Practical Electronics: Components and Techniques

John M. Hughes

Beijing · Cambridge · Farnham · Köln · Sebastopol · Tokyo

Practical Electronics: Components and Techniques

by John M. Hughes

Copyright © 2015 John M. Hughes. All rights reserved.

Printed in the United States of America.

Published by O'Reilly Media, Inc., 1005 Gravenstein Highway North, Sebastopol, CA 95472.

O'Reilly books may be purchased for educational, business, or sales promotional use. Online editions are also available for most titles (*http://safaribooksonline.com*). For more information, contact our corporate/institutional sales department: 800-998-9938 or *corporate@oreilly.com*.

Editors: Brian Sawyer and Mike Loukides
Production Editor: Nicole Shelby
Copyeditor: Rachel Monaghan
Proofreader: Amanda Kersey

Indexer: Ellen Troutman
Interior Designer: David Futato
Cover Designer: Ellie Volckhausen
Illustrators: John M. Hughes and Rebecca Demarest

March 2015: First Edition

Revision History for the First Edition
2015-03-10: First Release
2015-04-10: Second Release

See *http://oreilly.com/catalog/errata.csp?isbn=9781449373078* for release details.

While the publisher and the author have used good faith efforts to ensure that the information and instructions contained in this work are accurate, the publisher and the author disclaim all responsibility for errors or omissions, including without limitation responsibility for damages resulting from the use of or reliance on this work. Use of the information and instructions contained in this work is at your own risk. If any code samples or other technology this work contains or describes is subject to open source licenses or the intellectual property rights of others, it is your responsibility to ensure that your use thereof complies with such licenses and/or rights.

978-1-449-37307-8

[LSI]

Table of Contents

Preface.	xiii
1. Electrons in Motion.	1
Atoms and Electrons	2
Electric Charge and Current	3
Current Flow in a Basic Circuit	5
Ohm's Law	8
Power	8
Resistance	9
Example: Building a Voltage Divider	10
Summary	12
2. Fasteners and Adhesives.	15
Screws and Bolts	16
Screw and Bolt Sizes	17
Screw and Bolt Drive Types	19
Screw and Bolt Head Styles	20
Selecting Screws and Bolts	21
Washers	23
Self-Tapping Screws	25
Rivets	25
Adhesives and Bonding	26
Glues, Epoxies, and Solvents	26
Working with Wood and Paper	28
Working with Plastic	28
Working with Metal	29
Special-Purpose Adhesives	30
Summary	30
3. Tools.	31
Screwdrivers	31
Pliers	32
Wire Cutters	33
Wire Strippers	34
Crimping Tools	35
Socket and Hex Drivers	36
Clamps	38

Vises	39
Rotary Tools	41
Grinders	42
Drills	43
Drill Bits	44
Taps and Dies	45
Small Hand Saws	45
Miniature Power Saws	46
Specialty Metalworking Tools	47
Tweezers	49
Soldering Tools	49
Magnifiers and Microscopes	50
Workspaces	52
Summary	53

4. Tool Techniques ... 55

Working with Fasteners	56
Screwdriver Sizes and Types	56
Self-Tapping Screws	59
Hex-Socket-Head Fasteners and Hex Wrenches	59
Hex-Head Fasteners and Socket Wrenches	61
Adjustable Wrenches	63
Wrenches (Spanners)	64
Rivets	66
Dealing with Stubborn Fasteners	68
Soldering and Desoldering	70
Solder Types	70
Soldering Technique	72
Desoldering Wires and Through-Hole Parts	78
Surface-Mount Soldering	80
Surface-Mount Desoldering	84
Cutting	85
Rod and Bar Stock	85
Sheet Stock	87
Drilling	89
Selecting A Drill Size	89
Drilling Speed	91
Drilling Thin Sheet Stock	92
Lubricants	92
Punches and Pilot Holes	94
Using a Step Drill	94
Common Drilling Problems	94
Taps and Dies	96
Modification Cutting	102
Jeweler's Saw	102

Rotary Tool	103
Summary	104

5. Power Sources... 107
Batteries	107
Battery Packages	107
Primary Batteries	108
Secondary Batteries	111
Miniature Button/Coin Batteries	113
Battery Storage Considerations	116
Using Batteries	117
Battery Circuits	119
Selecting Batteries	120
Power Supply Technology	121
Wall Plug-in DC Power Supplies	122
Bench DC Power Supplies	125
Modular and Internal DC Power Supplies	125
Photovoltaic Power Sources	126
Fuses and Circuit Breakers	128
Fuses	128
Circuit Breakers	130
Summary	131

6. Switches... 133
One Switch, Multiple Circuits	134
Switch Types	135
Toggle	135
Rocker	137
Slide	137
Rotary	138
Pushbutton	139
Snap-Action	139
Slide and Rotary Switch Circuits	140
Switch Selection Criteria	141
Switch Caveats	143
Summary	143

7. Connectors and Wiring................................... 145
Wire and Cable	146
Wire Gauges	147
Insulation	149
Twisted Pairs	149
Shielding	152
Multi-Conductor Cables	153

Stripping Wire Insulation	155
Connectors	157
Connector Termination	157
Connector Types	161
Assembling Connectors	169
Soldered Terminals	169
Crimped Terminals	170
Connector Backshells	171
IDC Connectors	172
Ethernet Connectors	173
Summary	174

8. Passive Components 175

Tolerance	176
Voltage, Power, and Temperature	176
Packages	178
Resistors	178
Physical Forms	179
Fixed Resistors	180
Variable Resistors	185
Special-Purpose Resistors	191
Resistor Markings	192
Capacitors	194
Capacitance Values	195
Capacitor Types	195
Variable Capacitors	197
Surface-Mount Capacitors	198
Chokes, Coils, and Transformers	199
Chokes	199
Coils	200
Variable Inductors	200
Transformers	200
Packages	201
Summary	201

9. Active Components 203

How to Read a Datasheet	204
Datasheet Organization	204
Datasheet Walk-Through	206
Collecting Datasheets	208
Electrostatic Discharge	209
Packaging Overview	210
Through-Hole Parts	211
Surface-Mount Parts	211

Using Different Package Types	212
Diodes and Rectifiers	212
Small-Signal Diodes	213
Rectifiers	214
Light-Emitting Diodes	216
Zener Diodes	217
Exotic Diodes	217
Diode/Rectifier Axial Package Types	218
Diode/Rectifier Surface-Mount Packages	219
LED Package Types	219
Transistors	220
Small-Signal Transistors	221
Power Transistors	222
Field-Effect Transistors	222
Conventional Transistor Package Types	223
Surface-Mount Transistor Package Types	225
SCR and TRIAC Devices	225
Silicon-Controlled Rectifiers	226
TRIACs	226
Heatsinks	226
Integrated Circuits	228
Conventional IC Package Types	229
Surface-Mount IC Package Types	230
High-Current and Voltage Regulation ICs	232
Summary	233

10. Relays 235

Relay Background	235
Armature Relays	235
Reed Relays	236
Contactor	236
Relay Packages	237
PCB Relays	237
Lug-Terminal Relays	239
Socketed Relays	239
Selecting a Relay	240
Relay Reliability Issues	241
Contact Arcing	241
Coil Overheating	242
Relay Bounce	242
Relay Applications	242
Controlling Relays with Low-Voltage Logic	243
Signal Switching	244
Power Switching	244
Relay Logic	245

Summary 246

11. Logic 249
Logic Basics 250
Origin of Logic ICs 252
Logic Families 252
Logic Building Blocks: 4000 and 7400 ICs 253
    Closing the TTL and CMOS gap 253
    4000 Series CMOS Logic Devices 255
    7400 Series TTL Logic Devices 256
    CMOS and TTL Applications 256
Programmable Logic Devices 257
Microprocessors and Microcontrollers 259
    Programming a Microcontroller 260
    Types of Microcontrollers 261
    Selecting a Microcontroller 262
Working with Logic Components 263
    Probing and Measuring 263
    Tips, Hints, and Cautions 264
    Electrostatic Discharge Control 264
Summary 265

12. Discrete Control Interfaces 267
The Discrete Interface 268
    Discrete Interface Applications 269
    Hacking a Discrete Interface 270
Discrete Inputs 272
    Using a Pull-Up or a Pull-Down Resistor 273
    Using Active Input Buffering 274
    Using Relays with Inputs 274
    Optical Isolators 274
Discrete Outputs 277
    Current Sinking and Sourcing 278
    Buffering Discrete Outputs 278
    Simple One-Transistor Buffer 278
Logic-Level Translation 280
    The BSS138 FET 280
    The TXB0108 281
    The NTB0101 281
Components 282
Summary 282

13. Analog Interfaces 283
Interfacing with an Analog World 284

	From Analog to Digital and Back Again	284
	Analog-to-Digital Converters	289
	Digital-to-Analog Converters	290
	Hacking Analog Signals	291
	Summary	293

14. Data Communication Interfaces 295

Basic Digital Communications Concepts		296
Serial and Parallel		297
Synchronous and Asynchronous		298
SPI and I2C		299
SPI		300
I2C		302
A Brief Survey of SPI and I2C Peripheral Devices		306
RS-232		310
RS-232 Signals		313
DTE and DCE		314
Handshaking		315
RS-232 Components		316
RS-485		317
RS-485 Signals		318
Line Drivers and Receivers		318
RS-485 Multi-Drop		318
RS-485 Components		319
RS-232 vs. RS-485		320
USB		320
USB Terminology		321
USB Connections		322
USB Classes		322
USB Data Rates		323
USB Hubs		324
Device Configuration		325
USB Endpoints and Pipes		325
Device Control		326
USB Interface Components		327
USB Hacking		327
Ethernet Network Communications		328
Ethernet Basics		328
Ethernet ICs, Modules, and USB Convertors		333
Wireless Communications		334
Bandwidth and Modulation		334
The ISM Radio Bands		336
2.45 GHz Short-Range		337
802.11		338
Bluetooth®		340

Bluetooth Low Energy (BLE)	342
Zigbee	344
Other Data Communications Methods	345
Summary	346

15. Printed Circuit Boards..................................... 349

PCB History	349
PCB Basics	350
Pads, Vias, and Traces	351
Surface-Mount Components	351
Fabrication	352
PCB Layout	353
Determine Dimensions	353
Arrange Parts	354
Place Components	355
Route Traces on the Solder Side	355
Route Traces on the Component Side	356
Create the Silkscreen	356
Generate Gerber Files	356
Fabricating a PCB	357
PCB Guidelines	359
Layout Grid	359
Grid Spacing	360
Location Reference	360
Trace Width for Signals	361
Trace Width for Power	361
Trace Separation	361
Via Size	361
Via Separation	362
Pad Size	362
Sharp Corners	362
Silkscreen	362
Summary	362

16. Packaging..................................... 363

The Importance of Packaging	363
Types of Packaging	363
Plastic	364
Metal	364
Stock Enclosures	365
Plastic Enclosures	366
Cast Aluminum Enclosures	368
Extruded Aluminum Enclosures	368
Sheet Metal Enclosures	369

Building or Recycling Enclosures	371
Building Plastic and Wood Enclosures	371
Unconventional Enclosures	373
Repurposing Existing Enclosures	375
Designing Packaging for Electronics	375
Device Size and Weight	377
Environmental Considerations	378
Thermal Considerations	379
Sources	379
Summary	380

17. Test Equipment... 381

Basic Test Equipment	381
Digital Multimeters	381
Using a DMM	383
Oscilloscopes	385
How an Oscilloscope Works	387
Using an Oscilloscope	389
Advanced Test Equipment	391
Pulse and Signal Generators	391
Logic Analyzers	392
Buying Used and Surplus Equipment	394
Summary	396

A. Essential Electronics and AC Circuits..................... 397

B. Schematics... 455

C. Bibliography... 467

D. Resources.. 471

E. Components Lists... 481

Glossary... 493

Index.. 517

Preface

So, how much electronics do you need to know to be able to create something interesting, or creatively modify something that already exists? Well, that depends on where you start in the creative process. It also depends on your willingness to seek out new knowledge and acquire new skills as you go along.

The primary purpose of this book is to give you a reference for some of the more arcane (and possibly mundane) but essential aspects of electronics. These include things you would typically learn on the job and from years of experience, such as how to read the datasheet for an electronic component, determining how many things can be connected to an interface pin on a microcontroller, how to assemble various types of connectors, how to minimize noise and interference on a signal interface circuit, how to determine the resolution of an analog-to-digital converter, how various types of serial and network interfaces work, and how to use open source tools for schematic capture and PCB layout. And, of course, we will also cover the tools used in electronics work and how they are used, and we'll examine what's available in terms of test equipment beyond the garden-variety digital multimeter.

We'll start off with an introduction to the underlying physics of electricity that dispenses with the water-flowing-in-a-pipe analogy and gets right to the heart of the matter with a look at how atoms pass electrons around. We'll then examine the basic concepts of voltage and current. For those readers who might need or want a more detailed discussion of basic electrical theory, it can be found in Appendix A.

I should point out that this book is not intended to be an in-depth tutorial on electronics theory. There are already many excellent books on that topic, and to repeat that here would just be a pointless exercise in killing trees. So, while there is some introductory material to set the stage, so to speak, the primary intent of this book is to provide you with a reference for topics that aren't usually covered in an electronics text or a step-by-step project book.

With this book, perhaps one or two of the suggested reference works in Appendix C, and your own enthusiasm and ambition, you should be able to create that gadget or system you've been wanting to build and have it work as you intended. And remember, it's not the end of the world if you accidentally convert an electronic component into charcoal. It happens all the time; it's called *learning*.

Who This Book Is For

This book is for anyone with a desire to build an electronic device of some sort, but, to the maximum extent possible, I have made no assumptions about your skill level. What I *have* assumed is that you might not be familiar with the hardware, components, tools, and techniques that are used in electronics, or perhaps you already know something about electronics but could use some help with some of the more arcane aspects of the craft.

With this book as a workbench reference and guide to more detailed sources of information, you should be able to get started on building a nifty gadget and avoid some (hopefully, most) of the pitfalls that await the unwary. I've made the assumption that you will follow the pointers given to learn more about the various topics this book covers, and it covers a lot. It's simply not possible to cover all the topics presented in this book at more than just a surface level; the resulting tome would be huge. In lieu of a lot of details, I've tried to provide enough information to give you a basic understanding of the topics and a foundation to build upon.

So, if you've been thinking about something you'd like to build but aren't sure how to go about it, or you already know a fair amount about electronics but perhaps need some help putting it all together, then this book is for you.

How This Book Is Organized

Each chapter is devoted to a specific topic, ranging from hardware (screws, nuts, and bolts) to tools, and from switches, relays, and passive components to active solid-state parts. Each chapter is designed to allow you to easily find a specific subject and get quick answers to your questions:

Chapter 1: Electrons in Motion
    The first chapter provides a high-level "top-of-the-waves" look at electronics, using the notion of electrons in motion as the key to concepts such as voltage, current, and power.

Chapter 2: Fasteners and Adhesives
    Often overlooked or taken for granted, fasteners and fastening methods are essential to a successful project. The choice of fasteners can also have a major effect on the aesthetics of a project, so getting the right parts for the job can make the difference between elegant and clunky.

Chapter 3: Tools
    This chapter describes the basic tools needed to work with electronics (diagonal cutters, flush cutters, pliers, screwdrivers, etc.), along with some tools not commonly discussed in other texts, including things like crimp tools, rotary tools, step drills, professional grade soldering stations, and magnifiers and microscopes for surface-mount work.

Chapter 4: Tool Techniques
  Short sections for each tool discuss its uses and applications, including the correct use of sockets, wrenches, and screwdrivers; how to solder various component types, including surface-mounted components; and how to correctly size the holes needed to mount components like switches, lamps, or printed circuit boards in a chassis or panel.

Chapter 5: Power Sources
  An overview of power supplies for both DC and AC current, ranging from batteries to Variac-type devices, this chapter gives special attention to inexpensive DC power supplies in the form of plug-in modules (so-called *wall warts*). It also presents a discussion of fuses and circuit breakers and offers guidance on how to select an appropriate rating for these essential protection devices.

Chapter 6: Switches
  This chapter is a survey of the types of switches available and where they are typically used. This covers conventional switches, such as toggles and panel-mount pushbuttons, along with other types, such as PCB-mounted pushbuttons and membrane-type switches.

Chapter 7: Connectors and Wiring
  In electronics, almost everything connects to something, somewhere. This chapter describes the various types of connectors available, where they are commonly used, and how to assemble some of the more common types, such as DB-9, DB-25, high-density terminal blocks, and the 0.1-inch grid pin connectors found on Arduino, Raspberry Pi, and BeagleBone boards. It also covers related topics, such as soldering, crimping, and insulation displacement (IDC) techniques for connector assembly. This chapter deals mostly with those connectors that a typical human being can easily assemble without resorting to a microscope and tweezers, or a special tool that costs hundreds of dollars.

Chapter 8: Passive Components
  Passive components are the framework on which circuits are built. This chapter describes commonly encountered passive components such as resistors, capacitors, and inductors, including both through-hole and surface-mounted types. It also describes how to read component markings and how to understand component ratings for power, temperature, and tolerance.

Chapter 9: Active Components
  This chapter covers various types of active components, from diodes to ICs, with photos and package outline drawings to illustrate the various types. It also discusses key points to bear in mind when working with active components, such as static sensitivity, heat damage from soldering, and some of the package types available for surface-mount components.

Chapter 10: Relays
  Relays might be an old technology, but they are still essential in electronics. This chapter covers the various types of relays available and their typical applications. It describes types ranging from low-current, TTL-compatible reed relays

to high-power types used to control AC current. It also covers techniques for controlling a relay from a low-voltage circuit.

Chapter 11: Logic
Along with a condensed description of basic logic components (OR, NOR, AND, NAND, etc.), logic families (TTL, CMOS), and some examples of combinatorial logic circuits, this chapter also presents an introduction to microprocessors and microcontrollers, in terms of what is currently available and what you might need or encounter in your own activities.

Chapter 12: Discrete Control Interfaces
This chapter covers the basics of using a discrete signal (a single logic I/O port) to control things in the physical world. It also includes a discussion on the use of buffers, using both individual transistors and ICs, along with a discussion of current sink and sourcing considerations.

Chapter 13: Analog Interfaces
This chapter describes the basics of analog interfaces, both input and output, and includes discussions on resolution, speed, and the effects of quantization. It also covers aspects of analog I/O, such as voltage range, buffering, and circuit design considerations to reduce noise and improve performance.

Chapter 14: Data Communication Interfaces
Topics include common interfaces, from board-level SPI and I²C to RS-232, RS-485, USB, and Ethernet. This chapter also covers wireless interfaces, such as generic 2.45 GHz devices, 802.11 wireless networking, ZigBee, and Bluetooth. Serial and parallel, the two primary interface families, are introduced, followed by a discussion of synchronous and asynchronous modes of operation. The remainder of the chapter is organized into sections that cover each topic with a high-level technical discussion, and representative component part numbers are provided where applicable.

Chapter 15: Printed Circuit Boards
This chapter is an overview of PCB design and layout, with a focus on technique rather than specific tools. The chapter starts off with an introduction to PCB technology and concepts, including circuit board substrate materials and circuit trace (or track, if you will) pattern etching and plating techniques. An example from a real project is used to demonstrate the basic steps involved in creating a double-sided PCB layout. The chapter wraps up with a collection of general guidelines and tips.

Chapter 16: Packaging
A guide to the various options available for physically housing electronics, this chapter includes a discussion of plastic versus metal, sources for chassis components, and the use of unconventional enclosures to create unique packaging prototypes. Examples are given for commercial off-the-shelf packages in the form of small plastic enclosures, metal enclosures using both aluminum and steel sheet metal, extruded aluminum packages, and heavy-gauge kits for more demanding applications.

Chapter 17: Test Equipment
   A short tour of inexpensive test equipment, this chapter starts with the ubiquitous digital multimeter and moves on to oscilloscopes, signal generators, and logic analyzers. The examples include readily available, low-cost devices such as single- and dual-channel pocket digital oscilloscopes from China, and a multi-waveform signal generator module for the Arduino. The intent is to give you some suggestions that don't involve breaking the bank to purchase high-end test equipment (not that there's anything wrong with high-end gear—it's generally excellent; it just happens to be rather expensive).

Appendix A: Essential Electronics and AC Circuits
   For anyone interested, or anyone who could benefit from it for their projects, this appendix presents a terse, high-level overview of basic electronics theory beyond what Chapter 1 provides. Topics covered include capacitance, series and parallel resistor and capacitor circuits, basic AC circuit theory, inductance, noise, impedance, and grounding techniques.

Appendix B: Schematics
   This appendix defines the basics of schematic drawings, with examples of commonly encountered symbols. Light on text but heavy on graphics, this appendix is intended to be a place where you can quickly find the definition for a particular symbol. It also describes some available open source tools for creating schematic diagrams.

Glossary
   The glossary provides definitions of many key terms and acronyms used in this book.

Appendix C: Bibliography
   This appendix provides a bibliography of the suggested reference texts presented throughout the book, organized by topic.

Appendix D: Resources
   This appendix includes URLs for electronics distributors, sources for mechanical components, and vendors of surplus components of various types, as well as a brief discussion of buying electronics components and other items from vendors on eBay, with some guidance and caveats.

Appendix E: Components Lists
   This appendix lists most all of the IC components and modules mentioned in this book. While this collection is by no means comprehensive, it does contain enough representative parts from each category to provide a solid starting point for a new design.

Conventions Used in This Book

The following typographical conventions are used in this book:

Italic
   Indicates new terms, URLs, email addresses, filenames, and file extensions.

`Constant width`
   Used for program listings, as well as within paragraphs to refer to program elements such as variable or function names, databases, data types,

environment variables, statements, and keywords.

Constant width bold
Shows commands or other text that should be typed literally by the user.

Constant width italic
Shows text that should be replaced with user-supplied values or by values determined by context.

 This element signifies a tip or suggestion.

 This element indicates a warning or caution.

Safari® Books Online

Safari Books Online is an on-demand digital library that delivers expert content in both book and video form from the world's leading authors in technology and business.

Technology professionals, software developers, web designers, and business and creative professionals use Safari Books Online as their primary resource for research, problem solving, learning, and certification training.

Safari Books Online offers a range of plans and pricing for enterprise, government, education, and individuals.

Members have access to thousands of books, training videos, and prepublication manuscripts in one fully searchable database from publishers like O'Reilly Media, Prentice Hall Professional, Addison-Wesley Professional, Microsoft Press, Sams, Que, Peachpit Press, Focal Press, Cisco Press, John Wiley & Sons, Syngress, Morgan Kaufmann, IBM Redbooks, Packt, Adobe Press, FT Press, Apress, Manning, New Riders, McGraw-Hill, Jones & Bartlett, Course Technology, and hundreds more. For more information about Safari Books Online, please visit us online.

How to Contact Us

Please address comments and questions concerning this book to the publisher:

O'Reilly Media, Inc. 1005 Gravenstein Highway North Sebastopol, CA 95472 800-998-9938 (in the United States or Canada) 707-829-0515 (international or local) 707-829-0104 (fax)

We have a web page for this book, where we list errata, examples, and any additional information. You can access this page at *http://bit.ly/practical-electronics*.

To comment or ask technical questions about this book, send email to: *bookquestions@oreilly.com*.

For more information about our books, courses, conferences, and news, see our website at *http://www.oreilly.com*.

Find us on Facebook: *http://facebook.com/oreilly*

Follow us on Twitter: *http://twitter.com/oreillymedia*

Watch us on YouTube: *http://www.youtube.com/oreillymedia*

Endorsements

There aren't any endorsements in this book, at least not intentionally. I've made reference to many different component manufacturers, suppliers, and authors, but I've tried to be evenhanded about it. Any trademarks mentioned are the property of their respective owners and appear here solely for reference. As for the photography, I tried to use my own tools and other items as much as possible, and although an image may show a particular brand or model, that doesn't mean it's the only type available. It just happens to be the one that I own and use in my own shop. In some cases, I've used images with permission or from sources such as the Library of Congress, and this is noted as appropriate.

Acknowledgments

This book would not have been possible without the enduring patience and support of my family. In particular, I would like to acknowledge the photography and organizational assistance of my daughter, Seren, who put up with my fussiness and took yet another picture of something or other in the light tent for me when I didn't like the pose or lighting of the first (or second, or third) attempt. And, of course, my lovely wife, Carol, who would bring me things to eat in my shop and fret about me losing sleep.

Special thanks to Mike Westerfield for his technical review and input. It's always good to have more than one pair of eyes on the details, and Mike pointed out some rough spots that needed some editing and clarification. The end result is a better book, and it just goes to show why review is a crucial part of any development process.

The feedback from readers of the early release has been invaluable. Special thanks to those who suggested additions for the bibliography in Appendix D (you know who you are) and for the many helpful comments and constructive criticisms.

I would also like to thank the editorial staff at O'Reilly for the opportunity to work with them once again, especially Brian Sawyer for his willingness to put up with me in general, Mike Loukides for giving me this opportunity to write another book for O'Reilly, and the Atlas team for responding to my technical issues in a timely and helpful manner.

CHAPTER 1

Electrons in Motion

THE FIELD OF ELECTRICAL THEORY AND electronics is huge, and it can be somewhat daunting at first. In reality, you don't need to know all the little theoretical details to get things up and running. But to give your efforts a better chance at success, it is a good idea to understand the basics of what electricity is and how, in general terms, it works. So that's what we're going to look at here.

The main intent of this chapter is twofold. First, I want to dispense with the old "water-flowing-in-a-pipe" analogy that has been used in the past to describe the flow of electrons in a conductor; it's not very accurate and can lead to some erroneous assumptions. There is, I believe, a better way to visualize what is going on, but it does require a basic understanding of what an atom is and how its component parts work to create electric charge and, ultimately, electric current. It might sound rather like something from the realm of physics (and, to be honest, it is, along with chemistry), but once you understand these concepts, things like fluorescent lights, neon signs, lightning, arc welders, plasma cutting torches, heating elements, and the electronic components you might want to use in a project will become easier to understand. The old water-flowing-in-a-pipe model doesn't really scale very well, nor does it translate easily to anything other than, well, water flowing through a pipe.

Second, I'd like to build on this atom-based model to introduce some basic concepts that will come up later as you work on your own projects. By the end of this chapter, you should have a good idea of what the terms *voltage*, *current*, and *power* mean and how to calculate these values. If you need more details on a lower level, you'll find them in Appendix A, including overviews of serial and parallel circuits, and basic AC circuit concepts. Of course, numerous excellent texts are readily available on the subject, and I encourage you to seek them out if you would like to dig deeper into the theory of electronics.

If you are already familiar with the basic concepts of electronics, feel free to skip this chapter. Just don't forget to take advantage of Appendix A and the suggested references in Appendix C if you run into a need for further details somewhere along the line.

Atoms and Electrons

In common everyday usage, the term *electricity* is used to refer to the stuff that one finds inside a computer, in a wall outlet, in the wires strung between poles beside the street, or at the terminals of a battery. But just what *is* this stuff, really?

Electricity is the physical manifestation of the movement of *electrons*, little specks of subatomic matter that carry a negative electrical charge. As we know, all matter is composed of *atoms*. Each atom has a nucleus at its core with a net positive charge. Each atom also has one or more negative electrons bound to it, each one whipping around the positively charged nucleus in a quantum frenzy.

It is not uncommon to hear of the "orbit" of an electron about the nucleus, but this isn't entirely accurate, at least not in the classical sense of the term *orbit*. An electron doesn't orbit the nucleus of an atom in the way a planet orbits a star or a satellite orbits the earth, but it's a close enough approximation for our purposes.

In reality, it's more like layers of clouds wrapped around the nucleus, with the electrons being somewhere in the layers of the cloud. One way to think of it is as a probability cloud, with a high probability that the electron is somewhere in a particular layer. Due to the quirks of quantum physics, we can't directly determine where an electron is located in space at any given time without breaking things, but we can infer where it is by indirect measurements. Yes, it's a bit mind-numbing, so we won't delve any deeper into it here. If you want to know more of the details, I would suggest a good modern chemistry or physics textbook, or for a more lightweight introduction, you might want to check out the "Mr. Tompkins" series of books by the late theoretical physicist George Gamow.

The nucleus of most atoms is made up of two basic particles: *protons* and *neutrons*, with the exception of the hydrogen atom, which has only a single positive proton as its nucleus. A nucleus may have many protons, depending on what type of atom it happens to be (iron, silicon, oxygen, etc.). Each proton has a positive charge (called a *unit charge*). Most atoms also have a collection of neutrons, which have about the same mass as a proton but no charge (you might think of them as ballast for the atom's nucleus). Figure 1-1 shows schematic representations of a hydrogen atom and a copper atom.

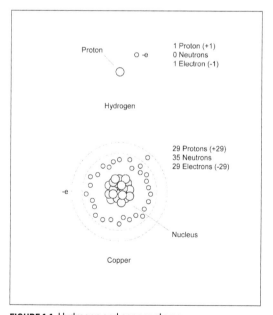

FIGURE 1-1. Hydrogen and copper atoms

The +1 unit charges of the protons in the nucleus will cancel out the −1 unit charges of the electrons, and the atom will be electrically neutral, which is the state that atoms want to be in. If an atom is missing an electron, it will have a net positive charge, and an extra electron will give it a net negative charge.

The electrons of an atom are arranged into what are called *orbital shells* (the clouds mentioned earlier), with an outermost shell called the *valence shell*. Conventional theory states that each shell has a unique energy level and each can hold a specific number of electrons. The outermost shell typically determines the chemical and conductive properties of an atom, in terms of how easily it can release or receive an electron. Some elements, such as metals, have what is considered to be an "incomplete" valence shell. Incomplete, in this sense, means that the shell contains fewer than the maximum possible number of electrons, and the element is chemically reactive and able to exchange electrons with other atoms. It is, of course, more complex than that, but a better definition is way beyond the scope of this book.

For example, notice that the copper atom in Figure 1-1 has 29 electrons and one is shown outside of the main group of 28 (which would be arranged in a set of shells around the nucleus, not shown here for clarity). The lone outermost electron is copper's valence electron. Because the valence shell of copper is incomplete, this electron isn't very tightly bound, so copper doesn't put up too much of a fuss about passing it around. In other words, copper is a relatively good conductor.

An element such as sulfur, on the other hand, has a complete outer shell and does not willingly give up any electrons. Sulfur is rated as one of the least conductive elements, so it's a good insulator. Silver tops the list as the most conductive element, which explains why it's considered useful in electronics. Copper is next, followed by gold. Still, other elements are somewhat ambivalent about conducting electrons, but will do so under certain conditions. These are called *semiconductors*, and they are the key to modern electronics.

This should be a sufficient model for our purposes, so we won't pry any further into the inner secrets of atomic structure. What we're really interested in here is what happens when atoms do pass electrons around, and why they would do that to begin with.

Electric Charge and Current

Electricity involves two fundamental phenomena: electric charge and electric current. *Electric charge* is a basic characteristic of matter and is the result of something having too many electrons (negative charge), or too few electrons (positive charge) with regard to what it would otherwise need to be electrically neutral. An atom with a negative or positive charge is sometimes called an *ion*.

A basic characteristic of electric charges is that charges of the same kind repel one another, and opposite charges attract. This is why electrons and protons are bound together in an atom, although under most conditions they can't directly combine with each other because of some other fundamental characteristics of atomic particles (the exceptional cases are a certain type of radioactive decay and inside a stellar supernova). The important thing to remember is that a negative charge will repel electrons, and a positive charge will attract them.

Electric charge, in and of itself, is interesting but not particularly useful from an electronics perspective. For our purposes, really interesting things begin to happen only when charges are moving. The movement of electrons through a circuit of some kind is called *electric current*, or *current flow*, and it is also what happens when the static charge you build up walking across a carpet on a cold, dry day is transferred to a doorknob. This is, in effect, the current (flow) moving between a high potential (you) to a lower potential (the doorknob), much like water flows down a waterfall or a rock falls down the side of a hill. The otherwise uninteresting static charge suddenly becomes very interesting (or at least it should get your attention). When a charge is not in motion, it is called the *potential*, and yes, we can make an analogy between electrical potential and mechanical potential energy, as you'll see shortly.

Current flow arises when the atoms that make up the conductors and components of electrical circuits transfer electrons from one to another. Electrons move toward things that are positive, so if you have a small light bulb attached to a battery with some wires (sometimes also known as a *flashlight*), the electrons move out of the negative terminal of the battery, through the light bulb, and return back into the positive terminal. Along the way, they cause the filament in the lamp to get white-hot and glow.

Figure 1-2, a simplified diagram of some copper atoms in a wire, shows one way to visualize the current flow. When an electron is introduced into one end of the wire, it causes the first atom to become negatively charged. It now has too many electrons. Assuming a continuous source of electrons, the new electron cannot exit the way it came in, so it moves to the next available neutral atom. This atom is now negative and has a surplus electron. In order to become neutral again (the preferred state of an atom), it then passes an extra electron to the next (neutral) atom, and so on, until an electron appears at the other end of the wire. So long as there is a source of electrons under pressure connected to the wire and a return path for the electrons back to the source, current will flow. The pressure is called *voltage*, which "Current Flow in a Basic Circuit" on page 5 will discuss in more detail.

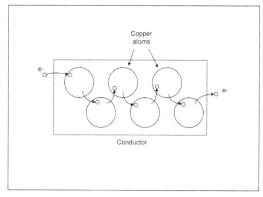

FIGURE 1-2. Electrons moving in a wire

Figure 1-3 shows another way to think about current. In this case, we have a tube (a conductor) filled end to end with marbles (electrons).

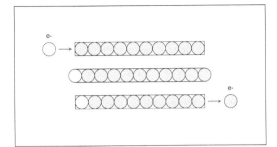

FIGURE 1-3. Modeling electrons with marbles in a tube

When we push a marble into one end of the tube in Figure 1-3, a marble falls out the opposite end. The net number of marbles in the tube remains the same. Note that the electrons put into one end of a conductor are not necessarily the ones that come out the other end, as you can see from Figures 1-2 and 1-3. In fact, if the conductor is long enough, the electrons introduced at one end might not be the ones that appear at the other end, but electrons would appear, and you would still be able to measure electron movement in the conductor.

Current Flow in a Basic Circuit

Electricity flows when a closed circuit allows for the electrons to move from a high potential to a lower potential in a closed loop. Stated another way, current flow requires a source of electrons with a force to move them, as well as a return point for the electrons.

Electric current flow (a physical phenomenon) is characterized by four fundamental quantities: voltage, current, resistance, and power. We'll use the simple circuit shown in Figure 1-4 as our baseline for the following discussion. Notice that the circuit is shown both in picture and schematic form. For more about schematic symbols, refer to Appendix B.

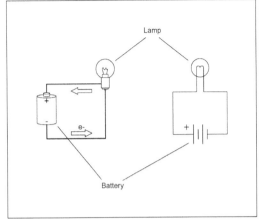

FIGURE 1-4. A simple DC circuit

A few words about the term *current* are in order here. The word has more than one meaning in electronics, which can be confusing at first. In one sense, current refers to the flow of electrons through a conductor of some kind. It is a reference to the movement of charge carried by the electrons. In the other sense, current refers to the number of electrons moving through the conductor. In this sense, it specifies the volume of electrons moving past some point in the circuit at some point in time. In other words, the measurement of current is the determination of the quantity of electrons in motion.

One way to think about current is to remember that it cannot be measured without movement, so when you see or hear the word *current*, it is usually referring to movement. To make the distinction clear, the term *current flow* is often used to mean movement of electrical charges. Static charges, even if just at the terminals of a common battery, have no current flow and hence no measurable current.

Current that flows in only one direction, as in Figure 1-4, is called *direct current* (DC). A common battery produces DC, as does the DC power supply in a typical computer system. Current that changes direction repeatedly is called *alternating current* (AC). AC is what comes out of a household wall socket (in the US, for example). It is also the type of current that drives the loudspeakers in a stereo system. The rate at which the current changes direction is called the *frequency* and is measured in cycles per second in units of Hertz (abbreviated Hz). So, a 60 Hz signal is made up of a current flow changing direction 60 times per second. When AC is used to drive a loudspeaker, a signal with a frequency of 440 Hz will be A above middle C to our ears.

By convention, DC is described as flowing from positive to ground (negative), whereas in reality, electrons flow from the negative terminal to the positive terminal of the power source. In Figure 1-4, the arrows show the electron flow. Basically, the discrepancy stems from an erroneous assumption made by Benjamin Franklin, who thought that electrons had a positive charge and flowed from positive to negative terminals. He guessed wrong, but we ended up with a convention that was already well ingrained by the time physicists figured out what was really going on. Hence we have conventional current flow and electron current flow. Although you should be aware of this discrepancy, from this point onward, we'll use conventional current flow, since that is what most of the electronics industry uses.

A *volt* (V) is the unit of measurement used for electric potential difference, electric potential, and electromotive force. When the term *voltage* is used, it usually refers to the electric potential difference between two points. In other words, we say that a static charge has a value of some number of volts (potential), but there is a certain amount of voltage between two points in a circuit (potential difference).

Voltage can be visualized as a type of pressure, or driving force (although it is not actually a force in a mechanical sense). This is the electromotive force (emf) produced by a battery or a generator of some type, and the emf can drive a current through a circuit. And even though it may not look like a generator, a power supply (like the one that plugs into the wall socket to charge a cell phone) is really nothing more than a converter for the output of a generator at a power plant somewhere.

Another way to think of voltage is as the electric potential difference between two points in an electric field. It is similar to the difference in the potential energy of a cannonball at the top of a ladder as opposed to one at the top of a tall tower. Both cannonballs exist in the earth's gravitational field, they both have potential energy, and it took some work to get them both into position. When they are released, the cannonball on the top of the tower will have more energy when it hits the ground than the cannonball dropped from the top of the ladder, because it had a larger potential energy due to its position.

These two descriptions of voltage are really just opposite sides of the same coin. In order to create a potential difference between two points, work must be done. When that energy is lost or used, there is a potential drop. When the cannonball hits the ground, all of the energy put into getting it into position against the pull of gravity is used to make a nice dent in the ground.

The main point here to remember is that a high voltage has more available electrical energy (pressure) than a low voltage. This is why you don't get much more than a barely visible spark when you short out a common 9-volt battery with a piece of wire, but lightning, at around 10,000,000 volts (or more!), is able to arc all the way between a cloud and the ground in a brilliant flash. The lightning has more voltage and hence a larger potential difference, so it is able to overcome the insulating effects of the intervening air.

Whereas voltage can be viewed as electrical pressure, current is the measure of the quantity, or volume, of electrons moving through a circuit at some given point. Remember that the term *current* can have two different meanings: electron movement (flow) and the volume of the electron flow. In electronics, the word *current* usually means the quantity of electrons flowing through a conductor at a specific point at a single instant in time. In this case, it refers to a physical quantity and is measured in units of amperes (abbreviated as A).

Now that we've looked at voltage and current, we can examine some of the things that happen while charge is in motion (current flow) at some particular voltage. No matter how good a conventional conductor happens to be, it will never pass electrons without some resistance to the current flow (superconductors get around this, but we're not going to deal with that topic here). *Resistance* is the measure of how much the current flow is impeded in a circuit, and it is measured in units of ohms, named after German physicist Georg Simon Ohm. "Resistance" on page 9 has more details about the physical properties of resistance, but for now, let's consider how resistance interacts with current flow.

You might think of resistance as an analog of mechanical friction (but the analogy isn't perfect). When current flows through a resistance, some of the voltage potential difference is converted to heat, and there will be a voltage drop across the resistor. How much heat is generated is a function of how much current is flowing through the resistance and the amount of the voltage drop. We'll look at this more closely in "Power" on page 8.

You can also think of resistance as the degree of "stickiness" that an atom's valence shell electrons will exhibit. Atoms that can give up or accept electrons easily will have low resistance, whereas those that want to hold onto their electrons will exhibit higher resistance (and, of course, those that don't readily give up electrons under normal conditions are good insulators).

Carbon, for example, will conduct electricity, but not as easily as copper. Carbon is a popular material for fabricating the components called resistors used in electronic circuits.

Chapter 8 covers passive components, such as resistors.

Ohm's Law

As you may have already surmised, there is a fundamental relationship between voltage, current, and resistance. This is the famous equation called *Ohm's law*. It looks like this:

$$E = IR$$

where E is voltage (in volts), I is current (in amperes), and R is resistance (in ohms).

This simple equation is fundamental to electronics, and indeed it is often the only equation that you really need to get things going. In Figure 1-4, the circuit has only two components: a battery and a lamp. The lamp comprises what is called the *load* in the circuit, and it exhibits a resistance to current flow. Incandescent lamps have a resistance that varies according to temperature, but for our purposes, we'll assume that the lamp has a resistance of 2 ohms when it is glowing brightly.

The battery is 1.5 volts, and for the purposes of this example, we'll assume that it is capable of delivering a maximum current of 2,000 milliamps (or mA) for one hour at its rated output voltage. This is the battery's total rated capacity, which is usually around 2,000 mAh (milliamp-hour) for a typical alkaline AA type battery. A *milli* is one-thousandth of something, so 2,000 mA is the same as 2 amps of current.

Applying Ohm's law, we can find the amount of current the lamp will draw from the battery by solving for I:

$$I = E/R$$

or:

$$I = 1.5/2$$

$$I = 0.75 \text{ A}$$

Here, the value for I can also be written as 750 mA (milliamperes). If you want to know how long the battery will last, you can divide its capacity by the current in the circuit:

$$2/0.75 = 2.67 \text{ hours (approximately)}$$

Power

In the simple circuit shown in Figure 1-4, the flow of electrons through the filament in the lamp causes it to heat up to the point where it glows brightly (between 1,600 to 2,800 degrees C or so). The filament in the lamp gets hot because it has resistance, so current flows less easily through the filament than it does through the wires in the circuit.

Power is the rate of doing work per unit of time, and is measured in watts. One watt is defined as the use or generation of 1 joule of energy per second. In an electrical circuit, a watt can also be defined as 1 ampere of current moving through a resistance at 1 volt of potential, and when charges move from a high voltage to a low voltage (a potential difference) across a resistive device, the energy in the potential is converted to some other form, such as heat or mechanical energy.

We can calculate power (P) in a DC circuit by multiplying the voltage by the current:

$$P = EI$$

In the case of the simple flashlight circuit, the power expended to force the current through the filament is expressed as heat, and subsequently as light when the filament gets hot enough to glow. If you want to know how much power the light bulb in our circuit is consuming, simply multiply the voltage across the bulb by the current:

$$P = 1.5 \times 0.75$$

$$P = 1.125 \text{ watts, or } 1.125\text{W}$$

Let's compare this power value with the rating for a common incandescent light bulb with a 100W rating. An old-style 100W light bulb operating at 110 VAC (volts AC, typical household voltage in the US) will use:

$$I = PE$$

$$I = 100/110$$

$$I = 0.9\text{A}$$

Amazing! The large light bulb consumes only a bit more current than the tiny light bulb connected to a battery! How can this be?

The difference lies in the voltage supplied to the light bulbs and their internal resistance. Now that you have an estimate of the amount of current flowing through a 100W bulb, you could easily work out what its internal resistance might be. You should also be able to see why leaving lights on (or using old-style light bulbs at all) is wasteful. The current adds up, and each watt of power costs money.

Resistance

Now let's look at the phenomenon of resistance more closely, since it is such a fundamental aspect of electronics. Formally stated, 1 ohm is equal to the resistance between two points of a conductor when a potential of 1 volt produces a current of 1 ampere. This is, of course, the relationship defined by Ohm's law, discussed in "Ohm's Law" on page 8.

Resistance is a key factor in electric circuits, which is why it is one of the three variables in the Ohm's law equation. As stated earlier, every circuit has some amount of resistance, except for things like exotic superconductors. Even the wires connecting a battery to a device have some intrinsic resistance.

Switches have internal resistance, as do connectors and even the copper traces on a printed circuit board (PCB). Figure 1-5 illustrates this by showing a simple DC circuit and its resistance equivalent.

You might notice in Figure 1-5 that even the battery has some internal resistance. Appendix A discusses series and parallel resistances, and how to calculate their values, but the point here is to show how nothing is free in the world of circuits. Resistance is everywhere, as far as electrons are concerned.

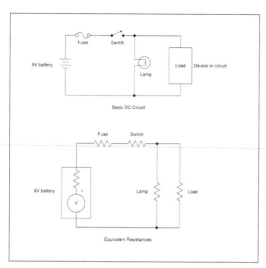

FIGURE 1-5. Circuit resistance example

Normally, this intrinsic resistance is ignored, as it tends to be small and doesn't really impact the overall operation of a device. However, if the device is a low-current one intended to run for a long time without the battery being changed, then it starts to become something to consider. Resistance to current flow means that energy is being expended pushing electrons through the resistive element, and that energy is dissipated as heat. Unless you are intentionally using a resistance as a heater (which is what electrical heating elements do), it is being wasted.

In electronics, the passive components called *resistors* are probably the most commonly used parts. Resistors come in a range of values and power ratings, from ultra-tiny little flecks for surface-mount use to huge devices used in diesel-electric locomotives to dissipate excess energy created during dynamic braking. Figure 1-6 shows a typical 1/4-watt carbon composition resistor. See Chapter 8 for more information about resistors and other passive components.

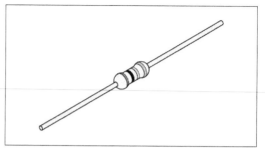

FIGURE 1-6. A typical resistor

Resistors can be used to limit current, reduce voltages, and supply a specific voltage at a particular location in a circuit. Resistance plays a big role in analytical applications such as network analysis (electrical networks, not data networks), equivalent circuits theory, and power distribution modeling.

Example: Building a Voltage Divider

You can do a lot with just a power supply of some sort, a couple of resistors, and Ohm's law. For example, let's say that you wanted to supply a circuit with 5V DC from a 9V battery. Provided that the circuit doesn't draw very much current (perhaps a few milliamps or so), and you are not too concerned about how stable the 5V supply will be, a simple thing called a *voltage divider* (shown in Figure 1-7) will do the job.

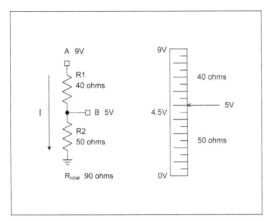

FIGURE 1-7. A simple voltage divider

We want the voltage at point B to be 5V when we apply 9V to point A. I've selected resistor values that will result in 100 mA of current flowing through both of the resistors. I've not taken into account the current consumed by the circuit connected to point B, but since the assumption here is that it will draw very little current, it won't have that big of an effect on the voltage level at point B.

Notice that the two resistors in the voltage divider of Figure 1-7 aren't the same values. One is 40 ohms; the other is 50 ohms. If both R1 and R2 were the same value, the voltage at point B would be 4.5V, not the 5V we want.

So how did I get those values? First, we determine the total resistance of the divider circuit. Since we already know the input voltage and the amount of current we want to pass through the resistors, the solution looks like this:

R = E/I

R = 9/0.1

R = 90

And, since there are two resistors in the divider, the sum of their values must be equal to the total resistance:

R1 + R2 = 90

If we use the current and the target output voltage of the divider (point B), we get the value of the second resistor, R2:

R2 = 5/0.1

R2 = 50

R1 is just whatever is left over:

R1 = 90 - R2

R = 40

The ratio between R1 and R2 and the resulting voltage at point B is illustrated graphically in Figure 1-7 by the vertical scale on the right side of the figure.

Another way to do this doesn't require any knowledge of the current through the divider, but instead uses the ratio of the two resistors:

$V_{out} = V_{in} \times (R2/(R1 + R2))$

Now, how long will the 9V battery last? A typical garden-variety 9V alkaline battery has a capacity rating of about 550 mAh. We can apply the same math used with the simple lamp circuit earlier. If we divide the battery's capacity rating by the current consumption of the voltage divider, we get this:

550/100 = 5.5

So, with this circuit, the battery will last for about 5.5 hours in continuous use.

As an exercise, calculate how much power this simple circuit will dissipate. Since resistors are rated in terms of both resistance and power dissipation, it should be quickly obvious that the two components will need to be rated for around 1 watt each. This circuit would overwhelm a small 1/8 watt component.

Also, I mentioned earlier that I was assuming that whatever was connected to the divider at point B wouldn't be drawing very much current. You could probably increase the values of the resistors by an order of magnitude (× 10), thereby reducing the total current to 10 mA, and still have enough margin to provide a very small current at around 5V. This would increase the battery life to 55 hours or so and significantly reduce the power rating requirement for the resistors. When you are using a voltage divider to produce a reference voltage for an active component in a circuit, the current draw is often very small (perhaps in the microamps range), since it's the voltage that matters. In cases like this, the values of R1 and R2 can be very large to further reduce the amount of current consumed by the divider.

This little exercise should make a few things readily apparent. First, you really don't want to use a voltage divider to try to create the equivalent of a power supply. Active regulators do a much better job and don't waste lots of energy as heat without doing any meaningful work. We will take a look at power supplies in Chapter 5 and active components like voltage regulators in Chapter 9.

Second, with multiple variables to work with, there is a lot of room to seek out solutions, some better than others. Don't settle on the first solution that pops up, because there might be a better way. Lastly, batteries are great for portability, but they really don't last long in continuous use when significant current is involved.

Summary

In this chapter, we've looked at the basics of atomic structure and how that contributes to how electrons move. We've also looked at the basic concepts of voltage, current, power, and resistance. In the process, we discovered that something rated for 100 watts of power at 110 volts uses only slightly more current than something at 1.25 watts at 1.5 volts, with the voltage being a major factor in the power difference.

With what you've seen so far, you should be able to determine how much power an electronic device is dissipating and determine how long a battery will last in a given situation, so long as you know the amount of current the battery is called upon to supply.

That should be enough basic theory to get things moving along, and later chapters will introduce additional concepts as necessary. If you really want to dig into the theoretical end of things to gain a deeper understanding, I would suggest one of the excellent reference works listed in Appendix C. Also note that Appendix B contains a listing of various schematic symbols commonly encountered in electronics work, as well as a write-up on using a schematic capture tool to create neat and tidy drawings of your circuits.

CHAPTER 2

Fasteners and Adhesives

CREATING SOMETHING NEW OFTEN ENTAILS attaching something to something else, and modifying something to adapt it to another purpose can entail attaching things where nothing was attached before. Then there's the issue of making sure things will stay attached. Fasteners and fastening techniques are key to creating reliable attachments, or building a solid, reliable chassis and enclosure for a circuit or electro-mechanical device. As a prelude to the upcoming chapters on tools and their uses, this chapter presents some of the more common hardware and adhesives used to fasten two or more things together. In Chapter 16, we'll look at enclosures and other packaging topics and see how fasteners and adhesives are used.

Fasteners come in a range of types and sizes. Some, such as screws and bolts, are familiar to just about everyone. Other types, such as rivets, are not as common but are widely used in a variety of applications. Still other types are designed for specialized applications. But regardless of the type or size, all fasteners are intended to do just what their name implies: fasten something to something else.

Although you might first think of nuts, bolts, and screws, a fastener isn't always a metallic part. Some reusable snap-on plastic fasteners are available for use with cardboard materials and are excellent for building disposable toys and play sets for children (among other applications), and screws and bolts are available in various nonmetallic materials, such as nylon, Teflon, wood, and ceramics.

Note that the word *fasteners* does not exclude things like adhesives (i.e., glues, silicon rubber, and other chemical compounds), and this chapter also includes a brief discussion of the various types of adhesives available. Adhesives are a handy way to attach things, and if done correctly, the bond is as strong and reliable as those made with screws or rivets. The concept of fastening can even be extended to include brazing, welding, and soldering, but this chapter won't cover those topics. Soldering is covered in Chapter 4 and elsewhere in this book. Brazing (a technique like a form of high-temperature soldering using an oxy-acetylene torch) and welding are art forms unto themselves, and there are numerous excellent reference and how-to books available. Also, many community colleges and vocational schools offer classes in both techniques.

Screws and Bolts

Various standards describe threaded fasteners, but there is no generally accepted definitive distinction between a *screw* and a *bolt*. Some sources define the difference on the basis of how the part is used, with bolts being mounted through something and tightened with a nut, and screws being inserted into a part with preformed threads or cutting their own threads as they are driven into position (*self-tapping* screws). Size also matters for naming threaded fasteners, with small parts often referred to as screws and larger pieces called bolts.

In any case, both bolts and screws use a spiral thread cut or pressed into a rod (usually metal but nylon, plastic, and wood have been used as well) to exert an axial force, which in turn is applied to hold two or more things in a fixed relationship to one another.

Figure 2-1 shows a sample of the various kinds of screws and bolts that are available. Just bear in mind that what might be a screw in one context could also be referred to as a bolt in another. This book will apply the terms *screw* and *bolt* based on the general criteria of size and usage stated previously.

FIGURE 2-1. Various screws and bolts

The variety of available screw types is staggering, and Figure 2-1 shows just a small sample. Fortunately, you don't need to be familiar with every type to make sensible choices; nor do you need to have a stockroom with stacks of containers full of fasteners to do useful work.

Distinguishing bolts from screws

The apparent ambiguity in the designation of a fastener as either a bolt or a screw is reflected in the official publication ICP013, "Distinguishing Bolts from Screws" (*http://bit.ly/bolts-screws*) from the US Customs and Border Protection division of the Department of Homeland Security. This document attempts to provide some guidance regarding how to determine whether a fastener is a bolt or a screw, but in the end, it still basically comes down to size and usage.

As the document puts it: "if it doesn't meet the primary criteria (and of course, if it doesn't conform to a fastener industry standard for a bolt), then it probably is a screw." In this case, the "primary criteria" is that a bolt is big and used with a nut, whereas a screw is small and not used with a nut. Except in those cases where a screw might be used with a nut (a machine screw). Or it's a really big bolt with a sharp tip that can cut its own threads. It gets confusing sometimes.

You can generally do most of the necessary fastener work you might encounter with a selection of machine screws in sizes 2, 4, 6, and 8, with a selection of lengths for each gauge, ranging from 1/4 inch long to 2 inches. A list of suggested sizes and types of machine screws to have on hand is shown in Table 2-1. The following section examines common gauge sizes in detail.

TABLE 2-1. A suggested inventory of screws (and associated washers and nuts)

Size	Type	Drive	Length
4-40	Pan	Phillips	1/4"
4-40	Cap	Hex Socket	1/4"
4-40	Pan	Phillips	1/2"
4-40	Cap	Hex Socket	1/2"
6-32	Pan	Phillips	1/4"
6-32	Cap	Hex Socket	1/4"
6-32	Pan	Phillips	1/2"
6-32	Cap	Hex Socket	1/2"
6-32	Pan	Phillips	3/4"
6-32	Pan	Phillips	1"
8-32	Pan	Phillips	3/8"
8-32	Pan	Phillips	1/2"
8-32	Pan	Phillips	3/4"
8-32	Pan	Phillips	1"

TIP: For definitions of the head types and drives, see "Screw and Bolt Head Styles" on page 20 and "Screw and Bolt Drive Types" on page 19.

If you purchase the suggested hardware in Table 2-1 as stainless steel parts in boxes of 100 each, you'll spend between $2.25 to $6.50 per box, and 14 small boxes don't take up much room at all. Also be sure to get nuts, flat washers, and locking washers for each size. "Washers" on page 23 discusses the various types of washers that are available. Of course, you can always just buy small quantities of what you need from a well-stocked hardware store, but you'll pay less per piece by buying parts by the box.

TIP: One of the best ways to become familiar with what fasteners are available and how they are used is to disassemble some electronic devices and observe what was used to hold things together (and, as a bonus, you might find some interesting bits to use in your own projects). You will find everything from cap head to Phillips, UTS/ANSI and metric, self-tapping types for both plastic and metal, and even some odd-looking things with star-shaped drive holes or even three slots (*Y* or *tri-wing* types, popular with some Asian manufacturers). If you have a military or industrial surplus outlet nearby, these can be a goldmine for hardware.

SCREW AND BOLT SIZES

Fortunately, unless you have a specific need to use an uncommon type of screw or bolt, you can do almost everything with five or so different sizes.

The Unified Thread System (UTS) commonly used in the US defines screws in terms of both diameter and thread pitch. The UTS is controlled by ANSI, the American National Standards Institute, and throughout this book, I'll use UTS and ANSI interchangeably when referring to bolt and screw sizes other than metric. Typical sizes encountered in electronics include diameter gauges of 2, 4, 6, 8, and 10. Common thread pitches include 40, 32, and 24. The *pitch* refers to the number of threads per unit of length in inches. So, a UTS screw or bolt size is defined as *gauge-pitch* (e.g., 4-40 or 6-32). Table 2-2 lists some common UTS screw and bolt sizes. The fractional diameter is the nearest value.

TABLE 2-2. UTS/ANSI machine screw diameter sizes

Gauge	Diameter (inches)	Decimals
0	1/16	0.06
1	5/64	0.07
2	3/32	0.08
3	7/64	0.09
4	7/64	0.11
5	1/8	0.12
6	9/64	0.13
8	5/32	0.16
10	3/16	0.19
12	7/32	0.21
14	1/4	0.24

Just because a gauge size is defined in Table 2-2, that doesn't necessarily mean that hardware in that size can be easily purchased. Fasteners in sizes 2, 4, 6, 8, and 10 are readily available. You can find other sizes (#1, for example) if you're willing to look hard enough, but if you need to special-order a gauge, a supplier won't be too interested unless you are willing to commit to purchasing a large quantity of the parts. Stick to the common sizes if at all possible. It's much easier that way.

Also be aware that with sizes of 1/4 inch or larger, the parts are often specified in diameter-pitch nomenclature, rather than gauge-pitch. In other words, while saying you want to use a 14-20 is technically correct, you may find that 1/4-20 is how the parts are stocked at the distributor or hardware store.

You should keep in mind that the gauge size numbers used with machine screws and bolts are also used with self-tapping screws, both sheet metal and wood. Be aware, however, that the actual diameter of a part can vary somewhat from the ideal value given in Table 2-2. The amount of variance depends on the tolerances applied by the manufacturer and the process used to create the threads, but it is generally no more than +/− 0.01 inches. This is important to remember when sizing holes and selecting drill bits, as discussed in Chapter 4.

In the metric-speaking world (which is almost everywhere outside of the US) a large range of metric screw and bolt types is available. These are defined in the international standard ISO 68-1, and the ISO 262 standard specifies a number of predefined sizes. Based on the standards, a metric screw with a shaft diameter of 1 mm may have a coarse thread of 0.25 mm or a fine thread of 0.2 mm. So, to specify a 1 mm screw with a coarse thread, you would use M1x0.25.

The remainder of this book will largely stick to UTS/ANSI nomenclature except when discussing things like tapping and clearance holes, but generally, everything you might

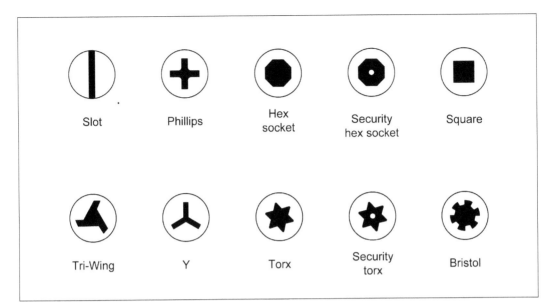

FIGURE 2-2. Screw and bolt drive styles

say about using a UTS screw or bolt also applies to its metric equivalent.

> **TIP** It seems inevitable that eventually the entire world, the US included, will go to the metric system. In industry this has already begun, with many US manufacturers shifting to metric parts to accommodate a global market and the use of parts and subassemblies from foreign suppliers. If you want to employ some "future-proofing" into your designs, you might want to consider learning how to work with metric parts, and metric units of measurement in general.

SCREW AND BOLT DRIVE TYPES

The two most commonly encountered screw drive types are slotted and Phillips. Figure 2-2 shows the drive types you might regularly encounter, but other specialty types are also in use. These are the common names; other names are also encountered, such as cross for Phillips, and Robertson for the square drive. Browsing through the website or catalog of a supplier such as McMaster-Carr, Microfasteners, or Amazon-Supply (formerly SmallParts) can be very informative.

Some types of fasteners are available with a six-sided (hex) hole for use with a hex wrench (also called a hex key or an Allen wrench). These *hex socket* types are usually found with button and cap head styles (see "Screw and Bolt Head Styles" on page 20) and are common in robotics and in scientific and metrology equipment such as interferometers, optical spectrometers, and telescopes. A hex socket drive allows for greater torque without tool slippage, which is a problem with both slot and Phillips drives.

Larger bolts often have a hex head suitable for use with a socket tool (i.e., the entire head of the bolt is the drive), and small sizes in this style are also available. The large sizes will be familiar to anyone who has ever worked on an automobile, and the small sizes can sometimes be found in scientific and military-grade equipment.

A small hex head part can be difficult to drive without a special socket wrench made specifically to seat flush around the head. Most common sockets sold at auto supply and home improvement centers will not work reliably, because they have a slightly rounded edge that will prevent the socket from fully seating on the sides of the head. The tool will likely slip and damage the head of the fastener. Chapter 4 discusses how to modify a socket to fit flush on a small hex head fastener.

If you elect to use a hex head screw or bolt, it is worth bearing in mind that while the part can be tightened (torqued) to a greater degree without tool slippage (or tool breakage) than a Phillips or a hex socket head part, it can be awkward to use in tight places where the socket to drive the screw or bolt won't easily fit. In other words, consider the tool that will need to be used to deal with the fastener in its eventual location.

Drive types such as tri-wing, Y, Torx, and other security variations are intended to discourage unauthorized access. They require special tools made expressly for that drive type, and the tools usually aren't available at the local home improvement center. Avoid them if at all possible.

> **TIP** You can defeat a security Torx or hex socket type drive by carefully knocking out the small pin-like column in the center of the drive hole using a pin punch and a ball-peen hammer. Alternatively, you can sometimes drill out the security pin with a good drill press if you first securely clamp down the work piece and use a drill bit not much larger than the security pin. Then again, if you expect to work with these types of fasteners on a regular basis, the tools to deal with them aren't that expensive and you can find them on eBay.

SCREW AND BOLT HEAD STYLES

Figure 2-3 shows some common head types for screws and bolts. The hex head is typically found on bolts (larger than common screws), and the button and cap head types typically have a hex socket and are used with a hex wrench. Notice that the cap head is shown with knurling, which allows for hand-tightening if the head is large enough or your fingers are small enough. The other head types can have either a slot or Phillips drive, and even a hex socket drive variation is available for the flat and oval head styles.

The round and pan head styles are most common around the home. Woodworkers routinely employ flat-head wood screws to keep the screw head flush with the surface of the wood. Consumer electronics tend to use a lot of flat-head machine screws and self-tapping types, again to keep things smooth and flush.

Be aware that the flat and oval head types require a countersunk hole to seat correctly. These are typically used in situations that require low-profile (or flush, in the case of the flat head types) screws or bolts. Chapter 4 covers drilling a countersunk hole, but basically, it involves creating a hole in the material, larger than the hole for the shaft of the screw, for the screw head to seat into.

SELECTING SCREWS AND BOLTS

Selecting the appropriate screw or bolt for a particular application involves taking into consideration several factors. If you want to fasten a plastic back onto a plastic case, a self-tapping screw would do the job. The back plate of a metal chassis box is usually fastened with self-tapping sheet metal screws. The screws used to mount things to a chassis, panel, or other surface can be 8-32, 6-32, or 4-40, depending on the size and mass of what is being mounted.

As a general rule of thumb, the larger the screw or bolt, the more load it can bear. Small screws (#2, for example) are fine for lightweight tasks, but a larger #6 or #8 screw should be considered for items or situations where the screw must resist a force (as with hinges or attaching a metal panel).

Also, the use of multiple screws or bolts can increase strength and reliability, but at the cost of extra weight and expense. Without knowing the type of materials being used and the forces expected, it's impossible for me to give specific recommendations for maximum loads, but Table 2-3 can serve as a rough guide.

TABLE 2-3. Machine screw clamp loads

Size	Clamp load
4-40	200
6-32	350
8-32	500
10-24	700
1/4-20	700

Clamp load is the amount of load applied to a bolted joint to hold the parts together and avoid relative motion (shear slippage). In Table 2-3 this is a maximum de-rated value derived from various references, and it assumes the lowest grade of part. It does not

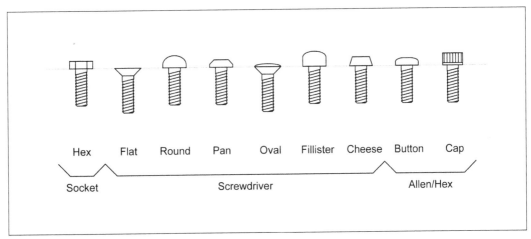

FIGURE 2-3. Common screw and bolt head types

take into account any additional stresses that may be placed on the fastener.

In general, parts with a nut tend to resist pull-out (axial failure) better than screws in a tapped hole, but the amount of the screw's threads engaged by the nut or the inside threads of a tapped hole plays a significant role. Likewise, the material the screw or bolt is used with has a big effect on the ability of the part to resist pull-out. Consider the case of a #4 stainless steel screw in a tapped hole in a piece of soft 1/10-inch (10-gauge) aluminum. Odds are, the inside threads in the hole in the aluminum will probably fail before the screw itself does, particularly if the hole was threaded using a 50% thread instead of a 75% cut. See Chapter 4 for more on threads and how to make them.

Shear strength is another consideration. If a screw or bolt is used to attach two pieces that may have a tendency to move parallel to each another, then there's a distinct risk that the fastener can be sheared off between the pieces. One way to compensate for this is to use the largest screw or bolt size possible. Another approach is to set hardened steel alignment pins into the pieces so that when they are mated, the pins will help take the stress if the pieces shift.

MIL-HDBK-60 from the US government contains a lot of useful information about threaded fasteners, along with definitions and formulas for various applications. You can download it for free at EverySpec (*http://bit.ly/fasten-hndbk*). Fastenal also provides a good online fastener guide (*http://bit.ly/tech-ref-guide*). Although it mainly addresses large bolts and nuts, the information can be easily scaled down for smaller fasteners.

The drive type is another important selection criteria. The two most common choices are Phillips and hex (i.e., Allen) drives. When selecting a Phillips drive, be careful not to get confused with a JIS cross-drive type screw head, which looks a lot like a Phillips but doesn't behave the same way (it won't "cam out," for example). See Chapter 4 for more on the JIS drive type.

Hex drives are popular in aerospace subsystems, military gear, and scientific instruments. Generally, you'll find hex drive screws and bolts in applications where there is a need to apply a significant and specific amount of torque to a fastener. A Phillips drive part will determine how much maximum torque you can apply by virtue of its design, while a hex drive will not stop you from twisting a hex wrench into two peices if you're not careful (or shearing the head off a bolt or screw). For this reason, hex drive fasteners are often installed with a torque wrench.

> **TIP** Never use a slotted-head screw or bolt if you can avoid it. Screwdrivers tend to slip out of the slot, and it doesn't take much to damage the slot to the point where it is unusable. For this reason, the Phillips drive was invented about 80 years ago to minimize damage to fasteners by "camming out" when the screw stalled at maximum tightness. A Phillips or hex socket drive is a much better choice than a single-slot part.

Use a screw that is just long enough to do the job; self-tapping screws can cause damage if they drive too far, especially in the case of plastic self-tapping types. Holes that are sized and tapped for a particular screw, such as the mounting holes in the aluminum frame of a hard disk drive, will accept only a screw of a specific maximum length. A screw that is too long either won't drive in completely, or it might drive through into something that it will damage (such as a circuit board in the path of the screw).

When using a nut, also try to use the shortest possible screw. A screw or bolt that protrudes out beyond the nut can interfere with other components and might get bent. A bent screw with a nut on it can be very difficult to remove gracefully without resorting to a cutting tool of some type.

WASHERS

Washers are essential when using fasteners to create a reliable load-bearing mounting point and help prevent the screw from working itself out and coming loose over time. A flat washer under the head of the screw or bolt helps to distribute the force applied by the screw. With metal, a washer slightly wider than the head of the screw is usually sufficient. If you are attaching to something thin or soft, such as polystyrene plastic, use a larger flat washer to help spread out the stress on the material. The same reasoning applies to soft materials like wood.

Figure 2-4 shows some of the available washer styles, and Figure 2-5 shows the typical assembly order for a screw and nut with washers.

FIGURE 2-4. Various types of washers

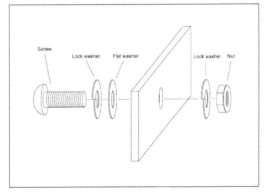

FIGURE 2-5. Using washers

A flat washer is also sometimes used as a spacer, although it is not a good idea to use a stack of washers to try to compensate for a screw or bolt that is too long to begin with. Occasionally an assembled bolt stack will include a flat washer under the lock washer beneath the nut. Although this does somewhat reduce the effectiveness of the lock washer, it also helps to prevent marring the underlying surface when the lock washer "bites" into the material.

A lock washer helps prevent a nut from becoming loose. These types of washers come in three basic styles: split-ring, inner-tooth, and outer-tooth. Figure 2-6 shows how a split-ring lock washer works.

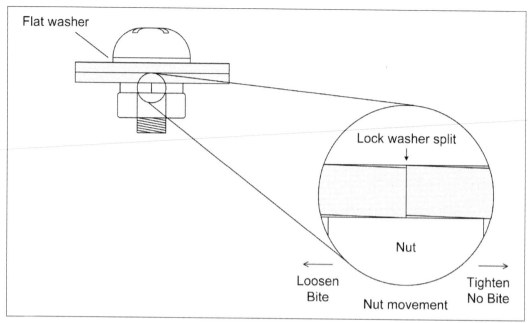

FIGURE 2-6. Split-ring lock washer

The edges of the split are bent slightly so that the nut will move over them smoothly when turned in the tightening direction, but the sharp edges will catch (or bite) on both the nut and the underlying surface if the nut moves in the loosening direction. This isn't enough to prevent the removal of the nut, but it is enough friction to help keep the nut secure. Be careful not to overtighten a split-ring lock washer, as it can be pressed flat, effectively disabling the locking capability of the washer.

The inner- and outer-tooth washers work in basically the same way, except that with these types, each tooth is bent slightly to bite into the nut and surface material if it starts to loosen. Figure 2-7 shows a selection of toothed washers. Toothed washers are often used with soft materials, such as plastic or sheet aluminum. Toothed washers are also used for establishing a ground connection, as the teeth can cut through paint or an anodized finish into the underlying metal.

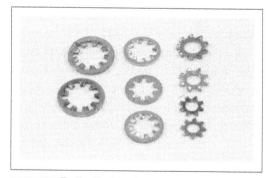

FIGURE 2-7. Toothed lock washers

When you are using a screw in a pre-threaded hole, there is, of course, no nut to lock, so the lock washer is placed under the head of the screw. A flat washer may also be used to protect the underlying surface, if necessary, but the smooth flat washer may reduce the effectiveness of the lock washer.

SELF-TAPPING SCREWS

Sometimes it makes sense to let a screw cut its own threads, and that is just what a self-tapping screw is designed to do. If you've ever worked with wood or wallboard (sheetrock), you've probably encountered a self-tapping screw. A self-tapping wood screw can cut its own threads as it is driven into place in a soft material such as wood.

Figure 2-8 shows a selection of different self-tapping screws.

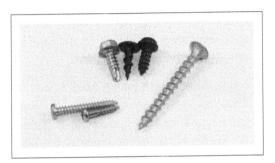

FIGURE 2-8. Self-tapping screws

Plastic is another material that works well with self-tapping fasteners, and they can also be used with soft or thin sheet metal. The small enclosures and electronics chassis sold by companies such as Hammond and LMB often use self-tapping sheet-metal screws to affix covers and panels to the main box or chassis.

As handy as they might appear, you should avoid self-tapping fasteners if possible. If you are working with something already built with self-tapping screws, sometimes you can't avoid them, but once a self-tapping screw goes in and comes out, it will almost never go back in as securely as it did the first time. This is especially true of sheet metal, because the original hole in the metal will be deformed by the screw when it is first installed. Chapter 4 describes a handy trick you can use when reinstalling a self-tapping screw.

Rivets

Rivets are commonly used to fasten sheet material, such as the aluminum skin sections of aircraft, fiberglass panels on a golf cart, and aluminum canoes. With suitable washers or purpose-made rivets, rivets can also be used to attach sheet plastic to a metal framework.

It is interesting to note that at one time (about 100 years ago), rivets—in the form of large red-hot chunks of metal shaped like a threadless bolt—were used to build the frameworks of skyscrapers, bridges, and the *Titanic*. In fact, until the widespread acceptance of arc welding and other modern methods of attaching one piece of metal to another, rivets and large bolts were the primary fasteners for large structures.

In electronics applications, rivets come in handy for creating small metal enclosures, attaching a bracket to something like, say, a metal can for use as a 2.45 GHz antenna, or even attaching a small metal enclosure to a section of metal pipe (like a light-beam sensor or sender on a chain-link fence post).

The type of rivet most commonly used for electronics work is relatively small, about the size of a #6 screw. Figure 2-9 shows some of the sizes available. Figure 2-10 shows an example of the tool used to install rivets.

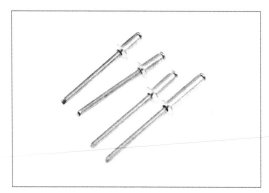

FIGURE 2-9. Small blind rivets

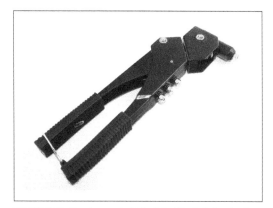

FIGURE 2-10. Hand-operated blind rivet installation tool

The items shown in Figure 2-9 are so-called *blind rivets*, which means that they are designed to be installed from one side of the work without requiring someone with another tool to apply pressure on the opposite side. They are sometimes also called *pop rivets*, after the brand name POP originally used for these products.

Blind rivets are available in a variety of head styles, including domed, flat, flush, and countersunk types. They also come in a range of sizes, lengths, and materials.

Blind rivets are installed with a tool that pulls a metal rod through the rivet body to cause the rivet to expand in a hole. The tool then trims off the excess length of rod flush with the exposed seat of the rivet. Chapter 4 covers riveting tools and how to use them. Chapter 16 discusses how blind rivets can be used to fabricate electronics enclosures or modify existing equipment or devices.

Adhesives and Bonding

Another way to fasten two (or more) things is with adhesives. Numerous types are available, ranging from cyanoacrylates (sometimes called "super-glue"), to single- and multi-part epoxies, to silicon rubber formulations like the adhesive used to glue the heatshield tiles onto the space shuttle. Adhesives are a science unto themselves, and some companies specialize in nothing but adhesives for special applications.

If you think you want or need to use an adhesive, it would be wise to do some research and see what is available. If nothing else, just reading the packages at the hardware store can be informative. When you are applying an adhesive, a pair of latex or nitrile gloves can save your hands (but make sure the adhesive won't attack the gloves!), and a wooden popsicle stick or medical tongue depressor can easily be trimmed with a sharp knife to make a disposable spreading and mixing tool.

GLUES, EPOXIES, AND SOLVENTS

Water-based glues work by creating a rigid matrix of linked chemical bonds in the adhesive material between two parts to be attached. Generally speaking, when the glue is "wet" it is in a liquid or semiliquid form, and the molecules in the glue can move and slide around quite easily. Wet glue has very little internal cohesive strength. It also has the ability to flow into the microscopic pits, bumps, and pores of the parts it is applied to. Some examples of water-based glues are the common white glue found in school classrooms or the wood glues used by carpenters.

For a water-based adhesive to form a tight bond, it needs to interface with the parts being bonded. With materials like wood or paper, this isn't a problem, since these are porous materials. Water-based adhesives usually don't work well with nonporous materials such as plastic or metal.

When a glue "dries" through evaporation or chemical reaction, the molecules in the glue can no longer move, and the internal cohesive force increases. Some types of glues also shrink, thereby pulling the glued parts closer to one another, while others expand slightly. The main point, however, is that the glue forms a hard interface between two parts that both adheres to the parts and is internally cohesive so it won't break apart under normal stress.

An *epoxy* is a type of adhesive that utilizes a chemical reaction (curing) to create internal cohesive bonds. Epoxy adhesives are based on an epoxy resin, which may be any of a number of compounds from what is called the *epoxide functional group*. The word *epoxy* actually refers to the cured form of epoxy resin.

An *epoxy resin* is a type of polymer that consists of chains of molecules. When the resin reacts with a hardener agent, either contained within the resin itself or applied as an additive, the reaction causes a chemical reaction involving cross-linking that is referred to as *curing*.

Epoxy adhesives come in various forms. There are one-part formulations that use light (typically UV) to start the curing action, while others work on contact with the air. A two-part epoxy consists of a resin and a hardener. These are mixed just prior to use. The shelf life of two-part epoxy is long, often on the order of years. Once the parts are mixed there is a period of time, typically between 5 and 30 minutes, when the epoxy can be worked before it starts to set and become too stiff to manipulate. Full curing can take up to 24 hours, depending on the formulation.

There is a vast range of applications for epoxy-based adhesives, and more for non-adhesive applications. Epoxies are known for excellent adhesion, good chemical and heat resistance, good-to-excellent mechanical properties, and excellent electrical insulating properties. With the appropriate formulation, epoxies can be used to bond materials such as metal, glass, wood, ceramics, plastics, and other resin-based materials (e.g., fiberglass and carbon-fiber materials). Some types of epoxies feature high thermal insulation, while other types offer thermal conductivity combined with high electrical resistance.

Unlike adhesion, *plastic bonding* is the process of causing two parts to partially dissolve at the point where they meet, and then allowing that joint to re-harden so that the two different parts actually become one. This applies only to plastics, and it is the plastic equivalent of welding two metal parts to one another. Bonding can be accomplished using heat generated by focused ultrasonic vibrations, applied by a heated metal tool, or chemically using a solvent. Here we will be looking at the chemical approach to plastic bonding; and if you have ever built a plastic model car or airplane or assembled PVC plumbing or conduit, then you are probably already familiar with plastic bonding. The ever-popular "clamshell" packaging is an example of ultrasonic thermal bonding, and cheap plastic toys are sometimes assembled with hot flat-tip tools that press a molded stub down into another piece and weld the two at that point.

Adhesives that utilize bonding work by attacking and literally melting the plastic to create a welded connection, and the plastic can react to the solvent very quickly. One chemical of this type is known as methyl ethyl ketone, or MEK, also referred to as butanone. It is particularly effective with polystyrene plastics. It also works with polyvinylchloride (PVC) and clear acrylic plastic. MEK can be purchased in small amounts at most hardware and home improvement stores. It is typically a thin liquid that will evaporate rapidly, and it must be used with good ventilation.

WORKING WITH WOOD AND PAPER

You can glue wood easily using adhesives specifically formulated for that purpose, although you can also use a general-purpose epoxy and get good results if you are careful and pick the correct one. Depending on how strong you want the joint to be, and whether or not it needs to be waterproof, you can use standard white glue, general-purpose glue, or specialty carpenter's wood glue. Even hot glue will work for some applications.

The same general caveats for wood apply to working with paper in all its various forms. A wide range of paper products is available, from heavy poster board to corrugated cardboard. These are useful for assembling a prototype enclosure or creating a scale model of something. Common white glue works well with paper (as any creative child can testify), and most wood glues will also work. Hot glue is popular with the arts-and-crafts crowd because it adheres reasonably well in the short term and is easy to apply.

Although it might tempting because of its convenience, resist the urge to use hot glue for anything except cardboard, fiberboard, and craft projects. Hot glue can be very unreliable; it's somewhat brittle when cool, and its adhesive properties on nonporous surfaces like metal or plastics are rather poor. It's great for making throwaway holiday decorations, but for long-term applications, not so much. Hot glue can also deliver some nasty burns if it comes into contact with bare skin while still in the molten state.

WORKING WITH PLASTIC

Plastics are a good place to consider using adhesives, but you need to be aware of just what type of plastic you are working with. Polyethylene, for example, is often heat-fusion-bonded to seal different pieces together, as standard adhesives and resins don't adhere to it very well. If you are working with something like polystyrene or PVC, you have a number of adhesive choices available.

> **TIP** There are multiple methods for identifying plastics, ranging from burning a sample sliver to laboratory spectroscopy. Another way is to take a sliver or sample piece and apply the adhesive you want to use. If you are using MEK or a MEK-like solvent-based adhesive, then the sample will show signs of a reaction with the adhesive (sagging, melting, deformation, softness, etc.). If it does nothing, then you will need to consider a different adhesive formulation, like a two-part epoxy.

Epoxies are a good choice, provided that the epoxy is formulated for the materials to be bonded. The downside to epoxies is that they tend to take some time to properly cure, so if you are in a hurry, you might want to consider something else. Also, because plastics are nonporous, the glued joint may be prone to breaking if it is overstressed by being bent or twisted. If it does break, it will most likely be at the place where the epoxy meets the material; the epoxy itself is tough and usually stays intact.

When you're working with polystyrene, PVC, or ABS materials, the best choice is to use a solvent-type bond (also known as *solvent welding*), unless there is some compelling reason not to. If you purchase MEK (described in "Glues, Epoxies, and Solvents" on page 26), you might want to consider making your own "glue" rather than try to work with the MEK in its liquid form. To make it thicker and easier to use, you can dissolve some bits of scrap polystyrene into it. The resulting goo is basically what you get from a tube of model cement from the hobby store, and it's a whole lot cheaper, too.

Note that a solvent like MEK won't work with some plastics. PEX, for example, seems to be unaffected by MEK, as is nylon. For these, you'll need to select a different adhesive, or resort to bolts, nuts, and brackets.

WORKING WITH METAL

Joining metal can be challenging, because the smooth, nonporous metal surfaces don't really offer much for the adhesive to grip. Water-based glues such as white glue and wood glue won't work, and solvent-based bonding methods are useless with metal.

Some types of specialty epoxies will grip metal surfaces, if the surfaces are properly prepared. When using an epoxy to join metal parts, be sure to follow the manufacturer's directions to the letter. As with plastics, an epoxy joint with metal parts is susceptible to shear forces. In other words, it might have good tensile strength, but it can break if twisted or bent.

Another method involves the silicon rubber adhesive mentioned earlier. It comes in a variety of types and colors, with some types useful for things like caulking a shower stall, and other grades suitable for use with the gaskets on automobile engines. Some high-temperature formulations are also available (such as what was used with the space shuttle). The downside is that silicon rubber works best for attaching large, flat surfaces. It doesn't work well for something that is small or narrow, such as when you're trying to glue one plate to another at a right angle.

In reality, the two best methods for attaching metal are to use fasteners or some kind of welding process. Epoxies come in third, but there is the issue of shear weakness. Silicon rubber works if there is a lot of surface area to work with, but otherwise it might not be a good idea.

SPECIAL-PURPOSE ADHESIVES

Be careful when working with cyanoacrylate adhesives. These glues work quickly and can create strong bonds, but they should be used with caution. Cotton or wool materials can react with cyanoacrylates in an exothermic reaction, and the heat generated can be high enough to cause a fire to break out. Cyanoacrylates also tend to have low shear strength, so while you might have trouble pulling a bond apart, applying sideways force will typically break it loose. These adhesives also tend to have a short shelf life, on the order of a year in an unopened package, and less than a month after they've been opened.

Summary

There are numerous ways to attach one thing to another. The best method depends on the material, the necessary strength, the desired reliability, and how much effort you want to put into it. Starting with bolts and screws, which are very strong fasteners when used correctly, we moved on to look at rivets, and finally adhesives. We did not cover gas, arc, or spot welding here.

With the information presented in this chapter, you should be able to make informed decisions about the types of fasteners that are suitable for your projects, and also be able to identify some of the less common types when modifying or re-purposing an existing device.

Although this chapter has mentioned in passing some of the tools used, and provided some warnings about selecting the right tool for a particular fastener, Chapters 3 and 4 provide further details about tools and their correct usage. Chapter 16 presents some examples of how to select and use various fastening techniques to create finished packages.

CHAPTER 3

Tools

THERE IS MUCH MORE TO TOOLS FOR ELECtronics than just screwdrivers and pliers. While most of the common tools can be found at a local hardware or home improvement store, many are unique to the electronics industry. These specialized tools have evolved over many years, in some cases starting out as modified versions of common hardware store types, and in other cases designed from the outset to fulfill a specific need.

For the most part, you shouldn't need to spend a lot of money on odd-ball tools if you stick to the common hardware described in Chapter 2 and avoid things like surface-mounted components with ultra-fine pitch leads. If you need to use an integrated circuit (IC) with something like 144 leads with hair-width spaces between the leads, then you should probably consider paying someone to mount it for you using screened solder paste and a reflow soldering system. For just a single project, it might not be worth the expense of acquiring a decent bench microscope and a fancy surface-mount soldering station and then learning to use it.

This chapter is a survey of some of the common tools you should consider owning for working with modern electronics. It is not intended to be a definitive or comprehensive guide. There are hand tools, power tools, and bench-mounted tools for tightening, cutting, drilling, and trimming. Other tools are used for soldering, inspecting, and finishing. I would suggest obtaining a selection of catalogs from companies such as Digikey and Mouser and perusing the tool sections. If you have a good electronics supply outlet nearby, it might be useful to browse its display racks to get an idea of what's available and examine the tools in person.

> Some of the tools described in this chapter can severely injure you if used incorrectly or carelessly. Always wear safety glasses when working with power tools, and always read and follow the manufacturer's safety precautions provided with the tool.

Screwdrivers

For every screw type there is a screwdriver. For most tasks, a basic selection of screwdrivers, such as the ones shown in Figure 3-1, is all you'll need.

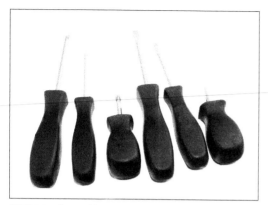

FIGURE 3-1. Screwdrivers

However, if you plan on disassembling consumer electronics or a toy, then you might also need some rather odd screwdriver types. A set of miniature and specialty screwdrivers, such as the one shown in Figure 3-2, is essential for these types of situations. You can find sets like this on eBay. Just bear in mind that these imported tools are generally not made from the highest-quality metal (that's why they are so inexpensive), and they can be easily ruined if used incorrectly.

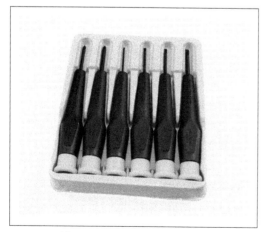

FIGURE 3-2. Miniature screwdrivers

Combination driver sets are available that use a common handle and a selection of driver bits. Figure 3-3 shows a set with slotted, Phillips, hex, Y, and other styles.

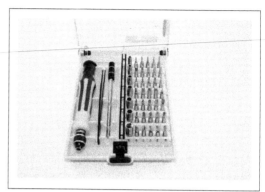

FIGURE 3-3. Combination driver set

> **TIP** Although inexpensive combination sets might seem like the answer to all your driver needs, you should bear in mind that you get what you pay for. The tool bits in these sets aren't always made from the best metal and tend to be brittle. The handles can also become loose or even break if stressed too much. That being said, sometimes the only place to find that oddball driver you really need is in one of these imported combination kits.

Pliers

The pliers available from a hardware store or other locations are acceptable for many tasks, but they are not always ideal for working with electronics. Figure 3-4 shows a selection of typical tools you might find at an auto parts or home improvement store.

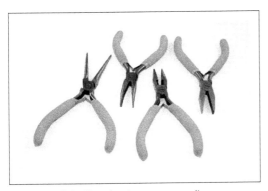

FIGURE 3-4. Selection of various common pliers

The large jaws of the common pliers are good for gripping stubborn bolts or holding a stiff spring while maneuvering it into position. But those same large jaws cannot really deal with things like resistor leads. For that type of task, you need a different tool.

Specialty pliers are available with narrow tips, and even with a 90-degree bend. Needle-nose pliers, shown in Figure 3-5, are a common tool in any electronics toolbox. But, as with any tool, they are intended for a specific set of applications, which are discussed in Chapter 4.

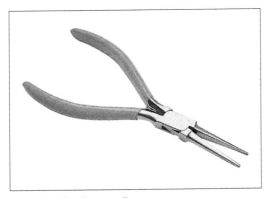

FIGURE 3-5. Needle-nose pliers

So-called *lineman's pliers* are a familiar tool for anyone who deals with household or industrial electrical wiring. They were originally developed for use by electrical linemen, hence the name. These tools are rugged and versatile and can be used to bend large-gauge wire, cut screws and small bolts, and pull cable through narrow channels or conduit, and some types have cut-outs to crimp lug-type connectors. They are sometimes used to hammer a screw or concrete anchor into a starting position, earning them the nickname of *electrician's hammer*. Figure 3-6 shows a typical example. You can find them at hardware and home improvement stores, online suppliers, and most electrical supply outlets.

FIGURE 3-6. Lineman's pliers

Wire Cutters

As with pliers, the typical wire cutters from the hardware store are suitable for cutting wires for home wiring and automotive work, but they are not designed for electronics. Specialty cutters are available with blades designed to cut flush against a surface to trim component leads on a printed circuit board (PCB) as close as possible, and some types have built-in retainers to prevent cut leads and wires from flying off. Figure 3-7 shows the so-called *flush cutter* type, which is most commonly used in electronics.

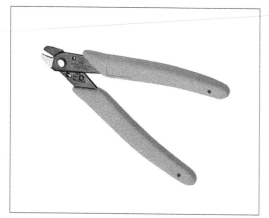

FIGURE 3-7. Flush wire cutters

The diagonal cutters shown in Figure 3-8, along with a pair of end cutters (also known as *nippers*), are common types of wire-cutting tools. As mentioned earlier, these are not designed specifically for electronics work, but they can, and should, be used for tasks that are too demanding for the flush cutters.

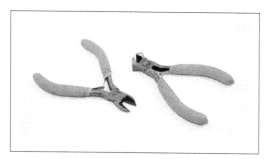

FIGURE 3-8. Diagonal wire cutters and end cutters (nippers)

Diagonal cutters come in a range of sizes, from small ones like the tool shown in Figure 3-8 to large cutting tools used by electricians and in industry. The end cutters are useful for nipping off wires close to a surface and can be used to (carefully!) remove small brads or finishing nails, provided that you are careful not to apply excessive force and put a notch in the blades.

There is one important thing to keep in mind when you are using wire cutters intended for electronics work: do not cut hard items that can create a nick or notch in the blades. In other words, use a pair of lineman's pliers or heavy cutters for things like clothes hanger wire or spring steel, diagonal cutters for large-gauge insulated wire, and flush cutters for the leads of components and thin wire only. Once your flush cutters have been nicked, that portion of the blade is useless (except perhaps for stripping small-gauge wires, but there are better tools for that).

Wire Strippers

Trying to strip the insulation from wire using something like a pair of flush or diagonal cutters is risky, at best. Unless you are very, very good, there is a distinct possibility that the wire will be nicked, and when that happens, the nicked spot can cause the wire to break. Some types of pliers include built-in wire strippers, but they don't always work that well, and they are fixed for one size of wire. A better option is a tool made specifically to strip wires, like the one shown in Figure 3-9.

FIGURE 3-9. Simple manual wire strippers

You can adjust the wire strippers shown in Figure 3-9 using the set screw seen on the lower handle. If set correctly, they will do a good job of removing most types of insulation without damaging the underlying wire. Manual wire strippers like these can be adjusted to any wire size from 10 down to 24 AWG (American wire gauge), but they can be hard to use on larger wire gauges.

There are, of course, fancier wire strippers available, such as the automatic strippers shown in Figure 3-10.

Automatic wire strippers cut and remove up to 1 inch of insulation in one step, repeatedly removing the same amount of insulation each time. The tool shown in Figure 3-10 handles 8- to 22-gauge wire, and a replacement blade set is available for 16- to 26-gauge wire.

As you might expect, there are also electric versions of automatic wire strippers. These are expensive heavy-duty tools intended for production line work, and they aren't something normally found in a typical small shop.

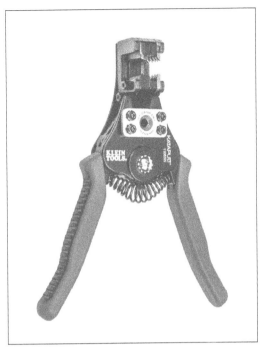

FIGURE 3-10. Automatic wire strippers

There is also a style of tool that incorporates lug crimper, machine-screw cutter, wire cutter, and wire stripper capabilities into the same tool. These are common in hardware stores, but I don't recommend them as wire strippers. When this type of tool is used to strip wire, it has a tendency to pull on the insulation rather than cut it cleanly, and sometimes the tool just doesn't have the right stripper hole for the wire. With the two types of stripping tools shown here, you can either set the tool for exactly the right size or you can rely on the blade set and grabber jaws to do a clean job without requiring you to wrestle with the wire.

Crimping Tools

If you are working with connectors that utilize crimped terminals, then a crimping tool is essential. There really is no other way to make a good connection with these types of connectors. Rectangular connectors that use insertable socket terminals are readily available, and they come in a variety of styles and sizes. Crimping tools range from simple things that look like pliers to aerospace-grade ratcheted devices with interchangeable crimping parts, called *dies*, for different contact sizes. Figure 3-11 shows a relatively inexpensive tool for working with crimp contacts like those used in rectangular connectors. The tools range in price from around $30 to well over $500.

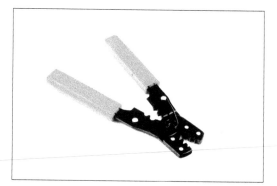

FIGURE 3-12. Spade lug crimping tool

Note that the crimping tool in Figure 3-12 has other things going on besides the crimping points. A tool like this can cut and strip wires, as well as trim machine screws. It can come in handy in an electronics shop on occasion, but it might not see a lot of heavy use.

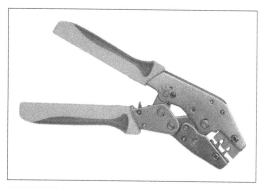

FIGURE 3-11. A crimping tool for small contacts

> TIP: Try to use the correct tool made for specific contacts or terminals. You may end up with a collection of crimp tools (that's typical), but when you need a tool, only the right tool will do the job correctly.

Do not try to use a crimping tool from an auto supply or hardware store for miniature electronic connectors. It won't work. Those tools, like the one shown in Figure 3-12, are made to be used with connectors such as the spade lugs and barrel splices found in automotive wiring, and for those types of applications they are fine. They just won't work with the extremely small terminal parts used in miniature connectors.

Lastly, if you plan to work with things like the "F" connectors used with cable TV wiring or make your own Ethernet cables, then you'll need crimp tools for those applications. You can find tools for video and telephone connectors at a well-stocked home improvement store, and there are multiple sources of RJ45 Ethernet-cable-making kits available online.

Socket and Hex Drivers

You can find a basic socket set at any auto parts, hardware, or home improvement store. These sets typically have socket sizes ranging from 1/4 inch to 3/4 inch, with some going smaller and some larger. It depends on how much you are willing to spend. Sockets come in metric sizes as well. Figure 3-13 shows a typical kit in a plastic carrying case. But, like many other common tools, these sets are not intended for the electronics industry. Rather, they are designed for automotive and other heavy-duty applications. Still, it is a good idea to have a decent socket set around.

FIGURE 3-13. Typical ratchet and socket set

Socket drivers made specifically for electronics work usually don't come with the ratchet. Rather, the kit is a set of tools that plug into a common handle, or each tool has an integrated handle, like the set shown in Figure 3-14.

Also note that the combination tool kit shown in Figure 3-3 comes with seven metric sockets, and still other kits are available with long (i.e., deep) sockets that can fit over a protruding screw or bolt shaft.

For hex-socket-head screws and bolts, you need a hex wrench or hex key, also called an *Allen wrench*. These come in both ANSI/ ASME (i.e, English) and a typical small set in a holder is shown in Figure 3-15.

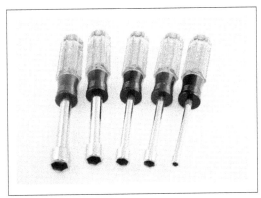

FIGURE 3-14. Socket tool set

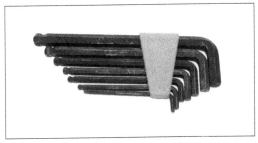

FIGURE 3-15. A standard set of hex wrenches (hex keys)

These hex wrenches have ball-type ends, and thus are sometimes referred to as *ball drivers*. This is a handy feature that allows the wrench to apply torque to a fastener without having to be directly aligned with the axis of the bolt or screw. As you can also see, this particular set already has some milage on it, but that's all right. Tools don't have to be pretty to work well.

I would not recommend the so-called *T-handle* tools, such as the example shown in Figure 3-16, for electronics work, mainly due to the expense, but that's largely a personal choice. This type of tool is popular in the optical sciences, where it is used to make fine adjustments to lens mounts and mirrors, and it is sometimes found in aerospace fabrication environments. These tools are less frequently seen in electronics labs or shops, however.

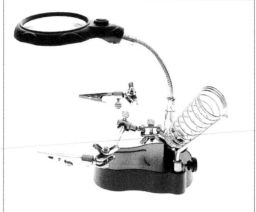

FIGURE 3-17. A "third-hand" fixture for holding work with small clamps

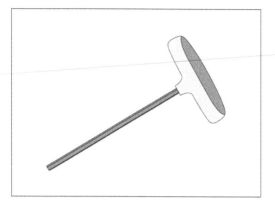

FIGURE 3-16. An example of a T-handle hex wrench

I would recommend getting sockets and hex wrenches in both ANSI/ASME and metric sizes, especially if you plan to hack an existing consumer or industrial device of some type. Most consumer electronics these days are assembled using metric fasteners, but a lot of industrial equipment made in the US is still ANSI/ASME.

Clamps

Clamps are designed to exert pressure to hold something, whether that is a single piece of wire, an electronic component, or two pieces of metal. Clamps may be locked or screwed into position, and they retain the pressure on whatever they're gripping until released. Technically, even a common alligator clip is a type of clamp, and it can be found in that role as part of so-called "third-hand" gadgets like the one shown in Figure 3-17.

Some types of clamps look like thin jaws with scissor-like handles. Also known as *hemostats*, these are just repurposed medical tools. Figure 3-18 shows some of the various types that are available for purchase from multiple sources. In the past, tools like this were used as heatsinks for soldering things like transistors into a point-to-point circuit. With the widespread adoption of printed circuit boards and temperature-controlled soldering irons, it became possible to solder parts onto the PCB without worrying too much about thermal damage (assuming, of course, that the person wielding the soldering iron has a good technique and doesn't loiter too long on the connection).

A hemostat is useful when you need to hold some parts in place for soldering, such as two pieces of wire or component leads. They are also useful for holding things while an adhesive sets or for just keeping something out of the way.

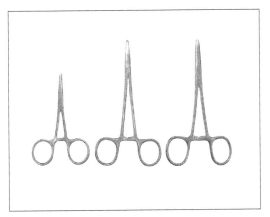

FIGURE 3-18. Hemostats (clamps)

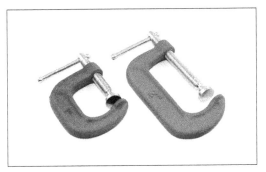

FIGURE 3-19. C-clamps

For larger jobs, there are various types of clamps available, from miniature C-clamps to plastic spring-loaded devices that look like clothes pins on steroids. Many of the clamps employed for woodworking can also be used for electronics work, so long as you keep in mind that things can get hot, and plastic clamps intended for wood might not fare well if the work pieces they are holding get too warm.

For dealing with metal, the ever-popular C-clamp is often a good first choice. Figure 3-19 shows some of the types available. I recommend keeping several of the smaller ones in your toolbox. You may or may not ever need the larger sizes, but having a couple around is not a bad idea.

Vises

A small vise is an essential tool in any shop. Figure 3-20 shows one type of bench vise commonly used in electronics work. Unlike its larger cast and forged cousins, this vise is lightweight, the head can rotate and swivel into various positions, and the jaws are padded with plastic strips to prevent damage to delicate items. This one happens to be made by PanaVise, and it consists of a model 300 base and a model 301 vise head.

The small vise shown in Figure 3-20 can also be incorporated into a compact workstation, as shown in Figure 3-21. In addition to the vise itself, it includes a soldering iron holder, circuit board holder, and a heavy base. This is a PanaVise model 315 circuit board holder with a model 300 base, all of which is attached to a larger base that has holders for a soldering iron and a spool of solder. It also also recessed tray spaces in the base to hold soldering-iron-tip cleaning sponges.

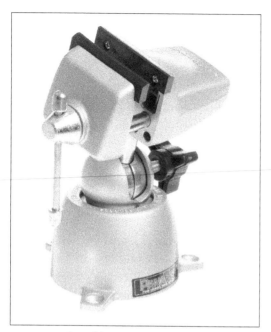

FIGURE 3-20. A small bench vise for electronics work

FIGURE 3-22. A large bench vise

big downside to a tool like this is that it really needs to have a permanent location—a very solid permanent location. So unless you think you might need to do some light metalwork, you can probably forgo the heavy-duty vise.

A somewhat smaller version of Figure 3-22 is shown in Figure 3-23. This model is designed to clamp to the edge of a table or workbench, and it's suitable for many lightweight tasks.

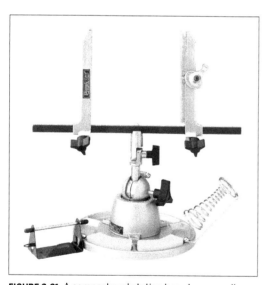

FIGURE 3-21. A compact workstation based on a small electronics vise

The vise shown in Figure 3-20 isn't suitable for bending metal or hammering out a stuck steel alignment pin. For that you need something like the one shown in Figure 3-22. A

FIGURE 3-23. A clamp-on bench vise

Rotary Tools

A good rotary tool is one of the most versatile tools you can own. There are many types available, ranging from the very cheap and somewhat flimsy to substantial tools suitable for production-line use. Some models come with a selection of speeds, and some have continuously variable speed control.

A rotary tool is extremely useful for cutting small square holes in a plastic box, trimming a slightly oversized printed circuit board to fit into an enclosure, drilling holes in a PCB, and performing other tasks that require a small tool with a lot of attachment options. Figure 3-24 shows a typical unit with variable speed.

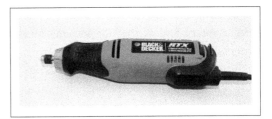

FIGURE 3-24. Rotary tool

You can also purchase a selection of various attachments for rotary tools in the form of kits from most hardware stores and home improvement centers, as shown in Figure 3-25. Of all of these, the most useful attachments are probably the miniature cut-off disks, sanding drums, and cutting tips. I purchase the cut-off disks in bulk packs, as I tend to go through them rather quickly. If you are building highly detailed model ships or custom jewelry, the other attachments might also be useful, but they aren't really essential for electronics work.

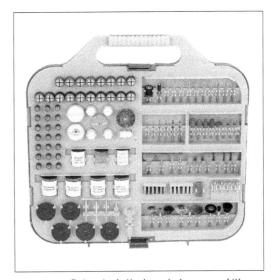

FIGURE 3-25. Rotary tool attachments (accessory kit)

One thing to keep in mind when considering a rotary tool is that the various attachments all use the same shaft size, typically 1/8 inch, and the chuck (the part that grips the shaft) is designed for only that size. In other words, you can't grab a 9/64 drill bit and expect it to work. It won't, not without an adapter or a different chuck. While a drill comes with a chuck that can be adjusted to accommodate bits of different diameters, a rotary tool usually doesn't.

> **TIP** A rotary tool is not really a drill. With the right bit, it can be used as a drill for soft materials and printed circuit boards, but it does not have the torque of a real drill. It is easy to burn out a rotary tool by using it for something it wasn't intended for, so if you need to drill holes, you really should reach for a drill.

Grinders

A small bench grinder is a handy thing to have. Grinders in general are useful tools to have, and they can save you a lot of time when you need to shape the tip of a screwdriver, take the corner off a bracket so it will fit, or clean up the edge of a piece of aluminum or clear acrylic. You can pick up one like the unit shown in Figure 3-26 from Harbor Freight for around $40, and some models come with a detachable flex-cable rotary tool, like the one shown.

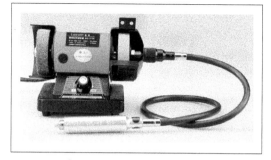

FIGURE 3-26. Small bench grinder with rotary tool attachment

Unless you plan to sharpen lawnmower blades or undertake some other heavy-duty activity, you probably don't need a heavy-duty grinder. These devices have high-power motors, often come with built-in work lamps, and are available in both bench and floor-mount versions.

A tool that I've found to be extremely useful is a right-angle grinder, like the one shown in Figure 3-27. You might not always need it, but when the need does arise, there is really nothing else that can do the job as quickly and efficiently as this tool. You can use it to cut small-diameter metal extrusions and tubing. You can also use it to remove the end of a machine screw or bolt that is protruding too far past a nut, or to remove the head of a blind rivet or ruined screw without resorting to drilling. It can also be used to slice up an aluminum chassis if you want to use part of it for something else, and, if you use it carefully, you can even cut square or rectangular access panels in a metal box.

FIGURE 3-27. Hand-held right-angle grinder

The downside to the right-angle grinder is that it is a loud—very loud—and powerful tool. It is also rather dangerous, and it can inflict serious injuries very quickly if you let it get away from you. Always use the side grip handle, wear heavy gloves, use eye protection, and never try to operate the tool single-handed. These tools aren't well suited to doing fine, detailed work (use a rotary tool for that), but they can slice through metal tubes, pipes, or extrusions with ease.

Drills

Electric hand drills are useful for a lot of things, but drilling a precise hole typically isn't one of those things. A small drill press is essential for drilling holes for screws, switches, LED indicators, or connectors. While you might be able to do a passable job with a common electric hand drill, the chances of slipping, creating an off-axis hole, or accidentally making the hole too large because of tool vibrations are great. If you want a clean, precise hole, use a drill press.

That being said, a hand-held electric drill is an essential tool for the shop. When you just need to make a quick hole, drill out a rivet, or use it with a driver bit to drive in a screw, an electric hand-drill is handy to have around. I recommend a battery-powered type like the one shown in Figure 3-28.

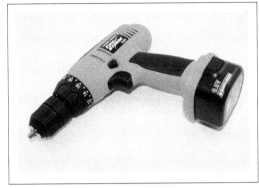

FIGURE 3-28. Cordless hand drill

For doing precision drilling, a drill press is essential. You can get a decent bench-top drill press like the one shown in Figure 3-29 for around $70 (this one came from Harbor Freight), and there are also rigs that allow a standard hand-held electric drill to be clamped into a drill-press-type fixture. These are all right for light jobs, but they do require care when you're mounting the drill. If you need to remove and replace the drill often to handle different tasks, then a dedicated drill press would probably be a better alternative; it's less tedious to deal with and it saves wear and tear on the hand-held drill.

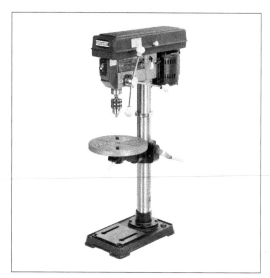

FIGURE 3-29. Small bench-top drill press

Drill Bits

The drill, be it hand-held or some type of drill press, is only a mechanism for spinning a tool. The working tool in a drill is the *drill bit*. Sets of drill bits in various sizes are widely available, in a variety of cases and holders. These range from small kits, like the one shown in Figure 3-30, to large sets found in machine shops and industrial settings.

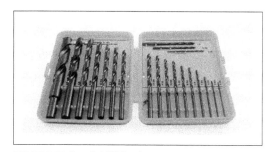

FIGURE 3-30. Small drill bit set

Not all drills are created equal, and most of the low-cost drill bit sets contain bits that will not stand up to extended or heavy use. The metal used to make the drills simply isn't that great, and a couple of attempts to make a hole in something like steel will quickly dull the tip. There are ways to drill a hole with lubricants that can help to reduce the wear and tear, but they won't stop it. Chapter 4 discusses recommended ways to use drills.

Some drill bits are intended for special-purpose applications, such as drilling the holes in a printed circuit board. Figure 3-31 shows a selection of these types of drill bits. Notice the color-coded plastic collars on the bit shanks. There are standard twist drill bits available with 1/8-inch shanks that can be used with a rotary tool, and you can typically find them wherever rotary tools are sold.

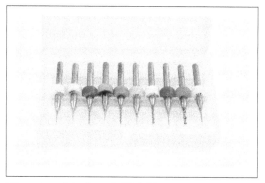

FIGURE 3-31. Selection of PCB drill bits

You can also find PCB drill bits listed as surplus items, mainly because the materials used to make PCBs tend to be hard on drill bits and they quickly get too dull to meet manufacturing standards. But, as always, caveat emptor. You may end up with a collection of dull or broken drills that are essentially useless. Better to spend a bit more and get one of the imported sets of new bits.

They might not be top quality, but with careful use, they will do the job.

Note that small drills bits like those shown in Figure 3-31 are useful for soft materials like plastic or wood, and of course they will make a hole in a PCB if they are sharp. In general, these types of drill bits will not work well with hard metals, but they can be used with soft metals like lead, silver, and gold.

Taps and Dies

Sometimes you really need a threaded hole, or maybe you want to make a special-purpose threaded shaft. Tapping is the process of cutting threads into an appropriately sized hole with a tool called a *tap*. A *die* is a tool for cutting threads into a blank rod to create a threaded shaft, and it is also sometimes used to repair damaged screw or bolt threads. Figure 3-32 shows an inexpensive tap and die kit.

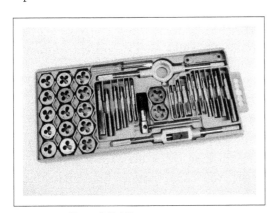

FIGURE 3-32. Tap and die kit

A tap and die kit is handy to have around, but it must be used with care to avoid damage to the tools. It's easy to break off a small tap in a hole that was drilled to the wrong size. Chapter 4 discusses some ways to use both taps and dies to get successful results.

Small Hand Saws

A small saw is useful for cutting things like tubing and sections of sheet materials. Specialty saws, such as a jeweler's saw, are extremely useful for creating odd-shaped holes in things.

Figure 3-33 shows a type of miniature hacksaw that is readily available at hardware and home improvement stores.

FIGURE 3-33. Miniature hacksaw

These tools will accept a standard hacksaw blade, and they are useful for getting into tight places. They do not have the same degree of stability that you will get with a regular hacksaw, however, because the blade isn't tensioned at both ends.

For doing detail work or creating small holes with odd shapes, something like the jeweler's saw shown in Figure 3-34 is the way to go.

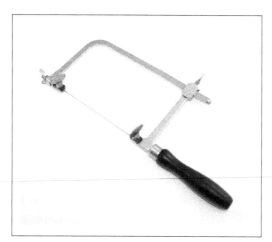

FIGURE 3-34. A typical jeweler's saw

A jeweler's saw uses a very narrow blade. This allows it to make sharp turns without binding, but it also means that it easy to snap the blade if you apply too much force while cutting. Figure 3-35 shows a pack of blades for a jeweler's saw.

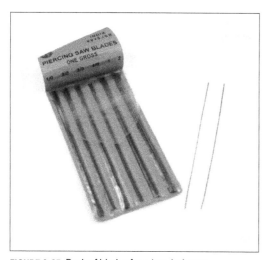

FIGURE 3-35. Pack of blades for a jeweler's saw

Miniature Power Saws

Since electronics often involves working with things that are small, there usually isn't a need for something like a full-up table saw or cut-off saw. However, you will find that there is often a need to cut plastic or metal tubing, aluminum extrusions, or small plastic or soft aluminum panels. For jobs like these, miniature tools are available, like the little table saw shown in Figure 3-36.

FIGURE 3-36. Miniature table saw

This handy device uses a standard 4-inch blade, and although it's somewhat clumsy to adjust, it does have a height adjustment. The blade is fixed in the vertical position, so it won't do angle cuts. You can find it for about $40 or so at places like Harbor Freight. Note that the table area is small, so it isn't suitable for anything larger than about 12 inches in size. You could, with a little effort, build table extensions for it to handle larger items, but then you would be treading into the territory of full-size table saws.

For cutting tubing, extrusions, rods, or other long, thin items, a cut-off saw with a grinder blade is a handy tool. Also known as a *chop saw*, a cut-off saw will produce a nice, clean cut, without the roughness and jagged edges that can result when a hacksaw is used. It's a lot quicker, as well. Figure 3-37 shows one of the various types of miniature cut-off saws that are available. They range in price from around $40 to over $300, depending on the brand, quality, blade size, and number of additional features. Note that some of these types of tools will work only with a grinder wheel, not an actual saw blade.

FIGURE 3-37. Miniature cut-off/chop saw

The primary advantage of small saws like these is their convenience and ability to deal with small work pieces. Just keep in mind that, like any powered tool, these are not toys. Although they are miniature versions of the full-size tools found in metal and woodworking shops, they can still cut you badly. Losing a finger with careless use is a distinct possibility, just as with their larger cousins.

Specialty Metalworking Tools

If you will be working with metal, there are some tools that you really should consider having in your toolbox. Metal can be a frustrating material to deal with, and having the right tool for the job can make the difference between annoyance and satisfaction.

When you are drilling or cutting a hole in sheet metal, the result often has sharp edges or little bits of leftover material called *burrs*. This is particularly true when you're working with the soft, untempered aluminum used for sheet-metal chassis parts. Steel pieces don't suffer quite as much from this, but drilling a 1/2-inch hole in a piece of 18-gauge steel can sometimes result in a somewhat ragged hole. This most often happens when you attempt to make a hole without going through a series of step-up holes first (see Chapter 4 for drilling techniques), which is a common mistake of the impatient.

A deburring tool, as shown in Figure 3-38, is essentially a swivel blade set in a handle. It is used in a circular motion to trim the inside of a hole or cut-out, removing burrs and helping to smooth out the cut. These tools are inexpensive and readily available from a variety of sources.

Creating a starting point is a good idea when working with metal. Even when you're using a drill press, a location point can help get the drill started to make a clean hole. The automatic punch, shown in Figure 3-39, is designed to take the place of a hand-held punch and ball-peen hammer. To use it, you simply push down until the internal spring-loaded mechanism releases, which causes the tip to create a small indentation in the material.

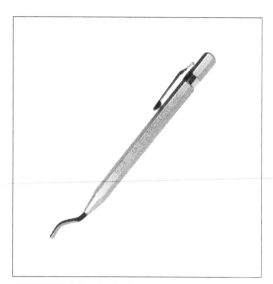

FIGURE 3-38. Deburring tool

FIGURE 3-39. Automatic punch

The concept of step-up drilling was mentioned earlier and is discussed in Chapter 4, but there is a tool that makes the process easier (which should not be surprising, since there seems to be a tool for every conceivable use). This is called a *step drill*, and a typical step drill set is shown in Figure 3-40.

FIGURE 3-40. Step drill set

A step drill is best suited for soft materials, such as plastic, soft aluminum, or mild steel as found in electrical boxes. Step drills are often used by electricians to make conduit holes in electrical enclosures. They can be used with hard materials if the tool is suitably hardened and rated for that kind of application, but many of the low-cost step bits won't take much abuse like that before they become useless.

Brave and bold electricians notwithstanding, I would suggest that a step drill be used with a drill press if possible, since it can potentially bind in the hole and twist the drill out of your hands. An electric drill can spin around and hit your hand before you can move it out of the way, and you could end up with a broken bone or two. At the very least, you can get a nice bruise if a drill gets away from you.

A good set of step drills tends to be rather pricey, running upward of around $100 for a set of three, but single drills can be had for a low as $10 each. When buying a step drill, get the best one you can afford.

Tweezers

A good pair of tweezers is invaluable for working with small parts that needle-nose pliers won't hold safely or reliably. Tweezers for electronics work come in a range of styles, as shown in Figure 3-41.

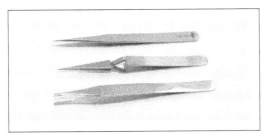

FIGURE 3-41. An assortment of tweezers

In addition to tweezers that operate like the ones found in a home medicine cabinet, there are also self-closing types, where you have to force the tips apart rather than force them together. It's not a bad idea to have at least one of both types in your tool kit.

Soldering Tools

A decent soldering iron or a soldering station is absolutely essential for working with electronics. Avoid the cheap soldering irons, as they won't hold up to heavy use and they don't hold their tip temperature reliably. Please don't even consider a soldering gun for electronics work (even though old Heathkit manuals show them being used to assemble vacuum tube equipment). A soldering gun is typically used for connecting heavy gauge wires and copper tubing, not for working with components on a PCB. In fact, if you happen to own a soldering gun, I would suggest hiding it so you won't ever be tempted to grab it in a hurry. Better yet, throw it away, donate it, or convert it into a low-cost spot welder. Figure 3-42 shows a low-cost temperature-controlled soldering iron.

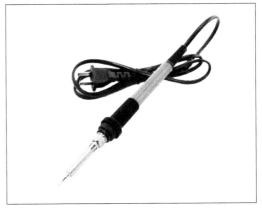

FIGURE 3-42. Soldering iron

Companies such as Weller make excellent industrial-grade soldering stations with continuous temperature control and grounded tips to reduce possible damage from static discharge. A variety of tip sizes and styles is available, ranging from something like a slot screwdriver to very fine points for surface-mount technology (SMT) work. Figure 3-43 shows a low-cost soldering station sold by Velleman, which includes the tool holder and a control unit to regulate the tip temperature.

Working with surface-mount parts requires soldering tools that are capable of working with small parts and closely spaced leads. Soldering stations for surface-mount work can be rather pricey, particularly for the stations that also include a hot-air attachment. The good news is that a soldering station like the one shown in Figure 3-43 will handle a lot of SMT tasks if used with a fine tip and the appropriate temperature. Chapter 4 discusses soldering temperature in more detail.

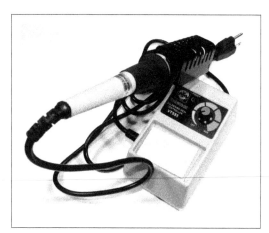

FIGURE 3-43. Soldering station

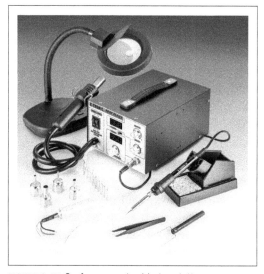

FIGURE 3-44. Surface-mount soldering station

Figure 3-44 shows a soldering station specifically designed for working with surface-mount parts (this one is X-TRONIC model number XTR-4040-XTS and is available through Amazon.com). In addition to the soldering iron with a fine-point tip, it also has a hot-air blower with a selection of nozzles. The hot air is used to desolder or rework a surface-mount part. The kit comes with the magnifying light shown. In Chapter 4, we'll look at how to work with surface-mount parts in more detail.

Magnifiers and Microscopes

If you plan on working with surface-mounted components, and you value your eyesight, you should consider purchasing some type of magnifier or low-power microscope. A true stereo microscope is best, but of course it is also more expensive than a simple single-objective type. Figure 3-45 shows a low-cost stereo microscope.

The types of cheap microscopes sold as toys for children are useless for electronics work, as are more high-end laboratory microscopes used in medical and biology work. The image quality of toy microscopes is usually rather bad, and both types typically have too high a level of magnification to be usable. An industrial microscope for electronics work is designed to provide a decent level of magnification (between 5X and 10X is typical, and some are adjustable) while still maintaining a relatively wide field of view. You won't be able to look at microbes in pond water with one of these, but you will be able to clearly see the leads on a TQFP144 surface-mount IC package.

The stereo microscope in Figure 3-45 was acquired for about $100 as a surplus (new overstock) item. You can find similar bargains on eBay or by checking out some of the optical surplus companies found online.

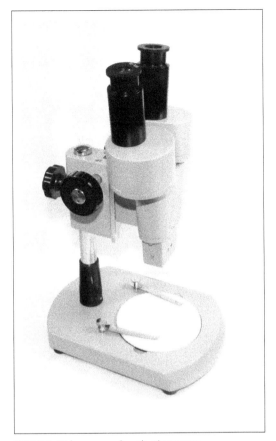

FIGURE 3-45. Low-power bench microscope

FIGURE 3-46. Bench work holder with magnifier

Another alternative—a handy item mentioned in "Clamps" on page 38—combines a magnifying glass with a pair of clips, as shown in Figure 3-46. It doesn't provide the same level of magnification as a microscope, but for many applications it's just fine.

Lastly, there are bench lamps with built-in magnifying lenses, like the one shown in Figure 3-47. Some are large enough to allow you to look through the lens with both eyes at once, so you can get a sense of depth.

FIGURE 3-47. Magnifying bench lamp

As a last resort, there are the jeweler's loupe and pocket magnifiers, like the one shown in Figure 3-48. The pocket magnifier, with its multiple lenses, is popular with geologists and rock hounds. It's also useful for electronics work, but it does require a hand to hold it and it needs to be fairly close to whatever you want to look at.

FIGURE 3-48. Pocket magnifier

Workspaces

Having a lot of tools on hand is great, but only if you also have a place to store them and use them. If you have tight quarters, a surplus combination cabinet and shelf unit like the one shown in Figure 3-49 might be just the thing. This particular item looks like it came out of a dormitory, or maybe an efficiency apartment, and it was purchased for about $40.

This setup has a board with rubber feet under it holding a small vise and an articulated work holder (with magnifying glass). Above that is a shelf with various supplies, and the whole front table folds up when it's not needed.

FIGURE 3-49. Combination cabinet and shelf unit with fold-down table opened

Things like the drill press, the miniature table saw, the cut-off saw, and the soldering station really need to live on a workbench of some sort. You can purchase relatively inexpensive workbenches from places like Sears or Harbor Freight, or you could spend some serious money and get an industrial-grade, metal-frame workbench for several hundred dollars.

Another alternative is to repurpose old metal office and dormitory furniture. If you have a local college or university that holds periodic auctions, it is possible to find some good bargains. Some manufacturers also hold auctions occasionally, and these are a good place to pick up used industrial-grade fixtures (if you don't mind scuffs, dings, acid burns, and other minor defects).

A good toolbox is also essential. Figure 3-50 shows a small, but overstuffed, toolbox. Larger roll-around toolboxes, like the ones used by auto mechanics, are nice if you have the space for them. A good one that will hold up to years of use is not cheap, however. You can get a roll-around cheap, but you'll more than likely end up with just a cheap roll-around that won't last very long.

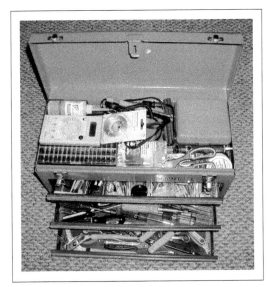

FIGURE 3-50. Overstuffed toolbox

Lastly, you can build overhead shelving for a table or desk for very little money. If you use standard shelf wood from a home improvement store, all you'll need to do is cut the shelf planks to the correct lengths and assemble each shelf with some brackets, some screws, and a suitable amount of wood glue, and you'll have a handy place to store books, parts, and tools while leaving the table or desk surface open. Paint is optional, but it does look nicer.

Summary

From what has been covered here, you should be able to gather up a minimum set of tools for your projects. As a suggested minimum, you will need screwdrivers, pliers of various types, flush and diagonal cutters, wire strippers, lineman's pliers, hex wrenches, sockets, soldering tools, a drill (and drill bits, of course), and a good set of files. A good pocket knife is also an incredibly handy thing to have.

In addition to checking electronics distributors, home improvement centers, hardware stores, electronics suppliers, and online sources, you might also want to look around your area to see if there are any surplus or second-hand tool shops. You can find a wide variety of things at low prices, although sometimes they do need to be cleaned up or at least readjusted before they are once again usable. If, for example, a shop has a bench-top band saw for $30 that really only needs to have the blade guides adjusted, then that's something that might be worth buying. You can also find buckets full of used screwdrivers, pliers, and wrenches at second-hand tool shops.

If you were to purchase everything covered in this chapter, you could end up spending less than $1,000, depending on where you purchased it and what level of quality you could tolerate. With some careful shopping, you might even be able to get some of tools used and get the total down to under $500.

A good set of tools can make all the difference between success and failure, but knowing how to use them and gaining experience is the other key ingredient. In Chapter 4, we will look at some of the techniques used with various tools.

CHAPTER 4

Tool Techniques

THIS CHAPTER DISCUSSES TECHNIQUES THAT can be used with some of the tools introduced in Chapter 3 and ties in with the discussions on fasteners found in Chapter 2. Not all of the tools from Chapter 3 are covered here in detail, because the uses for many of them are rather obvious. The objective is to expand on what was described in Chapter 3, discuss some of the little details that are often overlooked, and check out some of the less common tools.

In addition to the correct, and incorrect, use of common tools such as screwdrivers, this chapter will also cover some ways to use and modify sockets and wrenches. We'll look at how to solder various component types, with a special focus on surface-mounted components. We'll also cover riveting and dealing with stubborn fasteners. Drills and drill bits get a close look, along with the basics of cutting threads using taps and dies. An overview of cutting methods for dealing with sheet, bar, and rod materials is provided, which includes cut-off saws, hacksaws, the jeweler's saw, and rotary tools. We'll wrap up with a look at a technique for using a rotary tool that might surprise you.

This chapter is by no means a comprehensive discussion of tool usage. It is merely a summary of some useful techniques and things to look out for, along with a generous amount of advice gleaned from my own experience and the experiences of others. What works well for one person might not work well for another, and experience is the only way to develop your own techniques. This chapter is intended to be a starting point for acquiring that experience, or perhaps learning something new to add to what you already know.

> ⚠ Some of the tools described in this chapter can severely injure you if used incorrectly or carelessly. Always wear safety glasses when working with power tools, and always read and follow the manufacturer's safety precautions provided with the tool. Use lubricant when working with hard metals, and never reach into a running tool of any kind to clear out metal or plastic chips. Shut the tool down first, and then clean it out. Remember that a blob of molten solder can burn a deep hole in you, and even something as seemingly innocent as a pocket knife can do some real damage.

Working with Fasteners

The fasteners described in Chapter 2 are used with a variety of tools, which were covered in Chapter 3. Choosing the correct screwdriver, hex wrench, or socket is more important than you may think. When working with any kind of fastener, you must apply a significant amount of force on the head to rotate the screw or hold it in place while threading a nut onto it. The wrong size or type of tool can slip, ruining the part. Using a socket with a small socket-head bolt or nut where the socket doesn't seat correctly can also result in a ruined fastener.

The upshot here is that the correct tool will fit the fastener snugly, with little or no play, and it will turn or hold the fastener without damaging it. Inexpensive screwdriver and hex wrench sets are readily available that contain a selection of tools of reasonable quality, and the few dollars you'll spend is well worth it to avoid the aggravation of not having the correct tool or needlessly damaging a fastener. Small sockets can be easily modified to seat flush on a small nut or the head of a bolt, and both box and open wrenches (also known as *spanners*) can be modified to fit into narrow spaces or work with low-profile bolt heads and nuts.

SCREWDRIVER SIZES AND TYPES

Screwdrivers, like the ones introduced in Chapter 3, come in a range of standard sizes. Table 4-1 lists commonly available screwdriver sizes for UTS/ANSI screws, for both slot and Phillips types.

TABLE 4-1. Suggested screwdriver sizes for standard screw sizes

Screw gauge	Slot blade size	Phillips size
0	3/32"	0
1	1/8"	0
2	1/8"	1
3	5/32"	1
4	3/16"	1
5	3/16"	2
6	7/32"	2
8	1/4"	2
9	1/4"	2
10	5/16"	3
11	3/8"	3
12	3/8"	3
13	3/8"	3
14	3/8"	3
16	3/8"	3
18	3/8"	4
20	1/2"	4
24	1/2"	4

For the #4, #6, and #8 screws commonly encountered in electronics, the #1 and #2 Phillips are the appropriate screwdriver sizes to use. A #3 Phillips is sometimes used to mount equipment in a 19-inch rack using 10–24 screws, and there are cases where a heavy component (such as a power transformer, for example) is mounted to a chassis using a 10-24 or 10-32 bolt or screw. The #4 size tools are sometimes called for when you're dealing with things like large electrical distribution panels and automotive components, but they aren't used much with electronic devices. Most screwdriver sets include #1, #2, and #3 sizes, and the #0 or #4 sizes can be purchased separately if they are needed.

Be careful not to confuse JIS (Japanese Industrial Standard) cross-drive screws with Phillips drive parts. There is really nothing wrong with JIS screws and JIS screwdrivers; they're just not fully compatible with Phillips-style hardware. JIS fasteners are commonly found in electronics imported from Asia. Some tool vendors will sell imported JIS cross-drive screwdrivers as Phillips, when in reality they are actually JIS tools.

The main difference between JIS and Phillips is that a JIS screwdriver will handle a Phillips screw, but a Phillips screwdriver can wreck a JIS screw. So, if you've been using a JIS driver on JIS screws without realizing it, and then reach for a Phillips one day on the assumption that they are all the same, you may be in for a rude surprise in the form of a ruined JIS screw that you'll have to drill out or cut off to remove (this is discussed in detail in "Dealing with Stubborn Fasteners" on page 68).

Figure 4-1 shows the difference between the JIS and Phillips screw heads. Note the rounded corners of the Phillips screw, whereas the JIS has sharp right angles at the intersection. Figure 4-2 shows the difference between JIS and Phillips drivers. Note that the blades on the JIS tool are slightly shorter than the Phillips one, because the slots on a JIS screw are not as deep as on a Phillips screw.

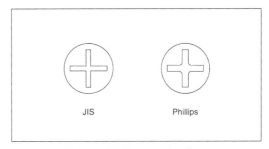

FIGURE 4-1. JIS versus Phillips screw heads

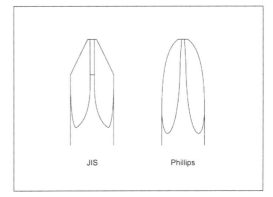

FIGURE 4-2. JIS versus Phillips drivers

The rounded corners at the slot intersection on the Phillips screw, along with the rounded blades on the driver, cause the tip of the screwdriver to "cam out" when the screw is fully seated and no more motion is possible. It's an intentional design feature. The JIS tool, on the other hand, is not designed to lift out of the screw head. A Phillips screwdriver will not seat correctly in a JIS-type screw, and it might round off the corners of the slot intersection on a JIS part.

Figure 4-3 shows how to check the size and fit of a Phillips screwdriver for a given screw, in this case a 6-32. Note that the #2 screwdriver on the left fits the screw, whereas the #1 tool on the right is a little too small. It might work, but the risk of damaging the screw head is much higher.

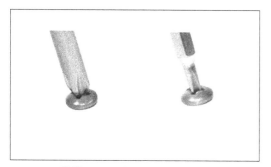

FIGURE 4-3. Phillips screwdriver sizing

Here's why all of this matters: a Phillips screw with the slots twisted out of the head can be a real hassle to remove. Figure 4-4 shows a damaged Phillips screw on the right, with an undamaged part on the left for comparison. The parts are identical in terms of size and threads, but one is now completely ruined. The damage shown in Figure 4-4 can occur when a screwdriver tip is too small for the screw head, or the screwdriver was allowed to bounce in and out of the head slots after it would have normally cammed out. This is always a potential risk when you are using some type of powered driver, but not so much when you're driving a screw by hand.

FIGURE 4-4. Damaged Phillips screw

Because a tool that is too large won't fit correctly in the first place, it should be readily apparent that continuing would be a bad idea (although I have seen people try it and, as expected, end up with a ruined screw head). The importance of correct tool sizing also applies to slotted head fasteners, but as stated elsewhere, these really should be avoided if at all possible.

If you are using a power driver with screws, you should be particularly careful. I have a used Ingersoll-Rand industrial driver that I'm quite fond of, but unless the torque is set correctly, it can ruin the head of a screw in the blink of an eye. A cordless drill can also damage a screw (as anyone who has ever used one for drywall installation or woodworking can attest), and even a small hand-held driver like the one shown in Figure 4-5 can ruin the head of a screw if used carelessly.

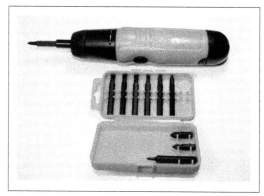

FIGURE 4-5. Battery-operated electric screwdriver

If used carefully, a battery-operated screwdriver like the one shown in Figure 4-5 can make your life a lot easier. If you do decide to purchase one, get a good one, preferably rechargeable. The own shown here uses AA size batteries, which occasionally will need to be replaced.

SELF-TAPPING SCREWS

Chapter 2 covered self-tapping screws, but it's worth mentioning again that they can be problematic to work with. Once a self-tapping screw has been removed, it's easy to drive it back in and destroy the threads it cut the first time. This is called *cross-threading*, and it's generally a bad thing.

There is a trick you can try to get a self-tapping screw to line up with the threads it cut into the material when it was first installed. First, put the screw into the hole without tightening it. Then, holding it in place with a slight amount of pressure on the screwdriver, turn the screw the "wrong" way (i.e., counter clockwise), as if you were backing it out. At some point, you should feel a slight "bump" when the start of the thread is encountered (if you are working with metal, you might even hear a faint click).

Now, immediately after the bump or click, turn the screw the correct direction to tighten it (clockwise), and it should pick up the threads and go in without cross-threading. Do not apply a lot of pressure when reseating a self-tapping screw; just let the existing threads pull it in and apply only moderate pressure when it stops to tighten it.

This little trick works best with plastics and thick sheet metal. With thin sheet metal, you might not have much of anything in the way of threads to work with, but it's still worth a shot.

HEX-SOCKET-HEAD FASTENERS AND HEX WRENCHES

A hex wrench (also called a *hex key* or *Allen wrench*) doesn't have a tendency to climb out of the hex socket in the way a screwdriver will with a Phillips or slotted screw. Figure 4-6 shows a hex wrench being used with a small cap-head bolt. Note that getting the wrench end fully seated in the hex socket is the secret to success.

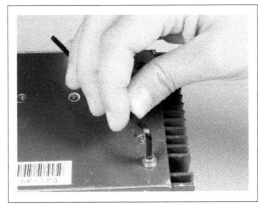

FIGURE 4-6. Using a hex wrench

> **TIP** A screwdriver is not a hex wrench, and trying to use one to tighten or loosen a fastener with a hex socket in the head will not do it any favors. Once the interior edges of the socket are damaged, a hex wrench will no longer seat correctly and the part is essentially useless. Inexpensive hex wrench sets appear often enough on the "bargain table" in hardware stores that there really isn't a good reason not to own one.

Typically, both ends of a common hex wrench are usable, as shown in Figure 4-7. Using the "short" end allows you to apply greater torque to the fastener by applying pressure to the long arm of the tool. Inserting the long end of the wrench into the hex socket won't allow for as much torque, but it does make it easier to spin the tool between your fingers and get the screw or bolt moving a lot faster.

ple concept, to be sure, but one that might not be intuitively obvious to those who haven't worked with hex wrenches before.

If you examine different sets of hex wrenches, you might notice that some of them have flat ends, whereas others have a rounded shape like a ball. The wrench shown in Figure 4-6 is a ball-end hex wrench, and Figure 4-7 shows a flat-end type. Figure 4-8 shows a side-by-side comparison of ball- and flat-end hex wrenches.

FIGURE 4-8. Ball-end and flat-end (straight shaft) hex wrenches

Flat-end (or *straight-shaft*) wrenches can tolerate higher amounts of torque and make more complete contact with the insides of the hex socket. A ball-end wrench is convenient for getting into tight spaces, but the ball doesn't make as much physical contact with the inner walls of the socket and the narrow section (the waist) of the wrench end might break under stress. Some people like to use the flat-end wrenches to initially loosen a part and then use a ball-end wrench

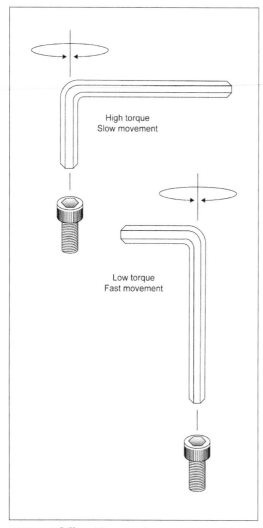

FIGURE 4-7. Different torque and movements possible with each end of a hex wrench

One way to take advantage of both ends of a hex wrench is to use the short end to break a fastener free and then flip the tool and use the long end to spin the fastener out. A sim-to turn it out. Conversely, a ball-end wrench can be used to turn a fastener into position, and then a flat end wrench can be used to fully tighten it.

When using very small hex wrenches, be careful not to apply too much force. The wrench shaft might snap under excessive torque, and while this won't usually damage the fastener, it does destroy a tool. Along the same lines, never use a hex driver in a power tool, such as an electric screwdriver or a cordless drill, unless the power tool has an adjustable torque limit. A hex wrench can apply enough torque to a fastener to twist the head off, leaving you just a threaded stub to deal with.

HEX-HEAD FASTENERS AND SOCKET WRENCHES

Socket wrenches come in both ratcheted and nonratcheted (direct-drive) forms. The use of the ratchet drive makes the tool less tedious to operate, and it also permits its use in restricted spaces. The socket wrench set shown in "Socket and Hex Drivers" on page 36 is a typical example. Some socket kits also include a screwdriver-type handle, which is handy for getting into tight spots, but it can't really apply the same amount of torque that a ratchet can supply.

Another variation is a direct-drive handle, which employs the mechanical advantage of a lever but does not incorporate a ratchet. Figure 4-9 illustrates the difference between a ratchet drive and a direct-drive socket tool handle. From left to right, we have a ratchet, a swivel-end direct-drive handle, a sliding bar handle, and a six-inch extension. These are all designed to work with sockets that accept a 1/4-inch drive. Many socket sets include both 1/2-inch and 1/4-inch drive sockets, along with a 1/4-inch drive handle that looks like a screwdriver, and a 1/2-inch to 1/4-inch adapter.

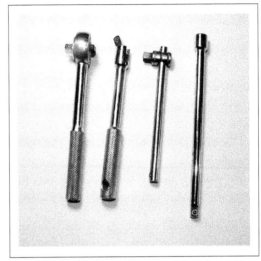

FIGURE 4-9. 1/4" drive socket handles and extension

An extension is useful if you need to get at a bolt that is beyond the reach of the wrench. Figure 4-10 shows an extension adapter for a 1/2" drive tool. It's the same idea as the one shown in Figure 4-9, only larger.

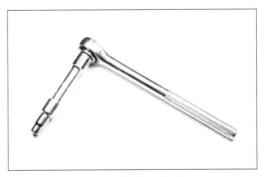

FIGURE 4-10. Socket wrench with extension

> Using pliers of any kind on a screw or bolt intended for use with a socket wrench is a sure-fire way to round off the edges of the head so that no socket will ever work with it again. Some types of pliers have a toothed section supposedly for use on hex-head parts. Don't use it unless it's an emergency.

A socket wrench can apply considerable force to a fastener. Be careful not to get too carried away and over-torque the part. The part might be difficult to remove at some future point in time, or even worse, the head of the bolt could shear off, leaving you with a headless bolt and a real problem. If there is a lock washer under the head or the nut, excessive torque can reduce its effectiveness.

A socket should fit snugly on the head of a hex-head bolt or screw, with little or no play (or wiggle). Chapter 2 discussed how a small hex-head fastener, such as a bolt or machine screw, can be difficult to drive if a socket does not seat flush around the head. *Small*, in this case, means 9/32 of an inch or 7 millimeters, and smaller, and sockets in this size range are usually 1/4-inch drive types.

Most common inexpensive sockets have a slightly rounded edge and angled faces at the entry of the socket that will prevent the socket from fully seating on the sides of the head of a small fastener or bolt with a low-profile head. Consequently, the tool could easily slip and damage the head. Figure 4-11 illustrates this situation. Note how an unmodified (stock) socket doesn't seat flush on a small hex-head bolt or screw. In some cases, it can be even more extreme than what is shown here.

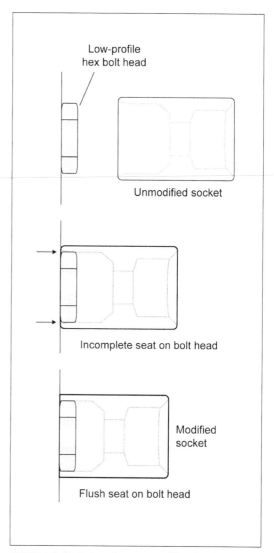

FIGURE 4-11. Socket seating on a small hex-head fastener

You can modify sockets by grinding off about 1/16 to 3/32 of an inch of the end of the socket, or up to where the interior angles end. The leaves a flat surface at the entry of the socket for the head of the fastener to fit into. Figure 4-12 shows sockets before and after such a modification.

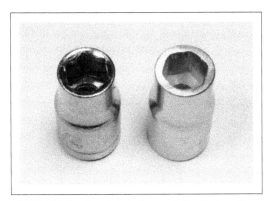

FIGURE 4-12. Small sockets before and after flush seating modification

Now the modified socket will seat flush on the head of the fastener, as shown in the bottom illustration in Figure 4-11. This isn't something I'd recommend for sockets larger than about 9/32 inch or 7 millimeters in size. Since the heads on larger fasteners tend to be taller and have more available surface area for the socket to contact, the modification really isn't necessary.

ADJUSTABLE WRENCHES

In general, an adjustable wrench is a bad idea, unless you are doing plumbing. But sometimes there is just no other way to get the job done due to space constraints or lack of other available tools. If used correctly and only occasionally, an adjustable wrench isn't necessarily bad, but it can be used badly.

If you are going to keep an adjustable wrench in your tool kit, make sure it is a good one. This is one tool you don't want to pick up from the bargain bin, and you should have a selection of sizes. In addition to single tools, adjustable wrenches often come in sets of four with 6-, 8-, 10-, and 12-inch wrenches. Specialty outlets carry 4-inch wrenches, as well. The wrench should have minimal play in the adjustable jaw, operate smoothly without sticking or binding, and look like some care went into its manufacture (forged with smooth machined surfaces).

Figure 4-13 shows the parts of a typical adjustable wrench. The worm drive moves the adjustable jaw in a track cut into the fixed jaw. The adjustable jaw is made so that the seat area will expand or contract while maintaining the correct angle to allow a hex-head bolt to seat correctly. Should you happen to encounter an adjustable wrench that does not maintain the correct angle in the seat, I would recommend tossing it into the recycle bin and getting a better tool.

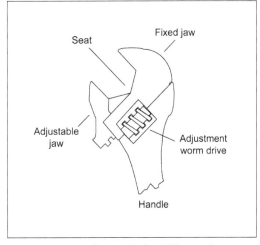

FIGURE 4-13. Parts of a typical adjustable wrench

If you do elect to use an adjustable wrench, you should make sure that the jaws are tight on the part, and be prepared to stop periodically to check that they remain tight. An adjustable wrench that slips is a fast way to ruin a hex nut or hex-head fastener. Figure 4-14 shows a typical adjustable wrench being used on a large bolt.

FIGURE 4-14. A typical adjustable wrench in use

There's a problem with how the wrench in Figure 4-14 is being used: it isn't fully seated against the bolt head. Notice that there is a gap between the head of the nut and wrench seat. The bolt head or nut should be as far back against the seat of the wrench as possible.

As with just about any tool, there is a right way and a wrong way to use it. The adjustable wrench is no exception, as Figure 4-15 shows.

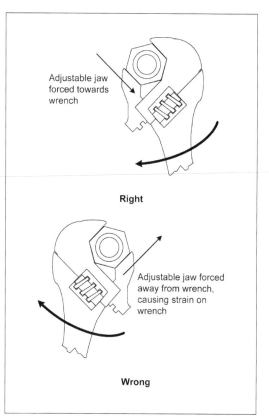

FIGURE 4-15. Right and wrong ways to use an adjustable wrench

WRENCHES (SPANNERS)

In electronics work, there is usually not much call for a box or open wrench, but as described in Chapter 3, small tools are available. Figure 4-16 shows a set of small open wrenches, ranging from 13/64 to 5/16 inch in size. These are old, so finding a set like this again might take some digging around in a used tool shop.

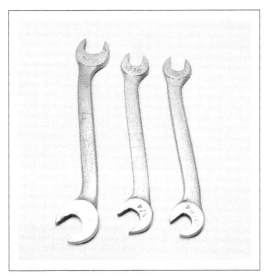

FIGURE 4-16. Set of small dual-size open wrenches

Figure 4-17 shows a more modern set. These are more conventional combination wrenches, with an open wrench on one end and a box wrench on the other. These range from 1/4 to 5/16 inch in size. Small metric sizes are also available.

FIGURE 4-17. Set of small combination wrenches

When you are working with a small wrench, it helps to keep pressure on the tool while it is in contact with the head of the fastener. A finger can be used for this purpose, although a small wood dowel or wood block will also work. Because the working faces of a box or open wrench typically have slightly rounded edges from the forging process used to manufacture them, they might not seat completely on the head of a small fastener. The width of the tool might also prevent it from reaching into tight spaces, such as when you are making an adjustment to something.

The set shown in Figure 4-16 is unique in that the wrenches are already thin and relatively flat, but this is more the exception than the rule. You can modify (hack, if you will) a small wrench to make it thinner by carefully grinding both sides of the open end of the wrench. Figure 4-18 shows a wrench prior to modification.

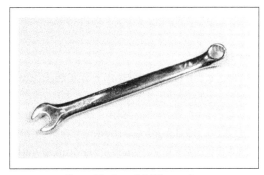

FIGURE 4-18. Unmodified small 1/4" wrench

A bench grinder was used to flatten the surfaces of the wrench shown in Figure 4-19, and the ground surfaces were then smoothed with a semi-abrasive buffing wheel. This technique can also be applied to the box end of the wrench. In case you're curious, this was done to a cheap 1/4-inch wrench that happened to be lying about, not doing much.

TOOL TECHNIQUES | 65

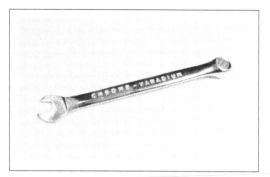

FIGURE 4-19. Modified small 1/4" wrench

FIGURE 4-20. Installing a blind rivet: setting the rivet

FIGURE 4-21. Installing a blind rivet: rivet set and stem cut

There's nothing wrong with hacking tools to make them fit your needs. Just be aware that, once modified, the tool will never be the same. But that's all right if you really need it and don't mind potentially sacrificing a tool. I generally keep a number of sacrificial tools on hand that I pick up from the bargain bin at the hardware store or scrounge up at one of the local second-hand tool stores. I wouldn't do this to an expensive tool unless there was a desperate need.

RIVETS

Using a blind rivet is simple and the technique is relatively easy to master. In addition to different diameters, blind rivets also come in different lengths, called the *grip depth*. The thickness of the parts to be joined determines the necessary grip depth. Figure 4-20 shows a blind rivet being set into a steel panel, and Figure 4-21 shows the rivet in place with the stem (or tensioning rod) cut free by the riveting tool. The tensioning rod is still in the barrel of the tool and will fall out when the handle is released.

There are four basic steps involved in installing a blind rivet:

1. Drill holes in both pieces to be joined at the location where the rivet will be placed.
2. Insert the rivet into the riveting tool.
3. Insert the rivet body though the holes in the pieces to be joined.
4. Pull the handle until the tool trims off the rivet's tensioning rod.

Figure 4-22 shows what is going on when a blind rivet is set. It also shows the two main measurements of a rivet: diameter and grip. The size of the hole in the material to be

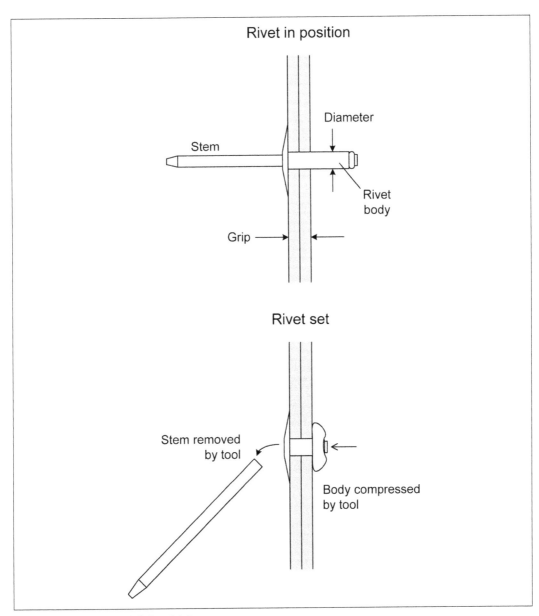

FIGURE 4-22. Blind rivet installation details

joined depends on the diameter of the rivet to be installed, which itself depends on how much load the riveted joint will need to bear. It is essential to make a hole that is just large enough for the rivet body to pass through, but not so large that the rivet is loose in the hole.

The *grip depth* is how thick the material can be for a given rivet size. Note that the grip depth is always less than the length of the rivet body. Blind rivets have a specified minimum and maximum grip. The minimum grip is the point at which a shorter length rivet may be used. A given rivet can be used with material that is less than the minimum grip depth, but it will cause the rivet body to "bunch" on the blind side. This might create a problem with other parts, and it can result in a less-than-optimal joint. In some cases, the tool might not be able to pull the tensioning rod out far enough to snap it off, which can be a hassle to deal with. If the grip depth is less than the minimum for a particular rivet, consider using the next shorter size.

The *maximum grip* is the thickest combination of materials the rivet can reliably and properly hold together. Do not use a rivet with a maximum grip that is less than the actual depth. When the rivet is set, it will not expand the blind side sufficiently to solidly and reliably join the pieces.

Some blind rivet tools come with a handy hole-sizing chart, along with a selection of rivets to start off with. Small rivets won't take as much stress as larger parts, and there is always the chance that a small rivet might be torn loose under extreme duress. A blind rivet can be used with soft materials such as plastic or even canvas if a backing washer is used to distribute the force exerted by the rivet. Some types of rivets are made with extended flanges for these types of applications. The rivets are cheap, so if you have a blind rivet tool, you really should have a good selection of rivets in different diameters and grip depths to go with it. Make sure you also have the correct drill bits to make the holes for the rivets.

DEALING WITH STUBBORN FASTENERS

Work with fasteners long enough, and at some point you will inevitably encounter a stubborn fastener that simply refuses to move. Here are some techniques for dealing with these situations.

Solvents

Using the solvent WD-40 can sometimes loosen parts that are bound by corrosion. There are also other solvents sold for this purpose. Use a cotton swab or small eye dropper to apply it to the base of the head of the stuck fastener.

> **TIP** Bear in mind that WD-40 is not really a good general-purpose lubricant. It is primarily a penetrating oil and was originally developed for water displacement and metal protection for military equipment, such as missiles and aircraft. It's also good for noisy door hinges, but it will dissolve heavier lubricants like motor oil and grease. Using it on something like a bicycle chain or a bearing is not a good idea.

Parts that have been sitting for a long time sometimes become "frozen" or "seized" in a threaded hole. One way to break them loose, in addition to applying a solvent, is to tap them. Do this using a small punch and a ball-peen hammer. A few taps might be all that is needed to break the part loose and get it moving. Just be careful not to damage the drive slots or the hex-socket hole in the head of the screw or bolt.

Torque

When you are dealing with a tight screw, it sometimes helps to apply a quick, sharp "snap" of force to the tool to break the bolt or screw loose. This requires a solid fit between the tool and the screw. Be careful not to apply an excessive amount of force, or you could be left with either a broken tool or a headless bolt or screw.

Temperature

In some cases, you can loosen a stuck fastener just enough to remove it by cooling it relative to the surrounding material. A can of so-called *freeze spray* (used in electronics to find temperature-sensitive parts) can suit this purpose. The trick is to apply the coolant directly to the head of the fastener and the metal immediately surrounding it after first applying a solvent. Cool the fastener and the surrounding area until it begins to develop frost, and then use the appropriate tool as quickly as possible. The objective is for the fastener and the hole it is in to contract; the fastener will become slightly smaller in diameter and the diameter of the hole will increase slightly (this might be counterintuitive, but it does work).

Drilling and Grinding

As a true last-resort measure, you can sometimes remove a stuck fastener by drilling out the head or grinding it off. If it is a flat-head screw, then drilling is about the only option. The trick is to drill out enough of the old screw head to pull the pieces it is holding apart, but not so much as to dig into the countersink below the head. Figure 4-23 shows how this is done.

FIGURE 4-23. Drilling out the head of a counter-sunk screw

The top image in Figure 4-23 shows the drill cutting into the head of the stuck screw. Notice that the diameter of the drill bit is slightly larger than the diameter of the screw shaft. In the bottom image, you can see what is left of the head of the screw on the drill bit. The operation was a success. You can remove what is left of the screw using pliers. Since the objective is to get the old screw shank out and discard it, you can be as rough with it as necessary.

For screws and bolts with raised heads (pan, cap, and such), it is possible to grind off the head of the fastener using a tool like a right-angle grinder (such as the one shown in "Grinders" on page 42). Just be careful not to let the grinder make contact with the underlying metal any more than necessary, and even then, some sanding and buffing might be needed to clean up the area. The end result of both procedures is the same: the shank of a now headless screw or bolt will be left that can then be removed and discarded.

Soldering and Desoldering

Soldering is the quintessential activity of electronics. One of the first odors you might notice when walking into an active electronics shop or lab is the smell of the hot flux from the solder wire. Almost every component used in electronics can be connected using solder in one form or another.

SOLDER TYPES

Solder is an alloy of two or more metals. In the past, the typical formulation used tin and lead, although nowadays other metals are found in alloys for lead-free solder or special applications requiring high conductivity or strength. The ratio of the component metals varies, depending on the application. In the past, the most common alloy for electronics work was a 60/40 blend of tin and lead (Sn60/Pb40). Another common type has a ratio of 63/37, which has a slightly lower melting point. For plumbing applications involving copper tubing and copper pipes, the ratio is usually 50/50, with a much higher melting point that requires a torch or a substantial soldering tool.

Early on, soldering was used to join copper tubing and the seams of tin cans, and then made the transition to electronics because of its ability to easily create strong, electrically conductive bonds between the metallic leads of components. Because of its lead content, solder was phased out of use in plumbing in the 1980s. Since 2006, lead-based solder has been phased out of most consumer electronics production, as well, although it is still common in high-reliability, aerospace, military, and hobbyist applications. Lead-free formulations are usually some variation of tin-copper (Sn-Cu) or tin-silver-copper (Sn-Aq-Cu). The lead-free types have different melting points than Sn-Pb formulations, although modern soldering irons can easily deal with this. The main issues with RoHS (restriction of hazardous substances—i.e., lead-free) solders involve ease of use and long-term reliability.

Although many people worry about the lead in solder, it's really not as big a hazard as some might think. In the solder alloy form, the lead is quite stable and poses a minimal hazard if handled properly. The primary concern is ingesting lead. Fermilab (the Fermi National Accelerator Laboratory) has a web page on lead solder hazards (*http://bit.ly/ solder-hazard*). The main points are to never eat or drink anything while working with solder, and be sure to wash your hands after using it. In fact, it's always a good idea to wash your hands after doing anything on the workbench, since the lead in solder isn't the only potentially hazardous material found around electronics.

In an editorial in the journal *ECN* (*http:// bit.ly/lead-free*), Jon Titus provides a quick overview of the impacts of making the transition to lead-free solders, most of which have been either neutral or negative. One major point is that to date there has been no data to indicate that lead-based solder is a major contributor to levels of free lead in the environment. The second major point is that lead-free solder joints can be less reliable and and are prone to introducing problems such as "tin whiskers," which can lead to long-term reliability issues. The original advocates of lead-free solder admit that this is indeed the case. Another interesting editorial, appearing in *Electronic Design News* (EDN) in 2007, listed and debunked some of the myths surrounding RoHS solder (*http://bit.ly/ROHS-myths*).

If you want to use RoHS solder, there are two main types available. A tin-copper alloy (99.3% Sn, 0.7% Cu) with a flux core is readily available in spools of various sizes. The flux used is the same for both tin-lead and RoHS solders, being either rosin-based or water-soluble. Sn-Ag-Cu (tin-silver-copper) solders are used by a majority of Japanese manufacturers for reflow, wave, and hand soldering in production environments. Kester has a good write-up on the use of RoHS solder for hand soldering (*http://bit.ly/hand-solder*).

The RoHS solders tend to not flow as smoothly as tin-lead (Sn-Pb) solder, and most types require a slightly higher working temperature (between 5 to 25 degrees C higher than with Sn-Pb solder). For this reason, some sources recommend that the beginner start off with a Sn-Pb solder and then make the transition to RoHS after gaining some soldering skill.

Although RoHS solder has seen improvements over the past few years, there are still some potential issues with tin whiskers, and because of the inherent properties of the metals involved, it will always be slightly more difficult to work with than lead-based solder. What type you choose to use is up to you.

> **TIP** When you are working with any solder, there will be some smoke. This is not the solder itself, as the soldering temperatures are not high enough to vaporize the metals in the solder. The smoke is coming from the flux. It's still not a good idea to breathe a lot of it, so having a small fan on the workbench will disperse it. Don't waste your money on a soldering iron with a "fume collector" attachment. These usually don't work (except for the really expensive industrial-grade tools), and in any case, a cheap fan from the local drugstore will blow the fumes away and keep you cool at the same time.

Tin-lead solder is still readily available and easy to work with and, provided that you don't ingest any, relatively safe to use. For electronics work, the most common form of solder is a wire with a hollow core filled with a flux material, typically some type of rosin (similar to what violinists use on their bows). Figure 4-24 shows a one-pound spool of rosin flux core solder. RoHS solder is also available in spools like the one shown in Figure 4-24.

FIGURE 4-24. Spool of solder wire

FIGURE 4-25. A tube of solder paste

For working with surface-mounted components, *solder paste* is essential. This is basically ground solder with flux and a binding agent that can be applied to the component mounting locations on a circuit board using an applicator. Solder paste is available in RoHS formulations as well as conventional tin-lead. Some solder paste comes in tubes that allow you to squeeze out what you need, or you can find it in small containers with screw-on lids. Figure 4-25 shows a tube of solder paste with a set of spare applicator tips.

SOLDERING TECHNIQUE

Chapter 3 covers soldering tools—namely, the soldering iron and soldering stations. If you want or need to work with lead-free solder, then investing in a good temperature-controlled soldering station would be a good idea. A selection of different tip sizes is also wise, since in some cases a fine tip is necessary (e.g., for surface-mount work). In other cases, you might need to apply much more heat (e.g., soldering to the terminals on a large toggle switch), so a large point or chisel tip would be appropriate.

Temperature

One of the questions that often comes up regarding soldering is how hot the tip should be. A general answer might that the tip of the soldering iron should be as hot as necessary to melt the solder and create a solid joint without damaging the part being soldered (or the underlying PCB). Fortunately, soldering has been around for a while, so there is a sizable body of knowledge (and advice) available to draw upon.

The melting point of 60/40 tin-lead solder is around 374°F (190°C), but the tip of the soldering iron needs to be much hotter than that, because it will cool rapidly when applied to the parts to be soldered. However, if the temperature is too high (as is sometimes the case with cheap, unregulated irons), the solder might not flow correctly, and the tip of the iron might not wet (take a thin initial layer of solder) properly.

Some sources recommend a tip temperature of around 700°F (370°C). For some work, you might want it hotter—for example, when soldering 18-gauge wire to a large switch or high-current connector of some type. If you are using RoHS solder, the tip will need to be hotter to adequately melt the solder. A hot iron will allow a connection to be made more quickly, which means the tip of the iron is in contact with the work for a shorter period of time than if the temperature were lower.

Another approach is to set the temperature to around 550°F (260°C) and increase it until you obtain the desired result. But bear in mind that if the temperature is too low, you risk damaging something because of the extended time necessary to heat the joint to the point where the solder will flow. The objective is to find the temperature range that works for you in typical situations. If you have a soldering station with a digital readout, I suggest keeping a list of temperatures for different tips and types of soldering activities, which you have determined for yourself, taped up near it.

Soldering Wires

Soldering can take some practice to master. Unfortunately, it is not as simple as just applying some heat and daubing on some solder. When you are soldering, you are, in effect, brazing, albeit at a low temperature. The idea is to get the work (the parts to be soldered) hot enough to let the solder melt and flow across and between them. The solder forms a bond with the metal of the components being soldered and becomes hard after it has been heated. Figure 4-26 shows a wire being attached to the connector lug of a small pushbutton switch.

FIGURE 4-26. Soldering wires to a switch

After the wires have been soldered to the switch, the solder lugs are covered with heat-shrink tubing, as shown in Figure 4-27. This protects the connections and prevents things from accidentally making contact with them. Using heatsink on a bare connection like this is always a good idea, even though it is an extra step.

FIGURE 4-27. Finished switch wiring with protective heatshrink tubing

Soldering a PCB

Soldering parts to a printed circuit board can be challenging. Unlike wires, the components on a PCB tend to be heat sensitive, so leaving the soldering iron on the connection for too long can potentially do some damage. Figure 4-28 shows a component being soldered to a PCB.

FIGURE 4-28. Soldering a component to a PCB

The place on a PCB where the lead of a component goes through is called a *pad*. A hole that exists in a PCB solely for the purpose of connecting one layer to another (like a vertical jumper of sorts) is called a *via*. PCBs come in single-layer, double-sided, and multilayer forms. Chapter 15 discusses the design of PCBs.

In Figure 4-28, you can see that the solder is applied to the point where the tip of the soldering iron meets both the component lead and the connection pad on the PCB. This happens to be a single-sided PCB, which can be slightly more difficult to solder than a double-sided PCB. In this case, the flux core in the solder was sufficient to help create a solid connection, but sometimes additional flux helps. Figure 4-29 shows the result for two components (a resistor and a small diode on the other side).

FIGURE 4-29. Component soldered to a single-sided PCB

A single-sided PCB can be tricky to solder, since there is little reason for the solder to wick down into the hole alongside the component lead. This can result in an incomplete connection, where only a part of the component lead is actually soldered and the solder flows around the rest of the lead without bonding.

It is not uncommon for the solder on a single-sided PCB to form a slightly rounded mound. This is because the solder does not flow down into the hole alongside the lead, but instead tends to form a "puddle" on the PCB pad. You can reduce this (if it bothers you) by using some additional flux and increasing the heat of the soldering iron.

With a double-sided (or multi-layer) PCB with plated-through holes in the pads, the solder should flow through the hole and form a fillet, or sloped fill, on each side of the PCB. The fillet will be largest on the side where the solder was applied and might not be visibly obvious on the component side of the PCB, but the solder should still flow through and fill the hole. Figure 4-30 shows a diagram of what a PCB solder connection

Using Heatshrink Tubing

Heatshrink tubing, also known as just *heatshrink*, is amazing stuff. It comes in a wide range of sizes, colors, and types. It's easy to cut, simple to use, and it does just what the name implies: it shrinks. It is used in everything from electric golf carts to computers. Anywhere that electrical tape might have been used in the past is a possible application for heatshrink tubing. The only somewhat tricky part about using heatshrink is getting it hot enough to shrink without damaging it or the components and wires around it.

As a general rule of thumb, it is safe to assume that a peice of common heatshrink will shrink to 1/2 of its original diameter, although some types can shrink to 1/4 of the original diameter or more. When heatshink is hot, it will self-adhere, and it has been used to create things like a temporary cap for a low-pressure gas line by crimping the open end of the heatshrink while it is still hot and pliable. It can also be used to seal an assembly inside. Chapter 12 describes a homemade optical isolator encased in a section of heatshrink.

As shown in Figure 4-27, a section of heatshrink was placed on each of the wires before soldering started. After the wire was soldered to the terminal of the switch and the solder joint was cleaned up, the heatshrink was moved down over the soldered terminal. A hot-air gun was then used to activate the heatshrink. Notice how the heatshrink extends from the terminal by about 1/3 of an inch or so along the wire. Since heatshrink becomes rigid after it has been heated, this serves as a strain relief for the wire. It also forms a seal around the wire to keep dirt and moisture out.

Using a cigarette lighter or a hot soldering iron is not a recommended way to work with heatshrink. Special hot-air blowers are available for just this type of application, and the desoldering/reflow blower on an SMT (surface-mount technology) workstart will also do a good job if used carefully. Too much heat can cause the tubing to split or melt, and it can do some serious collateral damage. Most common household hair dryers won't get hot enough to do the job. If you plan to work with heatshrink on a regular basis, you might want to consider purchasing an inexpensive hot-air blower. The hot-air blowers used for sweat soldering by plumbers will also work, but you have to pay close attention, as they run really hot.

The best way to learn how to work with heatshrink, and perhaps get some novel ideas for ways to use it, is to buy some and play with it. Heatshrink is not very expensive. Kits with various diameters and types are readily available online, and a local electronics outlet should have some available, as well.

should look like. Your results may vary, but the general idea is to have the solder flow through the plated hole in the pad along the component lead.

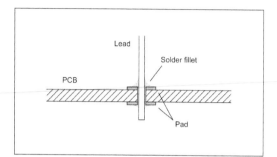

FIGURE 4-30. Through-hole solder joint for a double-sided PCB

> **TIP** Although some sources state that you should heat the pieces being soldered before applying the solder, it helps if the tip of the soldering iron is "wetted" with some solder before it is applied to the work. Without a good thermal connection, the work will take too long to heat up, and that heat will be transferred into something that could be damaged. Practice will allow you to experiment and see what works best for you, but I recommended using a wetted iron tip and applying the solder to the point where the tip meets the parts being soldered. The solder will melt and flow as soon as the tip and the work pieces reach an appropriate temperature.

When you are working with a PCB, and a single-sided PCB in particular, it helps if the PCB is clean—really clean. Wiping it down with 90% isopropyl alcohol will remove any residual oil or other contaminants. It also helps if the component leads are clean. Lastly, applying paste or liquid flux can make things go more smoothly, but it does leaves a mess that has to be cleaned up when the soldering is complete. Figure 4-31 shows a small container of paste flux that can be applied with a wood toothpick or a thin stick made from a wood-handle cotton swab (use the bare end, or just cut off the swab). I use the Puritan brand 6-inch wood shaft swabs, which can be purchased online through Amazon in packs of 1,000 for about $15 or less.

FIGURE 4-31. A small container of paste flux

After soldering, it is a good idea to go over the connections with a cleaning solution or just with 90% isopropyl alcohol, particularly if additional flux was used. A small brush, called an *acid brush*, is commonly used for this task. It is typically used to apply paste flux compounds for soldering copper tubing and pipes, but it is also used in the electronics industry for PCB cleaning applications. You can purchase acid brushes at electronics suppliers, hardware stores, and plumbing supply stores. They are also available in packs of 6, 12, or more from various online sources. I like to trim about 1/4 to 1/3 of an inch off the end of the brush to make it stiffer.

It is a good idea to remove leftover flux from the PCB after you have finished soldering, because flux can attract moisture and dirt. Also, if the humidity and temperature are just right, leftever flux residue can conduct a weak current, which introduces noise into a circuit and possibly cause erratic behavior.

> Alcohol and other cleaning solvents tend to leave a white residue on a PCB if it isn't flushed off completely. This is just flux residue that was deposited when the solvent evaporated. It is a good idea to remove it, and one way to do that is to go over the board with a dry acid brush. Another way is to wash the board in distilled water while brushing it. So long as there are no parts, such as potentiometers or switches, that might have problems with trapped water it won't hurt anything, and distilled water is nonconductive. Just make sure the board is completely dry before you apply power.

Soldering Defects

If the solder is disturbed while still molten, it can disrupt the solder's internal structure and create a rough-looking connection, or joint. Also, if the work is not hot enough, the solder won't flow properly, and the result will be a joint that is weak and might not conduct current reliably. This is called a cold solder joint, and an example is shown in Figure 4-32.

FIGURE 4-32. Cold solder joint*

When applied correctly, the solder should flow (literally) across the work, leaving a smooth, shiny joint. If the solder is applied quickly enough, the risk of thermal damage to components can be minimized. When the surface of the solder is grainy or dull, it indicates that the component lead may have moved before the solder cooled sufficiently, or the soldering iron wasn't hot enough. Although a cold solder joint might look like it is solid, it really isn't, and can lead to a broken joint down the road.

Although not as common as cold solder joints, broken solder joints can also occur. This happens when the lead of a through-hole component breaks free of the surrounding solder on a pad or terminal, often taking some of the original solder with it. This type of defect is typically seen when a circuit board has been subjected to shock or vibration. It was once quite common in portable electronics with single-sided PCBs but is less common with double-sided or multi-layer PCBs and does not occur with surface-

---

* Photo by coronium, CC-BY-SA-3.0, *http://bit.ly/cold-solder*

mounted components. A cold solder joint can be the precursor to a broken solder joint.

This type of problem can be difficult to track down, as it might be intermittent. Figure 4-33 shows what to look for on a PCB if you suspect that you might have a broken solder joint.

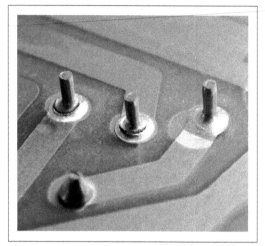

FIGURE 4-33. Broken solder joints*

To fix a cold solder joint or a broken joint, I recommend that you use some paste or liquid flux and apply a small amount of additional solder while reheating the joint. The flux will help to remove any existing contamination and oxidation, and the new solder will help to melt and reflow the original solder. Figure 4-33 illustrates an additional problem: the pad on the right has lifted from the PCB (delaminated). Since the solder joint appears to be solid, this can possibly be saved with the application of a small amount of epoxy. If this is a double-sided PCB, the opposite side will likely need some rework, as well.

DESOLDERING WIRES AND THROUGH-HOLE PARTS

There are basically two ways to remove solder and free a part: wicking and suction. Wicking involves the use of a copper braid, called (as you might guess) *solder wick*. Figure 4-34 shows a spool of solder wick.

FIGURE 4-34. Solder wick

The suction method involves a tool of some sort capable of pulling molten solder away from a pad or terminal using a temporary vacuum. The tools available include a squeeze bulb, a spring-loaded device called a *desoldering pump* (known colloquially as a *solder sucker*), and soldering repair and rework stations with built-in vacuum pumps. Figure 4-35 shows a typical hand-operated

---

* Photo by coronium, CC-BY-SA-3.0, *http://bit.ly/brokenjoints*.

desoldering pump (and, as usual with my tools, it's seen some use over the years—but it still works just fine).

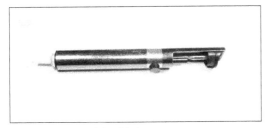

FIGURE 4-35. Typical desoldering pump

I don't consider the squeeze bulb to be of much use, so I won't discuss it here. If you want to try it, by all means do so. The technique is basically the same as that used with a spring-loaded desoldering pump, but it's been my experience that the bulb is just not as effective in terms of creating a quick vacuum to pull off the old solder. I'm also not a big fan of most soldering rework stations with a built-in vacuum pump. The concept is good, but often the execution leaves something to be desired. There are some good rework stations on the market, but they tend to be really expensive. The wick and the solder pump will work for almost every desoldering job that might come up, and they are cheap and easy to find.

Solder wick is relatively easy to use, but it does take a bit of practice to get it right each time. The first step is to make sure the solder joint is clean. Use some isopropyl alcohol to remove any residual dirt or grease. Next, apply the soldering iron and maybe a wee bit of solder to remelt the connection. Then, while the joint is still hot, place the wick on it as shown in Figure 4-36. The soldering iron is pressed down on top of the

wick and the molten solder should flow up into it. Sometimes it helps to apply a little bit of liquid or paste flux to help get the solder flowing smoothly. This is what the gooey stuff around the pads happens to be, and in this case it was liquid flux from a tube.

FIGURE 4-36. Desoldering with solder wick

The spring-loaded desoldering pump has been around for a long time. The early types didn't always work very well, but newer models have better seals on the internal plunger and more durable tips. The tip is made from a high-temperature plastic, so while it is possible to melt it, it takes some effort to do so.

To use the desoldering pump, first cock it by pushing down on the plunger until it locks. Then simply heat the solder connection until the old solder melts, quickly place the tip of the pump over the joint, and release the plunger. Figure 4-37 shows a desoldering pump in action.

FIGURE 4-37. Using a desoldering pump

As with the solder wick, using a desoldering pump is not something you can expect to master the first time (or second, or third). It take practice to develop a "feel" for when the solder is the right temperature and the best angle to use for placing the tip of the tool on the solder joint. In Figure 4-37, the tip of the pump is being held just above the pad while the iron melts the solder. It is then immediately placed on the molten solder and the trigger is pressed to release the plunger. Figure 4-38 shows the result.

FIGURE 4-38. Result of using a solder pump on a large pad

In case you're wondering, the images in this section came about because a potentiometer was mounted upside down on a PCB. It needed to be removed and reinstalled on the other side of the board. The operation was a success and it works correctly now.

SURFACE-MOUNT SOLDERING

Soldering surface-mount technology components (or SMDs, surface-mount devices) correctly and reliably by hand requires a considerable amount of practice and skill, but it can be done for many types of SMT components. You will need the correct type of soldering iron and tip, solder paste, a magnifier of some sort, tweezers, and a steady hand. Chapters 8 and 9 describe SMT component packages.

The first step is to apply the paste where the component will attach to the PCB, as shown in Figure 4-39. It doesn't really take much paste to do the job, but a little extra doesn't really hurt anything. It just makes things a bit messier and makes more work for the clean-up step.

FIGURE 4-39. Applying solder paste

Next, position the part (in this case, a resistor) in the correct location, setting it on and slightly into the paste. The binder in the paste acts as a weak glue, holding the part to the PCB, as shown in Figure 4-40.

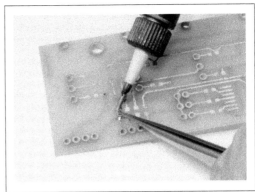

FIGURE 4-41. Soldering one end of an SMT resistor

When complete, the final result should look like Figure 4-42. Your part might have one end slightly raised, but that's all right for a manual soldering job. The main objective is for the part to be securely soldered to the PCB with bright, shiny solder joints.

FIGURE 4-40. Setting a part on the solder paste

While holding the part down with the tweezers (or something similar), apply the tip of the soldering iron to each end of the part. Be sure to touch the tip to both the part and the underlying PCB so that they will both heat at the same time. The paste should melt and flow onto and under the part, as shown in Figure 4-41.

FIGURE 4-42. Soldered surface-mount resistor

Soldering a surface-mount integrated circuit (IC) to a PCB is trickier than a single resistor. The first step is to apply solder paste to all of the connection points on the PCB, as shown in Figure 4-43.

TOOL TECHNIQUES | 81

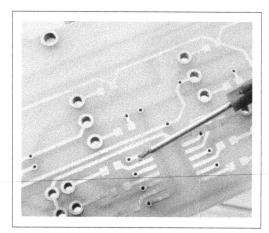

FIGURE 4-43. Applying solder paste for a surface-mount IC part

You want to apply just enough paste to make a good solder joint, but not so much that the part "floats" on the paste. In other words, it should settle close to the PCB when it is placed on the paste-covered pads, with little extra paste oozing out from under the leads of the part.

The reason for the close initial fit is to avoid having to force down the leads of the IC to meet the underlying PCB during soldering. If this happens, there is a definite possibility that they will spring back up when the soldering iron tip is removed, undoing the soldering job and making for a situation where you might spend considerable time trying to figure out why something works only intermittently.

To solder the IC, apply a fine-tip soldering iron to each lead of the IC, holding it only long enough for the paste to melt and flow. In Figure 4-44, a tool made of high-temperature plastic is being used to hold the part and keep it aligned on the PCB pads. The first pin is now finished and the tip of the iron is being pulled away. This will be repeated for each of the pin on the IC. Notice that another part has already been soldered to the board. It looks nice because it was cleaned up after soldering with alcohol and a brush.

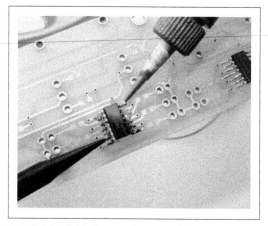

FIGURE 4-44. Soldering a surface-mount IC part

When the soldering is finished, you can use a razor knife with a fine tip to test each lead for a solid connection. None of the leads should lift when pried gently with the tip of the razor knife.

Figure 4-45 shows what a finished surface-mount IC should look like when it is soldered to the PCB correctly. It might be necessary to clean up around the IC using a brush and a cleaning solvent, as discussed in "Soldering a PCB" on page 74.

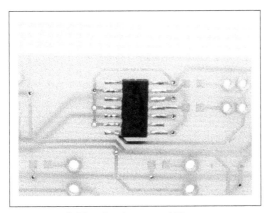

FIGURE 4-45. Soldered surface-mount IC

The SMT workstation shown in Figure 3-44 has both a temperature-controlled soldering iron and a hot-air attachment (there is a small pump inside the unit to provide a constant flow of air). A surface-mount part can be soldered (and desoldered) with the hot-air tool once it is set into the solder paste. This actually makes for cleaner connections with less leftover residue to clean up, but it does require a somewhat pricey workstation, and a fair amount of practice. Figure 4-46 shows the hot-air tool with a tip mounted on the end. The tool comes with a selection of tips with orifices of various diameters.

FIGURE 4-46. Hot-air tool for SMT soldering and desoldering

This tool can be used to solder an SMD IC to a PCB, provided that the temperature and air flow are set correctly. It takes some practice and experience to be able to define just what "correctly" means in each situation and with different models of SMT soldering tools. The preliminary steps (cleaning the PCB and applying the solder paste) are the same as before with the soldering iron.

When using a hot-air system, you should still hold the part down, mainly to prevent the air flow from moving it once the solder starts to melt. Use metal tweezers or a high-temperature tool made specifically for this purpose, like the one shown in Figure 4-44.

When you are doing hot-air soldering, it is important not to use too much solder paste. Not only does this reduce the mess you will need to clean up afterward, but the less solder to be melted the better. The hot-air nozzle isn't as discriminating as a soldering iron, so it is possible to heat not only the leads of the part, but the entire part as well. If you do elect to purchase one of these tools, you should also be prepared to buy a couple of tubes of solder paste and several surface-mount practice kits, and spend some quality time with it.

There are some surface-mounted components that simply cannot be soldered by hand. These include ball-grid array (BGA) parts or other package styles with connection points underneath the body of the part. Some types of connectors and inductors are made like this, and they can be difficult to solder manually as well. In these cases, you should probably consider paying someone with an SMT reflow soldering system to mount the parts for you. If this person has what is called a *pick-and-place* machine available, he or she can also populate and solder an entire circuit board automatically. The set-up costs for such a process can be significant, however, so it might not be practical for a one-off board. Still, most fabricators that deal with surface-mount production have people on staff who can manually place the parts on one or two PCBs and then put them through the reflow soldering machine.

Manual surface-mount soldering can be tedious, and it requires patience and a steady hand. For some applications, it is unavoidable, particularly if the parts you want to use come only in surface-mount packages. On the other hand, if you are building a prototype or just a one-off where size isn't a big issue, then opt for the through-hole or point-to-point soldering techniques, if possible.

SURFACE-MOUNT DESOLDERING

To be perfectly honest, without a hot-air tool like the workstation shown in Figures 3-44 and 4-46, desoldering any surface mount part with more than two connection points can be a major challenge, if not nearly impossible. The reason for this is that a SMD with multiple leads really needs to be desoldered all at once. Solder wick or a pump can get some of the solder, but it is difficult to get at the solder under tiny gull-wing or J leads. One technique involves heating and lifting each lead using a fine-tip razor knife, a large pin (e.g., a hat pin), or needle-point tweezers. This is tedious, to say the least, and with a J-lead part it can effectively ruin an otherwise reusable part. If the part uses metalized connection points under the body of the part, then using a razor knife, pin, tweezers, solder wick, or a pump is not a viable option.

Figure 4-47 shows the hot-air attachment for an SMT soldering station in use as a desoldering tool. You use the air flow to heat all of the part's leads at once by moving it around the part in a tight circular motion while applying pressure under the part with needle-point tweezers. There are also tools and attachments for the hot-air nozzle made specifically for this purpose that are not shown here.

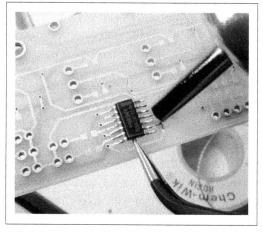

FIGURE 4-47. Desoldering a surface-mount IC with hot air

Once the solder under all the leads has melted, the part can be lifted off and set aside to cool. The PCB will probably need some cleanup with alcohol and perhaps the removal of leftover solder with solder wick.

Provided that the part wasn't heated too severely, it might still be usable, although excessive heat can stress the silicon chip inside the part and possibly cause some of the ultra-fine wires that connect it to the outside leads on the package to become unreliable. I keep a special parts bin just for "suspect" parts that have been removed from a PCB. They might work, and they might not, but if I'm in a hurry and willing to gamble, and component failure isn't a big deal, then I can dig through the bin and see if there's something I can use.

Cutting

The term *cutting* covers a lot of ground, ranging from dealing with things like rod, bar, and sheet stock, to cutting something to modify it. There are techniques for each situation, some better than others.

ROD AND BAR STOCK

Cutting small-dimension metal rod and bar stock, in the 1/8-inch to 1-inch range, is easily accomplished with a miniature cut-off tool like the one shown in Figure 4-48 (also described in Chapter 3). This type of tool is sometimes called a *saw*, but it is actually a type of grinder. It is relatively inexpensive, and it produces a clean, even cut through the material. The shield has been raised in Figure 4-48, so you can see the cutting wheel and the clamp.

FIGURE 4-48. Miniature cut-off saw in action

Notice that the miniature cut-off saw has a clamp built into its base. It is a good idea to use it—always. If you have a saw like this without a clamp, it would be wise to get one or improvise something. The reason is that, although it is small, this saw can produce a lot of torque, and it can easily throw a loose work piece across the room (or at you!). It can also remove fingers without too much effort, so you really don't want your hand near the blade trying to hold something down while you cut it.

There are different types of cutting wheels available for small cut-off saws. These aren't really blades, but rather thin grinding wheels. The level of abrasiveness is specified as the *grit*. A coarse grit, between 40 and 50, will cut heavier material, but it won't make a really fine cut. A finer grit, about 60 or so, will make a smoother cut, but it can overheat if used to cut thick material.

Some small cut-off tools are made to work with both blades and abrasive wheels, while others, like the one in Figure 4-48, use an abrasive wheel only. Since this tool is used only for cutting metal, there is no need for a toothed blade. There are other, larger, tools available with 7.25-, 8-, or 10-inch-diameter toothed blades for cutting non-metallic materials.

Alternatively, you can use a bench vise and a hacksaw, as shown in Figure 4-49. The secret to a clean cut is to make sure that the material is firmly clamped, the saw blade is sharp, and you have applied some form of lubrication. Trying to cut something by holding it down by hand on the bench might work with soft plastics, but it doesn't always

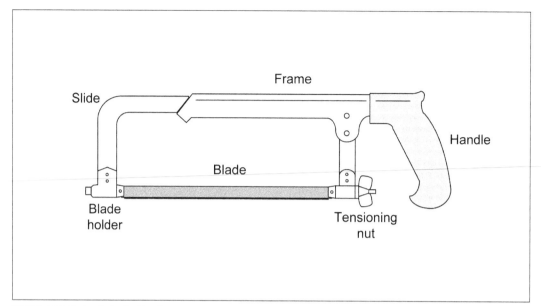

FIGURE 4-50. Parts of a typical hacksaw

work so well with metal. The hacksaw blade will tend to bind, bounce, and "chatter" on hard materials, resulting in a rough cut that might require additional work (grinding and sanding) to make it usable.

FIGURE 4-49. Cutting with a hacksaw

Move the blade of the hacksaw through the material with even strokes, and use the entire length of the blade. Let the teeth on the blade do most of the work for you. If cutting metal, go slowly—no more than about one stroke per second. The saw blade can get quite hot, even with a lubricant, and excessive heat can ruin the blade. The saw blades of the impatient have many dull and flattened teeth.

For a typical hacksaw blade, the cutting action occurs only on either the push or pull stroke, depending on how the blade was mounted in the hacksaw frame. I prefer to have it cut on the inward stroke (i.e., when pulling the saw toward you), somewhat like a Japanese woodworking saw. You might want to try it both ways and see which is best for you. Figure 4-50 shows the common names for the various parts of a typical hacksaw. Note that this type differs from the one shown in Figure 3-33, which is a more compact version.

Hacksaws like the one shown in Figure 4-50 have a sliding section of the frame that allows for different length hacksaw blades. The slide usually has two or three stop positions. In the US, blades are readily available in 10-inch (250 millimeter) and 12-inch (300 millimeter) lengths. The number of teeth per inch (or centimeter) is another specification, and you can get blades with 14, 18, 24, and 32 teeth per inch (TPI). The coarse blades are good for cutting thick metal, while the finer pitches are useful for making finer cuts and working with thin materials.

Using a lubricant when cutting metal with a hacksaw is always a good idea. A light oil works well. Table 4-7 lists lubricants for drilling, but these can also be applied to cutting with a hacksaw. The only exception might be the use of alcohol with aluminum. In that situation, I would recommend mineral oil or some other light oil. I like to use a bottle of oil with an extendable spout tube.

Lastly, make sure the blade is correctly tensioned in the saw frame. There is usually a wing nut at one end for this purpose. A loose blade might bind or create a ragged cut, so the blade tension should be tight, but not too tight. It is tight enough when the wing nut becomes too tight to easily turn by hand. Don't use a tool to make it tighter, as this could cause the blade to break or damage the frame.

SHEET STOCK

You can cut material such as sheet plastic with a pair of heavy-duty shop shears like the tool shown in Figure 4-51. These are useful for cutting thin material such as sheet brass or copper, soft plastics, and wire mesh.

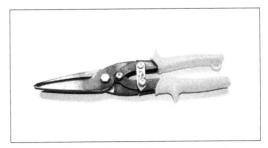

FIGURE 4-51. Heavy-duty shop shears (straight cut)

Generally speaking, plastic sheet, such as the polystyrene in Figure 4-52, is easy to cut, so long as it is not too thick. The shears will split the material and lift one side up along one of the blades, but thick material will not lift as easily, and the shears could bind. If the plastic sheet is too thick for the shears to deal with, then it's time to consider another approach, such as the rotary tool (see "Rotary Tool" on page 103), or the bench shear shown in Figure 4-55.

FIGURE 4-52. Using shop shears on sheet plastic

Sheet metal, unless it is very thin, can be a real pain to cut, even with special-purpose, aircraft-type sheet-metal cutters like those shown in Figure 4-53. These are also known as *snips* in some circles. These tools come in three forms: left, right, and center. Each type is designed to cut a curve in a particular direction (or no curve at all).

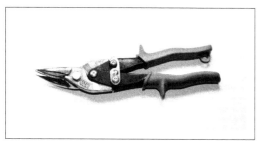

FIGURE 4-53. Aircraft-type metal shear (aviation snip)

Because metal is a stiff material, some effort might be required to make the cut. As with the shop shears shown in Figure 4-51, the metal shears will push up one side of the material as it is cut. It is easier to cut off a small section of the metal sheet, around 1/4-inch wide, and work into the desired depth in a series of shallow cuts. Trying to cut directly through a sheet of aluminum or thin steel from a starting position several inches into the material might be difficult, depending on the thickness of the material. In Figure 4-54, the cut was started deeper into the sheet because it was thin. If it had been thicker, it would have been easier to make a series of shallower cuts to get to the desired width.

Hand-held shears, or nippers, are handy for quick jobs or trimming something to make it fit, but the best solution for cutting both plastic and metal sheet stock is a bench shear, like the one shown in Figure 4-55. If you can justify purchasing one, then by all means do so. A good bench shear can save a lot of time and give the end result a professional appearance. This particular tool can also serve as a brake, which is useful for bending sheet metal into complex shapes.

FIGURE 4-55. Small bench shear/brake tool

A nibbler-type tool is yet another way to cut sheet metal. These tools come in manual, electric, and pneumatic forms. Figure 4-56 shows a manual nibbler tool.

FIGURE 4-54. Using aircraft-style metal shears

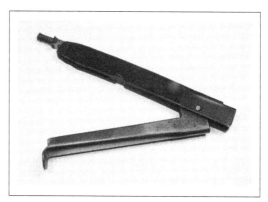

FIGURE 4-56. Manual nibbler tool

A nibbler works by literally nibbling away the material, one small bite at a time. For thin materials and short cuts, this tool can do a decent job, but for anything more substantial, you might want to consider using an electric or pneumatic-type nibbler tool. A manual nibbler tends to be hard on the hand after just a short time. You should also be prepared to clean up the cut with a file or a grinding attachment on a rotary tool. Nibblers don't always make clean, even cuts, but with some practice, they can do a decent job.

Last, but not least, is the rotary tool. For short cuts in thin material, it can do an acceptable job. Refer to "Rotary Tool" on page 103 for suggestions and examples of ways to get better control of a rotary tool and produce clean cuts. If you need a long, straight cut (6-inch or more), then you really need a bench shear like the one shown in Figure 4-55. The alternative is to make the best cut possible with the appropriate tool and then clean it up with a file, a grinder, or a bench-mounted belt sander.

Drilling

There's more to making a good hole than just eyeballing a drill bit and grabbing the power drill. By "good hole," I mean one that will fit the part correctly, be it a screw, a bolt, an LED, a switch, or anything else that needs to go through a hole. If the hole is too small, it will need to be drilled again, which is not a disaster, but can be an annoyance. If the hole is too large, the part will simply fall through it, and there are not many options available to recover from this kind of mistake.

SELECTING A DRILL SIZE

The appropriate size (diameter) of a drill bit depends on what you want to use the hole for. If you want a clearance hole (where

Fractions and Decimals

You might notice that many of the dimensions and drill sizes in this chapter, and in fact throughout this book, are given in decimal inches, not fractions. There are a couple of reasons for this. First, when you're selecting a drill bit for a tap hole, there usually just isn't a fractional drill size that will work as intended. Instead, we need to switch to the drill index system and select a drill bit on the basis of its decimal size. Second, when you're adding or subtracting dimensional values, the decimal representation is a lot more convenient than working with fractions.

Fractions do, however, get a lot of use in a typical machine shop. If you look at a standard drill size table, you might notice that the index numbers act as intermediate steps between the primary fractional dimensions. The fractional values increase in increments of 1/64 (or 0.0156): 1/32 is 2/64, 3/32 is 6/64, and so on. If you don't have a drill size table tacked up somewhere, you really should consider doing so.

A typical drill set from the hardware store or your local home improvement center will usually contain a selection of common drill sizes, from perhaps 1/16 inch up to 5/8 inch. What it won't contain are the drill sizes listed in Table 4-2. These are index sizes, and there are no fractional equivalents. If you plan on doing a lot of tapping, you'll need to also consider assembling a collection of index drills specifically for tapping, or purchasing a large drill bit set with both the standard fractional sizes and the index sizes.

something passes through it without interference), the bit needs to be slightly larger than the parts you want to use. If, on the other hand, you intend to tap the hole so that it will accept a threaded part, it needs to be sized correctly to allow for the tap to cut threads inside the hole to the correct depth.

For example, a clearance hole for a 6-32 machine screw would need to be no larger than 0.1495 inch, which is a #25 drill. If the hole will be tapped to hold a 6-32 screw, then it should be 0.1160 or 0.1065 inch in diameter, depending on the depth of the thread cut you want.

> **TIP** The tap drill sizes shown here are for both 50% and 75% taps. A 50% tap means that the threads will engage to 50% of their depth. This is usually more than enough for most uses, and because the tap tool doesn't cut as deeply, there is less risk that it will bind and break off in the hole during tapping. For very shallow holes, you might want to consider using a 75% tap. These are harder to create, but they will hold the fastener more securely when only a few thread turns are engaged. Also be aware that some drill and tap size tables give the 75% drill size, not the 50% size.

Table 4-2 shows a basic tap drill size chart for UTS/ANSI sizes. The drill index refers to the standard numbering system used for drill bits. Note that there are no fractional equivalents for these drill sizes.

TABLE 4-2. Drill and tap size in inches

Thread size	50% drill index	50% drill size	75% drill index	75% drill size
2-56	49	.0730	50	0.0700
4-40	41	.0960	43	0.0890
6-32	32	.1160	36	0.1065
8-32	27	.1440	29	0.1360
10-24	20	.1610	25	0.1495

Table 4-3 shows UTS/ANSI clearance drill sizes. These are mainly recommendations, but they are used throughout industry. If you don't have access to the specified index drills, choose a fractional drill size that is large enough to allow the screw or bolt to slide through but not so large that the head of the fastener falls through as well. The index sizes listed will produce what is called a *free-fit hole* (i.e., it will be somewhat loose). Most of the fractional sizes listed are closer to the tight clearances for each screw size. In the case of the 6-32 screw, there is no fractional size drill that is close, so a 9/64 drill might create a hole that is a little too tight, and the 5/32 might be too loose. Start with the 9/64 and see if it works for you. You can always make a hole larger with a file or a reamer, but you can never make a hole smaller.

TABLE 4-3. Free (loose) fit clearance hole drill sizes in inches

Thread size	Drill index	Drill size	Nearest fraction
2-64	41	.0960	3/32
4-40	30	.1285	1/8
6-32	25	.1495	9/64 or 5/32

Thread size	Drill index	Drill size	Nearest fraction
8-32	16	.1770	11/64
10-24	7	.2010	13/64

Table 4-4 shows some common tap drill sizes for metric hardware and Table 4-5 lists some common metric clearance drill sizes.

TABLE 4-4. Metric tap drill sizes

Screw size	50% metric drill size	75% metric drill size
1.5 × .35	1.25	1.15
1.6 × .35	1.35	1.25
1.8 × .35	1.55	1.45
2 × .4	1.75	1.6
2 × .45	1.7	1.55
2.2 × .45	1.9	1.75
2.5 × .45	2.2	2.05
3 × .5	2.7	2.5
3 × .6	2.6	2.4
3.5 × .6	3.1	2.9
4 × .7	3.5	3.3
4 × .75	3.5	3.25
4.5 × .75	4.0	3.75
5 × .8	4.5	4.2
6 × 1	5.4	5.0
7 × 1	6.4	6.0
8 × 1.25	7.2	6.8
9 × 1.25	8.2	7.8
10 × 1.5	9.0	8.5

TABLE 4-5. Metric clearance drill sizes

Screw size	Metric drill size	Closest American drill index/size
1.5 × .35	1.65	52
1.6 × .35	1.75	50
1.8 × .35	2.00	5/64
2 × .4	2.2	44
2 × .45	2.2	44
2.2 × .45	2.4	41
2.5 × .45	2.75	7/64
3 × .5	3.3	30
3 × .6	3.3	30
3.5 × .6	3.85	24
4 × .7	4.4	17
4 × .75	4.4	17
4.5 × .75	5.0	9
5 × .8	5.5	7/32
6 × 1	6.6	G
7 × 1	7.7	N
8 × 1.25	8.8	S
9 × 1.25	9.9	25/64
10 × 1.5	11.0	7/16

DRILLING SPEED

The appropriate rotational speed of a drill bit depends on the size of the drill, the type of drill, and what it is cutting.

For example, for a standard twist drill (the most common type) used in a drill press with aluminum or steel, the nominal speed varies from 3,000 RPM for a small (1/16 inch) drill down to 600 RPM for a 5/8-inch drill cutting steel. In general, the larger the drill, the slower the speed. Table 4-6 shows suggested speeds for various drill sizes and materials.

TABLE 4-6. Recommended drill-press speeds

Drill size	Acrylic	Aluminum	Steel
1/16 – 3/16	2,500	3,000	3,000
1/4 – 3/8	2,000	2,500	1,000
7/16 – 5/8	1,500	1,500	600

Controlling drill speed with a hand-held variable speed power drill is difficult, at best. Most hand-held cordless drills top out at around 1,500 RPM, while corded drills can reach 2,500 RPM. When drilling a small-diameter hole with a hand-held drill, you typically want to run the drill as fast as it will go and keep the drill bit well lubricated.

DRILLING THIN SHEET STOCK

When drilling holes in thin sheet stock (like the cover panels on some types of electronics enclosures) you should use a drill press, and the panel to be drilled should be sandwiched in between two sheets of heavier material, like thicker aluminum sheet or 1/2-inch-thick pieces of hardwood. Clamp it all down securely using C clamps. Figure 4-57 shows the concept behind this technique.

The benefits of this technique are that the hole will be cleaner (less jagged fringes) and the sheet metal won't be distorted by the pressure of the drill bit. You can also use just a bottom support to help reduce distortion. The downside with both approaches is that you will have to sacrifice a piece or two of material to your drill press. I recommend keeping a box of scraps near the drill press and just continue reusing them until they are too full of holes to keep around any longer.

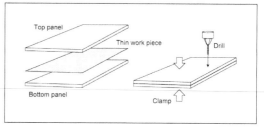

FIGURE 4-57. Drilling thin sheet metal between sacrificial panels

If this sounds like a lot of work just to drill a hole, well, it is. But even if the hole will be filled and covered by something like a switch or a knob, thin sheet metal can still distort a significant distance outward. The ideal solution is not to use thin sheet metal panels, but there may be times when that just can't be avoided.

LUBRICANTS

If you are cutting through hard metal, such as steel or tempered aluminum, a lubricant is a very good idea. Overheating a drill bit is a sure way to ruin it.

Some drill presses found in machine shops have a built-in lubricant pump and reservoir tank, but for the rest of us, a can of machinist's lubricant will do. If you are using a drill press and the work piece is clamped securely, you can apply the fluid yourself; otherwise, you'll need an assistant to help with the lubricant while you hold the drill.

You can pick up a can of tapping or cutting fluid at any well-stocked hardware or home improvement store, or even purchase some online. WD-40 also works well as a drilling lubricant, as does kerosene or rubbing alcohol when you're working with aluminum.

Just remember to continuously apply it to prevent overheating due to excessive friction.

A large syringe is sometimes useful in this case, as is a 1/2-inch or 1-inch-width brush along with a small metal can to hold the lubricant. In the past, it was common to find a cut-off soup can sitting next to a machine tool with kerosene or cutting oil in it and a small brush resting in the fluid.

Always check to make sure the cutting oil or lubricant you want to use is compatible with the material to be drilled.

Table 4-7 lists some drilling lubricant possibilities for different types of materials.

TABLE 4-7. Drilling lubricant suggestions

Material	Suggested lubricant/cutting fluid
Aluminum	Kerosene
	Isopropyl alcohol
	Mineral oil
Copper	WD-40
	Light machine oil (sewing machine or turbine oil)
	Mineral oil
Brass	WD-40
	Light machine oil
	Mineral oil
Iron	SAE 10 or SAE 20 oil
	Light machine oil (with flow)
	Purpose-made cutting fluid
Steel	SAE 10 or SAE 20 oil
	Light machine oil (with flow)
	Purpose-made cutting fluid

The SAE 10 and SAE 20 grade oils should be nondetergent motor oil. Do not use standard motor oil as a cutting oil, as it contains detergents and other agents. Also avoid the multi-viscosity oils (i.e., the 10W-40 type).

When Table 4-7 states "with flow," it means that the lubricant should be constantly flowing over the drill as it cuts into the material. The flow rate for something like the light machine oil doesn't need to be heavy, but drilling through iron or steel can generate a lot of heat, so using a brush might not be sufficient. Softer materials (such as aluminum, copper, and brass) are easier to drill without creating a lot of friction heat, so long as they are lubricated.

In general, lighter lubricants need to flow constantly when used with hard materials, and the lighter the lubricant, the more that needs to flow. This helps to prevent the build-up of heat (which is the main purpose of the lubricant), and lighter fluids will tend to burn off or evaporate faster than heavier lubricants. In any case, don't let the drill bit run dry, or you could end up with a dull or ruined bit.

Some machine tools use water-based cutting lubricants (called *emulsified coolants*) that flow constantly from a nozzle on a flexible tube that can be aimed at the tool. Other machines, like large multi-axis computer-controlled tools, operate inside an enclosure in a constant spray of coolant. For most tasks in a small shop, a brush or wash bottle and a selection of easy-to-obtain lubricants is usually fine.

Most of us don't have fancy machine tools with built-in lubricaton systems, but from the preceding discussion and Table 4-7, you can see that sometimes just drilling a hole with a standard drill press can use a lot of lubricant. The question then arises: where does all that lubricant go? I use an old rectangular cake pan for my miniature drill press; the entire tool is small enough to sit in it. For the other tools that don't need massive amounts of oil or kerosene, I use paper towels and place them where they will catch the drips. The large drill press uses a plastic tub that sits on the floor under the work table of the press. Some tools, such as manually operated vertical mills and lathes, have catch basins built into the frame of the machine for just this purpose.

If you are interested in the technical details of the effects of lubricants, and cutting fluids in particular, a paper from Michigan Technical University (*http://bit.ly/fluid-prop*) discusses the effects of cutting fluids on heat transfer.

PUNCHES AND PILOT HOLES

When you are drilling metal, it is a good idea to start with a punch point. This reduces the chance of the drill bit "walking" away across the surface of the material by giving it an indentation to settle into. The automatic punch described in "Specialty Metalworking Tools" on page 47 is an ideal tool for this. A center punch and a ball-peen hammer also works to create a starting point for drilling.

When you are drilling large holes, a pilot hole is a good way to align a larger drill. Typically you would want your pilot hole to be about 1/4 to 1/3 the diameter of the drill to be used to make the final larger hole. This provides enough of a hole for the larger drill to seat into. When you're using a hand-held drill, if the pilot hole is too small, the larger bit can jump out and go for a walk across the surface of the material. If you are drilling a very large hole, then you may need to create a sequence of progressively larger pilot holes.

USING A STEP DRILL

Using a step drill, like the ones shown in "Specialty Metalworking Tools" on page 47, correctly is an incremental process. This is particularly true when you're drilling into metal. The first step is to create a pilot hole large enough to accommodate the smallest step on the drill bit. Then, using generous amounts of lubricant, work each step of the drill bit into the material with a pause in between to allow the bit to cool.

Using a step drill bit in a hand-held drill is somewhat problematic, depending on the material you are drilling into, as a step drill might have a tendency to bind and good control of the drill is necessary to prevent this. If you do elect to use a step drill bit in a hand-held drill, make sure to maintain a tight grip on the drill at all times and don't plunge the bit. Let the bit do the work for you as much as possible.

COMMON DRILLING PROBLEMS

Drilling a hole sounds like a simple thing, and for the most part it is. But there are some traps for the unwary even with something this simple. The biggest things to watch out for are overheating, chatter, binding, and walking. Most of what is discussed in the following sections applies to drilling metal. Wood and plastic have their own set of similar issues, but metal is the most challenging.

Overheating

Attempting to drill through hard material like steel can result in a lot of friction. This, in turn, creates a lot of heat. In some cases, the resulting heat can damage the material being drilled, and it can definitely ruin an otherwise good drill bit. If you don't have any lubricant handy, then try to make the hole in small increments, rather than have at it all in one go. When drilling multiple holes in thin sheet steel, pause between each hole to allow the drill bit to cool off.

Chatter

Chatter is a serious problem with a hand-held drill, but not so much with a drill press. Chatter can occur when you attempt to drill a large hole without first drilling a pilot hole. For holes larger than about 1/8 inch in diameter, always drill a small pilot hole first. If, for example, you are trying to drill a 1/2-inch hole in a piece of 1/8-inch-thick tempered aluminum plate, start with a 1/8-inch drill, then go to a 1/4-inch, followed by a 3/8-inch drill, and then use the 1/2-inch drill as the last step. As the bit gets larger, the drill will be harder to control, so large holes are best done on a drill press, if at all possible.

Binding

A close relative of chatter, binding can occur when you're drilling a large hole and the bit attempts to take out more material than the drill motor can handle. On a drill press, this can stall the spindle and cause the drive belt to slip. It is particularly dangerous with a hand-held drill, because when the bit binds, it can rip the drill out of your hands and break fingers. Use the same step-up technique as described for chatter to help reduce the chance of a bind and possible injury.

Walking

A spinning drill bit will have a tendency to move across a surface unless there is a starting dimple (via a punch) or a pilot hole to help keep it in place. This is called *walking*, and with a hand-held drill it can result in a marred surface.

Dull Drill Bits

It seems that people (myself included) are constantly buying new drill bits to replace those that have become dull and useless. When a drill bit becomes dull, it will no longer cut correctly and will instead just spin and get extremely hot. Even bits used just for wood will dull over time. This is more of a problem with hard woods, although soft woods also take a toll. Composite materials like MDF (medium-density fiberboard), MDO (medium-density overlay), and Masonite can also wear down a drill bit. The materials used for printed circuit boards tend to be hard on drill bits, which is why PCB fabricators will replace their drill bits on a regular basis.

When you're working with metal, the first line of defense against dulling is to use lubricant. For woods and other materials, there isn't too much you can do about it, except to occasionally sharpen dull bits and try to avoid overheating. In the past, a drill bit was sharpened with a file and a vise, and the process was tedious and time consuming. It also took a fair amount of practice to get it right. Another method uses a large bench or stand grinder, possibly with a fixture to hold the drill bit to the grinding wheel at the correct angle. Doing this by hand is possible, but not recommended.

There are different types of drill-bit sharpeners available, some better than others. This is an area where you definitely get what you pay for, as the cheap drill-bit sharpeners are better at leaving a drill bit in worse shape than they are at actually sharpening a bit correctly. If you find yourself hunting through a collection of drill bits for one that is still sharp, or purchasing new bits on a regular basis, then you may want to consider buying a good sharpener.

Here are a few tips to avoid dulling drill bits:

- Do not overheat the bit. Cut a little, back off a little, then cut some more. Don't try to plunge all the way through thick material in one go. If the bit starts to smoke, it's definitely too hot.
- When drilling through metal, use a lubricant or a cutting oil. Use more lubricant for harder metals such as iron, steel, and tempered aerospace-grade aluminum.
- Keep your drill bits clean. Leftover bits of metal and wood (and wood sap) can act as abrasives and wear down the cutting edges of the bit.
- Use the appropriate drill speed for the material. Table 4-6 lists some suggested speeds for various drill sizes and materials.

TAPS AND DIES

If you are using screws or bolts with matching nuts or prethreaded holes, you don't need to worry about tapping a hole to create an internal thread. But if you need to attach something like a bracket to a heavy piece of material and there is no easy way to get behind the hole to attach a nut, then tapping the hole is a viable approach.

> **TIP** When you are drilling a hole to be tapped, it is essential to select the appropriate drill bit, as discussed in "Selecting A Drill Size" on page 89. If the hole diameter is too small, the tap might bind and break off in the hole, and it is extremely difficult to remove a broken tap. Conversely, if the hole diameter is too large, the threads will not be deep enough to reliably grip the screw or bolt, and it might pull out under stress.

Tapping a hole is fairly straightforward. Once a hole of the correct size has been drilled, the appropriate tap tool is screwed into the hole with the tap handle provided with the tap and die kit. If you drilled for a 50% hole, then the tool should go in without too much resistance. A 75% hole will require more force, since it is cutting through more material to create the threads inside the hole.

Use plenty of lubricant when tapping threads. Never try to tap a hole if the tool is dry. Even with a 50% hole, there's a small chance the tool might bind if there is no lubricant, and once that happens, there is a distinct risk that you might break the tap tool off in the hole while trying to extract it. A broken tap is extremely difficult to extract. In some cases, it's better to make a new hole and simply grind the broken tool flush with the surrounding surface.

As an example, let's look at the process of creating a tapped hole for a 6-32 machine screw. Figure 4-58 shows the piece of aluminum to be used, along with an automatic punch immediately after the drill point mark is created. Figure 4-59 shows the work piece clamped to the table of the drill press, and a #34 drill bit is being used to drill the hole. I elected to use a #34 drill bit instead of a #32 (50%) or a #36 (75%), mainly because I didn't have a #32 handy. The result will be somewhere between a 50% and 75% tap cut.

FIGURE 4-59. Drilling a hole for tapping with a drill press

In Figure 4-60, the hole is drilled, and tapping can commence. Note that I used a thin lubricant during drilling—rubbing alcohol (isopropyl), in this case. In Figure 4-61, the work piece has been moved away from the drill chuck with the clamps still in place (the work table of most drill presses can swivel on the main post of the press) and the tap has been mounted in the tap handle. This special tool usually comes with a tap and die kit, but can also be purchased separately (as this one was).

FIGURE 4-58. Preparing to drill a hole for tapping

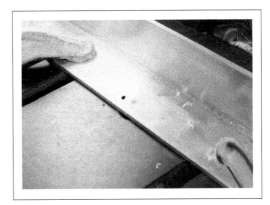

FIGURE 4-60. Drilled hole ready to tap

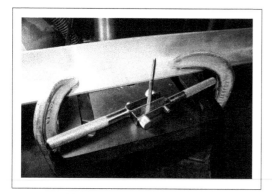

FIGURE 4-61. Tap mounted in tap handle tool, ready to go

FIGURE 4-63. Checking the tapped hole with a 6-32 pan-head screw

Figure 4-62 shows the tap completely through the hole, and again notice that I've used plenty of lubricant, this time light oil. Last, in Figure 4-63, a 6-32 pan head screw has been inserted to check the tap. After the residue from the lubricant is cleaned off, the part is ready to use.

> **TIP** Make sure the axis of the tap stays aligned with the hole, and always try to make the tap vertically. Trying to tap a hole at an angle can bind the tool and cause it to break, and once the alignment is lost, it's very difficult to correct it using a hand-tapping tool. This is one of reasons that machine shops use tapping fixtures, as described next. In Figure 4-62, the tap appears to be nonvertical relative to the work piece, but it really is lined up vertically; it's the image that isn't vertical.

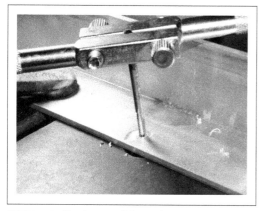

FIGURE 4-62. Tapping complete, ready to back the tool out of the hole

If you find yourself doing a lot of tapping, you might want to invest in a bench-top tapping fixture made specifically to hold the tap in constant alignment. Figure 4-64 shows a diagram of typical tool of this type. It is also possible to use the chuck of a lathe or drill press to align and hold a tap while you turn the chuck manually (power is off). There are also how-to videos and articles online that describe how to build your own hand-tapping fixture. The whole idea is to keep the tap in alignment with the hole.

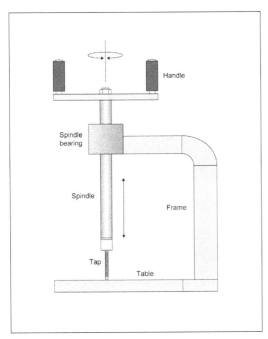

FIGURE 4-64. A bench-mounted tapping fixture

The fixture in Figure 4-64 uses the weight of the spindle assembly to apply a small amount of downward force to the tap, and the cutting threads of the tap itself will pull it into the work piece. It might take a small amount of pressure to get the tap started, but for the most part, all you need to do is turn the spindle using the handles. A hole in the table at the base of the fixture allows the tap to pass completely through the work piece, and there is usually some kind of clamp arrangement or perhaps a small vise (not shown here) to hold the work piece securely.

There are automated and semi-automated power-tapping tools available. These turn up mostly in machine shops and on assembly lines. Some people have also built their own power-tapping machines using cordless drills and screwdrivers. But for these tools to work correctly and reliably, everything has to be just right. The hole has to be exactly the right size and depth, and the work piece has to be held in exactly the right place. Lastly, the tool needs to be able to stop at the correct depth, reverse the tap rotation, and then back it out of the work piece. Commercial versions of these tools aren't cheap, so for most of us, it comes down to doing it the hard way by hand.

The counterpart of the tap is the die, and if you need a threaded rod that cannot be purchased at the hardware store, then this is the tool to reach for. For example, consider the situation in Figure 4-65. Here we have a steel rod with some threads at each end. This might be used as part of a linear position sensor (as described in Chapter 8 in the discussion of linear potentiometers), or perhaps it is a tie rod for the steering on a robot. While you could find a steel rod with full end-to-end threading, that might not be what you really want. With a die, you can make a rod like the one in Figure 4-65 to meet your requirements.

FIGURE 4-65. Steel rod with threaded ends

Using a die is relatively straightforward, but be forewarned that getting good results when cutting threads by hand is not easy. It takes practice and a fair amount of patience to get it right. In other words, it can be a real challenge.

TOOL TECHNIQUES | 99

Figure 4-66 shows a die for cutting 10-24 threads along with the holder it mounts into. Some people suggest grinding a slight taper into the end of the rod to allow it to feed into the die more easily, and I would agree with this. It's very difficult to convince a die to start to "bite" into a rod without a taper at the end.

FIGURE 4-67. Completed thread cut with a 10-24 die

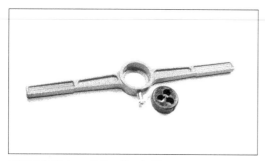

FIGURE 4-66. A small die with a holder

When you look closely at a die, you will see that the cutting teeth have a "lead-in" chamfer on one side. This is the side you start with on the rod to be threaded. This allows the tool to gradually work into the rod. The taper on the rod helps as well. You can see the taper in Figure 4-67. Make sure you mount the die in the holder with the entry side facing out, which is down in this case (the holder has a lip to hold the die; this is the exit side). If you try to use the holder with the die in backward, the force you might need to exert to start the cut will cause it to pop out of the holder.

The die can exert a lot of force on the rod, so it needs to be firmly clamped vertically in a vise of some sort, as shown in Figure 4-67. It is not possible to reliably hold the work piece in your hand and cut threads into it with a die. As with tapping, be sure to use a lubricant to reduce friction and prevent binding. I used a light oil for the lubricant in this case. Figure 4-68 shows the finished rod with about 1 inch of threads cut into one end. A 10-24 nut has been threaded onto the rod as a check.

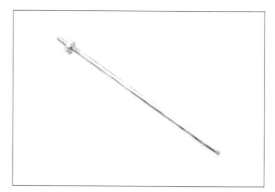

FIGURE 4-68. The finished rod with a 10-24 threaded section at one end

It is essential to always use a rod of the correct diameter for the threads you are trying to cut. Cutting an oversize rod with a die can result in the die becoming jammed onto the rod. If this happens, you might end up having to cut off the end of the rod, along with the jammed die, and then throwing it away. Forcing an oversized rod through a die can also dull or damage the die. Conversely, if the rod diameter is too small, the die's teeth won't cut deep enough threads and the threaded part probably won't work correctly with a standard size threaded hole or nut.

Screws and bolts have two basic measurements that define the diameter of a part. The first is the thread diameter (or *major diameter*). This is the diameter of the outside of the threads on the part. The second is the root diameter (or *minor diameter*), which is the diameter of the part across the bottoms (valleys) of the threads.

In general, for UTS/ANSI threads, the thread (major) diameter of a particular gauge less about 0.005 inch works well as a rod diameter. Table 2-2 lists the thread diameters for common machine screws. Some tool manufacturers may offer diameter suggestions that are slightly different than the standard for a particular screw or bolt diameter, but the differences aren't very much.

The dies found in an inexpensive kit are typically fixed in terms of the diameter of rod they will work with. Professional (and more expensive) tools have adjustments that allow the die to be sized. With these types of tools, one approach is to start off with the die opened up, and then bring it in to the final thread diameter over the course of multiple passes on the rod.

As stated earlier, cutting threads by hand with a die can be a major challenge. Perhaps the most challenging part is the amount of force needed to start the die while trying to keep it aligned at the same time. With a rod that has a diameter equivalent to the nut or threaded hole it will mate with, it can be really hard. Unless the part I'm making needs to be able to withstand significant stress, I typically use a rod with a diameter slightly less than what is listed in Table 2-2, just to make it easier on myself. If you find yourself cutting a lot of threads by hand, you may want to investigate the use of a small lathe or a special-purpose thread-cutting tool.

> **TIP** With both tap and die cutting, a general rule of thumb is that as the size (diameter) of the tap or die gets smaller, the harder it becomes to avoid breaking a tap, jamming a die, or twisting off the end of a rod with a die. Small-diameter screws and bolts typically have more threads per inch (or millimeter) than larger fasteners, and the threads are much finer. The result is that the tolerances are tighter and the chances for friction and binding much higher.

This section is just a cursory overview of the techniques used with taps and dies. This is an old subject, so there is a huge body of literature available. If you want to learn more, I would suggest investing in a machine shop reference book or two, such as *Machinery's Handbook* (see Appendix C). Even older books (from the middle of the last century, for instance) have valuable information, and I've found some real gems on the bookshelves at thrift stores. Taps, dies, and drills haven't changed much in the past 100 years, and neither have the techniques for using them.

Modification Cutting

Sometimes, you might need to cut something to modify it. For example, let's say you want to modify a device to install a new module or component, but it won't fit unless there's a new hole for it to fit into. This is a case of hacking in the truest sense, and there are several ways to do it.

JEWELER'S SAW

A jeweler's saw (introduced in "Small Hand Saws" on page 45) is designed to make small, precise cuts to create intricate patterns. Unlike a hacksaw, a jeweler's saw uses a very narrow blade, which can be turned easily while cutting. A coping saw is similar to a jeweler's saw but is used primarily in woodworking to create intricate patterns. Figure 4-69 shows a typical jeweler's saw in action.

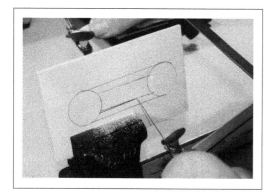

FIGURE 4-69. Using a jeweler's saw

Using a jeweler's saw takes a bit of practice. I recommend using guidelines, like the ones shown in Figure 4-69. The material is a piece of thin MDO (medium-density overlay, a type of plywood), but it could just as easily have been plastic or even thin sheet metal. The guidelines were made with a mechanical pencil. If you are cutting metal, you will probably want to use a fine-tip permanent marker (which comes off with rubbing alcohol, by the way).

The most common mistake is to apply too much force while trying to move the saw through the material to be cut. This usually results in a snapped blade. Let the saw do its job, and apply only enough pressure to keep it firmly in contact with the material. When turning a corner with a jeweler's saw, move slowly and cut in short, even strokes. The objective is to avoid binding the blade in the turn while still staying on the guideline. Be patient; these little saws can do a good job, but they don't do it very fast.

Note that the jeweler's saw is being held with two hands in Figure 4-69. One hand provides the cutting force, while the other steadies the saw and guides it. While it is possible to operate a small saw like this one-handed, I don't recommend it. It's easy for the saw to "get away" from you and head off in a direction you don't want it to go. Using two hands prevents this. You also need two hands to navigate corners.

I should also point out that the cut was started from the hole drilled to the inside of the circle on the right. The blade was removed from the far end of the saw frame, inserted through the hole, and then clamped back into the blade holder and retensioned. The near end of the blade (closest to the handle) was not removed. If you don't want to have an entry cut coming in from the side of the piece you are cutting, there really isn't any other way to do this. The blade is flexible and easy to remount once it is through the starting hole.

ROTARY TOOL

Modifying things is where a rotary tool really shines. Figure 4-70 shows a rotary tool with a cut-off disc attached. The cut-off disc is basically just a thin wafer of abrasive material made with a binding adhesive and formed under pressure, so it will wear down with use. If you are doing a lot of this type of cutting (or plan to), you might want to consider buying the cut-off discs in bulk packs.

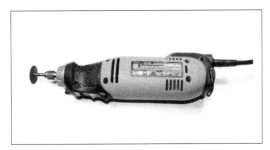

FIGURE 4-70. Rotary tool cut-off disc

It is a good idea to cut to the inside of the final dimensions of the hole, because holding a rotary tool steady while cutting can be a challenge. The edges of the hole can be finished out to the desired dimensions with a razor knife and a file, but a hole that is too large cannot be easily fixed.

You might find it difficult to hold the rotary tool steady while cutting a straight line. Here's a trick to help with that: clamp the rotary tool down to the workbench, not the work piece you are cutting. A small bench vise, with some padding if necessary, works well for this, as shown in Figure 4-71.

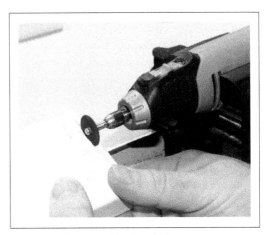

FIGURE 4-71. Rotary tool in a bench vise

Mounting the rotary tool in a vise allows you to move the piece you are cutting under the cutting wheel while the rotary tool is held firmly. The piece you are cutting is probably going to be much lighter and easier to maneuver than the tool. You can use something like a block of wood to brace your hand, or even support the work piece while cutting it. With a simple setup like this, you can get a much cleaner and straighter cut than when attempting to hold the tool in one hand and the work piece in the other.

Alternatively, you could use something like the X-Y cross-slide vise shown in Figure 4-72. In this case, I've used a cross-slide vise (it can be moved in the x and y axes) to hold the rotary tool. Once the tool and the piece to be cut are both clamped down, you can move the vise using the crank handles to produce a straight and clean cut. It is particularly useful for creating a clean notch in materials such as aluminum, which is what is being done in Figure 4-72.

Widening the notch just requires a slight turn on one of the crank handles to shift the cutting blade over a small amount, and the other crank is used to move the cut-off disc through the material once more. Repeat as necessary to make a clean notch of the desired width.

FIGURE 4-72. Holding a rotary tool in an X-Y cross-slide vise

A small bench vise (discussed in "Vises" on page 39) is holding the piece of aluminum to be cut. One downside to this setup is that there is no z-axis—in other words, no height adjustment. If the cut-off wheel of the rotary tool needs to move up or down relative to the work piece, then the work piece must be unclamped, adjusted, and then clamped back down again. But since this arrangement is primarily intended to make straight cuts through small items, it isn't really that big of an inconvenience. If being able to move a cutting tool in three axes is a requirement, you can use something like a small milling machine.

A small file can be used to clean up the finished notch and level out the bottom of the cut. Since the abrasive wheel on the rotary tool wears down rather quickly, a notch made with multiple passes will tend to have a slope at the bottom. Alternatively, you could adjust the work piece to raise it slightly, and then repeat the cutting passes from the high end of the bottom of the notch back towards the original starting end. Some filing will still be necessary if you want a really clean and square cut.

If you're curious as to what the cross-slide vise looks like, Figure 4-73 shows the model I happen to have. This is a 6-inch unit purchased from Harbor Freight for about $80 USD, and it does a fairly good job. For use with the rotary tool, I've used C clamps to hold it down, rather than bolting it directly to the workbench. This allows it to be moved if necessary, and it often ends up on the work table of a drill press. A handy thing to have; it gets around in the shop.

FIGURE 4-73. Cross-slide vise

Summary

In this chapter, we've covered a variety of tool techniques. There are, of course, more

types of tools available than what we've seen here and in Chapter 3, but to do justice to them all would require a set of books rather than a couple of chapters. With what is here, you should be able to get a basic idea of how the various tools are used and what can be done with them, as well as what to watch out for when using them.

Vertical Milling Machines

Don't let the cross-slide vise tempt you into thinking that your drill press can be a vertical mill. The bearings in most small-shop drill presses are not designed to take the side loads that milling can generate; they're made to take a load along the axis of the drill. Using a standard drill press as a mill with anything harder than soft plastic will almost certainly damage it, and replacing the bearings can be an annoying task. While there are drill presses available that are rated for use as light-duty milling machines, they really aren't an ideal substitute for a real vertical mill.

A *vertical mill* is a machine tool that is designed specifically to handle high side loads, and they look something like a drill press on steroids. It has a vertically mounted spindle to hold a cutting tool, a table that can move in X and Y like the cross-slide vise, and usually in Z (height) as well. A vertical mill is designed to take a lot of side force, and these tools can cut a 1/2" wide channel through a block of aluminum without breaking a sweat. If you don't know what a vertical mill looks like, just do a search for "vertical mill" in Google Images.

Designed primarily to cut metal, vertical mills use special cutting tools. The cutting tools come in a variety of styles, and they usually have a flat or blunt end without a sharp point like a drill bit. Some vertical mills can also be used as a drill press with conventional drill bits, albeit a rather large drill press.

If you want a small vertical mill for your shop, you might consider one of the imported machines from China to start with. They are relatively inexpensive, but you should be prepared to make some adjustments, and possibly even some minor modifications, to make the tool stable and accurate enough to do any serious work. Some of the sources for these are Harbor Freight, Enco, and Penn Tool Company. Another option is combination tools, such as the Smithy Midas that combines a mill and lathe into a single machine. Sherline also makes a 3-in-1 tool suitable for light work.

Lastly, there's always used equipment to consider. A number of interesting small milling machines appeared on the market from about 1940 to around 1970. If you look around enough, you might find an old Hardinge, Clausing, or even a Rockwell. These are heavy-duty little machines made for regular use in a working machine shop. Consequently, they are tough and very heavy. A special stand is required, as a typical workbench won't support the weight.

In addition to a huge number of books, both new and used, on machine shop practices, there is a lively and dynamic online community of enthusiasts, collectors, and hobbyists to turn to with questions. There are also websites devoted almost exclusively to various brands and types of machine tools, so getting parts and advice is usually not a problem.

We've also seen that there is nothing wrong with modifying a tool so that it will perform better in a particular application. While it is possible to purchase sets of precision tools for special applications already shaped to specific widths and lengths, you can save a lot of money by doing it yourself if you are patient and have the right tools. The "right tools" in this case means a small grinder like the one shown in "Grinders" on page 42, and possibly a combination belt/disc sander. A vise also comes in handy for bending and shaping metal items, like tools, and there is also a small anvil in my shop for times when I need to beat on something to get it into the shape I want. (Or just when I feel like beating on something.)

An important point to take away from this chapter is that even though a tool might seem conceptually simple, there is often more to correctly using it than you might first think. The key is to gain experience with the various tools, research them to acquire additional details, and talk to people who have experience with them. Or, to summarize it another way: use the right tool for the job, and use the tool the way it was intended to be used.

One of the best ways to learn how to use tools correctly is to practice. This is one reason why this book mentions disassembling old electronic equipment in many different places. Another approach is to purchase practice kits made specifically to use as training aids. There are several sources for SMT soldering practice kits, which usually consist of a double-sided PCB with a selection of mounting patterns and some parts to solder. What the parts happen to be isn't relevant, just the form factors. Try searching for "SMT practice kit" on Google to get some ideas of what is available and the typical cost.

Lastly, consider keeping a notebook for your own discoveries and thoughts regarding your tools and how you use them. While you probably don't need to take notes on using a screwdriver, if you modify a screwdriver, it might be helpful to understand why you did so later on. This would apply to any tool customization that might not be intuitively obvious, and it definitely applies to special-purpose custom tools.

CHAPTER 5

Power Sources

ALL ELECTRONIC OR ELECTRICAL AND ELECtronic devices, from the simplest flashlight to the control systems on a modern aircraft, need one thing in common: power. Without a source of electrons, it's all just a pile of inert metal, plastic, and silicon (among other things).

This chapter presents an overview of power sources for both DC and AC, ranging from batteries to linear and switching power supplies. Special attention will be given to batteries and inexpensive DC power supplies you can use to power your project that plug into a standard AC wall socket (so-called *wall warts*).

We'll wrap up the chapter with a look at fuses and circuit breakers, essential but often overlooked devices that can save the day if used correctly. To that end, there is a brief discussion of how to select an appropriate rating for these essential protection devices.

Batteries

Batteries are the simplest source of DC power, and they are essential when portability is a major consideration. Unfortunately, a battery will last only so long before it must be either replaced or recharged.

There are numerous types of batteries available, ranging from the tiny types used in things like hearing aids, all the way up to huge arrays that are found in solar power installations and in the back-up power systems for large computer installations. Some batteries are rechargeable, such as the types used in cell phones and automobiles, while others are single-use and must be disposed of when they are exhausted.

BATTERY PACKAGES

Batteries for general-purpose consumer applications are available in a variety of package styles, from the common cylindrical forms of AAA, AA, C, and D types used in portable devices, toys, and flashlights, to the square package used for 9V types. Table 5-1 lists the five most common general-purpose battery sizes.

TABLE 5-1. Common battery sizes

Type	Dimensions (mm)
AAA	10.5 × 44.5 mm
AA	14.5 × 50 mm
C	25.5 × 50 mm
D	34 × 61 mm
9V	48.5 × 26.5 × 17 mm

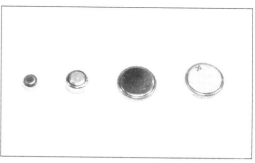

FIGURE 5-2. A selection of small "coin" batteries

Figure 5-1 shows a side-by-side comparison of the common battery sizes in the order listed in Table 5-1, from left to right.

FIGURE 5-1. A selection of common batteries

The so-called *coin* or *button* batteries found in some toys, watches, hearing aids, and other small devices do indeed look like small coins or buttons, and there are multiple sizes and ratings available for these, as well. Figure 5-2 shows a sample of some of the available sizes.

The batteries shown in Figure 5-2 are, from left to right, LR41, LR44, CR2032, and CR2025. See "Miniature Button/Coin Batteries" on page 113 for more details on the various types and sizes of coin-style batteries that are available.

Other package styles include the rectangular form used for sealed lead-acid batteries, such as those found in things like emergency lights and the uninterruptible power supply (UPS) devices used to keep computers and other devices running when the lights go out. Rectangular 6V lantern batteries with spring contacts are still used for some types of camp flashlights and portable fluorescent lighting units. The lithium-ion batteries found in cell phones and tablets come in a range of shapes, most of which resemble flat rectangular wafers about 3 to 10 millimeters thick. And, of course, there are the large, heavy lead-acid batteries used to provide starting power for cars, trucks, and boats.

In addition to the common sizes, batteries can be made in custom sizes for a particular product. Generally, however, it is a good idea to stick with something that is readily available at almost any grocery store if at all possible.

PRIMARY BATTERIES

Batteries with nonreversible electro-chemical reactions are referred to as *primary batteries*. Primary batteries are usually alkaline types, but other types (such as lithium, silver-oxide, and zinc-air) are also common in the form of coin batteries (see "Miniature Button/Coin Batteries" on page 113 for details about coin batteries). Large silver-oxide primary batteries have also been used in aircraft, submarines, and spacecraft.

Alkaline Batteries

The alkaline battery is the most common type of battery in use today. An alkaline battery is a chemical device that employs the reaction between zinc and manganese dioxide to create an electric current. Alkaline batteries are size interchangeable with the older zinc-carbon types.

For the common sizes listed in Table 5-1, each type of alkaline battery has its own standard voltage and capacity, as shown in Figure 5-3.

Note that the capacity values shown in Figure 5-3 are conservative and are provided for reference only. The actual capacity of a particular battery will depend on the manufacturer's formulation and production process.

A typical alkaline battery is a one-shot device. Once it is discharged, it is generally not a good idea to try to recharge it. There are some types of alkaline batteries that are somewhat rechargeable, but unless a battery specifically states that it is a rechargeable type, don't try it. There is a risk of a burst battery and possibly fire.

While lightly loaded (low current drain), an alkaline battery will produce a relatively constant voltage at a constant current. This is the amp-hour rating mentioned in Chapter 1 and shown as mAh (milliamp-hour) values in Figure 5-3. The battery's output is a function of the chemical reaction occurring inside the battery, which in turn is affected by the load on the battery. Over time, the battery loses its ability to react chemically, and the battery dies. When the battery starts to fail, its voltage will drop, and it will no longer be able to hold its rated voltage as it discharges. When heavily loaded, alkaline batteries start to fail much sooner than other types.

The usable capacity of an alkaline battery depends on the load on the battery. Figure 5-4 shows typical discharge curves for both an alkaline and a nickel-metal hydride (NiMH) battery under load. Notice how the NiMH battery is able to maintain a voltage above the cut-off level for longer than an alkaline type when both are loaded equally. From this we can conclude (correctly) that the capacity rating for an alkaline battery is more applicable to low-current-drain situations.

As you can see in Figure 5-4, the output of an alkaline battery over time is not a flat line up to some point where it starts to drop. Rather, it is more like a gradual descent into darkness over time because an alkaline battery will develop an internal resistance when the current drain is high. It drops below the usable voltage cut-off limit much sooner than the NiMH battery, even though both have roughly the same rated capacity. The cut-off voltage shown in Figure 5-4 depends on the device the battery is powering. Some can operate down to 1V, whereas others might give up at 1.2V (which means the use of an NiMH might not work out too well).

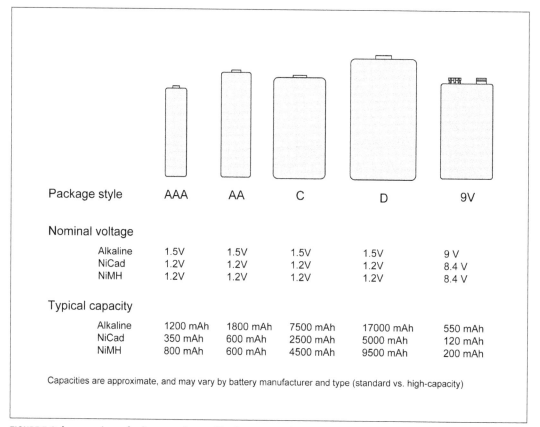

FIGURE 5-3. A comparison of voltages and capacities for common battery types

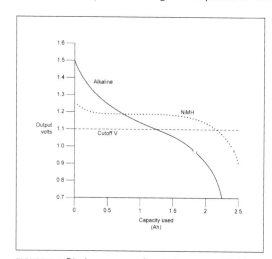

FIGURE 5-4. Discharge curves for alkaline versus NiMH batteries

Silver-Oxide

Silver-oxide batteries are primary cells typically found in the form of small button or coin-style batteries, due to their high energy-to-weight ratio. A silver-oxide cell also has a flatter discharge curve than an equivalent alkaline battery, and the nominal voltage per cell is only slightly higher (1.5V for alkaline versus 1.55V for silver-oxide).

Larger sizes are possible but limited in application due to the cost of silver. A button-style battery uses little silver, so it is not a major cost factor. Prior to the invention of lithium batteries, the silver-oxide battery had the highest energy-to-weight ratio (the energy density). Originally developed for aircraft, these types of batteries have been used in spacecraft and on submarines.

Lithium

Not to be confused with the lithium-ion (Li-ion) secondary batteries found in cell phones and other portable electronic devices, a lithium battery is a disposable primary battery type that uses some form of metallic lithium in the anode of the battery. A primary lithium battery has a high charge density, which equates to a long useful lifetime. Found mostly in the form of coin or button batteries, AAA, AA, and 9V sizes are available as well. They are also rather expensive, with a four-pack of AA-size lithium batteries going for around $15.

Output voltages for lithium batteries (individual cells) range from 1.5V to 3.7V. The current capacity can be as high as 3,000 mAh for a single AA cell, and the discharge curve is virtually flat right up until the battery is exhausted.

Zinc-Air

Zinc-air primary batteries are based on the oxidation of zinc when in contact with the air. For this reason, coin-type zinc-air batteries come with a seal that must be removed before the battery will produce any output.

Zinc-air batteries have a high energy density and are relatively inexpensive to manufacture. They are mostly found in the form of coin cells in hearing aids, medical devices, and pagers, although they have also been used in film cameras and, in large forms, as the primary power for electric vehicles.

SECONDARY BATTERIES

Batteries that are rechargeable (meaning that the electro-chemical state of the battery can be reversed), are referred to as *secondary batteries*. Secondary batteries include nickel-cadmium (NiCad), nickel-metal hydride (NiMH), lithium-ion (Li-ion) types, as well as the common lead-acid types found in vehicles and in large-scale power storage applications.

The oldest example of a rechargeable battery is the lead-acid type often found in automotive applications. A more modern variant, the deep-cycle battery, is sometimes used in solar power installations. Other more recent types include nickel-cadmium (NiCad), nickel metal hydride (NiMH), and lithium-ion (Li-ion).

NiCad: Nickel-Cadmium

Created in the late 1800s in Sweden, the NiCad battery has been used in things like portable two-way radios, emergency lights, cordless power tools, and electric vehicles. Manufactured in a variety of sizes, including AAA, AA, C, D, and 9V, NiCad batteries are readily available and provide decent performance. NiCad batteries do have a tendency to develop a sort of memory with repeated charge-discharge cycles, which prevents them from taking a complete charge until they have been subject to a deep discharge cycle to reset the charge memory.

NiCad batteries have a typical output voltage of 1.2V per cell, instead of 1.5V as found in alkaline batteries. Although many devices will operate fine with the lower voltage, some will not.

NiMH: Nickel-Metal Hydride

Since their introduction in the late 1980s, NiMH batteries have replaced NiCad in many applications. The charge capacity of NiMH is much better than NiCad, and NiMH batteries don't suffer from the charge memory effect that sometimes afflicts NiCad types. NiMH batteries do have a higher self-discharge rate than NiCad types, however, and like NiCad batteries, a NiMH cell output is typically 1.2V.

Figure 5-5 shows the battery pack from a cordless telephone before and after the pack was opened to reveal its contents. This pack has an output voltage of 3.6V at 500 mAh capacity, and it contains three 1.2V cells wired in series. Notice in Figure 5-5 that there is a fuse link incorporated into the battery pack, as indicated by the arrow.

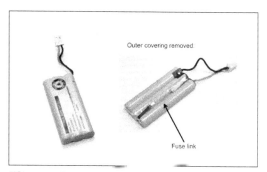

FIGURE 5-5. Small NiMH battery pack for a cordless telephone

These types of battery packs are readily available from various online sources and sell for about $5.

Li-ion: Lithium-Ion

Lithium-ion batteries have become common in consumer electronics such as cell phones, tablet computers, netbooks, and some types of electric tools and medical equipment. These batteries have good charge density and low self-discharge rates, and they do not suffer from memory effect.

Li-ion packaging is varied, ranging from conventional tubular packages in the standard sizes to custom-made flat rectangles or even circular shapes. Be aware that the output voltages from Li-ion batteries in standard sizes may not be the 1.5 or 1.2V you would expect. Some can range as high as 3.6V. It would be easy to damage an electronic device expecting, say, 6V (four alkaline batteries) but instead being supplied with almost 14.5V!

The batteries found in things like MP3 players, cameras, and cell phones are often packaged in flat, rectangular shapes like the one shown in Figure 5-6. These typically have three or four terminals at one end that make contact with a matching number of spring-loaded pins or perhaps leaf contacts.

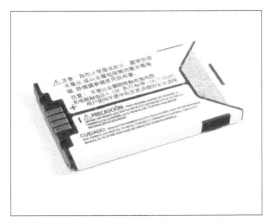

FIGURE 5-6. A typical Li-ion battery from a portable device

Although most Li-ion batteries are custom-made for a specific application, if the correct fixture can be purchased or fabricated, they can be an excellent compact source of power for a project. An old cell phone can provide such a mount if it is removed carefully.

You should be aware that Li-ion batteries can have a relatively high energy storage capacity for their size and will sometimes overheat if they're mistreated or happen to be defective. This is particularly true of the types used in mobile devices, and there have been reports of Li-ion batteries bursting into flames or even exploding. While not as big of a problem with more recent batteries as it once was with earlier types, it is still something to take into consideration.

Lead-Acid

Lead-acid batteries have been around for a long time, at least since the mid-1800s. They have low energy density for their weight, but they are relatively inexpensive and have the ability to deliver large surge currents, which makes them ideal for applications such as turning the starter motor in an automobile.

In the 1970s, the so-called *gel cell* was created. This is essentially a lead-acid battery with a silica gel mixed into the electrolyte. These types of batteries don't suffer from significant evaporation issues, and their larger usable temperature range makes them more suitable for extreme conditions. Due to the gel, these types of batteries do not have the same surge capability as a conventional lead-acid battery. So while a gel cell is great for energy storage applications, it is not a good choice as the starting battery for a vehicle.

MINIATURE BUTTON/COIN BATTERIES

We wrap up our brief survey of batteries with a look at coin or button types, such as the ones found in hearing aids, watches, laser pointers, miniature digital scales, and other small devices. They come in both primary and secondary forms, although we'll cover only the primary types in this section.

International standard IEC 60086-3 defines an alphanumeric coding system for coin-style batteries, although some manufacturers have their own naming system. This can make things confusing, but fortunately you can usually identify a replacement battery by cross-referencing the manufacturer-specific ID number to an IEC equivalent.

The full version of the coding scheme incorporates a letter code for the battery type followed by a letter code for the package type (which is always *R* for *round* when referring to coin or button batteries). These two characters are followed by two, three, or four digits that encode the physical dimensions (diameter and height) of the battery. The full form contains both the diameter and the height, like this:

- [type][package][diameter][height]

For example, a CR2032 (one of which is shown in Figure 5-2) is a round lithium cell with a diameter of 20 mm and a height of 3.2 mm. Another example is the LR736, which is another round lithium cell with a 7.9 mm diameter and 3.6 mm thickness. The 7.9 mm diameter is indicated by the value 7; the fractional part of the diameter is not included in the identification code. The standard diameter values are defined in Table 5-3.

An alternative form of the coding system uses a numeric value to specify the case size rather than the diameter and height:

- [type][package][size code]

LR41 is an example of this type of ID code. This would be a round lithium battery of size type 41, which is equivalent to an LR736. The size codes are listed in Table 5-4, which can be used as a cross-reference between the two types of codes.

Lastly, some batteries have just a numeric identification number. This is not part of the IEC standard but is an example of a manufacturer-assigned ID code, some of which have become de facto standards over time. Fortunately, most manufacturers follow the IEC case size specifications, so many batteries with unique ID codes are largely interchangeable with other types that use the IEC identification system. For example, the ID numbers 186, 301, 386, SR43, and SR1142 all refer to the same battery (a 1.55V silver-oxide type with a diameter of 11.6 mm and a height of 4.2 mm).

The battery type code refers to the chemistry of a battery and can be one of C, L, P, or S. Table 5-2 defines these type codes, along with the nominal output voltage for each individual cell of that type.

TABLE 5-2. Button/coin battery types

Type code	Chemistry	Output (volts)
C	Lithium	3
L	Alkaline	1.5
P	Zinc-air	1.4
S	Silver-oxide	1.55

The case dimensions or case code for a particular battery type follow the R code (a round case). In the full-form numbering system, the diameter is specified with either a one- or two-digit value indicating the diameter of the case in whole millimeters (rounded down). The height of the case is always a two-digit value that specifies the physical height in millimeters and tenths of a millimeter. Table 5-3 lists the diameter code values specified in the IEC standard.

TABLE 5-3. Button/coin battery diameter values

Diameter code	Nominal diameter (mm)
4	4.8
5	5.8
6	6.8
7	7.9

Diameter code	Nominal diameter (mm)
9	9.5
10	10.0
11	11.6
12	12.5
16	16
20	20
23	23
24	24.5

With this coding system, the battery ID is sufficient to physically describe the battery. So, for example, if you encounter an SR926, you'll know that it's a silver-oxide button type that is 9.5 mm in diameter and 2.6 mm in height.

The standard size-code system is a more compact scheme used to identify a particular type of button battery not only in terms of physical size, but also in terms of current capacity. Table 5-4 lists the size codes for alkaline and silver-oxide batteries and the typical capacity of each type. The capacity ratings are shown for both alkaline (L cap) and silver-oxide (S cap) batteries. The *x* in the IEC ID codes can be replaced with either an L or S, as appropriate. For the SR67 and SR68 types, no alkaline (L) equivalent is readily available.

TABLE 5-4. Button/coin battery size codes (capacity in mAh, dimensions in mm)

Code	IEC ID	L cap	S cap	D	H
41	xR736	25–32	38–45	7.9	3.6
43	xR1142	80	120–125	11.6	4.2
44	xR1154	110–150	170–200	11.6	5.4
45	xR936	48	55–70	9.5	3.6
48	xR754	52	70	7.9	5.4
54	xR1130	44–68	80–86	11.6	3.1
55	xR1121	40–42	55–67	11.6	2.1
57	xR926	46	55–67	9.5	2.6
58	xR721	18–25	33–36	7.9	2.1
59	xR726	26	30	7.9	2.6
60	xR621	13	20	6.8	2.1
63	xR521	10	18	5.8	2.1
64	xR527	12	20	5.8	2.7
66	xR626	12–18	26	6.8	2.6
67	SR716	n/a	21	7.9	1.65
68	SR916	n/a	26	9.5	1.6
69	xR921	30	55	9.5	2.1

Lithium coin-style batteries use the full-form IEC ID numbers. Table 5-5 lists some of the more common types you might encounter in the wild. The ANSI designation is also given, where applicable.

TABLE 5-5. Lithium button/coin battery size codes (capacity in mAh, dimensions in mm)

IEC ID	ANSI	Capacity	D	H
CR927		30	9.5	2.7
CR1025	5033LC	30	10	2.5
CR1216	5034LC	25	12.5	1.6
CR1220	5012LC	35–40	12.5	2.0

IEC ID	ANSI	Capacity	D	H
CR1225	5020LC	50	12.5	2.5
CR1616		50–55	16	1.6
CR1620	5009LC	75–78	16	2.0
CR1632		120–140	16	3.2
CR2012		55	20	1.2
CR2016	5000LC	90	20	1.6
CR2025	5003LC	160–165	20	2.5
CR2032	5004LC	190–225	20	3.2
CR2330		255–265	23	3.0
CR2354		560	23	5.4
CR2430	5011LC	270–290	24.5	3.0
CR2450	5029LC	610–620	24.5	5.0
CR2477		1,000	24.5	7.7
CR3032		500–560	30.0	3.2

BATTERY STORAGE CONSIDERATIONS

Alkaline primary batteries typically have a longer shelf life than a secondary battery of the same physical form factor and rating. For this reason, they are a good choice when the battery must sit for extended periods of time. The downside is that they cannot be recharged (in most cases, that is), and all primary batteries eventually deteriorate, so you will need a replacement schedule in place to install fresh batteries.

Three to four years is typical, with some types capable of a shelf life in excess of six years. The time interval depends on the chemistry of the particular battery and the thermal environment (a battery in a hot environment will tend to degrade faster than one kept in a cool location). Some people put batteries in the refrigerator, but there is no solid evidence that this will help extend their shelf life. It can, however, help prevent premature deterioration due to overheating. I would not recommended putting batteries of any type in a freezer. Overall, keeping spare batteries in the door of a refrigerator is not a bad idea, particularly if you happen to live in a hot environment. Just don't expect the batteries to last any longer than normal.

Primary batteries in coin cell form have shelf life durations from 1.5 to 10 years, depending on the chemistry of the battery. A lithium coin cell can have a shelf life of up to 10 years. A silver-oxide coin cell can hold out for about 1.5 to 2 years, and an older style mercury cell (now obsolete) can last for up to 3 years. A zinc-air type can stay viable for up to 6 years, so long as the seal remains in place (they are activated when a seal is removed and air can enter the battery).

A secondary battery has the advantage of being rechargeable, but it has a relatively short shelf life due to internal leakage currents (self-discharge). A secondary battery will significantly outlast a primary type if it can be recharged periodically. They are good choices when a reliable power source is available, such as a floor-sweeping robot that can be recharged from a wall outlet, or a device that uses a solar cell to keep its battery (or batteries) charged.

Storage of secondary batteries can be somewhat more involved than with a primary battery. For example, a lead-acid or Li-ion secondary battery needs to be stored with a minimum charge to prevent internal deterioration and possible damage. In the case of a lead-acid battery, it can accumulate lead sulphate crystals on the plates, which renders the battery useless. This is why the battery in a vehicle in storage should be charged periodically and why the batteries purchased from an auto parts supplier are already charged when you buy one. Lithium-ion batteries can also suffer internal damage if left in a completely discharged (0V) state for extended periods of time. Some sources recommend maintaining a minimum of 2 volts per cell by periodically "topping off" the battery. Ideally, a Li-ion battery should be stored with a partial charge of between 30% to 50%.

NiCd and NiMH batteries are somewhat more forgiving when it comes to storage, but they too have their own unique issues. NiCd batteries can be stored in discharged state, but they might require multiple deep discharge cycles to restore them to full capacity (this is the same technique used to erase the so-called *charge memory* effect found with NiCd batteries). NiMH batteries don't suffer from the charge memory of NiCd types, but they will self-discharge more rapidly.

Most manufacturers recommend storing secondary batteries in a 15°C to 20°C (59°F to 68°F) environment. It can be colder than this, with some types rated for storage down to −20°C. A fully charged lead-acid battery can withstand −35°C, but a fully discharged lead-acid battery will freeze at 0°C.

In summary, you can see that a primary battery is a relatively low-maintenance device that can be stored for extended periods of time before use, but it's a one-shot thing. It will need to be discarded and replaced at some point and, if in continuous use, that time will come sooner rather than later. A secondary battery, on the other hand, will last in continuous use for a considerably longer time than a primary type, provided that it can be recharged routinely. A secondary battery is much more particular about storage conditions and, in general, secondary types involve more storage maintenance considerations than primary batteries.

USING BATTERIES

Batteries require some form of fixture to hold them, unless they happen to have terminals like those found on automobile batteries and some gel cells. Battery holders are available in a variety of configurations, from a single AAA or button cell to fixtures capable of holding multiple D size cells. Figure 5-7 shows a sample of some of the types of battery holders that are available.

The standard 9V battery package has a pair of contacts at one end that are designed to mate with a set on either a snap-connector or a plastic holder. Figure 5-8 shows a snap-on connector with flexible wire leads.

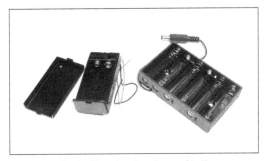

FIGURE 5-7. A selection of various types of battery holders

Avoid soldering wires directly to a battery unless the battery was specifically made to be soldered. Some of the small batteries found on computer motherboards have solder tabs. They are intended to be soldered into place once and then left in place to provide memory retention power for things like the real-time clock and configuration memory. Attempting to solder a wire to a conventional battery of any type that is not intended to be soldered can damage the battery, and it often requires the use of an acid-based flux (if the metal used for the battery's end caps will even bond with the solder in the first place).

FIGURE 5-9. Battery holder for a conventional 9V battery

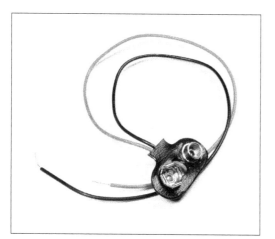

FIGURE 5-8. A snap-on connector for a 9V-style battery

Figure 5-9 shows a plastic holder made specifically for a standard 9V type of battery. This holder is intended to be attached to a flat surface using the four holes in the bottom of the plastic shell.

If you elect to use the snap-on connector, you'll also need to work out a way to keep the battery secure. Otherwise, it can float around inside an enclosure and do some physical damage to other components. Figure 5-10 shows one way to secure a battery holder. In this case, it's a holder for two AA cells that has a 9V snap-on connector. It could just as easily have been a bare 9V battery.

FIGURE 5-10. A battery holder secured with plastic zip ties

The secret to Figure 5-10 is revealed in Figure 5-12. A stick-on zip-tie anchor, like the one shown in Figure 5-11, has been affixed to the bottom of the enclosure. These plastic parts are typically used to secure and route bundles of wires and cables through a chassis, but they also work very well as an anchor point for situations like this.

FIGURE 5-11. A common plastic zip-tie anchor with adhesive backing

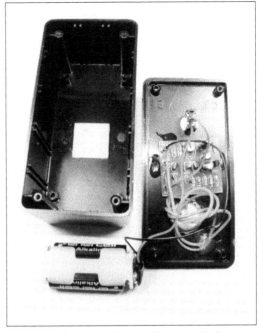

FIGURE 5-12. A zip-tie anchor used to secure a battery holder

In case you're curious, the device (i.e., gadget) shown in the preceding images is supposed to be an audio mosquito repeller. Its effectiveness is still in question, but it was fun for the kids to build and test it.

BATTERY CIRCUITS

Batteries can be connected in series or parallel to achieve voltages or capacities greater than what can be obtained from a single cell. Batteries in series are commonly used to increase the voltage, such as might be found in a large flashlight with multiple C or D size batteries. In other cases, you might want to increase the available current, and connecting batteries in parallel will achieve this, as shown in Figure 5-13.

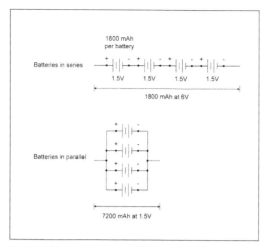

FIGURE 5-13. Series and parallel battery circuits

Of course, it is also possible to build a combination of series and parallel circuits to achieve both a desired voltage and current capacity. It just depends on how many batteries you are willing to use and how much space is available for the battery holders.

SELECTING BATTERIES

When selecting a battery for a given application, most people look first at the voltage and the physical size, but there are other factors to take into consideration, as well:

Physical size

How large is the available space for a battery? This includes not only the battery itself, but also the holder (if one is used, a PCB-mounted battery doesn't need a holder, just sufficient room for the battery package). If there isn't a lot of room available for something like a AA size battery, a smaller type could be used (an LR41 button type, for example), but the trade-off is battery capacity versus size, and a reduced battery capacity means that the circuit will not be able to operate for the same amount of time as would be the case with a larger battery.

Voltage

If a circuit uses 5V DC, you can use four alkaline or silver-oxide batteries to get 6V. Five NiCad or NiMH batteries will also provide 6V. If there is sufficient space available, a single lantern-type battery will provide 6V. In each case, the circuit will need a voltage regulator to provide the necessary 5V power.

Capacity

The capacity rating of a battery will determine how long the battery can supply a circuit with current at a constant discharge rate before the battery is exhausted. For small batteries, capacity is specified in mAh (milliamp-hours).

Environment

Many battery types do not handle heat or extreme cold very well, and their performance can suffer if operated outside of their rated temperature range. If a circuit will be used in an environment where the temperature of a particular battery type might exceed its range, then either select a different battery, or make provisions to keep the battery's immediate environment within its operating range.

Replacement

Will the batteries need to be replaced at some point? If so, then selecting a battery that can be easily removed is better than one that must be unbolted or desoldered. This depends, of course, on the size of the battery itself and its operating environment. Large batteries, such as some sealed lead-acid types, need to be physically secured with straps or brackets, and the terminals might employ screws or bolts.

Shelf life

It is important to bear in mind that an alkaline battery has a considerably longer shelf life than a rechargeable type. If the battery will sit for an extended period before it is used (such as an emergency flashlight in a wall-mounted bracket), it should be an alkaline type. If it will be used routinely, and it can be recharged routinely, then a NiMH rechargeable type might be a better choice.

The best source of information for a particular battery is, of course, the manufacturer. The reference data will (usually) specify the nominal capacity, output voltage, maximum suggested discharge rate, and environmental constraints. Some manufacturers also provide useful reference materials for download or as web pages. There are also many other sources of information available on the Internet. For example, the MIT electric vehicle team has created "A Guide to Understanding Battery Specifications" (*http://bit.ly/battery-spec*).

Power Supply Technology

The purpose of a power supply is to provide electrical power in a usable form through a conversion of some sort. In most cases, this involves converting AC at a high voltage (110 or 220 volts) to a much lower DC voltage. In other situations, there might be a need to convert from one DC voltage to another— say, from 5V to 3.3V. Conversely, some types of LCD displays use a high-voltage backlight,

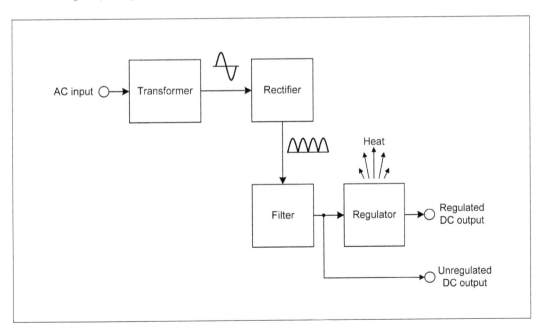

FIGURE 5-14. Linear power supply block diagram

so a low-level DC input needs to be "stepped up" to a much higher voltage.

There are two basic technologies used in DC power supplies: linear and switching. The linear type is the oldest, dating back to the time of vacuum types (which used high-voltage DC). A linear power supply takes the output of a step-down (or step-up, in some cases) transformer, rectifies it to convert the AC to DC, filters it to remove as much leftover ripple from the original AC source as possible, and sometimes regulates it to provide a constant voltage or current (or both) output. Linear power supplies are simple to understand and build, but they tend to be bulky and inefficient, particularly if they need to supply a substantial amount of current. Figure 5-14 shows a block diagram of a basic linear power supply.

A complete linear voltage regulator is available as an IC in a TO-92, TO-220, SOT-89, SOT-223, or similar type of package. These devices can be used to build a modular or bench type power supply, or they can be included into a circuit on a PCB to set the voltage level for the entire circuit or just a some portion of it. These devices are covered in more detail in Chapter 9.

The block labeled "Regulator" in Figure 5-14 could be single IC voltage regulator, or it might be a high-power transistor circuit. It depends on the application, how much adjustability is desired, and the amount of current the power supply can provide.

A switching power supply is much more efficient than the earlier linear type, and it is also much smaller for an equivalent power output rating. Most of the plug-in supplies used with things like notebook computers, DVD players, and external hard drives use switching power supply technology to keep the cost and size to a minimum. The power supply in a desktop PC is a switching supply.

Internally, a switching power supply works by directly rectifying the AC's main input, and then chopping it into pulses at a high frequency. The high-frequency pulses are applied to a small transformer and then converted to DC on the other side. Because a high frequency is used, the transformer can be very small, but the trade-off comes in terms of circuit complexity. A switching power supply is typically more complex internally than a linear supply, with isolated control and feedback circuits to regulate the output voltage and current. Figure 5-15 shows a block diagram for a simple switching power supply.

> The high-voltage side of a switching power supply might have voltages as high as 165V if it connects directly to the AC line voltage (and even higher if the mains are 220-230V AC). This is a definite shock hazard.

Although it is inefficient, a linear power supply does have the advantage of an adjustable output, in terms of both voltage and current. A switching power supply is not as easily adjustable. There are switching power supplies available with variable outputs, and as the technology improves, they are becoming more common.

WALL PLUG-IN DC POWER SUPPLIES

A simple wall plug-in power supply (also known as a *wall wart*) is an easy and conve-

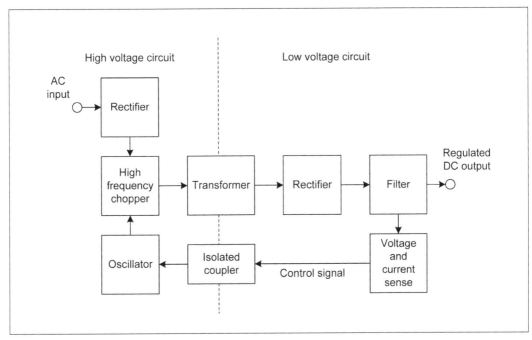

FIGURE 5-15. Switching power supply block diagram

nient way to provide DC power. These units are typically already UL and CSA approved, so that's a hurdle that doesn't need to be overcome if your project becomes a commercial product. Figure 5-16 shows a typical wall plug-in supply.

FIGURE 5-16. A typical wall plug-in power supply

These devices some in a variety of voltage and current ratings. Some provide only a few hundred milliamps of current, while others are fairly substantial with up to 3 or more amps of available current. Many of these types of power supplies use a circular coaxial connector like the ones shown in Figure 5-17, which shows four different sizes.

FIGURE 5-17. Coaxial DC power connectors

Two popular sizes are 2.5 mm diameter by 5.5 mm length, and 2.1 mm diameter by 5.5 mm length. There are other sizes in use in various types of consumer electronics and, unfortunately, there isn't a whole lot of standardization. A retail electronics outlet like Radio Shack will often have a large display of various shapes and sizes of replacement power connectors.

Be sure to read the output voltage specification on the power supply before you connect it to anything. There's nothing obvious to differentiate a 5V DC supply from a 24V AC supply with the same size of connector, except what is printed on (or molded into) the case. Also be sure to check the polarity of the connector. Some have the positive on the inner connector and ground on the outer ring. Others do it the opposite way. Check-

ing a power supply with a digital multimeter (DMM; see Chapter 17) before using it is a prudent step.

There's nothing that says you have to use the connector already attached to a wall power supply. You can cut it off and replace it with your own, or use no connector at all. The one thing you should do, however, is determine which of the wires is positive and then mark it. I like to use either a dab of white paint or a small 1/4-inch section of red heat-shrink tubing. I would not recommend using a 3.5-millimeter phone plug, like the types used for audio input and output. These can momentarily short while being removed or inserted.

It's also prudent to check the output of the wall supply to see how "clean" the output is. It may claim to be 5V DC, but there might also be a lot of "ripple" on the output if it's not well filtered. This didn't matter for some applications, because the device it was originally used with might have incorporated some type of internal filtering.

The frequency of the extraneous signal might be anywhere from 60 Hz up to several thousand (if it's a switching power supply). You can test for this by measuring the output of the power supply while connected to a load (something like a 100-ohm resistor will do; see Chapter 8 for information about resistors) and using the AC scale of a DMM or an oscilloscope (see Chapter 17). You shouldn't see any more than about 10 or 20 millivolts of ripple on the DC output. Any more than that and it might need some filtering.

If the output is "dirty," you'll need to use an RC (resistor-capacitor) filter to clean it up. Or, if you need 5V you can use a 9V or 12V DC supply and a voltage regulator IC to produce a clean 5V output. See Chapter 9 for more on voltage regulators.

BENCH DC POWER SUPPLIES

As the name implies, a bench DC power supply is intended for use on a workbench or in a test setup. Most of these types of supplies, such as the one shown in Figure 5-18, feature variable voltage and current control, and some come with fancy digital readouts, as well.

FIGURE 5-18. Variable-output bench DC power supply

You'd typically use a bench supply while testing a new device or circuit, before committing it to a final packaged form. You can also use it to power sections of an existing device in order to probe its inner workings. Most bench supplies don't provide a lot of current, and most of them are linear power supplies. A linear supply is easier to control than a switching supply and allows the voltage and/or current to be adjusted from zero to whatever the maximum rating happens to be for the supply.

MODULAR AND INTERNAL DC POWER SUPPLIES

The power supply in a typical desktop PC is an example of an integrated modular DC power supply. These types of power supplies are intended for use within another device or system, and they usually don't have convenient terminals like the bench supply shown in Figure 5-18. In some cases, they don't have terminals at all but instead have connection points on a PCB to attach wires.

Figure 5-19 shows a modular power supply. Modular power supplies similar to this one come in both closed and open frame styles. This one happens to be a closed-chassis type with an integral fan. It is a switching supply rated for 12V DC at 35 amps. An open-frame type would have just a simple metal frame, no cover, and holes in the frame for mounting hardware. Modular and open-frame power supplies are available as both switching and linear types.

FIGURE 5-19. Modular 12V DC switching power supply

The power supply shown in Figure 5-20 is also a switching unit, rated for 5V DC at 30 amps. Power supplies like the ones shown in Figures 5-19 and 5-20 are not adjustable, except by a small amount for fine-tuning the output voltage.

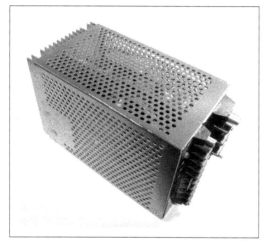

FIGURE 5-20. Modular 5V DC switching power supply

An important thing to note about a power supply of this type is that there are often two terminals labeled +S and −S. These are the *sense* inputs, and they are used to compensate for voltage drop over an extended wire between the supply and the load. In many cases, the power wires are short and the drop isn't a major concern, so a simple jumper is placed between +S and the positive output terminal, and between −S and the negative output terminal. But, in any case, unless the supply is designed to operate with the +S and −S inputs disconnected, these terminals must be connected back to the + and − output terminals at some point.

These are rugged, high-current power supplies intended to be integrated into a larger system. They are definitely not something you would want running around loose on a workbench, but they can be safely tucked away inside some type of enclosure. This type of power supply can be found in surplus shops and inside discarded medical, industrial, and scientific equipment. The downside is that, as mentioned earlier, they are not adjustable. Unless you add an adjustable regulator to the output, you get whatever the supply was designed to deliver.

Another thing to consider is the amount of current these types of power supplies can deliver. Unlike a low-current bench power supply with an adjustable current limit, some of these heavy-duty switching power supplies will just keep on pumping out the current, even into a short circuit (they go into a constant-current mode of operation that can tolerate a short circuit). Keep in mind that 30 amps can do a lot of damage, so it is essential to have a fuse or circuit breaker on the output of a high-current power supply.

Photovoltaic Power Sources

A photovoltaic cell (or *solar cell*), is a semiconductor device that converts the energy of photons to electric current by means of the photovoltaic effect. Primitive forms of photovoltaic cells existed in the early to mid-1800s, but it wasn't until the middle of the 20th century that these devices started to become efficient enough to make the transition from laboratory curiosity to usable devices.

On a sunny day with clear skies, up to around 1,000 W of solar energy falls on a square meter of the earth's surface, depending on the distance from the equator. If that energy could be converted directly into electricity with 100% efficiency, we would be able to power many of our homes and cities with it. Sunlight is, after all, essentially free, which is a major source of its appeal. Unfortunately, modern solar cells are not that efficient, and not everyone lives where there is intense sunlight for most of the year.

Today's commercial-grade solar cells are between 15 and 25% efficient, depending on how much you are willing to pay. New technologies either in research or entering production are boasting efficiencies of better than 40%, and hopefully it won't be long until these become cost-effective alternatives to conventional power sources. But even in their current form, solar cells are very useful for providing power where there are no outlets and routinely changing batteries is not an option.

Solar cells are rated by voltage output and available current for a given size (area) of a cell. These characteristics are in turn determined in large part by the efficiency of the cell. Basically, the efficiency of a solar cell is the measure of how well it can convert the light energy impinging on it into usable electric current. The higher the efficiency rating, the better the cell is at the conversion.

By itself, a solar cell is useful only when there is light falling on it, so it is useless at night. For this reason, solar cells are often used in conjunction with rechargeable batteries to provide power during darkness. You may have seen the emergency telephones alongside some major freeways, each with a small array of solar cells on the top of the pole to which the phone box is mounted. Inside the phone box is a radio transceiver, a charging controller circuit, and a rechargeable battery pack of some type.

Solar cells, like batteries, can be connected in series and parallel configurations, as shown in Figure 5-13. Solar cells connected in series will produce a higher output voltage, and when connected in parallel, the current output increases. Some solar cells have a built-in diode (called a *bypass diode*) to prevent reverse current flow when one cell is shaded and others aren't. The reason for this is that unshaded cells can drive reverse current through a shaded cell and possibly damage it. If the individual cells in an array do not have bypass diodes, they should be added. Figure 5-21 shows how bypass diodes are incorporated into a series array of solar cells.

Figure 5-22 shows an example of a low-cost solar-cell module. This is a Parallax 750-00030 rated for a maximum of 6V output at 1 watt, which is about 160 mA of current (0.167 A). This may not seem like much, but if the solar panel can produce 160 mA of current for six hours a day, it can be used to charge a battery during daylight hours.

Charging a battery from a solar-cell module like the one shown in Figure 5-22 might be as simple as placing a resistor in series with the solar cell to limit the current to the battery, or it might involve using a regulator of some sort if an array of solar cells is used that will produce more voltage or current than the battery can accept during charging.

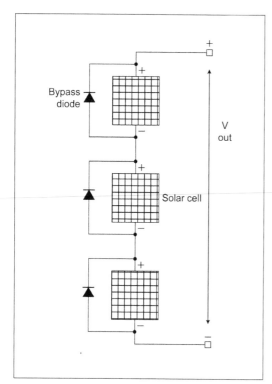

FIGURE 5-21. Bypass diodes in a series array of solar cells

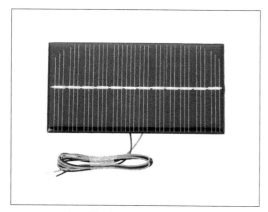

FIGURE 5-22. A small solar-cell module

If a large enough array of solar cells is used, it might be possible to power a device and charge a battery for nighttime use at the same time, similar to what occurs when a tablet or notebook PC is connected to its charger but it's still completely usable. When this isn't feasible, for whatever reasons, a charging control system might employ two batteries. The batteries alternate between charging and active use to ensure that the solar-cell module doesn't have to work to charge a battery that is also being used to power the electronics at the same time.

For any battery-assisted solar-powered system to work reliably, the current draw of the device using the battery must be low enough that it won't discharge the battery faster than the solar cell can charge it. This is a balancing act that requires careful consideration of things like the output current of the solar cell, the current consumed by the device, and the average number of days with enough sunlight to produce sufficient charging current.

For systems that don't need to operate at night, just a solar-cell array might suffice. Solar-powered calculators are an example of this approach.

Fuses and Circuit Breakers

Connecting anything that could be damaged or cause a fire hazard to a source of electrical power without some kind of protection is never a good idea. Fuses and circuit breakers exist to help prevent problems from occurring.

FUSES

Fuses are sacrificial devices with a low internal resistance. When the current flowing through the device exceeds a certain level for a specific period of time, the device will open (or *blow*) by melting the internal connection, called the *link*. In some cases this is gradual, and the link simply opens up, but other times it can be violent, leaving the inside of the fuse tube coated with vaporized metal.

Fuses come in various sizes and ratings, from tiny surface-mounted devices on a circuit board, to huge tubular objects with solid copper end tabs used in AC power distribution applications. Most glass cylinder fuses encountered in electronics fall into the AGW or AGC size category. In addition to standard fast-acting fuses, there are *slow-blow*, or *time-lag* types. A time-lag fuse is designed to withstand an over-current condition for a short period of time before it opens. Figure 5-23 shows the AGA-, AGW-, and AGC-type fuses, along with a GMA metric dimension fuse.

than a few milliseconds might need a time-lag type of fuse. With these types, the fuse will hold some percentage of current over its rated maximum for a specific period of time (which depends on the amount of current involved) before it opens the circuit.

AGC-type fuses, and similar styles, can be used with a fuse holder that is designed to mount through a hole in a panel, or with a fuse clip holder that has two metal clips to capture and retain the fuse. These tend to be large, bulky parts and are found mainly in large pieces of equipment. Figure 5-24 shows a clip-type fuse holder for an AGC-type fuse.

FIGURE 5-24. AGC clip-type fuse holder

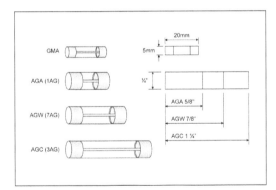

FIGURE 5-23. Common fuse types

Even though size is an important design consideration, it is also important to understand the differences between fuse types. A fast-blow fuse might not allow current spikes such as those encountered with electrical motors, whereas a slow-blow type will. Circuits that exhibit high current draw during turn-on (high inrush current) for more

PCB-mounted fuses are available with axial leads in conventional glass tube bodies similar to the AGC type, and as sealed components that resemble resistors. Figure 5-25 shows an example of a PCB-mounted fuse.

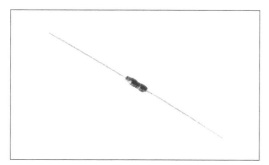

FIGURE 5-25. Printed circuit board-mounted fuses

Surface-mount types are also available in various sizes and ratings. Note that while it is easy to change out a blown fuse if it is mounted in a fuse clip or in a fuse holder, changing out a PCB-mounted fuse can be a bit of hassle for through-hole parts with leads and a downright pain for surface-mounted parts.

Fuse selection can be a complex process, involving things like inrush current profile curves, temperature derating, and surge duration. However, for most small solid-state DC circuits, a standard fast-blow fuse will suffice. Pick a fast fuse with the appropriate voltage rating; that is, rated to blow at approximately 30% to 50% greater current than the circuit will ever use. So, if you have a circuit that should never draw more than 750 mA at 5V, a low-voltage 1A fast-blow fuse would not be an unreasonable choice.

Note that this is simplistic in the extreme and does not take into account what might cause the circuit to draw more than 750 mA. Perhaps it's not a good idea to allow anything over the maximum. In that case, you would want to use a fuse with a lower current rating.

Selecting a fuse involves doing some research to determine the best type and rating for your application, and that's well beyond the scope of this book. There are numerous resources online, and most fuse manufacturers publish selection handbooks for their products.

CIRCUIT BREAKERS

As with fuses, circuit breakers come in a variety of shapes and sizes. The basic mechanism of a conventional circuit breaker relies on current flow to release a spring-loaded mechanical switch of some sort. This might be the result of electromagnetism, in the form of a solenoid, or it might also have a thermal component in the form of a bimetallic strip. A circuit breaker that relies solely on a solenoid is referred to as a *magnetic breaker*; one that uses only a bimetallic strip is a *thermal breaker*; and one that includes both a solenoid and a bimetallic strip is called a *thermal-magnetic breaker*. Figure 5-26 shows a small circuit breaker that operates on the thermal principle.

FIGURE 5-26. Small thermal circuit breaker

Inside a typical circuit breaker, the internal contacts are held closed by a spring-operated mechanism. When the solenoid is activated or the bimetallic strip deforms, a release is engaged and the spring pulls the contacts apart, thus opening the circuit. The breaker is reset manually with a button or a toggle-like handle.

Although not unheard of, electromagnetic or thermal-magnetic circuit breakers aren't often found in low-voltage DC electronic circuits, with the exception of some aerospace applications. They are more often encountered in AC power distribution and control systems, such as the breaker panel in a house. Many utility output strips also have a small thermal breaker built into the housing.

Summary

As discussed in Chapter 1, electrical current flow arises as a result of energy expended as a form of work to create an electrical potential. This chapter looked at three common sources for electrical potential: primary batteries (conversion of chemical energy to electromotive force), power supplies (conversion of mechanical energy via a generator somewhere to electromotive force), and photovoltaic cells (conversion of photon energy to electromotive force). The chapter wrapped up with a high-level tour of fuses and circuit breakers. All of these subjects are much deeper than what has been discussed here, but to do them full justice would require each to have its own book. For additional information, check out the books listed in Appendix C.

Batteries were examined from the perspective of type, physical form factor, and capacity, with an emphasis on those types most likely to be used or encountered in small electronic devices. Batteries come in a wide range of sizes, capacities, and voltages. The internal chemistry can utilize elements such as lithium, nickel, silver, and zinc, each with its own unique set of characteristics.

When you are selecting a battery, the primary characteristics to consider are size, voltage, capacity, and environment. There are potential trade-off decisions to be made between size and capacity, or between size and voltage. The environment the battery will operate within can also constrain the available choices, or it might require some clever solutions to keep the battery within its rated temperature range.

Power supplies are available with linear or switching modes of operation, with sizes and styles ranging from the common plug-in types for consumer electronics to modular units capable of delivering considerable amounts of current. For linear power supplies, the inherent inefficiency of the voltage regulation is compensated by the ability to easily incorporate a continuously adjustable output, which is why many workbench supplies are linear types. We also looked at the basic operating principles of a switching power supply, which boasts much better conversion efficiency than a linear type, but can be technically challenging to incorporate a variable output. For this reason, most low-cost switching power supplies have fixed output voltage and current ratings.

Photovoltaic solar cells are based on an old technology that is just now starting to come of age. For some applications, such as remote sensing or control, a solar cell is an ideal way to provide power during daylight hours. When coupled with a rechargeable secondary battery, solar cells can extend the useful life of a device to the service life limit of the battery, and possibly beyond.

If you plan on eventually making your creation widely available to others, or perhaps even commercializing it, it would be a good idea to do some further research on batteries and other power sources. Based on what was covered in this chapter, you should now have a good working idea of what is available in terms of batteries, power supplies, and solar cells and how they are used. Depending on what type of device you need to power, you should now have enough information to make an informed initial decision.

CHAPTER 6

Switches

SWITCHES ARE ALL AROUND US AND HAVE been for the past 150 years or so. Early switches were rather crude things, consisting mainly of one or more *blades* that could be moved into position between two terminals of some sort using a handle attached to the blade assembly. The old scary movies of the last century used these as set props for locations like Dr. Frankenstein's lab, and they can also be seen in old photos of Tesla's and Edison's actual workshops. In fact, they are still in use and are available for purchase. Figure 6-1 shows a modern version of this ancient switch type.

FIGURE 6-1. A knife-blade switch

Switches have come a long way since they made their first appearances in the 19th century. They are now available in a vast range of styles and types. Some are incredibly tiny, while others are huge. They can be push activated, sliding, or rotary. Some have a so-called *toggle* handle to operate them, whereas others have a rocker type of operation. Some are designed to be operated solely by a mechanism, rather than by a human being. Still others have built-in lamps of some sort to let the operator know when a circuit is active or a fault has occurred. A whole industry has grown up around the design and production of switches, and many engineers spend their entire careers designing and testing new types of switches and improving on the designs of older types.

A look at the switches section of the website for a major distributor such as Digikey or Mouser will reveal a mind-numbing variety to choose from. In this chapter, we'll look at some of the more commonly used types and pay special attention to things like physical mounting requirements, connection methods, current ratings, and durability.

Relays are a type of electromechanical switch, and they are discussed separately in

Chapter 10. Digital switches are discussed in Chapter 11. In this chapter, we'll stick to the old-fashioned mechanical types of switches. These devices are all around us in a multitude of types and forms, and they probably will be for a long time to come.

One Switch, Multiple Circuits

In its simplest form, a switch controls one circuit with one contact mechanism by being either open or closed. Figure 6-2 shows a simplified illustration of the internal components of a typical modern single-contact switch. This is a toggle switch, covered in more depth in "Switch Types" on page 135.

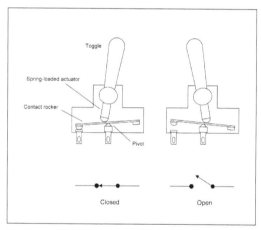

FIGURE 6-2. Simplified diagram of a toggle switch mechanism

The mechanism is simple, consisting of a contactor that can rock into one position or another. The end of the toggle handle inside the switch is a spring-loaded plunger (the actuator) that pushes the contactor to the desired position and then holds it there. As you can see in Figure 6-2, the contact rocker will either allow current to flow when in the closed position, or it will be open in the opposite position and no current will flow.

This is a common design in switches that incorporate a spring-loaded mechanism of some sort. When the switch is closed, it "snaps" closed and stays in that position. Applying force (via a toggle, rocker, or other means) overcomes the spring tension in the actuator and allows the switch to snap open and remain open. Figure 6-3 shows a typical commercial miniature toggle switch with this type of mechanism.

FIGURE 6-3. A typical miniature toggle switch

Some available switches have two, three, or more internal contacts operating in parallel, thus allowing a single device to control multiple circuits simultaneously. In switch terminology, a single contact is called a *pole*. Each pole can have one or more positions so that a multi-pole switch can control multiple circuits. The only real limitations to this concept are the physical constraints on the switch and the force required to operate it.

If the switch has only one active position (on or off), it is referred to as a *single-throw* (ST) type. This is the type of switch illustrated in Figure 6-2 and pictured in Figure 6-3. A switch with two active positions (ON-ON) is called a *double-throw* (DT) switch.

An ST switch is like a simple gate, whereas a DT type can route current between two different paths, A or B. Some varieties of DT switch have a mechanical "neutral" position in between, which is an ON-OFF-ON type of action.

The contactor (or *contact rocker*) shown in Figure 6-2 is also called a *pole* in switch terminology. If the switch has one pole, it is a *single-pole* (SP) type. If it has two poles, it is referred to as a *double-pole* (DP) switch. In a DP switch, the poles are mechanically linked and move in unison, allowing the switch to control two different circuits at the same time.

ST, DT, SP, and DP functions can be combined to create SPST, DPST, SPDT, and DPDT switches. For example, the knife-blade switch shown back in Figure 6-1 is an SPDT type. Figure 6-4 shows the schematic representations of the four common switch pole and position arrangements (note that Appendix B covers schematics in detail).

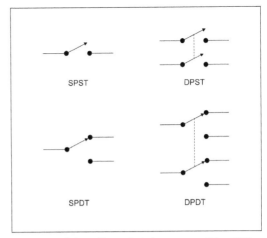

FIGURE 6-4. Schematic symbols for common SP and DP switch types

Notice that a dashed line is used here to indicate that the poles shown in the symbols are mechanically coupled. A schematic might also use labels, such as S1-A and S1-B, to denote that a switch has more than one pole, or it could use both methods. The current trend is to dispense with the dashed line, but you will likely encounter it in older schematics.

Switch Types

In this section, we'll look at toggle, rocker, slide, rotary, pushbutton, and snap-action switches. These cover most of the mechanical switch spectrum, and all are really just variations on a theme.

TOGGLE

The toggle switch is a common type that comes in a variety of styles. Figure 6-5 shows a heavy-duty toggle switch (you've already seen a miniature version in Figure 6-3).

FIGURE 6-5. A heavy-duty toggle switch

Toggle switches can be even smaller, with a micro form available, as shown in Figure 6-6. These are useful in applications where space is limited, but a toggle switch is still needed.

FIGURE 6-6. A mico-toggle switch

Toggle switches are readily available in SPST, SPDT, DPST, and DPDT forms. They can also be had in three-pole and four-pole versions. It is useful to remember that a SPDT switch with only the common terminal and one position wired is the same as a SPST switch.

Yet another variation is the center-off switch (the ON-OFF-ON type mentioned earlier). In these switches, the internal contacts might be SPDT or DPST, but the toggle handle has three positions. In the center position, the plunger at the end of the toggle actuator will settle into a detent in the middle of the contact rocker, leaving the contact rocker with both ends suspended, and neither of the terminals will be connected.

In some applications, a center OFF with a momentary ON in one of the positions is used for things like testing a subsystem (momentary ON) and enabling the subsystem (normal ON). The manufacturer achieves the momentary action by shaping the contact rocker such that it will not hold the toggle actuator, whereas the regular ON position will. Some designs also incorporate an internal spring to push the actuator back to the center position when it is moved to the momentary position and then released.

Toggle switches might not be the flashiest or most futuristic-looking switches, but they are rugged, easy to use, and ubiquitous. Older styles of common residential light switches are a type of toggle, as are some of the switches found on the front panels of musical instrument amplifiers. The control panels of spacecraft like the *Soyuz* or the space shuttle contain hundreds of them, and in early computers, toggles switches were one way to enter a program or data into memory. These days, new houses are often wired with a type of rocker switch, which we'll look at next.

ROCKER

A rocker switch employs a plastic or metal piece shaped in a shallow V so that when one end is up, the other is down. In other words, it rocks from one position to another. Internally, the mechanism is identical or similar to that of a toggle switch (or perhaps a slide switch, depending on the manufacturer; see "Slide" on page 137). Figure 6-7 shows a typical miniature rocker switch.

FIGURE 6-7. A generic SPST miniature rocker switch

All that is really different about a rocker switch is the physical means of changing the internal mechanical state of the switch. Most rocker switches use a pivoted contact rocker similar to that found in a toggle switch. Also note that some rocker switches come with built-in lamps. There are neon, LED, and incandescent bulb types available, and they are commonly encountered in aviation and industrial controls. As with the toggle switches, some older mainframe and minicomputer systems made extensive use of rocker switches to enter data into the machine. Also as with toggle switches, rocker switches can also be found in a center-OFF form, as well as a momentary action in one or both directions.

SLIDE

Slide switches are a convenient way to select from more than two circuits. A small tab or knob slides in a track to move the pole contactor (or contactors) between positions, and slide switches can have more than one pole. Figure 6-8 shows a typical small slide switch with solder eyelets. This particular switch is designed to be panel-mounted with screws.

FIGURE 6-8. A typical small slide switch

Other types of slide switches have pin-like legs for mounting on a PCB, as shown in Figure 6-9. These are useful for situations where a PCB can be used to support the switch, and they look a bit neater than a switch with screws. For a front-panel control, the slide tab protrudes through the panel and the PCB is mounted behind it to support it.

FIGURE 6-9. PCB-mounted slide switch

Some older radio communications equipment used slide switches with three or more poles to change how the internal circuitry behaves (tuning range, power outputs, etc.). Physically, the pole contactor slides across each of the contact positions while staying in contact with a bar or rail inside the switch. The track the pole contactor moves in usually has small indentations, and a small ball bearing is used to provide some tactile feedback in the form of a "click" or "bump" at each of the contact positions.

Slides switches are available in miniature and micro forms, along with surface-mounted types. PCB mounted versions can be found in both vertical and right-angle designs, and momentary actions are available.

ROTARY

Slide and rotary switches are closely related, in that a slide switch is similar to a rotary switch laid out flat with the contacts all in a row instead of a circle. The action of each type is essentially the same. As with slide switches, a rotary switch can have multiple poles. In some older types of test equipment, it wasn't uncommon to find rotary switches with upward of 10 or even 15 poles per switch.

Rotary switches are readily available and, like every other switch type, come in a variety of sizes and capabilities. Figure 6-10 shows a single-pole rotary switch with six positions. One solder eyelet is the pole, and the rest are the switch-position terminals.

FIGURE 6-10. Single-pole rotary switch

Rotary switches come in PCB-mount versions, both through-hole and SMT, and there are even some types, like the one shown in Figure 6-11, that require a small screwdriver to operate. These are typically used in places where there might be an occasional need to alter the behavior of a circuit, but not under end-user control. Testing is one situation that comes to mind, or perhaps infrequently changing the behavior of a device.

FIGURE 6-11. Surface-mounted micro rotary switch

FIGURE 6-12. Panel-mounted pushbutton switch

PUSHBUTTON

The pushbutton switch is about as ubiquitous as the toggle switch and can be found on everything from a cell phone to the ignition button on late-model cars with electronic key-lock systems. They come in a wide range of styles, some internally illuminated, some not. Figure 6-12 shows a panel-mounted pushbutton switch, and Figure 6-13 shows a PCB version.

Like the other switches we've reviewed so far, a pushbutton switch can have more than one pole, although SP and DP types are the most common. The common type of mechanism is a momentary action, but there are some that will mechanically stay in one position or another, like the emergency stop switches on elevators and industrial machines.

FIGURE 6-13. PCB-mounted pushbutton switch

SNAP-ACTION

A snap-action switch is typically found in the role of a sensor, rather than something that a human might operate, although there are some varieties that do incorporate a toggle or pushbutton mechanism. These switches are typically used to sense things like physical limits for the moving parts of a machine, the passing of a lobe of a cam, or if a device is resting on a surface or suspended in the air. Figure 6-14 shows a small snap-action switch. Various other types of input mechanisms are available, including rollers, plunge rollers, and pushbuttons.

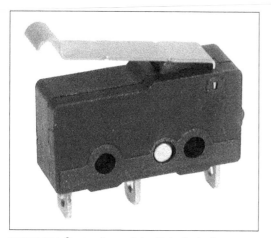

FIGURE 6-14. Snap-action switch with leaf actuator

A snap-action switch is usually intended for heavy use, and some are sealed to prevent accidental fires or even explosions in hazardous environments (such as might be found in grain elevators). As you might expect, a heavy-duty snap-action switch tends to be both physically large and somewhat expensive, although there are some miniature types available that are popular with the robotics folks.

Slide and Rotary Switch Circuits

As mentioned earlier, slide and rotary switches are similar in terms of functionality, and they differ mainly in their physical form. Electrically, the objective of both types is to select a circuit from multiple choices.

Figure 6-15 shows the schematic representation of a four-position double-pole slide switch, similar to Figure 6-9, but with one additional position.

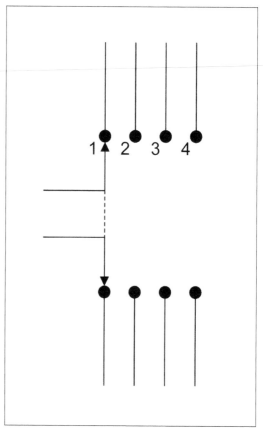

FIGURE 6-15. Diagram of a four-position slide switch

The schematic for a five-position rotary switch like the one in Figure 6-10 is shown in Figure 6-16.

If a slide or rotary switch has multiple poles, the old-style schematic convention is to show the physical connection between each of the poles with a dashed line, although in some schematics, each pole might be located in different parts of the drawing. More current schematics dispense with the dashed line and just use a numbering scheme (e.g., SW1-A, SW1-B, etc.) to indicate the poles of a single switch assembly.

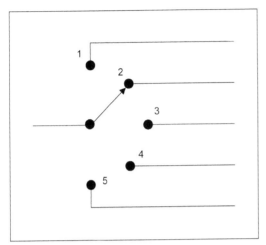

FIGURE 6-16. Diagram of a five-position rotary switch

Switch Selection Criteria

There are three primary criteria for selecting a switch: what the switch needs to do (electrical characteristics), how it needs to do it (physical form), and where it will be placed (mounting). As often happens in engineering, there are trade-offs that must be made. That really awesome little miniature toggle switch that would be perfect for a project might not be able to handle the current flowing through the circuit it is controlling. Conversely, you might find that the right switch for the voltage and current levels involved will be physically bigger than the rest of the project components combined. The goal is to find the middle ground and pick the best possible switch for the job.

Here are some suggested essential criteria to consider when selecting a switch:

How much voltage will the switch need to handle?

Consider the application and the range of voltages the switch could possibly encounter. Select one that is capable of at least the anticipated maximum value.

How much current will flow through the switch?

Consider how much current the switch will need to handle. Even at low voltage levels, such as 5V DC, there can be considerable current. While internal arcing at low voltage might not be a problem, with a large enough amount of current moving through the switch, its contacts won't last long. With a significant amount of current, the inherent resistance in the switch's contacts and metal bus components can be enough to cause the switch to overheat, or even burst into flames. If possible, selecting a switch with twice the capacity of the expected maximum current is a prudent choice.

How many circuits, or contacts, will the switch need to have?

Don't use a switch with more poles than is really necessary. The more poles a switch has, the more expensive it will be, and the more prone it will be to failure. Less is better.

Should the switch be a toggle, a slider, a rotary type, or some other mechanism?

Don't let aesthetics be the primary guiding criteria to switch selection. A nifty-looking switch might seem like a good idea, but will it be easy for the user to operate? Will its function be intuitively obvious? Will it be able to endure multiple operations over an extended period of time? Could someone wearing gloves operate the switch? These are consideration that often get a lot of scrutiny in engineering design meetings, and for good reason. Many a gadget has been built with cool space-age controls, only to fail miserably in actual use after a short period of typical user abuse.

Does the switch need to be small? How small?

This criteria is related to the previous consideration, but here the main concern is size versus cost versus available mounting space. Miniature switches that are also rugged aren't cheap, and they can be a pain to assemble into a design. Select the largest switch you can reasonably use, given the physical constraints of its intended mounting location and the manner in which it is physically connected to the circuit.

Should the switch be a low-force type, or does it need to have a stiff mechanism?

This might seem like an odd topic, but it's one that is sometimes overlooked in commercial designs, resulting in a product that is difficult (or sometimes even painful!) to use. An extremely stiff slide switch, for example, can leave a user with a sore finger if it must be operated routinely. A rotary switch can be frustrating if it is difficult to turn to a desired position. On the other hand, there are situations where a stiff switch really is the correct choice. The switches on heavy machinery or high-voltage equipment, where an incorrect switch position could result in major problems, are two possible applications. The upshot here is to select a switch that is suitable for the intended application and won't be easily bumped or jostled into an incorrect position, if that's a concern.

What are the mounting options? Panel, PCB, or something else?

Toggle, pushbutton, and rotary switches can be mounted in holes drilled into a panel or in some type of chassis. A slide switch will require a rectangular hole, with the length of the rectangle being proportional to the number of switch stop positions. The downside of panel or chassis-mounted switches with soldered leads is that wires are needed to connect the switch to the circuit it is controlling. Switches are also available for PCB mounting, like any other component on the PCB. With a PCB-mounted switch, the wires are eliminated, but now the switch is part of something else, which will have its own mounting requirements. A panel or chassis-mounted switch with PCB terminals can also be used to hold the PCB it is part of.

How rugged should the mounting be?

A switch mounted using a shaft nut or mounting screws will typically be more robust then one soldered onto a PCB, although this depends to a large extent on how well the switch is soldered to the PCB. Some PCB mount switches come with extra metal tabs as part of their body construction. These are intended to be soldered into holes in the PCB to secure the switch. There are protective guards and covers available for toggle switches to help protect them from both impact damage and accidental operation.

Switch Caveats

Being a mechanical device, a switch has certain limitations and behaviors that you should be aware of. With toggle, rocker, pushbutton, and snap-action switches, *contact bounce* can be a significant concern. If a switch is used to control power to a circuit or a device, this might not be an issue, but when a switch is used to generate or control a signal (such as, say, an input to a microcontroller that is used to count something), contact bounce can become a very big deal.

Contact bounce can be reduced electronically with a filter or a one-shot timer (see Chapter 11 and Appendix A) or, if the switch is connected to a microcontroller, it can be eliminated in software. It is beyond the scope of this book to address software topics, but check out the texts listed in Appendix C for details.

Contact bounce is not the main issue with slide or rotary switches. In a slide or rotary switch, you should be aware that some switches don't immediately switch from one set of contacts to another. In other words, they can be either *shorting* or *nonshorting*. A shorting switch will allow the pole contactor to span two contacts when moving between them. A nonshorting type will have a definite physical gap between the contacts. Why would you want to use a shorting type slide or rotary switch? They are commonly found in circuits that carry audio signals, since the shorting behavior reduces any "pop" that might occur when switching between, say, microphone A and microphone B, or from one filter setting to another. Never use a shorting switch in a circuit that is handling power, as opposed to signals, or evil things might occur, and in general, don't use a shorting type switch unless there is a definite need for it.

Summary

In this chapter, we've looked at some of the various types of switches. You should now have a good idea of what types and styles of switches are available and also have a basic understanding of how they work mechanically. We have reviewed toggle switches, slide switches, rocker switches, pushbutton types, and rotary switch mechanisms. We also took a brief look at snap-action switches, which are common in industrial environments and some robotics applications, but aren't usually seen in small electronic devices.

The main takeaway from this chapter is to select the switch that meets the needs of the application, unless, of course, you are simply throwing something together to see if it works. Spending some quality time online reviewing the various switches that are available is a worthwhile effort, and a trip to a local electronics outlet can also be informative.

CHAPTER 7

Connectors and Wiring

WIRES AND CONNECTORS ARE THE GLUE that binds electronic components, assemblies, and devices into cohesive systems. Wires can be the copper traces on a PCB or actual insulated wires from one point to another inside the chassis of a device. The concept of a wire extends to include things like USB cables, power supply wiring, Ethernet cables, and shielded cables carrying audio, video, or RF (radio frequency) signals.

This chapter covers the basics of wire sizes, stranded versus solid wires, and multiconductor cables. It also looks at shielding and how it is employed to reduce interference from external noise sources, as well as how twisted-pair wires work.

Connectors provide convenient end points for wires and allow the parts of a device or system to be modularized. This makes for easier testing, assembly, and maintenance. Without connectors, we would have to resort to soldering the wires that connect various parts of a device, or removing wires from a circuit if part of it needed to be replaced. At one time, this was indeed the case. If you ever get the opportunity to disassemble an old television set from the 1960s, you should. It is an eye-opening example of how to do things the hard way (but, in all fairness, there weren't a whole lot of affordable options back then).

Nowadays, connectors are ubiquitous. This chapter presents descriptions of some of the more common of the various types of connectors available and describes where they are typically used. It also covers the techniques used to assemble some of the more common types, such as DB-9, DB-25, high-density terminal blocks, and the 0.1-inch grid spacing pin connectors found on things like the Arduino, Raspberry Pi, and BeagleBone boards. Along the way, it also touches on topics such as soldering, crimping, and insulation displacement connector (IDC) techniques for connector assembly.

This chapter specifically focuses on those types of connectors that a typical human being can easily handle or assemble without resorting to a microscope and tweezers. It doesn't cover USB connectors, which are not something you would want to assemble by hand if you can avoid it. Nor does it delve into the world of high-reliability connectors used by aerospace and the military, or miniaturized connectors such as the types found in consumer electronic devices, as these typically require special crimping and assembly tools that cost many hundreds of dollars.

Wire and Cable

The terms *wire* and *cable* are sometimes used as synonyms, but in general you can think of a wire as a single conductor of some sort and a cable as a collection or bundle of two or more individual wires. For example, the shielded and insulated bundle of wires between a microphone and an amplifier is referred to as a *microphone cable*, never as a *microphone wire*. This isn't a hard-and-fast rule, however, and confusion sometimes arises when size is involved. Large wires, like those used to carry electricity between poles along the side of the road, are sometimes referred to as cables.

Wires come in specific sizes, with different types of insulation, or even no insulation at all. Wire is available as a single solid conductor or as a set of smaller wires in a bundle, called *stranded wire*. Figure 7-1 shows how smaller wires are twisted into a bundle to create stranded wire, whereas the solid wire is just a single conductor. Figure 7-2 shows cross-section views of both a solid and a stranded wire.

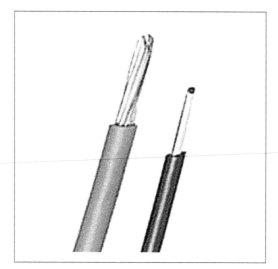

FIGURE 7-1. Stranded and solid wire

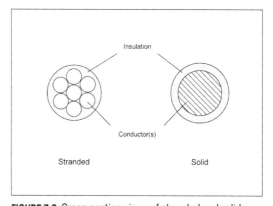

FIGURE 7-2. Cross-section views of stranded and solid wire

Stranded wire is generally preferable to solid wire because it is more flexible and less prone to breaking, but solid wire still has a role in some applications. Telephone patch bays, for example, contain what are called *punch-down blocks* and use solid wire. The punch-down blocks work by pushing a solid copper wire between metal blades. These slice through the insulation and make contact with the wire. Of course, there is a special tool for this, but unless you plan to do a lot of telephone-type wiring, it's not worth buying one.

Solid wire might be a valid option in applications where the wire will not experience any significant flexing and cost is a consideration (solid wire is typically less expensive than stranded wire). For small wire gauges, it is more common to find solid wire, and for applications such as wire-wrap circuit construction, solid wire is the only appropriate choice.

Figure 7-3 shows a selection of insulated wire on small spools, often referred to as *hook-up wire*. Kits like this can be purchased from electronics distributors and other online sources in a variety of gauges and insulation types.

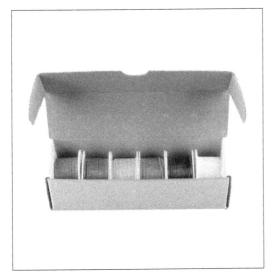

FIGURE 7-3. An assortment of hook-up wire spools in a dispenser box

WIRE GAUGES

Wire gauges are defined by the American Wire Gauge standard and are referred to as AWG sizes. With the AWG system, a higher number indicates a smaller diameter. This is historical and refers to the number of times a wire needed to be passed through a drawing die (a metal block with a hole in it) to reduce the wire to the desired diameter.

Both solid and stranded wires use the same gauge system, but there are some differences. An 18-gauge stranded wire has a larger diameter than an 18-gauge solid wire, because the spaces between the strands are not counted. Both wires will have an equivalent cross-sectional surface area of copper, however.

Table 7-1 lists the diameters and resistance for gauges 12 through 40. For electronics work, the most common type of wire encountered will be 32- to 16-gauge PVC insulated stranded wire. The most commonly used sizes seem to be 20, 22, and 24 gauge. Large diameter wire, such as 16 and 18 gauge, is sometimes used to carry large amounts of current (10A or greater) from a power supply.

TABLE 7-1. Common AWG solid-wire gauges

Wire gauge	OD (inches)	OD (mm)	R/1,000 ft	R/1 km
12	0.0808	2.053	1.588	5.211
14	0.0641	1.628	2.525	8.286
16	0.0508	1.291	4.016	13.17
18	0.0403	1.024	6.385	20.95
20	0.0320	0.812	10.15	33.31
22	0.0253	0.644	16.14	52.96
24	0.0201	0.511	25.67	84.22
26	0.0159	0.405	40.81	133.9

Wire gauge	OD (inches)	OD (mm)	R/1,000 ft	R/1 km
28	0.0126	0.321	64.90	212.9
30	0.0100	0.255	103.2	338.6
32	0.00795	0.202	164.1	538.3
34	0.00630	0.160	260.9	856.0
36	0.00500	0.127	414.8	1361
38	0.00397	0.101	659.6	2164
40	0.00314	0.0799	1049	3441

Table 7-2 shows a sample of some of the more commonly encountered AWG gauges for stranded wire.

TABLE 7-2. Common AWG-stranded wire gauges

Wire gauge	Stranding	OD (inches)	OD (mm)	R/1,000 ft	R/1 km
16	7/24	0.060	1.5240	3.67	12.04
18	7/26	0.048	1.2192	5.86	19.23
20	10/30	0.035	.8890	10.32	33.86
22	7/30	0.030	.7620	14.74	48.36
24	7/32	0.024	.6096	23.3	76.44
26	10/36	0.021	.5445	41.48	136.09
28	7/36	0.015	.3810	64.9	212.92

In Table 7-2 the "Stranding" column defines how the wire is organized internally. In the case of 26 gauge wire, for example, the table indicates that it comprises 10 strands of 36-gauge wire.

Note how the resistance of the wire over a given distance drops as the diameter of the wire increases (as shown in the R/1,000 and R/1 km columns for both solid and stranded wires). Also note that stranded wire conducts better (has less resistance) than solid wire of the same gauge. This is because, while the wire types might have the same cross-sectional area, the surface area of the stranded wire is greater than the solid wire.

As mentioned in Chapter 1, everything in a circuit has resistance, including the wires used to connect the circuit to a power source or another module. Since current is defined as the volume of electric charge moving through the cross-sectional area of a conductor in some unit of time, it stands to reason that the larger the cross-section area, the larger the current capacity of the conductor. This is one case where the analogy to a water pipe actually does apply fairly well. You can move a lot more water (in gallons/minute) through a 1-inch hose than you can through 1/4-inch tubing, so, by analogy, you can safely assume that 4-gauge wire will safely carry more current than 18-gauge wire.

For a realistic example, let's say you have an application where a sensor is remotely located 250 feet or so from the rest of the equipment, and it is connected by a cable with multiple 28-gauge stranded connectors. According to Table 7-2, each conductor in the cable will have about 16 ohms of resistance over that distance. Failing to take this into account could result in bad readings or an excessive power drop to the sensor. Furthermore, because even the common (ground or neutral) return line has resistance, the sensor will tend to "float" at a voltage higher than the actual ground at the local controller, and this can introduce all kinds of nasty side effects.

Using a cable with larger conductors will help to reduce the problems, but in reality, it's better to power a remote device with its own power source (perhaps a battery or solar panel, covered in Chapter 5) and use digital signaling and balanced twisted-pair wiring to communicate with it. Chapter 14 discusses techniques like these.

INSULATION

No discussion of wire and cables would be complete without a discussion of insulation. Some time around the late 19th century, with the widespread introduction of electrical devices such as the telegraph, most wires were either bare or were wrapped with fabric or paper. In some cases, a coating of a tar-like substance or varnish was also applied to help preserve the cloth or paper insulating material.

In some historic buildings, you can still see the bare wires used to route AC power between the rooms, hopping from one ceramic insulator to the next across rafters, under floor joists, and down the inside of a wall. In smaller devices, such as early radios, the wires were often left bare and simply soldered between terminal strips or tube sockets. Electromagnetic components with tightly wound coils of wires, such as relays and solenoids, used either layers of insulating paper or wires coated with a type of varnish or shellac. Many of these types of components are still made this way today.

Times have changed, and now wires and cables are available with a variety of insulation materials. The most common type is polyvinyl chloride (PVC). This is a relatively soft, low-temperature material that is flexible and easy to strip. It does have a tendency to melt, shrink, and sometimes char when it is close to a solder joint that has been overheated during soldering. Because heavier gauge wires require more time to solder than a thinner gauge, PVC insulation on wire gauges larger than about 18 gauge can be a problem. A better choice is Teflon, but it can be difficult to strip cleanly.

Very small wire gauges, such as the fine solid wire used to do wire-wrap construction, are usually coated with Kynar, a type of polyvinylidene difluoride (PVDF). It can be annoying to strip and requires a special tool and some patience to get it right. Most manual wire-wrap tools come with a small stripper tool of some sort. But unless you need to do wire-wrap construction (which is not covered in this book) or you need to solder patch wires on a PCB, you probably won't encounter wire-wrap wire.

TWISTED PAIRS

A wire is essentially an inductor (see Appendix A) and, as such, it can be subject to external interference from AC power lines, electric motors, lightning bolts, and even a nearby radio transmitter. The telescoping antenna on a portable radio is nothing more than a section of wire specifically intended to pick up the electromagnetic energy generated by a radio station. The conductive traces on a PCB can also inductively couple to one another, creating interesting and difficult-to-fix problems. Wires extending from a sensor back to a microcontroller can conduct more than just the sensor data. One technique to reduce the influence is to use twisted-pair wiring. Another involves shielding (see "Shielding" on page 152).

Twisted pairs, as the name suggests, are two wires that are twisted around each other. A twisted pair is always used for a single circuit, not two. Figure 7-4 shows both a single-channel and a multi-channel connection between two devices (device A and device B).

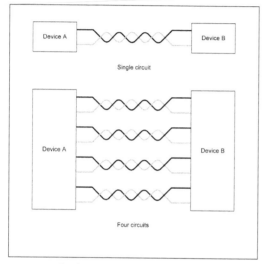

FIGURE 7-4. Single- and multiple-circuit twisted-pair connections

The concept of twisted pairs was invented by Alexander Graham Bell in 1881 to deal with noise on early telephone circuits. The early telephone system used existing telegraph lines, which involve just a single conductor with grounded batteries, keys, and clickers (sounders) at each end. The unshielded lines turned out to be excellent antennas for picking up things like electric street cars and noisy electrical motors.

The solution was to use twisted pairs. In a twisted-pair circuit, the wires carry the same signal with opposite polarity, so when one is positive, the other is negative. It is the difference between the wires that counts, and that difference is measured across a load at the end of the twisted pair. If external noise interacts with the pair and causes the same level of interference on both, it is ignored, because it will not induce a potential difference across the line load (this is called *common-mode rejection*). Chapter 14 discusses twisted-pair wiring in more detail, but the focus in this chapter is on what twisted pairs look like and how they are specified.

Figure 7-5 shows what a section of a twisted pair looks like. This was removed from a multi-conductor cable like the one shown in Figure 7-11. A twisted pair is specified in terms of wire gauge, wire type (solid or stranded), and twist rate (also called the *twist pitch*). In a multi-conductor cable with multiple twisted pairs, the twist rate is usually different between the pairs to minimize unwanted coupling (called *cross-talk*) between each of the pairs.

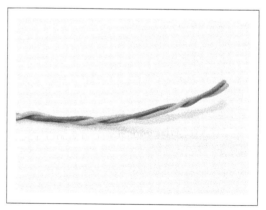

FIGURE 7-5. A section of twisted-pair wire from a larger cable

There is also a color code for twisted-pair wires. Typically, one member of the pair will be a solid color (blue, green, red, etc.) and the other will either be white with a stripe of the same color as its mate, or just white. Orange, blue, green, and brown are common colors found in Ethernet cables (they're actually specified by a standard, TIA/EIA T568B). Other color schemes can be found, depending on the manufacturer and, in some cases, what the customer specifies.

Twisted-pair cables come in unshielded (UTP), shielded (STP), and foiled twisted-pair (FTP) forms. Common Ethernet cables are UTP types with four internal pairs. The cable shown in Figure 7-11 is an STP cable. In an FTP cable, each pair has conductive foil wrapped around it, and the entire cable might have an outside shield (this would be an S/FTP cable). These are sometimes found in instrumentation applications where any external interference that might skew a measurement is unacceptable.

You can make your own twisted-pair wires by simply twisting two wires. This isn't quite as easy as it may sound, since the wires will have a tendency to end up with one wrapped around the other, instead of both twisted around each other. If you anchor the two wires at one end using something like a small vise, you can keep constant tension on them as you twist. This will help prevent wrapping.

Another way to make a twisted pair is to make a loop hook out of a piece of steel coat-hanger wire that can be used with a hand-held drill, and then drill two holes through a small block of wood large enough to pass the wire through. Cut off two lengths of wire at the length you need to twist, plus some extra (the twisting effectively shortens the wire a small amount). Knot the wires together at one end, and put the knot over the hook. Feed the loose wire ends through the holes in the wood block. While someone holds the drill and runs it at a slow speed, pull the wood block back along the wire as the drill creates the twist. Be careful to keep the wires separated before they enter the block. Figure 7-6 shows the guide board and what the homemade twist hook looks like in a drill. Figure 7-7 shows the result.

FIGURE 7-6. Improvised hook and guide board for making twisted-pair wire

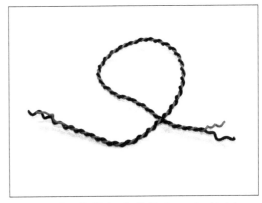

FIGURE 7-7. Result obtained with the improvised twist tools

Granted, the twist on the wire in Figure 7-7 might be a wee bit on the tight side, but it's still perfectly usable. It helps to have two people to do the twisting when you're using this technique. If you don't have another person available to help out, or you just want to avoid the issue completely, you can purchase ready-made twisted-pair wire as just a single pair on spools in various lengths and gauges. Special bench-mounted power tools can also make twisted-pair wire, but they aren't cheap. If you need just a little, I would suggest making it yourself. If you need a lot, buy it by the spool.

One more thing about twisted pairs: if two or more twisted pairs are run in a bundle, the degree of twist for each pair should be different. If each pair has the same degree of twist, they can cross-couple, which can create unwanted side effects. If you really need multiple pairs in a bundle, and it doesn't have to carry much current, consider using a section of stranded CAT5 Ethernet cable. I also keep a large box of old computer and instrumentation cables on hand, and when I need a short length of multi-conductor cable, I can fish one out, lop off the old connectors, and put it to use. Then again, spools of multi-conductor twisted-pair cables are available in various lengths.

SHIELDING

Shielding is a way to minimize the effects of external sources of electromagnetic interference (EMI) on the conductor in a wire or cable. The concept is similar to a *Faraday cage*, which is a grounded conductive enclosure that prevents external electrical and electromagnetic energy from entering an enclosed space. The EMI is shunted away to ground before it can have any effect on the conductor.

The two most commonly encountered forms of wire shielding are braid and foil, and it's not uncommon to find both in use in the same cable. Figure 7-8 shows an single conductor shielded cable, called a *coaxial* (or just *coax*) cable, which is used in video, CATV, and radio applications. Coaxial cable isn't new; it was patented in England in 1880 by Oliver Heaviside.

FIGURE 7-8. Shielded coaxial cable

Figure 7-9 diagrams the inner parts of a typical coaxial cable. The center conductor can be either solid or stranded; for RF use it's typically solid, and for audio and video applications, a stranded conductor is often used.

Figure 7-10 shows another interesting example of a shielded cable, part of an old wireless Ethernet range extender. The cable is small-diameter coax with a connector at one end for an antenna and another at the opposite end to plug into a connector on a PCB. This was repurposed for use with a wireless data link module like the ones described in Chapter 14.

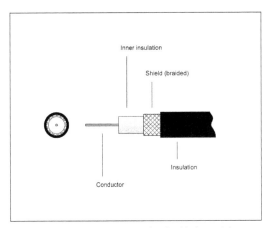

FIGURE 7-9. Inner construction of a shielded coaxial cable

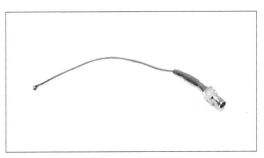

FIGURE 7-10. Small-diameter coaxial cable with antenna and PCB connectors attached

In some shielded cables with both a braided shield and a foil wrap, a drain wire is run along the length of the cable between the braid and the foil. This helps to ensure that the foil is grounded. The drain wire might be a solid bare wire or a bare twisted wire like the one shown in Figure 7-11. The drain wire doesn't have to be connected so long as the braid is connected to ground, but it's more convenient than pulling back the shield and using that as the shield ground connection.

MULTI-CONDUCTOR CABLES

As mentioned earlier, the term *cable* is usually reserved for bundles of two or more multiple conductors, or for very large single-conductor wires. This section focuses on the bundle definition.

Individual Conductors and Twisted Pairs

Some types of multi-conductor cables consist of individual wires or twisted pairs. They can be either shielded or unshielded, and the shielding might be a braid, foil, or both. Figure 7-11 shows one end of a multi-conductor cable that was used to connect a peripheral device to a PC.

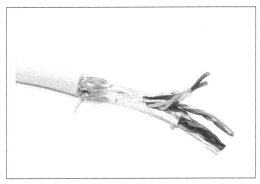

FIGURE 7-11. Multi-conductor shielded twisted-pair cable end

The cable shown in Figure 7-11 is a shielded twisted-pair type with both a braid and a foil shield around the entire wire bundle. The conductors consist of 10 pairs of 28-gauge stranded wires and two pairs of 20-gauge stranded wires, for a total of 24 conductors. The bare twisted strand sticking off to the side near the outer jacket insulation is the drain wire that runs along the length of the cable between the foil and the braided shield.

Unshielded multi-conductor cable is commonly used in applications where the extra protection against EMI is not necessary. Figure 7-12 shows an example of a multi-conductor cable that is suitable for DC power and control, such as connecting a remotely located relay bank or motor controller to a local microcontroller. This particular cable has two 20-gauge wires, two 22-gauge wires, and four 30-gauge wires, all stranded. There is no outer shielding.

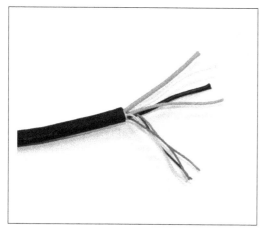

FIGURE 7-12. Multi-conductor unshielded cable

Ethernet cable is another type of unshielded twisted-pair multi-conductor cable. Because Ethernet uses balanced differential signaling, it is robust when it comes to rejecting external interference. So just the twisted pairs are sufficient.

Ribbon Cables

In electronics, you will often need to route a set of parallel signals from one place to another inside a device. Instead of connecting numerous single wires or using a multi-conductor cable with individual wires, you'll find that a better solution is to use something like the ribbon cable shown in Figure 7-13. Note that part of the cable has been split and pulled back, for a reason described shortly.

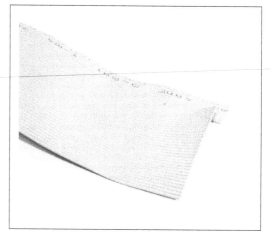

FIGURE 7-13. Ribbon cable

A *ribbon cable* (also sometimes referred to as *multi-wire planar cable*) consists of a set of conductors set side by side and molded into a common insulator. These flat cables come with various numbers of conductors and various conductor gauges. Cables are available with 4, 6, 8, 10, 14, 15, 16, 18, 20, 24, 25, 26, 34, 37, 40, 50, 60, 64, and 80 conductors. It just so happens that connectors are available for use with ribbon cables that have the same number of pins. These are described in "Insulation Displacement Connector" on page 158.

It is also possible to separate groups of wires in a ribbon cable into cables with some smaller number of conductors. Say, for example, you needed a ribbon cable with eight conductors, but all you have on hand is a roll of ribbon cable with 24 conductors. This is not a problem for most PVC-insulated ribbon cables. Just start the cut at the end of the larger cable for the number of conductors you want, and then pull it back. With a little care, you'll get a nice eight-conductor ribbon cable, as shown in Figure 7-13.

The wire used in a ribbon cable is typically stranded wire in 22, 24 or 26 gauge, although some specialty cables have larger or smaller gauge wires. There are even ribbon cables with solid conductors that are intended to be soldered directly into a PCB. These would typically be used to connect two PCB modules over a short distance.

An example of where you might find solid-conductor ribbon cable is an LCD display module attached to a larger PCB. If the display will never be removed during the lifetime of the device, and the physical constraints of the packaging limit the mounting options, then a soldered solid-conductor ribbon cable might be a good choice.

As you might expect, there are special tools available to cut ribbon cable, and other tools to attach connectors. We'll look at these a little later on.

Flex Cables

A flex cable is a relative of the ribbon cable, and is fabricated by bonding metallic conductors to the surface of a thin film and then applying another layer of film to seal the conductors inside. If you have ever opened up a cell phone, laptop computer, or portable DVD player, chances are, you've seen a flex cable. Figure 7-14 shows an example of a flex cable in its natural habitat. A flex cable is essentially a flexible PCB, and the cable also incorporates active components in some applications.

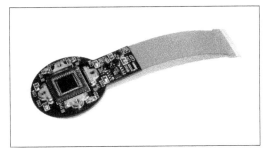

FIGURE 7-14. An installed flex cable

These types of cables are great for the production of miniature electronic devices, but not so good for prototypes or hacking. Still, you should probably be aware of them and have a general idea of how they are used.

STRIPPING WIRE INSULATION

Removing insulation from wire or cable isn't always as easy as it might appear. Using a knife or wire cutters might seem like a good idea, but it's easy to nick the underlying wire. The nick then becomes a weak point where the wire is more likely to break if flexed.

Chapter 3 describes special tools for stripping insulation from wires. It's a good idea to invest in a couple of different types. They aren't all that expensive and they can make the difference between having a reliable connection and a wire that breaks off during final assembly, leaving only a short stub on a connector or in a hole on a PCB.

An important point to keep in mind is how much insulation to remove. This varies, depending on how the wire will be attached. When dealing with crimped connections (see "Crimped" on page 160), you should follow the recommendations of the crimp contact manufacturer. Removing too much insulation defeats the purpose of the wire strain relief portion of the crimp contact, and leaving too much can result in no connection at all. Once a crimp contact is crimped, it cannot be easily undone.

For wires that will be soldered into a PCB, you must remove enough insulation to allow the wire to protrude through the hole in the PCB with about 1/4 inch of bare wire showing on the opposite side. Any more than this is a waste of wire. Once soldered in place, the excess wire can be removed with flush cutters (also described in Chapter 3).

As you might guess, specialty wire strippers are available for coaxial cable, ribbon cable, and multiconductor cables. These can range from cheap (and almost useless) to expensive tools designed for a production-line environment.

Figure 7-15 shows a semiautomatic wire stripping tool in action. This particular tool uses a set of blades with holes equal to specific AWG wire gauges. The main advantage of this tool is that it pulls the cut insulation away from the wire after the cut is made, and it maintains tension on the wire the whole time. It requires almost no effort to use, unlike a completely manual tool. If you elect to purchase a tool like this, also be sure to purchase an alternate blade set for other wire gauges.

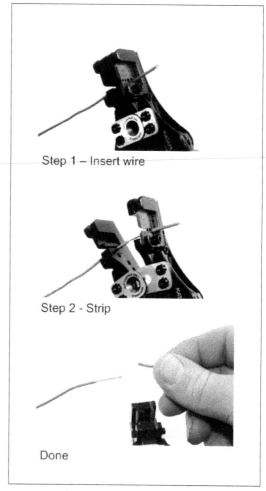

Step 1 – Insert wire

Step 2 - Strip

Done

FIGURE 7-15. A semiautomatic wire stripping tool in action

Sometimes, however, the manual tool is the better way to go, particularly in those cases where the semiautomatic tool might not have a position for a given wire gauge. The manual tool can be continuously adjusted to accommodate a specific wire gauge. It also allows the user to feel when the tool has cut through the insulation. The semiautomatic tool has no such feedback during operation.

Connectors

It's been said that most electronics failures are due to connectors. That may well be true, but connectors are essential, and all physical interfaces in electronic devices that aren't soldered directly to something will utilize a connector of one sort or another. The key to success is picking the right connector for the application and then assembling it correctly.

For any given interface, there's a connector to use for it. A look at the back of a typical desktop PC might reveal multiple types of connectors: a DB-9 for the serial port, DB-25 for the parallel printer port, a high-density DB-15 for connecting a VGA analog video monitor, an RJ-45 type jack for Ethernet, several type A USB sockets, and perhaps two or three 3.5 mm jacks for audio input and output.

Internal connections between modules and components in an electronic device often take the form of ribbon cables (see "Ribbon Cables" on page 154), coaxial cables, and bundles of wires with multi-terminal connectors at each end.

CONNECTOR TERMINATION

There are many ways to connect wires and cables, each one designed to meet a specific need for a particular type of wire or cable. Connectors can help to make assembly more efficient and improve the maintainability of a device or system. Soldering wires directly into a circuit board, while effective, can make it difficult to disassemble for repair without causing damage, and in some cases might actually be less reliable than would be the case if the appropriate connector had been used. This is particularly true in situations where the wiring might be subjected to flexing or vibrations. The point where a wire enters a PCB or attaches to a solder-lug on a part is a *flexure point*, which can weaken and break over time.

We'll start out with some descriptions of how to attach wires or PCB traces to a connector, because most of the connectors covered here are available in more than one mounting form.

Terminal Blocks

The venerable terminal block has been around for a long time, and it's still a good option for some applications. Figure 7-16 shows what is called a *barrier terminal block*. Notice that each terminal position (the screws) is isolated from adjacent terminals by a low barrier ridge, hence the name.

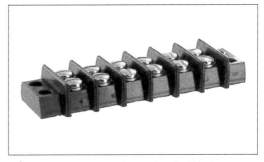

FIGURE 7-16. A six-position barrier terminal block

Miniature PCB mount types, such as the five-position part shown in Figure 7-17, are another variation on the terminal block. These are common on things like industrial controllers, motor driver modules for robotics, and lawn sprinkler timers. Generally, they can appear wherever there is a need to connect individual wires to a PCB without using an integrated connector of some type.

FIGURE 7-17. A five-position PCB-type terminal block

Insulation Displacement Connector

The Insulation Displacement Connector (IDC), shown in Figure 7-18, is a commonly used type of connector for working with small-wire-gauge ribbon cables.

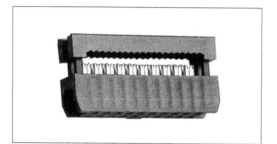

FIGURE 7-18. A typical IDC

An IDC generally consists of two parts: the connector body and a pressure plate. To install the connector, you insert the ribbon cable into the gap between the body and the pressure plate and compress the entire assembly using a special tool. You can also do this using a small vise, but if you are making more than just one or two cables, it's worth it to buy the tool. Figure 7-19 shows a ribbon cable assembly with an IDC at each end.

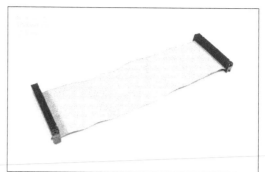

FIGURE 7-19. A ribbon cable assembly with IDCs at each end

IDCs are available in single-, double-, and even triple-row configurations, with a pin/socket spacing of 0.1 inch being the most common for low-voltage, small-signal applications. A mating PCB-mounted header connector is used to create board-to-board connections, as shown in Figure 7-20.

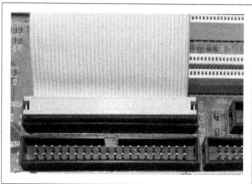

FIGURE 7-20. Ribbon cable with an IDC and a PCB-mounted header

IDC versions of DB-9, DB-15, DB-25, DB-37, DB-50, and HDB-15 connectors are also available, as well as the so-called Centronics-style connectors once used with printers. These are all still made today and are readily available. It is much more convenient to use an IDC DB connector than it is to solder or crimp each wire into the connector, but due to space constraints (ribbon cables and the associated connectors can take up a lot of room) and the fact that a ribbon cable is relatively easy to damage, they might not be an appropriate substitute for a soldered or crimped connector.

A properly assembled IDC is a fairly robust connector that lasts for many years (or even decades) in active service in an enclosed and protected environment. In fact, experience has shown that problems with IDCs don't typically involve the ribbon cable-to-connector coupling, but instead are due to the pin sockets and solder joints that couple the IDC into a circuit module. "IDC Connectors" on page 172 describes the assembly of an IDC connector.

Soldered

Some connectors are designed to be assembled using soldering techniques. These connectors typically have the rear of the pin or socket formed into a cup-like shape to accept a wire. In other cases, the connector might use insertable pins or sockets with a hole in the rear for a wire to be inserted and soldered.

Figure 7-21 shows a well-done solder connection per NASA specification NASA STD 8739.3 (marked as obsolete by NASA but still a good reference).

NASA STD 8739.3 (*http://bit.ly/nasa-sec*), marked as obsolete by NASA but still a good reference). The main idea is not to have an excessive amount of wire exposed after stripping the insulation, but not to have so little stripped off that it is melted or burned during soldering. "Soldered Terminals" on page 169 discusses connector soldering technique, and Chapter 4 provides additional information on soldering in general.

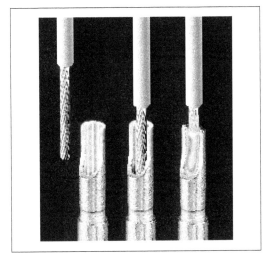

FIGURE 7-21. NASA-style solder-cup connection

Figure 7-22 shows a female DB-25 connector with solder-cup terminals. This is a fixed-terminal connector, meaning that the sockets (in this case) or pins (if it's a male connector) are permanently set in the connector body when it is manufactured. Also note that the body is made of plastic, so it may melt and the contacts may shift if too much heat is applied.

CONNECTORS AND WIRING | 159

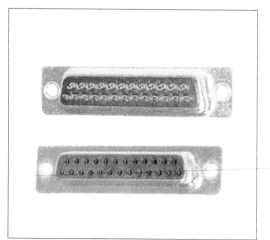

FIGURE 7-22. DB-25 connector with solder-cup terminals

Crimped

Connectors with crimp contacts offer the advantage of reliable and consistent connections (if done correctly), speed of assembly, and long-term durability. The downside is that there is usually an up-front cost in terms of tooling necessary to work with the crimp contacts used in the connector. In some cases, this is relatively minor, but for some connectors, the cost of the tool necessary to form the crimp connection can run into the hundreds of dollars, or even more.

Crimped contact connectors are available in DB forms, as circular connectors with anywhere from 2 to over 200 contacts, and as rectangular forms designed for use with PCBs to serve as module interconnects. They also come in a variety of sizes, from large, heavy-duty types for carrying large amounts of current to the tiny connectors found in things like DVD players.

Figure 7-23 shows two types of female crimp contacts, a barrel type (on the left) and a leaf type (on the right). Crimp contacts like these are designed to be seated in a plastic housing with one hole per contact and are typically used with a mating set of pins on a PCB (a header block or strip). They are held in place by small metal tabs that lock into a slot in the connector body, or by a plastic tab molded into the connector that captures and retains the contact.

FIGURE 7-23. Two types of female crimp contacts for rectangular connectors

The crimp contacts used in DB and circular connectors are in the form of pins and sockets. Low-cost versions are available, but the high-reliability types are fully enclosed like the ones shown in Figure 7-24. These require a special (and rather expensive) tool to assemble correctly.

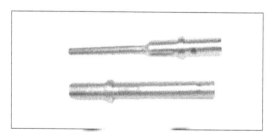

FIGURE 7-24. Male and female high-reliability crimp contacts

The so-called *lug* connectors used in automotive and industrial settings (among other places) are also a form of crimp contact. Wire lugs can be found in a number of styles, including spade lug and ring lug. The crimp tool shown in Figure 3-12 is designed specifically to work with connectors like the spade lug shown in Figure 7-25.

FIGURE 7-25. A typical spade lug

"Crimped Terminals" on page 170 describes some of the tools and techniques for working with crimp connectors.

CONNECTOR TYPES

The previous section discussed some of the ways that connectors are connected to wires, from ribbon cables to single wires in a cable or wire bundle, using solder, crimping, and insulation displacement methods. This section looks at some of the various types of connectors that are available, keeping in mind that, for any given type, one variation might use solder, another crimping, and still another IDC approach. In the end, however, it's still the same type of connector and should mate correctly with other connectors of that type.

Connectors come in a variety of types and styles, ranging from some that are so small that it is impossible to work with them without a low-power microscope, to others that almost need two hands to wrestle. Some consumer electronics manufacturers seem to be particularly fond of coming up with new and unique connectors for their products. Unfortunately, that means those products are incompatible with anything else. In some cases, this works out all right, as when the rest of the industry adopts the new connector and it becomes readily available. In other cases, the oddball connector will simply fade into history and perhaps end up as a curiosity on some item in a museum. Constantly changing connector types also contributes to the growing problem of electronic waste, as yesterday's trendy gadget with the oddball connectors becomes today's obsolete piece of junk, along with whatever once connected to it.

This section looks at the physical characteristics of common connectors that can be readily purchased from a major distributor or from your local electronics supply house. What is covered here is just the tip of a huge iceberg. There are many, many other types available, and this book can't possibly cover them all. But rest assured, for any given application, there is probably a connector available for it, somewhere.

DB Connectors

Figure 7-26 shows the female version of a DB-9 connector, and Figure 7-27 shows the male connector. These are commonly used for RS-232 interfaces, such as those found on some older desktop PCs. Because a DB connector is relatively easy to assemble and connect to a circuit, they are also used as DC power connectors, as RS-485 connectors (another type of serial interface), and as signal interfaces for gadgets of all types.

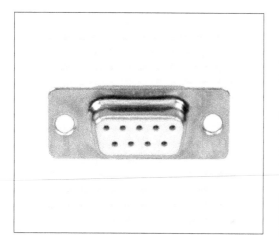

FIGURE 7-26. A female DB-9 connector

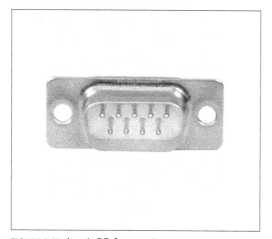

FIGURE 7-27. A male DB-9 connector

These connectors are not what one might call miniature types. They take up a serious amount of real estate, either in a chassis panel or on a PCB. However, if you want to put some things at the end of a long multi-conductor cable, a DB connector is an easy and rugged way to do it.

Both of the connectors in Figures 7-26 and 7-27 are panel-mount or shell types. In other words, they are designed to be bolted into a D-shaped hole in a chassis panel or other flat surface, or enclosed in a metal or plastic shell. Figure 7-28 shows a DB-9 connector and shell assembly. Internally, the DB connector might have solder-cup connections (like the part shown in Figure 7-22), crimped sockets (since this is a female connector), or even be an IDC type. "Connector Backshells" on page 171 covers backshell assembly.

FIGURE 7-28. DB-9 connector with backshell

DB connectors are readily available with 9, 25, 37, and 50 pins. A high-density version, the HDB-15, is commonly used for analog computer monitors and is still found on the back of many desktop PCs. DB connectors are also available in PCB mounting styles, both right-angle and vertical (upright) mount.

Figure 7-29 shows pin numbering for a DB-9 connector. Notice that the pins are reversed between the male and female connectors when viewed from the face of the connector.

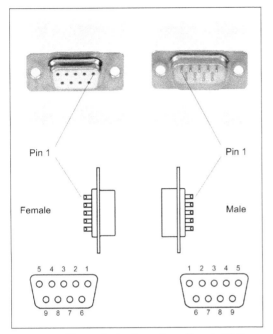

FIGURE 7-29. Pin numbering for male and female DB-9 connectors

If size isn't an issue, or you just want to put a connector on something while you are prototyping it, then a DB type can be a reasonable choice. They are handy for situations where a significant amount of current might be flowing, such as in the battery-charging circuit for a robot. Use four of a DB-9's pins for positive and the other four for the negative, with the ninth pin serving as a frame (earth) ground, and you can push many amps through the connector (between 5 and 10 amps per pin at 5 to 12V DC, depending on the type of pins and sockets used—solid machined pins can handle more current than stamped pins).

PCB Edge Connectors

Connecting one PCB to another can be accomplished in a variety of ways, but it might not be a good idea to use soldered connections to do the job. A connector of some type makes it much easier to disconnect a PCB and replace it if something breaks, or just take it out and work on it.

When printed circuit boards first started to appear about 50 years ago, there were some interesting notions about how they should be connected to the other modules in a system. After various methods were tried, a popular trend emerged that utilized the copper traces on the PCB itself as part of the connector. These special traces are colloquially called *fingers*, and they were laid out with the correct width and spacing to mate with what is known as an *edge connector*. A thin layer of gold is usually applied to help improve reliability.

Figure 7-30 shows a PCI (peripheral component interconnect) *riser* that has both connector fingers and a socket for a PCI circuit board. This type of add-on circuit board is commonly found in 1U (1.75 inches high) computer servers and allows for an additional board to be installed in the limited height available by orienting it sideways.

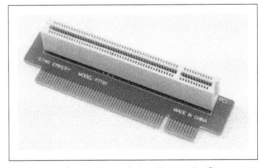

FIGURE 7-30. Connector socket and connector finger traces on a PCI riser PCB

The edge connector is still in use today, and it offers the advantage of simplicity and reduced cost, since only one actual connector is required, with the PCB itself serving as the other part. There are even edge connectors with IDC termination available so that you can use a ribbon cable to connect to a PCB using the built-in connection fingers. While this might sound a bit strange, it was once quite common.

One downside to the ribbon cable and edge connector approach is that the connector holds onto the PCB by friction, so it must be strapped or otherwise restrained to keep it from working off due to vibration. Edge connectors can also suffer from wear and corrosion, and over time they might start to exhibit problems. Occasionally cleaning the PCB fingers with a cotton swab and alcohol was once a common ritual for service technicians. Today, special chemicals are available that will lift off corrosion and help restore the surface without requiring excessive friction that might wear off the gold plating.

Pin Headers and Sockets

Another common type of PCB connector uses set of pin or socket contacts spaced at regular intervals. Figure 7-31 shows an example of both male and female versions of these types of connectors. A spacing (or *pitch*) of 0.1 inch (2.54 millimeters) is common, and other pitch spacings are also available, including 0.4 millimeters, 0.5 millimeters, and 1 millimeter. For a 0.1-inch pitch connector, the posts (pins) have a square cross-section of 0.025 inches (0.635 millimeters) per side.

FIGURE 7-31. Examples of 0.1" pitch headers

The use of pins and sockets to make connections is very old, and over the years, the concept has been extended to a multitude of shapes, sizes, and styles. For PCB applications, there are IC sockets that accept the pins of DIP-style IC packages (see Chapter 9 for IC package types); strips of machined sockets for the same purpose that will also work with crimp-on pins of the correct diameter; header blocks with one, two, or three rows of either pins or receptacle contacts; housings designed to couple with IDC connectors; and housings designed to accept crimp-on contacts with space for anywhere from 1 to 40 or more positions. Figure 7-32 shows a dual-row 16-position header mounted on a PCB.

FIGURE 7-32. Dual-row 0.1" pitch header mounted on a PCB

Connections made between cables are often referred to as *in-line*, and the pin-and-socket concept can also be found in the in-line connectors used in some types of vehicle wiring harnesses, the power wiring inside an assembly line machine, and in the precision high-density circular connectors used in aerospace applications. Some examples of the different types of crimped connectors used for pin-and-socket applications appeared earlier in "Crimped" on page 160.

An odd thing here is that, even though this is an old concept, no consistent naming conventions appear to be in use. One supplier might call a part a *socket strip*, while another might refer to a similar part as a *female header*. The receiving contact is sometimes called a *socket*, a *female contact*, or a *crimp terminal*, depending on how the part is manufactured and how it is intended to be used (soldered directly to a PCB or crimped onto a wire). A pin contact is almost always called a *pin*, however, although it will sometimes be referred to as a *male contact*.

For our purposes, I'll refer to the pin contacts as *pins* and the receptacle contacts as *socket contacts*. Rows of either type in a molded plastic block or housing will be called either a *pin* or *socket header*, as appropriate.

Figure 7-33 shows how these types of connectors are assembled. In this case, the headers and housings have a single row of eight contact positions, but they are also available in other patterns, from a single contact to three rows of contacts with up to 40 or more positions per row.

Note that, while it's not shown in Figure 7-33, a pin header strip will mate with a socket header, which can be used for board-to-board interconnects. You can find these on things like Arduino boards, for example.

Figure 7-34 shows another style of pin-and-socket connector. These are intended for connecting wiring to a PCB, as shown here. Notice the upright tab on the pin strip. This is used to grip the socket shell, providing a more reliable interface than you would get by just pushing a socket onto the pins.

FIGURE 7-34. PCB interconnect using pin-and-socket headers

The idea of using a part of the connector as a retainer can be extended to create pin-and-socket assemblies that mate so tightly that a tool is necessary to release the plastic latch (or latches, in some cases). Connectors with a heavy-duty latching mechanism are found, for example, in automobiles, soda dispensing machines, and in some PCs for the motherboard power cables.

These types of connectors are designed for use with individual wires, not ribbon cables. Although a pin header like the one in

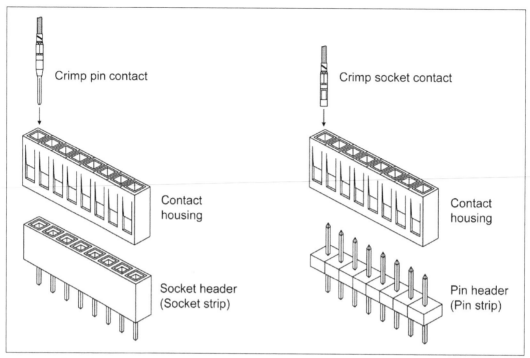

FIGURE 7-33. Pin and socket connectors and headers

Figure 7-32 can be used with a ribbon cable IDC, special-purpose headers are also available, like the one shown in Figure 7-20.

Other types of connectors for use with PCBs are available from multiple sources, such as Molex, Hirose, TE (formerly Tyco/AMP), and others. Some are miniaturized and require special tooling to assemble. Some are used for cable-to-cable connections, while others are intended for use as miniature PCB interconnects. There are through-hole types available, and others for surface-mount applications. Parts and tools are available from major distributors such as Allied, Digikey, Mouser, and Newark, so it's worthwhile to look through what's available. Lastly, make a point to disassemble an old portable CD player or digital camera and look at how things are connected. There's no better way to find ideas than to look at how someone else solved a similar problem.

2.5 and 3.5 mm Jacks and Plugs

Anyone who has ever used earbuds to listen to music with an MP3 player knows what a 3.5 mm stereo plug looks like. Some devices, like cell phones, use the smaller 2.5 mm version for things like a hands-free headset.

These plugs, and the matching jacks, are common and readily available. The jacks come in both panel-mount and PCB-mount styles, and some have a threaded barrel with a nut to hold the jack in place. Soldering either the jack or the plug can be a challenge, however. The plugs can be particularly troublesome, as they don't have much room inside the shell. The solder connections must be as minimal as possible and still be reliable, and the small size limits the size of the wire that can be used.

Another potential drawback to small plugs like these is the limited number of circuits available. A basic plug has only a tip and ring, as shown in Figure 7-35.

FIGURE 7-36. A three-contact (stereo) 3.5 mm plug

If you need to be able to connect headphones to a device, or if you want an easy way to implement patch cables, 3.5 mm plugs and jacks are a good solution. They are not really a good idea for supplying power, since there is a chance that a temporary short might occur when the plug is inserted or removed.

USB Connectors

In Universal Serial Bus (USB) terminology, there are *hosts* and *devices*. USB employs a master-servant type of communications protocol, where only the host can initiate a conversation and the device only responds (this is covered in Chapter 14). Hosts and devices have unique connector types, the shapes and sizes of which are defined by the USB standards.

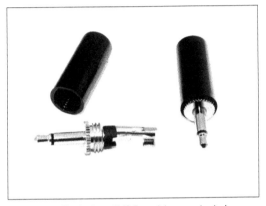

FIGURE 7-35. Typical small (3.5 mm) two-contact plug

A stereo plug is a three-contact version, like the plug shown in Figure 7-36. With this type of plug, there are two possible active circuits and a ground return. Although typically used for stereo audio applications, there is nothing that would prevent it from being used for another purpose, provided that the voltage and current are within the capabilities of both the connector and the wire used with it.

All USB connectors have four contacts: two for the data signals (D+ and D−) and the other two for power (+5V) and ground. Typically, these are premolded, so (hopefully) you won't need to assemble any of them. What you should be aware of are the four different types of connectors defined by the USB standard and shown in Table 7-3.

CONNECTORS AND WIRING | 167

TABLE 7-3. USB connector types

Type	Appearance	Application
Type A		Used primarily at the host or controller end of a USB connection.
Type B		Used primarily on servant devices such as USB hubs, printers, and cameras.
Mini		Commonly found on consumer digital devices such as cameras, toys, and some cell phones.
Micro		Similar to the mini USB connector but slimmer. Designed to be more resistant to wear than the mini type.

Type A connectors are common on desktop and notebook PCs that act as hosts. Some external devices use the B type, although the mini and micro forms are becoming common as devices (such as cell phones, miniature cameras, single-board computers, and toys) shrink in size. Avoid the temptation to create your own USB connector, because once you go down that path, you'll also need to make custom cables. It's easier, and in the long run cheaper, to just buy the PCB-mount connectors and pre-made cables.

Receptacles for all of the connector types come in PCB-mount styles, both through-hole and surface mount. Figure 7-37 shows both type A and type B receptacles with through-hole leads.

FIGURE 7-37. PCB-mounted type A and B USB receptacles

A word of caution: it is probably not a good idea to use a USB connector for anything other than a USB interface. While the venerable DB-9 and DB-25 connectors have been used for a variety of things besides RS-232 serial interfaces, USB connectors are part of an industry-accepted standard. A mini USB receptacle implies that someone could plug it into a PC and expect it to work. If the connector isn't wired for standard USB, the result could be disappointing (or worse).

Ethernet Connectors

Ethernet connectors have evolved over the years, from so-called *vampire taps* and tubular twist-lock coaxial connectors (known as BNC types) and shielded coaxial cable strung from machine to machine, to the present-day use of 8P8C connectors, also referred to as an *RJ45* type. Most Ethernet networks are arranged in what is called a *star configuration*, where a switch or hub distributes the signals to attached machines and other switches and hubs. An Ethernet switch—as its name implies—switches, or routes, the data between the primary cable to the connected PCs or other devices. The other devices can include Arduino or Raspberry Pi boards with Ethernet connectors, test equipment (newer types), video cameras, and even kitchen appliances.

Attaching something to an Ethernet network is usually as simple as plugging in a cable. Figure 7-38 shows an Ethernet cable with an RJ45 attached and a PCB module with the mating receptacle.

FIGURE 7-38. RJ45 Ethernet connector and receptacle

The same caution given for repurposing a USB connector also applies to an RJ45 connector. When someone encounters an RJ45, she expects it to be an Ethernet connection, not a DC power or control signal connection. You can devise your own connectors for Ethernet, but avoid using the standard RJ45 types for anything but Ethernet.

Assembling Connectors

Assembling a connector correctly is essential for durability and reliability. It is well known that a large percentage of failures in electronic devices are the result of connector failures, so getting it right from the outset can eliminate a lot of annoyance later on.

SOLDERED TERMINALS

Connectors available with solder terminals include the DB family, circular connectors (not covered here), 3.5-millimeter and 2.5-millimeter audio jacks, and 1/4-inch jacks like those used for musical instruments (also not covered here). Connectors such as Ethernet and USB typically come as complete parts that just need to be soldered to a PCB, or crimped onto a suitable cable in the case of an RJ45 connector.

As discussed in Chapter 4, the idea behind soldering is to heat both the connector and the wire with the tip of the iron so that the solder will flow smoothly onto both. In some cases, this might be tricky, since inexpensive connectors might have an injection-molded plastic body. If the soldering iron is in contact with the pin for too long, it might melt the surrounding plastic and ruin the connector.

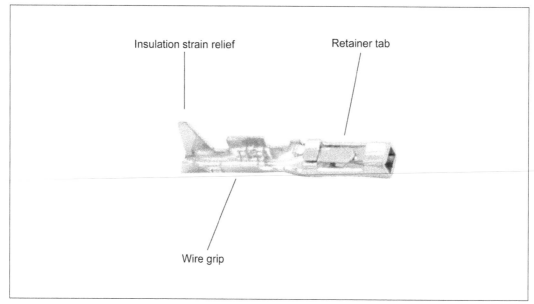

FIGURE 7-39. Main parts of a crimp contact

The key is to remove the iron immediately once the solder has flowed into the connector's cup. Blowing on it gently to cool it down can help reduce the risk of melting the plastic of the connector body around the pin. Placing the connector in a small vise or using a "helping hand" fixture (both described in Chapter 3) can help to avoid some aggravation. It also sometimes helps to apply a small amount of paste flux prior to soldering. This will help lift off contamination from the wire, the contact, or both, and the solder might flow more smoothly. Just be sure to clean off any flux residue after the soldering is complete, and make sure not to get the flux residue on the contact pins or in the socket contacts.

CRIMPED TERMINALS

When you are assembling a connector with crimped contacts, the first step is to crimp the wires into the contacts and crimp them securely, one at a time. Figure 7-39 shows three of the four main parts of a crimp contact (the fourth is the barrel or pin that mates with a matching contact or header pin).

As shown in Figure 7-40, the wire is gripped in two locations: the insulation strain relief and the wire grip that makes the electrical connection. The strain relief is essential, as it helps to prevent the wire from flexing and possibly breaking off at the wire grip point.

The key to making a good contact assembly is paying attention to the amount of insulation removed from the wire. Too much, and the strain relief won't grip correctly; too little, and the wire grip section won't be able to make a solid electrical connection.

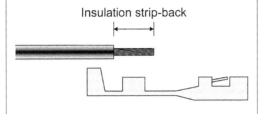

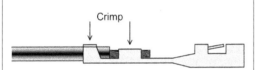

1. Strip back insulation to recommended length.

2. Place wire into contact as shown.

3. Use crimp tool to apply correct amount of pressure to both the strain relief and the wire grip parts of the contact.

FIGURE 7-40. Diagram of a correctly assembled crimp contact

> **TIP** Contacts that are designed to be crimped need a tool made specifically for that type of contact. Pliers or a lug crimper like the ones found at an auto supply store simply won't work correctly. You might be able to get something that sort of works, but it probably won't be as neat and reliable as it would be if the correct tool were used instead.

The tool for the job is something like the one shown in Figure 7-41 (described in "Crimping Tools" on page 35). This particular tool has provisions for two different contact sizes, and it's used primarily for DB connector pin-and-socket contacts.

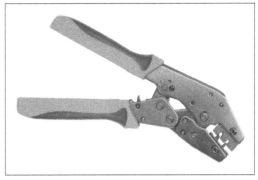

FIGURE 7-41. A multi-size contact crimping tool for small contacts

The crimp locations in the tool (or *die*, as they are sometimes called) actually have two areas for each that apply pressure to both the strain relief and the wire grip parts of the contact, and at the same time the fingers on the contact are rolled over the wire. The tool is ratcheted, so it can apply a significant amount of force without excessive hand force. It costs about $40.

In some cases, you might need a crimped contact on a single wire, but not an entire connector body full of them. This is often the case when you're working with things like an Arduino. You can purchase bundles of premade jumper wires, or you can make them yourself. If you do elect to make your own, remember to slip a piece of heatshrink tubing over the contact after it's been crimped, to prevent shorts and possible damage to the wire.

CONNECTOR BACKSHELLS

Some connectors, the DB series in particular, can be assembled with what is called a *backshell*. Figure 7-42 shows the various bits and pieces that make up a backshell like the one shown in Figure 7-28. A backshell protects both the connector and the wiring by covering the terminals and providing strain relief for the cable or wires attached to the connector.

FIGURE 7-42. The parts that make up a DB-9 connector backshell

The assembly is straightforward: two screws with matching nuts hold the shell together, two more screws with retainers are used to tighten the connector against a matching connector, and two threaded clamp straps and matching screws form the cable strain relief. The DB-9 connector seats into the front of the shell, and once the shell is assembled, the connector will sit securely with minimal movement (if any).

IDC CONNECTORS

Figure 7-44 shows an IDC before and after assembly. In some cases, there is also an additional, strap-like part that holds the ribbon cable to the pressure plate to provide strain relief and prevent the cable from working loose. The strain relief strap is attached to the connector after the ribbon cable has been inserted and clamped.

An IDC works by forcing metal blades with slits through the insulation of the ribbon cable, with one blade per wire in the cable. The conductors in the ribbon cable are forced into the slot in each blade and, if assembled correctly, they will form a gas-tight, cold-weld bond between each blade and the corresponding wire. You can see the blades in Figure 7-18.

When you are assembling an IDC, there are some points to keep in mind. First, the end of the ribbon cable should be cut straight and even. In other words, it's better to use a ribbon cable cutter like the one in Figure 7-43 than to try to work through the cable with flush cutters, although that is possible with some patience. The downside is that a tool like this can run about $100, or more. A sharp guillotine-style paper cutter will also do an acceptable job, but it won't stay sharp for long if used as a ribbon cable cutter.

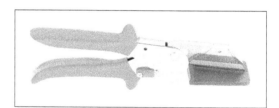

FIGURE 7-43. Ribbon cable cutter

Secondly, there cannot be any stray wire strands at the cut end of the cable, because they can cause all kinds of problems later on. Finally, the cut end of the ribbon cable should not protrude beyond the body of the connector, but should instead end up flush with the body and the pressure plate.

You assemble the IDC by pressing the parts together, and there is, of course, a tool for that as well. Figure 7-44 shows how an IDC compresses a ribbon cable to force the individual conductors into the contact blades.

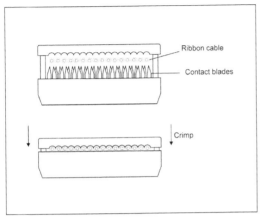

FIGURE 7-44. Before and after diagrams of IDC assembly

If you're making only a small number of ribbon cable assemblies on a sporadic basis, a small bench vise will work, as long as you're careful to keep things lined up while applying pressure. To avoid marring the connector body or damaging the socket holes, use a vise with soft jaw strips or place some thin strips of wood between the connector components and the metal surfaces of the vise jaws.

ETHERNET CONNECTORS

Unlike USB connectors, you can easily make your own Ethernet cables. Figure 7-45 shows an example of the type of tool used to do this. The crimping tools come in a diverse range of styles, from ultra-cheap (and not really worth buying) to robust, industrial-grade things made from machined steel. Avoid plastic crimping tools; they won't last long and may actually cause more problems by making substandard connectors.

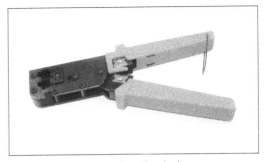

FIGURE 7-45. RJ45 connector crimp tool

Note that there are two basic cable types used for Ethernet: stranded wire and solid wire. Each uses a slightly different type of connector to accommodate the wire used in the cable. Typically, a cable with solid-core wire would be used in an application where the cable won't see much flexure, such as wiring in a wall or a cable tray. Ethernet cables that will be moved, such as patch cables, are typically made with stranded-type wires for flexibility. The connectors used for each type are different internally, so be sure to use the correct connector shell for the wire type. Most crimp tools and cable kits come with an instruction sheet or pamphlet, and it's a good idea to read it before using the tool.

An RJ45 connector works by forcing a thin metal blade or a slotted blade like those used in an IDC through each of the wires in a standard CAT5 or CAT6 Ethernet cable. This means that after the correct amount of outer insulation is removed from the cable, the internal wires must be separated (they are bundled as sets of twisted pairs) and then pushed into the connector in the correct order. This part usually takes a bit of practice to get it right each time.

Ethernet cable kits are readily available from a number of online sources, and even from some unlikely sources, such as large department and home improvement stores. Figure 7-46 shows an example of a kit that contains a cable cutter, wire stripper, and a crimping tool that will handle both RJ45 and RJ11 (telephone) connectors. It also contains a selection of RJ45 connector shells for both stranded- and solid-wire cables, as well as some RJ11 telephone-type connector shells. All that's needed is a spool of Ethernet cable (not shown).

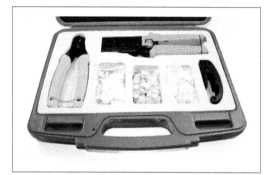

FIGURE 7-46. Ethernet and telephone RJ45/RJ11 connector kit

The proliferation of Ethernet has made it a common feature in many new homes. Many kits have RJ45 connectors for both solid- and stranded-wire cables, a cable insulation stripper, a ratcheting tool to crimp the connectors to the cable wires, and instructions. The cable can be purchased in spool lengths of anywhere from 25 to 1,000 feet.

Summary

This chapter has covered some of the more common connectors and wire types and shown you how to create complete connector and wire assemblies. With regard to connectors, an important point to remember is that any given type of connector might be available in a form that you can connect by soldering the wires, crimping the wires into individual contacts, mounting the connector on a PCB, or using an IDC method to deal with things like ribbon cables.

Selecting the right tool for the job is another important detail covered in this chapter. Connectors can be unforgiving, and the correct tool helps make the process less frustrating and produces better connectors. You might end up with a selection of different types of cutting and crimping tools, but if you use them more than just a few times, it's not a bad investment.

We didn't cover things like "F" connectors, BNC connectors, SMA and SMB miniature connectors, and DC power connectors. These are all relatively straightforward and easy to figure out, and there are a lot of sources of information available online. We also didn't go over circular connectors, which are common in aerospace and industrial applications. This is where purchasing and disassembling some old surplus electronic equipment can be a valuable educational experience and, as usual, there are plenty of sources of information available, including NASA, the US military (in the form of military specification, or MIL SPEC, documents), and the connector manufacturers themselves.

CHAPTER 8

Passive Components

PASSIVE COMPONENTS ARE THE FRAMEWORK on which circuits are built. One way to think of a passive component is as something that only responds to voltage or current; it doesn't exhibit any active control behavior. Stated more formally, a passive component either dissipates energy (resistors) or stores and releases energy (capacitors and inductors), but does not actively contribute energy.

Unlike transistors and integrated circuits, passive components don't require a power supply, just whatever happens to be going through the circuit of which they are a part. In other words, a resistor simply resists, and a capacitor or inductor just responds to changes in voltage and frequency to store and release electrons in a consistently predictable way. There is no way, short of physical manipulation, to alter the intrinsic behavior of a passive device (as you'll see, there are ways to physically manipulate a passive device and alter the behavior in a controlled way).

Something like a transistor, on the other hand, can be used with a control input to modify its response to voltage and current (i.e., it has gain), and active devices can exhibit nonlinear behaviors. An active device can also supply energy to a circuit via an external power source. For these reasons, transistors, and other semiconductor devices, are classified as active components (the subject of Chapter 9).

For an example of how important passive components are, a typical transistor circuit (perhaps a small headphone amplifier or an old-style portable radio) is composed primarily of passive components, with a few transistors scattered around. The passive components set voltage and current levels, couple signals from one part of the circuit to another, and filter out unwanted signals and AC currents. The transistors are the active gain elements that boost signal levels, produce regulated DC supply voltages and current, and serve as oscillators (in a radio circuit).

But passive devices can be more than just the "glue" of a circuit. It is possible to build functional and sophisticated electrical devices using only passive components by exploiting their passive behaviors. In fact, at one time, long ago, this was a popular approach, mainly because early active components like vacuum tubes tended to run hot, required high voltages, and weren't very reliable. Go back far enough and vacuum tubes didn't even exist, yet people were building fairly sophisticated electrical gadgets. For example, thumb through a book on Nikola Tesla and see how many vacuum tubes you can spot in photos of his labs. The answer is none, yet he was able to build a remotely controlled boat and created one of the first working radio-type devices. He did all of it using nothing more than resistors, capacitors, coils, and mechanical generators running at various speeds. All this in the late 1800s.

This chapter describes the physical characteristics of commonly encountered passive components such as resistors, capacitors, and inductors, including both through-hole and surface-mounted types. It also describes how to read component markings and how to understand component ratings for voltage, power, temperature, and tolerance.

Tolerance

Almost all passive components have values specified with a given tolerance, which is stated in terms of percentage. What this means is that a part with a 20% tolerance (for example) will have an actual value that is anywhere from −20% to +20% of the value stated for that part. So a resistor with a tolerance of 5% and a value of 10,000 (10K) ohms can be anywhere from 9,500 to 10,500 ohms and still be within tolerance. The same applies to capacitors and inductors.

It is important to bear in mind that precision is expensive. A part with a tolerance of 1% will cost more than a 5% part, and one with a tolerance of 0.1% can be very pricey. It is also important to remember the old axiom of electronics: if a circuit requires precision parts to work, it might not have been designed correctly.

There are only a few special cases where high precision is necessary. For the most part, electronic circuits are rather forgiving. This is particularly true when there is a knob for a user to turn, or some other means of adjusting the circuit during operation (as with an automatic gain control, for example).

Using a precision resistor in combination with a potentiometer is a classic example of a design that wasn't thought all the way through, but precision resistors used to produce a bias or offset voltage might make sense in some types of high-precision measurement circuits. In general, precision parts should not be used unless there is a real and compelling need to do so.

Voltage, Power, and Temperature

In addition to whatever value a component might have (within its tolerance), passive components also have ratings for working and peak voltages, power dissipation, and usable temperature range. So long as a component is used within these limits, it will exhibit a value (ohms, microfarads, or millihenries) within its tolerance range.

Voltage is a fairly obvious limitation. A small part might not be able to tolerate an extremely high voltage, whereas a larger part, made from special materials, might be able to withstand thousands of volts.

The working voltage of a resistor is determined by how the material that is used to fabricate the part will withstand a high potential difference across it. If the working voltage is exceeded, the carbon or metal film material in the part can start to break down, resulting in failure (it either goes open, or its value will change unpredictably). Another failure mode in an over-voltage condition is *arcing*. In other words, at some point, the potential difference might be great enough to cause the current to arc across (or through) the resistive material to the opposite terminal.

For a typical carbon film resistor, the maximum working voltage will vary between 100 and 350V DC, depending on the manufacturer. In general, resistors rated for high-voltage applications tend to be longer than typical 1/4- or 1/2-watt parts, but length isn't always a definite indication of high-voltage capability.

Capacitors have definite voltage limitations, since at some point, the dielectric material will cease to be an insulator and become a conductor. Most small ceramic capacitors are capable of withstanding 50V DC. Other types can work with voltages up to 500V or more, and still others that use oil as the dielectric can handle thousands of volts. Always check the voltage rating on an electrolytic or tantalum capacitor. If the maximum voltage is exceeded, the part can fail in a rather spectacular fashion (often involving a small ball of fire and flying sparks).

For most circuits operating in the 3V to 9V range, the voltage ratings of the components aren't really a major concern. It does become a concern when you're dealing with the input side of a switching power supply, a high-voltage charging circuit for a flash tube, or a vacuum tube circuit (yes, people still build and use these).

The power rating of a component is the maximum amount of power, in the form of heat, that the part can safely dissipate before it becomes something resembling charcoal. Recall from Chapter 1 that DC power is computed as the product of the voltage potential and the amount of current ($P = EI$). Axial lead through-hole resistors, for example, are rated at values of 1/8, 1/4, 1/2, 1, and 2 watts, with other ratings possible. Surface-mount parts can be found with power ratings ranging from 0.03 to 1 watt, depending on the size of part. For DC circuits, power ratings apply primarily to resistors. For more information about capacitors and inductors, see Appendix A or consult one of the texts listed in Appendix C.

Temperature can have a significant effect on the behavior of a passive component. Almost all resistors will exhibit some degree of temperature sensitivity, and the performance of a capacitor can be severely degraded if it is operated outside of its rated temperature range. Fortunately, many circuits don't experience temperature extremes, although heat can be a problem with things like solar panel tracking controllers or an engine controller for a vehicle. High-performance computers can also have thermal issues. Devices such as outdoor weather sensors might see some extreme high or low temperatures, depending on the local climate, but there are ways to deal with this using internal heating elements, insulated enclosures, or active cooling systems.

As a general rule, if you have a circuit that will utilize high voltages, operate at high power levels, or operate in an environment subject to temperature extremes, it would be a good idea to get the component specifications from the manufacturer. These so-called *datasheets* have all the parametric values for a given part (or family of parts). Chapter 9 contains a walk-through on how to read a datasheet. Appendix D lists component distributors.

On the other hand, if you are building a small robot, hacking a household appliance so it can join other appliances in a local network, or just modifying something that uses batteries or a wall power supply, you probably don't need to worry about temperature or voltage ratings too much. You still need to be aware of power ratings, since, as mentioned in Chapter 1, a AA battery can deliver an impressive amount of current.

Packages

Through-hole (non-surface-mount) passive components come in one of two basic package types: axial and radial. Figure 8-1 shows the difference between the two.

Other packages are used for special-purpose components, such as resistors in TO-220 style packages originally designed for transistors. In general, however, these are the two styles most often encountered with low-voltage, low-current electronic circuits.

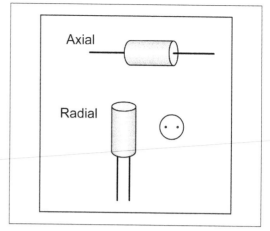

FIGURE 8-1. Axial and radial package types

Surface-mount packages are a subject unto themselves, and they are covered later in this chapter. Capacitors, inductors, and resistors are all available as surface-mount devices, and the package nomenclature is generally the same for the various component types.

Resistors

Resistors are the most ubiquitous type of electronic component and also one of the oldest. They come in both fixed-value and mechanically variable types with power-handling capability ranging from a less than 1/10W to hundreds of watts or more. Cadmium sulfide photocells are photo-sensitive resistors that have found applications ranging from electronic music synthesizers and guitar effects pedals (e.g., Jimi Hendrix's famous "Cry Baby" wah-wah pedal sound) to daylight detectors in street lamps and the light-beam intrusion detectors used in some burglar alarm systems.

Other types of resistors are temperature sensitive and are often found in thermostat control circuits, and one type is sensitive to humidity. Others are designed to serve as strain gauges, changing their resistance in response to mechanical deformation.

PHYSICAL FORMS

Resistors come in a variety of sizes and package types. Axial lead components, with wire leads protruding from either end, are common. High-power parts might have solder lugs or even screw terminals on both ends. Surface-mount parts have no leads at all, but instead are soldered directly to a printed circuit board. Figure 8-2 shows a selection of the various physical shapes available for fixed-value resistors.

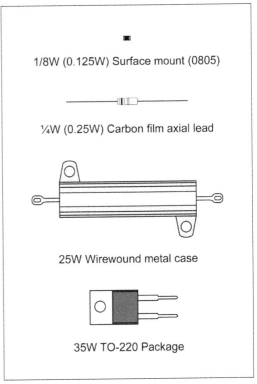

FIGURE 8-2. Examples of the various package styles available for resistors

> **TIP** In the technical data from component manufacturers, and in the catalog listings of many distributors, you will often find component wattages given in decimal form, while in everyday usage, we refer to 1/10 watt, 1/8 watt, 1/4 watt, and so on. The decimal values are necessary for performing calculations, and they can be easily displayed on a web page, whereas the fractional forms are more intuitive to many people and seem to appear more frequently when we are talking about the wattage of a part. In this chapter, for example, when discussing a 0.125W component, it will usually be written as 1/8W, not 0.125W. In tables and figures, it might appear as 0.125W or 1/8W, depending on the context. The reality is that you need to be able to switch from one format to the other as necessary, just as when dealing with dimensions (as covered in Chapter 4).

You should note that much of what can be said about axial lead fixed-value resistors can also apply to surface-mount parts. Figure 8-3 shows a scale comparison of axial lead and surface-mount resistors. Rulers are provided to give a sense of just how small some of these parts really are. Also, notice how much smaller the surface-mount parts are compared to the smallest axial lead component shown, the 1/8w (.125W) resistor. This size scale difference applies to most surface-mount components relative to their through-hole counterparts.

Another package is the *resistor array*, a set of resistors all of the same value arranged either as individual components in a

PASSIVE COMPONENTS | 179

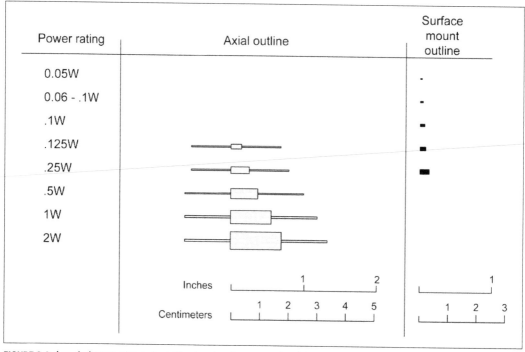

FIGURE 8-3. A scaled comparison of axial lead and surface-mount resistor sizes

common package or connected to a common lead. Figure 8-4 shows a single in-line package (SIP) with multiple resistor elements. Each lead is one end of a resistor, and the other ends are all connected to a common pin.

FIGURE 8-4. Single in-line package resistor array

Resistor arrays are also available in dual in-line packages (DIP) identical to those used for integrated circuits, and surface-mount packages as well. The array shown in Figure 8-4 is a set of eight 1 k ohm resistors with a common lead (hence the nine pins on the package). These parts are used for pull-up resistors on a parallel digital bus, current limiters for seven-segment LED displays, and as a way to conserve PCB real estate.

FIXED RESISTORS

Early resistors from around the early part of the 20th century that are still recognizable as such today were typically large clunky things consisting either of a ceramic tube with a resistive wire wound around it, or a solid rod of resistive material (carbon-based). Metal end caps and leads were used to make the connections, and the whole thing might be covered with some kind of ceramic or shellac coating. The color code for the value was painted on the body of the device, sometimes by hand. These devices were usually about the diameter of a common #2 pencil and around an inch or so in length. Some types had a tendency to emit copious amounts of smoke, or even burst into flames, if they were severely abused.

Modern fixed-value resistors come in a variety of forms, with the most common being the carbon film and metal film types. Other types include carbon composition and wirewound designs. The selection of the type of resistor used in a design will depend on factors, such as power-handling requirements, precision needed, and physical size constraints. In this section, we'll look at some of the common types you are likely to encounter. A look through a national distributor's catalog or website will reveal more exotic types of resistors, but for the most part, you shouldn't need to worry about them.

Carbon Composition

A carbon-composition resistor contains a solid core of resistive material with leads attached to either end. Figure 8-5 shows a couple of carbon-composition resistors.

The carbon-composition resistor was once common, mostly prior to the 1980s. They can still be found in older electronic devices, and if you plan to use old "junk" gadgets and appliances for parts, you will most likely see quite a few of them. They have since been replaced by carbon and metal film types, and while you can still purchase carbon-composition resistors, they are expensive compared to the film types. For a definition of the color codes, see "Resistor Markings" on page 192.

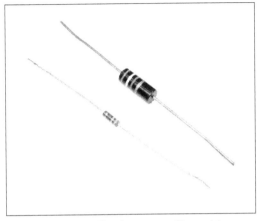

FIGURE 8-5. 1/4 and 1W carbon-composition resistors

Carbon Film

As the name implies, a carbon-film resistor consists of a thin layer of a carbon-based material deposited on an insulating substrate such as ceramic. The resistance is determined by the physical dimensions and thickness of the deposited film. Figure 8-6 shows a typical 1/4W carbon-film resistor.

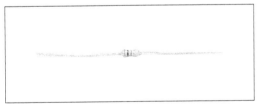

FIGURE 8-6. A typical 1/4W carbon-film resistor

Power ratings of 1/8 and 1/4 watt are commonly used, although carbon-film resistors are available with up to 5 watts of power dissipation capacity. Prices for 1/8-watt parts are typically around 1.6 cents each in quantities of 1,000, or about 30 cents each when purchased in small quantities at an electronics retailer. For a definition of the color codes, see "Resistor Markings" on page 192.

Metal Film

Metal-film resistors are typically used when the tolerance needs to be better than what can be obtained from other types. The film, usually an alloy of nickel, is first deposited and then any excess is physically removed to adjust the part to the desired value. Tolerances from 2% down to 0.5% are available, with the higher tolerance parts costing more, as you might expect.

Physically, a metal-film resistor might look a lot like a carbon-film type, except that it will have an additional color band to indicate a third significant digit for its value. Figure 8-7 shows a typical precision axial lead, metal-film resistor.

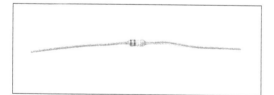

FIGURE 8-7. A precision metal-film resistor

Precision resistors are sometimes used in measurement circuits where you need to establish a definite voltage or current level for some type of sensor. For the most part, however, precision parts are not necessary for most circuits, and there's no compelling reason to pay extra for the unneeded precision. For a definition of the color codes, see "Resistor Markings" on page 192.

Wire-Wound

Modern wire-wound resistors are similar to their ancient ancestors and use the same concept of a resistive wire on a ceramic core form. These types of resistors are most commonly used for high-power applications wherein the part will need to dissipate many watts of power without self-destructing. Figure 8-8 shows a type called a *sandbox* resistor, which has a ceramic outer shell containing the wire-wound resistive element in a ceramic potting matrix. These types come in a variety of power ratings. This one happens to be a .47-ohm, 5-watt part. It is used as a current limiter in the output stage of a high-power amplifier.

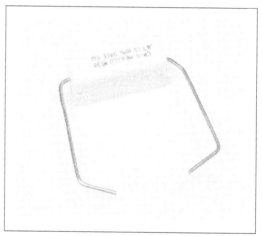

FIGURE 8-8. A "sandbox" wire-wound resistor

Another common style utilizes a metal shell, typically aluminum, with the wire-wound element encapsulated inside. Figure 8-9 shows a metal case wire-wound resistor. The advantage of the metal shell is that it can be attached to a larger surface that can act as a heatsink.

FIGURE 8-9. A metal case, wire-wound resistor

Applications for high-power wire-wound resistors include supplying heat in test chambers, dumping excess energy in a diesel-electric locomotive (the vents on the end of a locomotive engine aren't for a radiator; they are for the fans that cool banks of resistors), and serving as so-called *dummy loads* used to test audio amplifiers and radio transmitters.

Wire-wound resistors tend to be inductive, due to their construction. They can also exhibit capacitance between the windings. Manufacturers try to minimize these effects, but they can never be completely eliminated. For this reason, you will almost never see a wire-wound resistor in a circuit that deals with high-frequency signals, unless of course the effect of the resistor was taken into account when the circuit was designed. They are used to limit current through output devices like transistors and FETs (field-effect transistors), in linear power supplies, and in some battery-charger control circuits.

Precision wire-wound resistors in the low-power range (approximately 1W or less) use the same color code scheme as carbon composition, carbon film, and metal film resistors. For a definition of the color codes, see "Resistor Markings" on page 192. Larger parts (greater than 1W) typically have the value printed on the body of the part, as shown in Figure 8-8.

High-Power Packages

High-power resistors are available in packages other than the axial-lead forms shown in Figures 8-8 and 8-9. The type shown in Figure 8-2 in a TO-220 package was originally designed to house high-power transistors. Other types include devices up to several feet in length and many inches in diameter, with heavy mounting tabs at each end for the electrical connections.

High-power resistors are typically used to safely dissipate unwanted energy or provide a source of heat. One application for the TO-220 style package is when there is a need to heat a flat metal surface, such as in a thermal-vacuum test chamber. An array of TO-220 resistors can provide the necessary heat. Arrays of high-power resistors are also used to dissipate the energy produced by the electric motors in a diesel-electric locomotive during dynamic braking.

In more common settings, high-power resistors in nonstandard packages can be used to build things like load simulators for power supplies, dummy loads for audio amplifiers, a safe heat source for a small epoxy curing oven, or heaters to keep a small remote sensor from freezing when deployed in sub-zero temperatures.

Surface-Mounted Fixed Resistors

While axial lead resistors were once the norm for electronics, they are being replaced by surface-mount parts. This is both good and bad, depending on your perspective. It's good because surface-mount technology (SMT) parts are smaller, cheaper, and more easily incorporated into an automated production system. It's bad because SMT parts can be difficult to work with by hand. In fact, some parts are so small that it is almost impossible to work with them, even when using a microscope and specialized tools.

Surface-mount resistors are available in carbon composition, carbon-film, ceramic, and metal-film forms. They are available in other formulations as well, depending on the power rating and tolerance. Figure 8-3 shows some of the various package types in scale with axial lead components.

Figure 8-10 shows a surface-mount resistor, along with a 680 ohm, 1/10W part in a 0805 package. "Resistor Markings" on page 192 describes the numbering system used with surface-mount resistors.

Table 8-1 lists common package types for surface-mount resistors. Note that these are typical sizes; the actual sizes may vary slightly from one manufacturer to the next. Check the specifications before assuming anything about the dimensions of a particular part. Also note that the wattage may vary between parts of a particular size, so be sure to double-check the specifications before making any assumptions.

TABLE 8-1. Nominal surface-mount resistor sizes

US	Metric	Length	Width	Watts
0201	0603	0.024 (0.6 mm)	0.012 (0.3 mm)	0.05
0402	1005	0.039 (1.0 mm)	0.020 (0.5 mm)	0.03/0.063
0603	1608	0.063 (1.6 mm)	0.031 (0.8 mm)	0.063
0805	2012	0.079 (2.0 mm)	0.049 (1.25 mm)	0.1
1206	3216	0.126 (3.2 mm)	0.063 (1.6 mm)	0.125
1210	3225	0.126 (3.2 mm)	0.098 (2.0 mm)	0.25
2010	5025	0.197 (5.0 mm)	0.098 (2.6 mm)	0.25
2512	6332	0.25 (6.3 mm)	0.13 (3.1 mm)	0.5

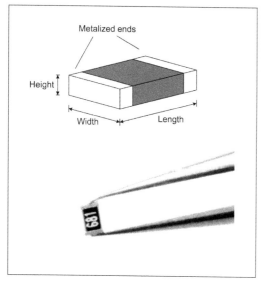

FIGURE 8-10. A 1/10W surface-mount resistor

You might notice a pattern in the package types codes in Table 8-1. The numbering system is based on the length and width of the part, so, for example, a resistor with a length of 0.039 inches (1.0 millimeters) and a width of 0.02 inches (0.5 millimeters) is given the package code 0402 and the metric code 1005. The first two digits in each code refer to the length, and the last two refer to the width.

Chapter 4 includes an example of soldering a surface-mount resistor onto a PCB. This was done with the same 0805 part shown in Figure 8-10, along with some solder paste to make the process easier. A skilled technician with a bench microscope, good tweezers, and an SMT soldering station can work with parts down to about the 0402 (1005 metric) package size. Anything smaller than that becomes increasingly difficult or impossible to do by hand, and yes, there are much smaller parts available.

VARIABLE RESISTORS
Variable resistors can be adjustable types with fixed or adjustable taps. Two-terminal variable resistors (typically wire-wound types) with a sliding contact arranged so that the device can be operated using a knob are called *rheostats*.

A three-terminal variable resistor designed so that a wiper slides across a resistive surface (usually a carbon-based material) is called a *potentiometer* (or just a *pot*). There is a terminal at either end of the resistive element and a third connected to the wiper. A three-terminal potentiometer is, in effect, a continuously variable voltage divider.

Rheostats

The rheostat was invented around 1845 by Charles Wheatstone, an English scientist and inventor who also developed the Wheatstone bridge, musical instruments, and a cipher method, among other things. The term *rheostat* combines the two Greek words *rheos* (meaning stream) and *stat* (meaning to regulate).

A rheostat is typically a two-terminal variable resistor, usually built around a coil of wire formed into a semicircle or wound along an insulating tube. A conductive material, such as a carbon-based ceramic, is also sometimes used as the resistive element. Rheostats that employ resistive wire are commonly used in situations where the device must dissipate significant amounts of power.

Figure 8-12 shows a generic version of a rheostat, like the high-wattage types that might be found in an old-style temperature control for a soldering iron or as a control in a lighting system. This one happens to have three terminals, but it's still classified as a rheostat. To make it a true two-terminal device, the center terminal is connected to one of the two terminals to either side. Similar types can be found in the passive filter module of a high-end loudspeaker, and they were once common in the power supplies for model trains. This one happens to be an Ohmite RHS50R, which is a 50-ohm part rated at 25 watts.

Rheostats like the one shown in Figure 8-12 are tough and durable, and it's not uncommon to find them still in perfect working order when disassembling some old piece of

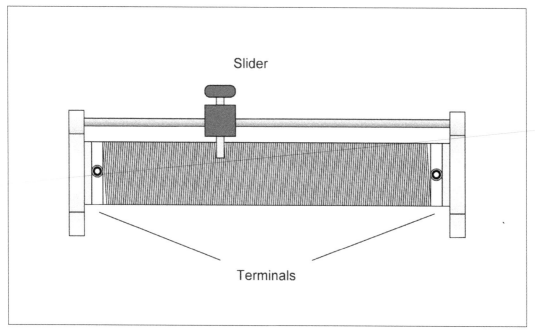

FIGURE 8-11. Diagram of an old-style sliding contact rheostat

electrical equipment. Note, however, that rheostats do tend to get hot when handling large amounts of current, so make sure to take this into consideration.

Linear slider types, like the sliding rheostat shown in Figure 8-11, are sometimes used in laboratories and in some industrial applications. These are similar in most aspects to the original devices created by Wheatstone. All that has really changed over the past 165 years are the materials used to build the device.

Note that, for low-power applications, a potentiometer (discussed in the next section) can be, and often is, wired to behave as a rheostat by having the wiper terminal connected to one of the end terminals.

Potentiometers

Potentiometers come in a range of sizes and power ratings. Some can handle several watts of power, while others are small and rated for only fractions of a watt. Some types are designed to mount in a hole using a nut and lock washer or mounting screws, while others can be soldered directly to a PCB.

One common type of potentiometer utilizes a semicircular strip of carbon-based resistive material with a metal wiper contact that slides across the surface as a shaft is turned. There are three terminals, as shown in Figure 8-13. One way to think of a potentiometer is as a continuously variable voltage divider (see Chapter 1 and Appendix A for information on voltage dividers), which is how these devices are typically used.

Side view

Bottom view

FIGURE 8-12. Example rheostat device

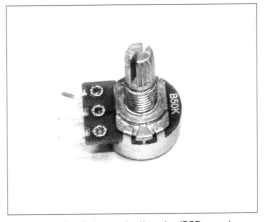

FIGURE 8-13. Single-turn potentiometer (PCB mount, threaded shaft)

Figure 8-14 shows a selection of trimmer potentiometers designed for PCB mounting. These potentiometers are designed to be adjusted with a small plastic tool similar to a screwdriver. These are often referred to as *trimmer* potentiometers because they are intended to be set once (or at least not very often) to adjust (or trim) a circuit and then left in the appropriate position.

FIGURE 8-14. Various sizes of trimmer potentiometers

In addition to a circular rotary form, potentiometers also come in linear, or slider, styles. The primary physical difference between a linear potentiometer and a rotary type is the physical arrangement of the resistive element and how the contact wiper moves across it. These are commonly found in things like the audio mixers used in recording studios, public address systems, and stage lighting control systems.

Figure 8-15 shows an example of four slider linear potentiometers mounted on a PCB. Small rectangular knobs press onto the ends of the shafts, although this is more a cosmetic aspect than a functional requirement (in this case, I think the knobs have been long lost). The front panel has four slots cut to allow the shafts to protrude through when the PCB is mounted.

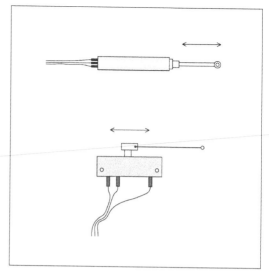

FIGURE 8-16. Linear potentiometers configured as position sensors

FIGURE 8-15. Linear slider potentiometers on a PCB

Linear potentiometers are also sometimes used as position sensors. Figure 8-16 shows a couple of examples of linear potentiometers configured as motion sensors. The top diagram is a tubular device with a shaft. The lower diagram is a variation on the type of slider potentiometer shown in Figure 8-15, except that this type mounts to a panel or bracket with small bolts.

Outside of position-sensing applications, the primary advantage of a linear potentiometer over a rotary type is that it provides an easily comprehended visual indication of its setting. Physically mounting a linear potentiometer is more involved than just drilling a hole, since it needs a slot for the contact arm to move within, and some types are designed to be attached to a panel with small screws at each end of the body of the part. The types intended for use as motion sensors have mounting arrangements specifically intended for motion sensing applications.

Multi-Turn Potentiometers

A multi-turn potentiometer is designed for situations that reaquire a fine degree of control. They are usually precision devices. Like single-turn potentiometers, they are available in a range of styles and sizes, from large devices suitable for panel mounting to miniature parts designed for mounting on a PCB. Figure 8-17 shows a miniature type. These are available in several different package types and are typically referred to as *trim* potentiometers because they are primarily used to adjust some aspect of an active circuit. Almost all of the miniature types like this are adjusted with a screwdriver or something similar.

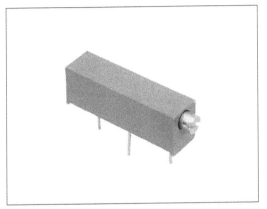

FIGURE 8-17. A miniature multi-turn "trim" potentiometer

FIGURE 8-18. Panel-mounted, multi-turn potentiometer

Larger multi-turn potentiometers, like the one shown in Figure 8-18, come with a shaft that can be fitted with a special knob that counts the number of full turns and incorporates a graduated circular scale, as shown in Figure 8-19. This allows an operator to "dial in" a specific setting with some degree of repeatability.

FIGURE 8-19. Counting dial for use with a multi-turn potentiometer

So where would a device like the one shown in Figures 8-17 and 8-18 be used? They are often found on test instruments and laboratory equipment. Ultrasonic metal crack detection equipment uses them, as do some medical devices. Basically, you'd use a multi-turn precision potentiometer wherever you need to be able to repeatedly set something to a relatively precise value or make a small precise adjustment to a circuit.

Surface-Mount Potentiometers

Surface-mount potentiometers and trimmers are internally the same as their larger cousins, only instead of PCB leads or a threaded barrel for mounting in a panel, these devices have small tabs or metalized areas that are soldered directly to the surface of a PCB. The types with extended leads are relatively easy to work with, but some types have the contact points under the body of the part, which can be difficult to solder manually.

Figure 8-20 shows two different types of trimmer potentiometers in surface-mount form. Be forewarned that these are tiny components, as shown in the relative scale diagram in Figure 8-21.

FIGURE 8-21. Relative size example for a surface-mount trimmer potentiometer

Figure 8-22 shows an example of a larger potentiometer in surface-mount form. This is more like the parts that you might mount in a panel or in a through-hole location on a PCB, except that it is designed to be surface-mounted and can be used with a pick-and-place machine for automated assembly.

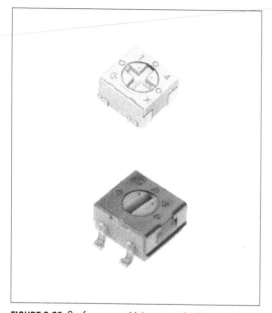

FIGURE 8-20. Surface-mount trimmer potentiometers

Figure 8-21 shows a 4.5 × 4.5 mm surface-mount trimmer potentiometer, a 1/4W resistor, and a US dime for size comparison. The surface-mount soldering technique described in Chapter 4 will work with these types of parts, but take care to ensure that solder paste or flux does not get into the part. The ones shown in Figure 8-20 are relatively well sealed, but other types have an open design and are susceptible to contamination.

FIGURE 8-22. Large surface-mount potentiometer

SPECIAL-PURPOSE RESISTORS

Resistors can be made so that they are sensitive to certain aspects of their environment, including light, humidity, and temperature. These devices have been used in a variety of applications where low cost and simplicity are important considerations. On the other hand, they often don't have the same level of precision and responsiveness of more expensive components. But if all the device needs to do is control a bathroom night light, measure the outside temperature and humidity, or determine if it is a sunny day or not, then a high level of precision probably isn't necessary.

Temperature Sensitive

A temperature-sensitive resistor is commonly known as a *thermistor*. Although temperature sensitivity is also present in many types of common fixed-value resistors, it is intentionally enhanced in a thermistor, and thermistor devices can be manufactured so as to exhibit consistent response behavior.

Thermistors can be made with either a positive or negative temperature response coefficient. In other words, a positive temperature coefficient (PTC) device will exhibit increased resistance in response to increased temperature. A device with a negative temperature coefficient (NTC) will exhibit decreasing resistance with increasing temperature.

NTC thermistors are commonly used in temperature-sensing applications because they have a relatively linear response over a given temperature range. Figure 8-23 shows a small bead type of NTC thermistor used for temperature sensing.

FIGURE 8-23. Small bead type of NTC thermistor

A PTC thermistor tends to act more like a switch, with a sudden increase in resistance once a certain specific temperature is reached. PTC-type thermistors are typically found as surge and inrush current protection devices.

Humidity Sensitive

The resistive humidity sensor, or *humistor*, is a type of resistor designed to respond to the amount of moisture in the gas surrounding it. A resistive humidity sensor measures the humidity level by measuring the change in the resistance of a hygroscopic element as it absorbs or releases moisture. The sensing element might be composed of an organic polymer (such as a polyamide resin, polyvinyl chloride, or polyethylene) or a metal oxide.

Light Sensitive

Light-sensitive resistors go by a variety of names, including *photoresistor*, *photocell*, or *light-dependent resistor* (LDR). A photoresistor works by decreasing its resistance in response to increasing light intensity. Figure 8-24 shows a typical cadmium sulphide (CdS photoresistor). These devices are cheap and readily available.

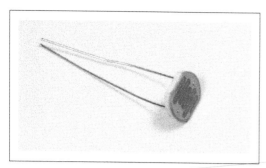

FIGURE 8-24. A typical CdS photoresistor component

Inexpensive CdS photoresistors can be found in many applications. They have been used as photographic light meters, both in handheld standalone devices and integrated directly into a camera. They can also be found in street lights, night lights, clock radios, and alarm systems.

A photoresistor is also useful in situations that don't involve sensing ambient light levels. These include electronic devices where a small lamp is used to alter the resistance of the device without a direct electrical connection. Audio signal processing is a common application, and CdS photoresistors have also been used to couple control signals between high- and low-voltage sections of a system. If the response time of the photoresistor is short enough, it can be used to build a simple communications device. Figure 8-25 shows a couple of possible applications, which are both really just variations on a theme.

RESISTOR MARKINGS

The standard color code for nonprecision through-hole resistors consists of three to four color bands printed around the body of the part. As shown in Figure 8-26, the first two bands are the most significant digits, the third band is the multiplier (in 10x steps), and the fourth band, if present, indicates the tolerance of the part.

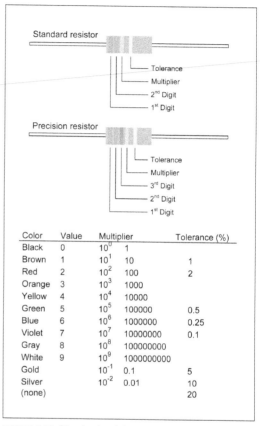

Color	Value	Multiplier		Tolerance (%)
Black	0	10^0	1	
Brown	1	10^1	10	1
Red	2	10^2	100	2
Orange	3	10^3	1000	
Yellow	4	10^4	10000	
Green	5	10^5	100000	0.5
Blue	6	10^6	1000000	0.25
Violet	7	10^7	10000000	0.1
Gray	8	10^8	100000000	
White	9	10^9	1000000000	
Gold		10^{-1}	0.1	5
Silver		10^{-2}	0.01	10
(none)				20

FIGURE 8-26. Standard resistor color codes

So if a carbon film resistor is marked with red, red, red, and gold, it is a 2,200 (2.2 k) ohm 5% part.

Precision resistors typically have a fifth band. In this case, the first three color bands are significant, with the third digit band for the decimal fraction part of the value. From this, we can see that a resistor with color bands orange, violet, red, orange, and green is a 47.2K with 0.5% tolerance.

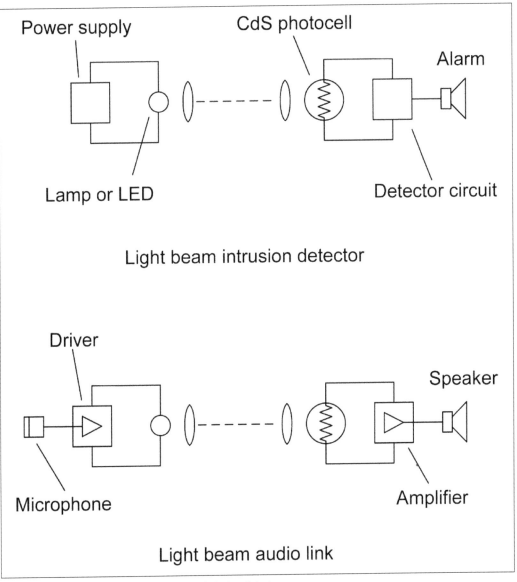

FIGURE 8-25. Two example applications for a CdS photocell device

Surface-mount resistors typically utilize numbers rather than colors. Standard parts (5 to 20% tolerance) will include three numbers. The first two are the significant digits, and the third is the multiplier. For example, a part marked with 474 is a 470,000 (470K) ohm resistor. In Figure 8-10, the resistor is marked with 681, which translates as 680 ohms.

There is one special type of resistor, which isn't really a resistor at all: it's a jumper that looks like a resistor, called a *zero-ohm link*. In the through-hole style, this will be a part with just a single black band. The surface-mount version has just a zero printed on it. These parts are mostly used in automated assembly systems, often because the machine that places the parts on the PCB for soldering can't easily handle wire jumpers, or because the manufacturer doesn't want or need to invest in a jumper-placing machine. If the zero-ohm link part is wide enough, one or more circuit traces may be safely routed under it without touching either of the part's connecting pads.

Capacitors

A capacitor is a device that stores and releases electrical charge. You can think of it as a type of reservoir, or as an impermeable but flexible membrane. A capacitor does not allow direct current to flow, but it will allow alternating current to pass. The value of a capacitor depends on how much charge it can store, and it also has a direct bearing on how the capacitor will respond to an AC signal (see Appendix A for more details).

Capacitors are often used to couple one circuit section to another to allow only the desired signal to pass but block any residual DC that may be present. In digital logic circuits, they are often found connected between the Vcc (V+) pin of a TTL part to ground, and they are used to "decouple" the transient current spikes that can occur when a logic gate changes state. They are also used extensively in passive filter circuits to remove noise and 60 Hz ripples from the output of a DC power supply, and they show up in sensing circuits to help remove unwanted RF interference (RFI). Capacitors are used to build high-pass or low-pass filters, and they are an integral component in tuned circuits such as those used in radios.

Early capacitors were as simple as two metal plates separated by a small gap or a piece of oil-soaked paper. Later, they evolved into oil-filled canisters, or they were made from strips of foil separated by a layer of oil-impregnated paper. In any case, the concept is simple: two plates for accumulating charge are separated by some kind of dielectric material (i.e., an insulator that can be polarized). The surface area of the plates and the space separating them determines the overall capacitance of the device. Figure 8-27 is a generalized diagram of a capacitor.

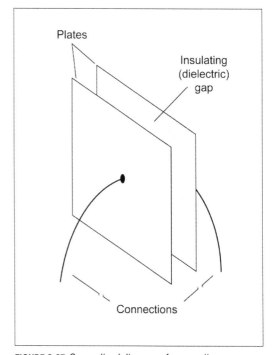

FIGURE 8-27. Generalized diagram of a capacitor

A capacitor stores energy in an electrostatic field that is generated by a potential difference across the plates. When a voltage is applied across a capacitor, one of the plates will accumulate a positive charge and the other will have an equal negative charge. The capacitance is largest when the distance between the plates is the smallest possible and the plate area is the largest possible. In other words, the capacitance increases as the area of the plates increases and the distance between them decreases.

There is a limit to how much of an electrical potential difference a capacitor can withstand before something breaks down. This is referred as the *breakdown voltage*. For some types of capacitors, this can be quite low, on the order of a few volts. Other types can withstand hundreds or even thousands of volts before breakdown occurs. The breakdown voltage limit is why electrolytic capacitors are specified with a maximum working voltage. Exceeding this voltage in operation courts disaster.

No capacitor is perfect, and after a charge is accumulated, it will immediately start to dissipate as soon as the potential source is removed. Capacitors with a dialectic made from polypropylene or polystyrene can exhibit extremely low leakage rates, whereas other types, such as electrolytics, tend to exhibit high leakage current.

CAPACITANCE VALUES

Capacitance is measured in farads (after Michael Faraday), with 1 farad being defined as a capacitance that produces a potential difference of 1 volt after being charged by 1 ampere of current flowing for 1 second (which happens to be the same as 1 coulomb).

Most capacitors encountered in electronics have values measured in millifarads (mF), microfarads (µF), nanofarads (nF), or picofarads (pF). For example, the ceramic decoupling capacitors mentioned in the previous section are typically 0.1 µF types. Parts with very small values, in the pF range, are often found in RF circuits, while power supplies will employ capacitors with values of 470 µF or higher in the output filter section of the circuit.

CAPACITOR TYPES

Modern capacitors can be based on a ceramic dialectic, on thin layers of a plastic such as polyester, or on thin sheets of the mineral mica. Some types, such as electrolytic devices, use aluminum or tantalum foil and employ an oxide layer to serve as the dialectic. In these capacitors, an electrolyte paste serves as the second electrode.

Capacitors come in polarized and nonpolarized types. Electrolytic types are usually polarized, and you must be careful not to apply a reverse polarity voltage to the device. The results of this can be somewhat spectacular (e.g., an exploding part, smoke, a ball of fire, etc.).

A special class of electrolytic capacitors, called *supercaps*, is starting to appear as a backup power source for flash memory, and some supercaps are capable of storing enough energy to replace small batteries for short periods of time.

This section describes only ceramic, electrolytic, and plastic film capacitors. There are other types available, and some, like silver mica types, were once common but are now generally considered to be obsolete. A look through a distributor's catalog will show you the wide variety of types and ratings available.

Ceramic

Ceramic capacitors are based on a ceramic wafer with a metal film applied to opposite sides. The area of the wafer and its thickness both work to determine the resulting capacitance of the part. Ceramic capacitors are inexpensive and relatively stable. Some types are large, such as disc forms, whereas others, like the monolithic types, can be very small. These types of capacitors are also available in surface-mount packages and share the same package numbering scheme as surface-mount resistors. Figure 8-28 shows both a disc and monolithic type of ceramic capacitor.

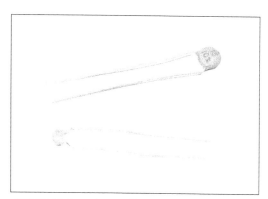

FIGURE 8-28. Ceramic capacitors

Most ceramic capacitors are marked using a three-digit scheme. The first two digits are significant and the third is the multiplier. Capacitors are marked in units of picofarads. Thus a part marked as 103 would be 10,000 pF, or 0.01 µF. Occasionally, you might come across an older ceramic capacitor that uses the color code shown in Figure 8-26, but these are rare nowadays.

Electrolytic

Electrolytic capacitors are capable of holding significantly more charge than other types, and for this reason they can often be found in power supplies and flash tube circuits. Electrolytic capacitors come in a variety of sizes and shapes, a sample of which is shown in Figure 8-29.

FIGURE 8-29. Some available electrolytic capacitor styles

Most electrolytic capacitors are polarized, so take care to make sure that a part isn't installed incorrectly. These parts are available in both axial and radial forms, and there is also a surface-mount version of the metal-can electrolytic capacitor. The radial forms are common and readily available, but sometimes an axial lead component makes more sense for a design where the height above the PCB is constrained or when you might improve the design of the PCB by running traces under the body of an axial lead capacitor.

Tantalum eletrolytic capacitors are common in miniature electronic devices, mainly because they are more physically compact than a conventional electrolytic of the same value. They also exhibit better performance than conventional electrolytic types, but they are more expensive. Figure 8-30 shows a pair of typical tantalum capacitors with radial through-hole leads, but they are also available in axial packages and surface-mount packages.

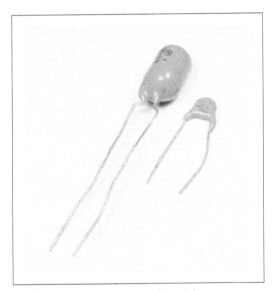

FIGURE 8-30. A generic example of a tantalum electrolytic capacitor

Plastic Film

As the name implies, plastic film capacitors utilize a thin plastic film as the dialectric material between the electrodes. Some common types of plastic used include polypropylene, polyester, polyenthylen naphthalate, polystyrene, and polycarbonate, to name just a few.

The primary advantages of a plastic film capacitor over a ceramic or electrolytic type are stability, low temperature dependence, and possibly a high capacitance value without being polarized. For this reason, they are often found in loudspeaker crossover filter circuits. These types of capacitors don't appear very often in low-voltage electronics utilizing solid-state components.

VARIABLE CAPACITORS

A variable capacitor is a rarity nowadays, but at one time, it was the primary means to alter the behavior of a tuned circuit. In other words, it was part of every radio receiver up until the late 20th century. As shown in Figure 8-31, a variable capacitor consists of a set of plates arranged such that one set can move in between another matching set. The extent to which the movable plates are within the fixed plates determines the capacitance.

FIGURE 8-31. Variable capacitor for radio tuning

This type of capacitor is still used in some types of radio transmitting equipment,

PASSIVE COMPONENTS | 197

although it has been largely replaced by solid-state devices in modern radio receivers. Variable capacitors are also sometimes found in RF test equipment. Figure 8-32 shows a simple radio kit that uses a variable capacitor in the tuning circuit.

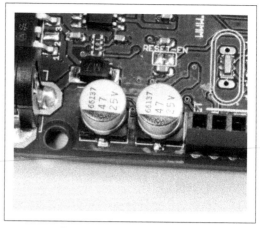

FIGURE 8-33. Surface-mount electrolytic capacitors

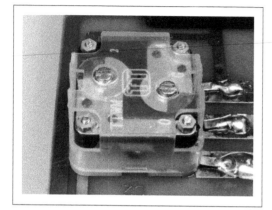

FIGURE 8-32. A simple AM/FM radio kit that uses a variable capacitor for tuning

SURFACE-MOUNT CAPACITORS

Surface-mount capacitors come in a range of sizes in rectangular packages similar to surface mount resistors. Some types are made with a can-type package, similar to the electrolytic capacitors shown in Figure 8-29, but with SMD mounting tabs or metalized areas rather than leads. Figure 8-33 shows some surface-mount electrolytic capacitors on a PCB.

Figure 8-34 shows another example of surface-mount capacitors, in this case identifying some of the capacitors in an Arduino Leonardo.

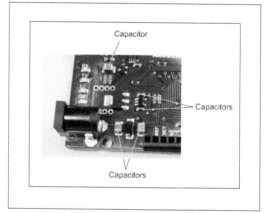

FIGURE 8-34. Surface-mount ceramic and tantalum capacitors on an Arduino Leonardo

Physically, a surface-mount capacitor can have a greater height than a resistor with the same length and width. This is particularly true for tantalum and electrolytic types.

Chokes, Coils, and Transformers

Inductors make up the third catagory of passive components and, like capacitors, they store and release energy, except that in the case of inductors, this is magnetic energy created by current flow through the inductor. Creating a magnetic field in an inductor requires electrical energy, and the inductor will initially oppose the incoming current and behaves like a load. When the initial current is removed, the collapsing magnetic field generates a current flow and the inductor behaves as a current source.

Recall that a capacitor will block DC but allow AC to pass. An inductor has the opposite behavior. An inductor impedes AC but does not impede DC, except for the inherent resistance of the wire in the inductor. In many cases, an inductor will appear like a short or an extremely low resistance to DC.

Inductors come in the form of chokes and coils. A *choke* uses a solid core of some type of ferromagnetic material—it is, in effect, a solenoid. A *coil*, on the other hand, is just an open coil of wire. The unit of measurement for an inductor, of any type, is the henry (abbreviated H), with common values being in the millihenry (mH) range.

This section is intended to provide only a quick look at inductors. For more on the theory behind inductors, refer to Appendix A.

CHOKES

All inductors exhibit the property of resisting changes to the current passing through them, but chokes are designed to take advantage of that behavior. Chokes get their name from their applications in blocking AC while allowing DC to pass relatively unimpeded. Some types are specifically designed to block low-frequency signals, such as audio signals. Others are designed to block radio frequencies. The nodule sometimes found on computer cables contains a choke of some sort to help prevent interference from passing between a device and whatever it is connected to. Figure 8-35 shows a choke of the type that might be found in a power supply, or in series with the DC input to a circuit.

FIGURE 8-35. An RFI choke

Chokes come in a range of styles and ratings. They are useful when you are dealing with circuits that might be susceptible to external radio frequency interference on the power supply, or when you are attempting to take DC readings from a sensor connected to very long wires. Chokes also appear in the filter stage of linear power supplies to block the residual AC ripple from the rectifier section and help smooth the output.

COILS

A coil, otherwise known as just an *inductor*, is typically used to create a tuned circuit such as an oscillator or a radio receiver, although it sometimes assumes the role of coupling components between the sections of a circuit in radio equipment.

Although inductors are available in a selection of values, it is often the case that a coil must be constructed specifically for a particular application. Ham radio enthusiasts have been doing this for a long time, and texts like the *ARRL Handbook* (refer to Appendix C) contain formulas to use in order to determine the wire guage, diameter, and number of turns needed to construct a coil with a specific value. Figure 8-36 shows the tuning coil of a simple radio.

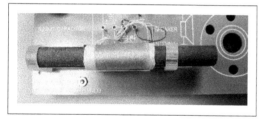

FIGURE 8-36. RF tuning coil for a simple radio circuit

For the most part, coils don't show up much in electronics that don't use radio frequency signals, except perhaps when a circuit needs to be filtered or tuned for a specific application.

VARIABLE INDUCTORS

Variable inductors are typically used in radio frequency applications. A common type of variable inductor contains a movable core, and you change the inductance by varying the depth of the core in the winding. When adjusting a variable inductor while the circuit is active, you use a nonmagnetic tool, somewhat like a plastic screwdriver, to keep from disturbing the behavior of the inductor. Figure 8-37 shows a common package type for a PCB-mounted variable inductor.

FIGURE 8-37. Typical PCB-mounted variable inductor

TRANSFORMERS

A common example of an inductive device is the *transformer*. In a transformer, energy is inductively coupled from one winding to another. A transformer can reduce an input voltage to a lower level (a *step-down transformer*) or increase it (a *step-up transformer*). In a linear power supply, the incoming AC line voltage is reduced by using a smaller number of turns in the output winding than in the input winding. A step-up transformer is just the opposite. The ratio of input to output windings, the number of windings in each coil, and the material used for the core of the transformer determine its operating characteristics.

Other types of transformers are used to couple signals rather than transfer power. Audio transformers, as the name implies, are used to couple an audio signal from an input stage to an output stage. The transformer might be 1:1, which does not change the voltage level but does isolate one part of the circuit from another.

Figure 8-38 shows a representative example of a small power transformer. The input winding, also called the *primary winding*, is rated for 230 VAC input. The output, or *secondary*, winding will produce 12.6 VAC (the secondary is actually center-tapped, so it can be used as two 6.3 VAC outputs). The core of the transformer is made up of layers of a material with high magnetic permeability to contain the magnetic field and improve inductive coupling between the windings.

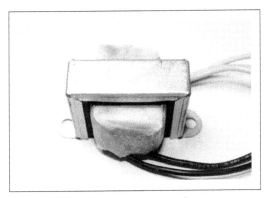

FIGURE 8-38. A typical small power transformer

PACKAGES

As with resistors and capacitors, inductors and transformers are available in a variety of package styles. In addition to the axial choke shown in Figure 8-35, inductors are availabe in small axial lead packages not much larger than a 1/2 watt resistor, in radial lead packages that resemble electrolytic capacitors, and in surface-mount packages. Some SMD inductors look like surface-mount capacitors or large resistors, while others have a square shape and a circular internal coil form to hold the wound wire. Transformers are available in chassis-mounted styles like the one shown in Figure 8-38, compact packages with leads for through-hole PCB mounting, and as surface-mount parts.

Summary

This chapter covered the three primary types of passive linear circuit components: resistors, capacitors, and inductors. Of these, resistors and capacitors are the components that will be the most commonly used in low-voltage analog circuits and microcontroller applications. Inductors might come into play for RFI suppression and will definitely be useful when you are dealing with RF circuits.

Passive components such as fixed resistors and capacitors come in one of three basic physical forms: axial, radial, and surface-mount. Variable resistors can take either a rotary or linear form, and variable capacitors are commonly built around a rotary mechanism.

What we haven't covered are large electrolytic capacitors, such as the types used in heavy-duty linear power supplies. The rise of switching power supplies has largely eliminated the need for capacitors the size of soup cans, although they can still be purchased. If you ever get the chance to strip down an old mini or mainframe computer system from the 1960s or 1970s, you will see these in the DC power supplies. However, electrolytic capacitors that are 30 years old probably aren't good any longer, since the internal electrolyte will likely be dried out.

We also didn't cover all the capacitor types available, such as mica and some of the plastic dielectric-based components. These are not commonly used with digital electronics or sensing circuits. They are more commonly encountered with low-noise audio, RF, and some types of laboratory-grade precision circuits that need the low noise and thermal stability they can provide.

Each of the three passive component types comes in a vast array of physical sizes, physical compositions, and electrical ratings. Each one would deserve its own book to really do it justice. Fortunately, those books have already been written, and some of them are listed in Appendix C. A look through the catalog of a major distributor or a search on the Web will reveal a massive amount of information, most of it free. Hopefully, with the overview provided by this chapter, you should be able to make some sense of this information.

CHAPTER 9

Active Components

This chapter covers some of the various types of active components you might encounter, from diodes to ICs—in other words, devices that utilize semiconductors, referred to as *solid-state components*. We will also discuss some of the package types available for both through-hole and surface-mount components, as well as how to prevent damage to solid-state components from static discharge.

The primary emphasis here is on available types of solid-state devices and the packaging used in their manufacture, not on how these devices work. That would be another complete book (or two) in itself. If you want to know more about the internal operation of active components, take a look at Appendix A or one of the excellent texts listed in Appendix C.

Another key point of this chapter is datasheets. These are the defining documents for electronic components (and many other things as well). Learning to obtain and read a datasheet is a crucial skill for anyone dealing with electronic circuits. To that end, "How to Read a Datasheet" on page 204 steps through the contents of a typical datasheet and also offers suggestions on where and how to obtain them.

The distinction between an active component and a passive one is based primarily on how a part deals with electrical energy. In Chapter 8, it was described as the difference between something that dissipates energy, or stores and releases energy, without the need for an external source of power, as opposed to something that could actively alter circuit behavior and use an external source of power.

Active components typically rely on an external power source and are able to alter current flow in both linear and nonlinear ways. A transistor, for example, can perform amplification by altering the current flowing through it in proportion to changes in a small input signal. The result is an output that is a larger version of the input. Its behavior is also nonlinear, in that a transistor's response to an input is based on thresholds and current flow direction. A resistor, on the other hand, will simply resist current flow, no matter how much, from zero up to its physical power dissipation limit. The response graph for a resistor is a straight line, meaning that it is linear. But for a transistor, the voltage-versus-current response curve has distinct curves, and is therefore nonlinear.

Diodes are also included in this chapter, simply because a diode is a semiconductor device with a single solid-state junction and nonlinear behavior. A transistor is a close cousin of the diode with a more complex solid-state junction arrangement. Some texts categorize diodes as passive devices, but because they do exhibit nonlinear responses to current and voltage, I think they can be placed here with the other active components.

How to Read a Datasheet

Before we delve into active solid-state components, it would be a good idea to take a look at the definitive source of information for these parts: the datasheet. Datasheets have become commonplace for things other than transistors, diodes, and ICs; even nuts and bolts have datasheets available. Reading them and making sense of what they contain is what this section is all about.

Datasheets have been around for quite a while. Sometimes called a *data sheet* or a *spec* (specifications) sheet, they began as short documents (one page or so) that listed the most important characteristics of a device, product, or mechanism. Over time, they have evolved into elaborate documents that describe the essential electrical, physical, and mechanical characteristics an engineer might need to know in order to use the product. While some datasheets are still only one page (there's only so much that can be said about a blind rivet, for instance), in other cases, they consist of tens of pages of detailed graphs, tables, and sometimes equations and example program code.

Once, before the Internet made it easy to move electronic documents around, electronic components manufacturers used to spend millions of dollars a year on creating, publishing, and distributing so-called *data books*, which were generally collections of datasheets for individual components. It was a big day in the lab when a sales representative from a distributor (or even a manufacturer) would show up with a pile of data books in the trunk of her car. It was common to find a wall of bookshelves filled with data books in a well-stocked electronics R&D lab. These days, almost all of those books can fit on a single 16 GB flash drive.

> **TIP** Datasheet authors assume that the reader is conversant with the concepts and terminology used in the datasheet. There is seldom any attempt made to define unique terms or explain concepts. One way to approach this is to get a collection of datasheets from different sources and compare them while having a glossary (such as the one in this book) handy. I haven't duplicated any datasheets in this section, due primarily to space and copyright considerations, but these documents are readily available from distributor and manufacturer web pages.

DATASHEET ORGANIZATION

Most datasheets for electronic components are organized into multiple sections. I've listed six possible sections here, but some may have fewer, and some may have more. It depends to a large extent on the complexity of the part and the internal datasheet style guidelines of each manufacturer:

Summary/Overview

May provide a paragraph or two of overview discussion, a list of features, suggested applications, short summaries of the process technology (how it is fabricated), and other salient points the engineer might want to know at a glance. May also include ordering information if the part is available in different package and temperature range variations, although this information sometimes has its own section.

General Specifications

Typically contains a table of absolute maximum ratings, such as maximum power supply voltage and maximum operating temperature, a table of thermal characteristics, and optimal operating conditions.

Electrical Specifications

May include connection diagrams and a description of the connection pins or input/output terminals, if applicable, and a table of electrical characteristics (min/max operating voltage, clock speed, rise/fall time, and similar data).

Functional Description or Diagram

Some datasheets include a description of the operation of the device from a functional perspective (what it does, not specifically how it does it), or there might be just a functional diagram to illustrate the primary internal functions of the device. If a diagram is included in this part of the datasheet, it can usually be considered to be more of an equivalent circuit diagram (see Appendix A for more on equivalent circuits) than an actual schematic.

Application Examples

In some cases, the designer needs to know some critical details about things like physical mounting, the use of a heatsink, and PCB trace layout considerations. This section might also include what is called a *reference design*, which is a circuit using the part that is known to work and which can be used as a jumping-off point for the engineer's own design.

Physical Specifications

This section provides details on available packaging (through-hole, surface-mount, etc.), including dimensions. This is where you look for details concerning correct PCB layout for the type of part you want to use.

Not every datasheet contains all of the sections listed here, or provides them in the order given, but even something as seemingly straightforward as a metal film resistor might have an 11-page datasheet. For example, see the Vishay metal film resistor datasheet (*http://bit.ly/vishay-data*). We will look at the datasheet for an IC in detail in the next section, but it is informative to see how datasheets for other types of components are written and organized.

DATASHEET WALK-THROUGH

The 555 is a popular (and old!) part first released in 1971 by Signetics as the NE555. It is a timer that uses an RC (resistor-capacitor) time constant (see Appendix A) to control a relatively simple but robust circuit. The 555 can be configured as an astable (continuous) oscillator, or it can be used as a monostable (one-shot) timer. It is an extremely versatile device, and entire books have been written about it (the *555 Timer Applications Source Book* by Howard Berlin, for example). For the truly ambitious, there is even a kit that allows you to build a working 555 circuit from discrete components. It's available from Evil Mad Scientist Laboratories (*http://bit.ly/ems-555*).

As an example, I'll use the datasheet for a 555 timer IC from Intersil, known as the ICM7555 (or ICM7556 for the dual-timer version). If you want to follow along, you can download the datasheet from Intersil's website (*http://bit.ly/intersil-timer*).

The front page contains the general description, a list of features, a list of possible applications, and the package pinout diagrams for both the single-timer and dual-timer versions of the part (the dual timer simply has two complete 555 timers in one package). Figure 9-1 shows a layout diagram of the front page.

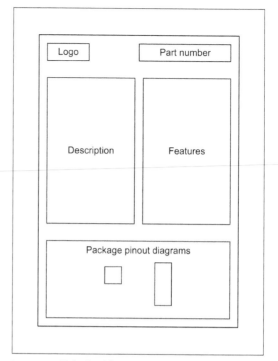

FIGURE 9-1. Datasheet front-page layout example

This is probably the most important page in the entire datasheet, since it is the one you would want to refer to for pinout information and general specifications, which are incorporated into the list of features in this particular datasheet. After reading this page, we know that the part can operate on a supply voltage from 2V to 18V, it can generate a pulse rate of up to 1 MHz, and it draws around 60 µA of current.

The second page contains ordering information. From this, we can see that the part is available in various package types and three different temperature ranges: 0° to +70°C, –25° to +85°C, and –55° to +125°C. The acronym SOIC stands for *small outline IC* (a surface-mount package), and PDIP refers to a plastic DIP (dual inline package) through-hole package. The Cerdip package is a ceramic package intended mainly for military and harsh environment applications. "Surface-Mount IC Package Types" on page 230 describes the SOIC family of packages.

The third page provides absolute maximum ratings and electrical specifications. You might notice that the maximum output current is 100 mA, which implies that the chip will drive an LED or relay without breaking a sweat. The fourth page continues the electrical specifications table and has a functional diagram, along with a truth table that defines the logical behavior for the timer.

But wait a minute, did you catch it? The features list states that the operating current was 60 µA, but the absolute maximum ratings state that the timer could supply 100 mA of output current. There is a big difference between 60 µA and 100 mA. What's going on here? The answer is shown on page 5 of the datasheet.

Page 5 includes the schematic diagram of the 555 timer. Notice that the output is driven by two MOSFET (metal-oxide semiconductor field effect transistor) devices, one connected to VDD (V+) and the other connected to ground. In other words, the 555 can swing its output from ground to almost full V+ (sometimes also referred to as *rail-to-rail*). The 100 mA limitation arises from the current-carrying ability of the internal output circuitry on the silicon chip itself. The rest of the timer (everything to the left of the two output MOSFETs) requires only 60 uA to operate.

The applications section, titled "Application Information," also begins on page 5. This is where you can find practical information about the use of the part. In this case, the datasheet shows three example circuits. Two of the circuits are for an astable oscillator, and the third shows how to use the 555 as a monostable timer. Let's say you wanted to operate an alarm for some specific period of time and have it turn itself off. You could use a 555 and a relay to accomplish this (see Chapter 10 for information on relays). The circuit shown in figure 3 on page 6 of the datasheet shows how to make a timer, and the text on the page describes how to calculate the values of the RC time constant components. Of course, you could just buy a prebuilt module (such as the one shown in Figure 9-2) that does this, using basically the same concept.

FIGURE 9-2. 555 timer module kit

The kit shown in Figure 9-2 is available from Apogee Kits (*http://bit.ly/kit-mkIII*) for about $5. Other kit suppliers sell similar items for about the same price, so there are many to choose from.

But back to the datasheet. Pages 7 and 8 contain graphs that show the performance of the part for things like minimum trigger

ACTIVE COMPONENTS | 207

pulse width, supply current versus supply voltage, frequency stability versus temperature, and so on. These are good to have, and they provide a way to illustrate the dynamic behaviors of the part as a function of change in various parameters and conditions. In particular, notice in Figures 14 and 15 that if you want to have a wide range of astable frequencies or monostable times, then you'll need to be able to change the values of the timing components (RA, RB, R, and C) to cover a particular range. A rotary switch can be used to achieve this (see Chapter 6). It's generally a good idea to at least look through the various graphs provided and see if any of them can tell you something about the part that the equations in an earlier section might not make immediately obvious.

Pages 9, 10, 11, and 12 show detailed package drawings and provide dimensions for each type. Notice that the package descriptions for the 8-pin and 14-pin PDIP parts use the same drawings, so don't let that fool you. Also, if you're going to do a PCB layout, this is where you would look to make sure the PCB layout software's component library has a predefined template that will work with this particular part (see Chapter 15 for a discussion of PCBs). While this isn't really a big issue with JEDEC (Joint Electron Device Engineering Council) standard through-hole parts, some library templates occasionally get it wrong when it comes to surface-mount components. It's always a good idea to do a quick check before committing the PCB design to fabrication.

COLLECTING DATASHEETS

Once you become familiar with datasheet formats and conventions, they aren't all that hard to decipher. Just remember that the most commonly referenced information is almost always located at the front, where the designer can get a quick sense of what the part does and what its essential limitations might be. The rest of the datasheet goes into deeper levels of detail, and each datasheet usually wraps up with a description of how the part is packaged.

Of course, none of this is cast in stone, and some datasheets are easier to read than others. Badly written ones are relatively rare, and lack of sales (and customer complaints) will usually get them fixed within a short period of time. Datasheets from major manufacturers are an essential sales tool. They are the first place to look to get information about a part, and almost every manufacturer makes them readily available. Most major distributors also provide links on their web pages to the original datasheets for the parts they sell, and there are websites that offer datasheets for many different parts from multiple sources for free.

After a while, you may find that you have collected a large number of datasheet PDF files. Personally, I like to print out the ones I refer to on a regular basis (double-sided, of course) and put them into a binder (or two, or three). That way, I can make margin notes on the printed pages, and use a highlighter on important sections in the document. But in any case, if you work with electronics long enough, you'll find that datasheets tend to accumulate. The days of bookshelves full of databooks and three-ring binders may be fading into the past, but the datasheet will be with us for a long time to come in electronic form.

Lastly, lest anyone think that the electronics industry lacks a sense of humor, in 1972, an engineer at Signetics created a datasheet for a fictional component called a write-only memory device (*http://bit.ly/womd-data*). Signetics was the first company specifically established in 1961 to produce integrated circuits, and it was fairly prolific, so a lot of datasheets for various products were generated each year (the company was purchased by Philips, now known as NXP, in 1975).

The joke datasheet caught on. According to some sources it was a hit with marketing at Signetics, and even made it into the databook. Some customers also liked it and a few apparently attempted to order the part as a joke. The story of how the 25120 Write-Only Memory datasheet evolved into its current apocryphal form is also interesting. You can read more about it at *http://www.sigwom.com*.

Electrostatic Discharge

Electrostatic discharge (ESD) is the bane of solid-state active components. Some devices, such as diodes and rectifiers, are less sensitive than parts like CMOS ICs and FETs (field-effect transistors). But, nonetheless, ESD is something that can wreck a solid-state part if it is handled and stored carelessly. Even worse, a part can sustain ESD damage and still appear to function normally, at least initially. The damage sometimes doesn't appear until the part is stressed in operation, or after some period of time has elapsed. When it does fail, it will do so without warning.

ESD is, as its name implies, an event that occurs when an accumulated static charge is suddenly discharged to ground (lightning is a form of ESD, albeit a rather dramatic one). The problem arises when the path to ground happens to be through an IC or other solid-state device, and most of the time it can't even be felt or seen. The first line of defense is to make sure that you are not the source of the static discharge, or part of the discharge path. The easiest way to avoid becoming an ESD source is to wear a wrist strap, like the one shown in Figure 9-3.

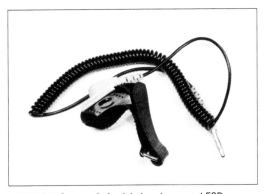

FIGURE 9-3. A grounded wrist strap to prevent ESD damage

Wrist straps don't connect directly to ground, but rather through a 1M ohm resistor to limit the current and prevent shock hazards. The resistor also serves to limit the effect of sudden discharge through a part should it be at a high potential when touched by someone wearing the ESD wrist strap. The discharge can work either way, from handler to part, or part to handler.

Another technique is to use an ESD mat. These mats come in a range of sizes and are made of a conductive material. They are designed to cover the surface of a workbench and they can be trimmed to fit a particular space. They have a small terminal in one corner that is connected to a ground point. They typically come in rolls and in a variety of colors.

Soldering equipment is also available in ESD-safe designs, and most better-quality soldering irons and soldering stations have this feature. Ungrounded soldering tools can accumulate a formidable amount of charge, and it can be released into a part as soon as the tip of the iron touches the PCB. For this reason, you should always use an ESD-safe soldering iron or soldering station when working with solid-state components, and never defeat the ground on the AC mains plug.

Lastly, there is the issue of component storage. Storage bins should be ESD safe, if possible. If that's not practical (or affordable), it is a good idea to have a supply of antistatic high-density foam on hand. Figure 9-4 shows some ICs and a few FETs inserted into a piece of anti-static foam.

The foam is made from a conductive material, usually impregnated with carbon, that acts as a high-value resistor. If you handle the foam with an ESD wrist strap on, and work with it on an antistatic mat, you should be fine. Just remember while wearing the wrist strap to touch the foam first, not the components themselves. This allows the strap to do its job and bleed away any high potential without damaging the parts. You can purchase this type of foam in sheets from most major electronics distributors and ESD-control product suppliers.

FIGURE 9-4. Anti-static foam carrying sensitive components

Packaging Overview

Active components come with anywhere from 2 to 144 leads (or more), depending on the device. Small through-hole components such as diodes and rectifiers are often packaged in axial lead forms, much like resistors. Transistors come in plastic and metal packages, with three or four leads. Radial lead and pin-grid array packages have the leads protruding from the bottom of the device, whereas surface-mount parts often have leads at the sides, or as contact points on the underside of the package. Some large ICs are available in a package style called a *ball-grid array* (BGA), with the contacts points arranged in a grid on the underside of the IC. If you've ever removed the CPU from a recent vintage PC motherboard, you've likely seen a pin-grid array intended for use with a socket of some type. Parts with a ball-grid array are soldered directly to the PCB. High-end graphics adapter cards often use a BGA package for the GPU (graphics processor unit) chip.

The Joint Electron Device Engineering Council (JEDEC) defines over 3,000 different package types for electronic components. This section gives some brief descriptions of the more common packages you might encounter when working with solid-state active components.

THROUGH-HOLE PARTS

A through-hole part is any electronic component that is designed to be placed on a PCB by way of a lead inserted through a hole in the PCB. In all fairness, this is a rather PCB-centric definition, since there is nothing to prevent you from using a so-called through-hole part with a solderless breadboard or wiring it into a circuit by soldering the leads to an old-style terminal strip. In fact, prior to the widespread adoption of PCBs, point-to-point wiring was exactly how it was done (see Chapter 15 for more on this and PCBs).

The timer module shown in Figure 9-2 uses through-hole components exclusively. For axial lead components such as diodes and rectifiers, the spacing between the holes (or pads) on a PCB for the leads is left entirely to the PCB layout designer—up to the maximum permitted by the length of the component's leads, of course. For radial lead parts such as LEDs and transistors, the spacing between the leads is fixed by the physical package used for a given part. ICs in through-hole packages generally follow the JEDEC standards for package body width and lead spacing.

SURFACE-MOUNT PARTS

Surface-mount packages (referred to as SMD, or *surface mount devices*, which are a form of SMT, or *surface-mount technology*) offer a significant savings in terms of PCB space. This is part of SMT's main appeal, along with the ease of incorporating automation into the fabrication process.

As the name implies, SMD packages mount to the surface of a PCB; no through-hole is necessary. Surface-mount diodes and rectifiers come in small packages similar to resistors and capacitors (see Chapter 8). For ICs, the spacing between the leads (also called the *pitch*) and the width of each lead depends on the package type.

Figure 9-5 shows the relative size difference between the PCBs for a circuit with four transistors in SMT and through-hole styles. The drawings are approximately to scale.

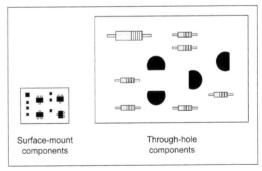

FIGURE 9-5. Comparison of surface-mount versus through-hole circuit boards

The surface-mount PCB in Figure 9-5 is a fraction of the size of the conventional through-hole board. If the through-hole PCB is 1 inch wide by 2 inches long, the SMT PCB is about 1/3 inch wide by 1/2 inch long. Granted, this is a somewhat contrived example (there are no connection points on the

PCBs, for example), but it does show the relative size difference possible with surface-mounted components as opposed to conventional through-hole types. "Surface-Mount IC Package Types" on page 230 covers surface-mount packages for ICs.

USING DIFFERENT PACKAGE TYPES

Note that it is always a good idea to use the manufacturer's datasheet dimensions when you're laying out a PCB or sizing a part for a design. Don't blindly trust whatever a PCB layout CAD package might have in its parts library, because there can be variations between manufacturers and even from year to year from the same source. For more about PCB layout, see Chapter 15.

In the following sections, each type of part has a description of the package types available, so we won't go into any more details here. The primary point is to gain an appreciation for just how small things can get when surface-mount devices are used. The downside is that they can be extremely difficult to work with, and the investment required to obtain the necessary tools and supplies is not trivial.

Diodes and Rectifiers

Diodes and rectifiers are semiconductor devices that only allow current to flow only in one direction. Both types of devices have the same behavior, so the distinction arises from how they are used. Diodes typically deal with small amounts of current, while rectifiers are designed to handle large (sometimes very large) currents. Rectifiers are commonly used to convert the AC from a transformer connected to 120 VAC house current to DC that can then be filtered and regulated into DC for electronic devices. Diodes are typically used to rectify signals, restrict current flow, and limit signal levels, and in applications like video switching circuits and digital logic. A diode will typically respond to a change in current more quickly than a rectifier. This is called the *switching time* of the device. Appendix A provides some details about how the semiconductor junction in a diode or rectifier works, and Figure 9-6 shows graphically what happens when an AC signal encounters a rectifier.

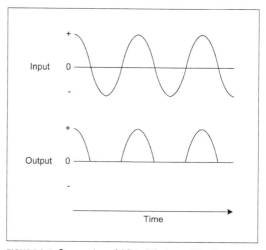

FIGURE 9-6. Conversion of AC to DC via rectification

Rectifiers have been around in one form or another for well over 100 years. Early rectifiers (late 1800s, early 1900s) used mechanical components such as vibrating armatures to rectify AC and convert it to DC. These were replaced by vacuum tubes in the early part of the 20th century, and also by rectifiers that were based on stacks of either selenium or copper oxide wafers. These early rectifiers tended to be large, bulky, and sometimes dangerous.

An early form of small-signal diode was the so-called *cat's whisker* used in some simple radio circuits. It was invented around the same time as vacuum tube rectifiers, but, outside of some kits and specialty applications, it was obsolete by the 1920s. This device typically used a crystal of galena (the crystal form of lead sulfide) and a wire "whisker" that could be moved around on the face of the crystal to find the best location to rectify a radio signal. In other words, these were the original crystal radios. Literally.

The solid-state diodes and rectifiers we know today became common with the advent of silicon and germanium-based semiconductors. Modern rectifiers and diodes employ a solid-state junction, and they can be fabricated directly onto a silicon wafer as part of an integrated circuit.

Some devices have behaviors similar to common rectifiers but also have characteristics that make them uniquely suited for certain applications. Among these are Zener diodes, light-emitting diodes (LEDs), and photo-sensitive diodes. Of course, there are even more exotic types available, but we won't delve into those, as their applications tend to be highly specialized. If you are curious, I encourage you to search out technical data on these exotic components, as some of them are quite fascinating.

Some parts are marked with a number printed on the body, while others have no markings at all except for the cathode band. For this reason, it is a good idea to keep diodes and rectifiers separate in small containers or envelopes with the part number written or taped to it. If they get jumbled, the only good way to sort them out is by using a test instrument called a *curve tracer*. While it's not hard to build a workable curve tracer from junk parts and an old oscilloscope, it's a chore that you can avoid by taking care to keep things neat and organized from the outset.

SMALL-SIGNAL DIODES

Small-signal diodes, as the name suggests, are used to rectify low-voltage, low-current signals. Some types are extremely fast, making them suitable for use in radio frequency circuits or in some types of high-speed logic circuits. In fact, early computers used large numbers of small-signal diodes in their circuits. These served as early forms of read-only memory, among other things.

Figure 9-7 shows a typical small-signal diode in a glass package.

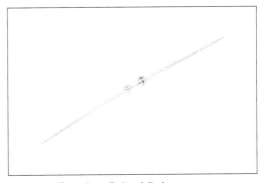

FIGURE 9-7. Typical small-signal diode

The JEDEC standard specifies that diodes use part numbers beginning with "1N." In the case of Figure 9-7, this is a 1N4148 part, which is common. It has largely replaced the earlier 1N914 diode for most small-signal applications.

Diodes are rated in terms of forward voltage, peak inverse voltage, switching time, and forward current::

Forward voltage
Also referred as the *forward bias*, the voltage potential applied to a solid-state P-N junction that will result in current flow through the junction.

Forward current
The maximum amount of current that the P-N junction can safely carry when it is conducting while forward biased.

Peak inverse voltage (PIV)
The maximum reverse polarity voltage the P-N function can withstand before it breaks down and destroys the device. Note that avalanche and Zener diodes are designed to take advantage of reverse bias.

Switching time
The amount of time required for the P-N junction to transition between non-conducting to conducting states.

Table 9-1 gives the typical specifications for a 1N4148 diode. These are the primary things to note when you are examining the datasheet of a small-signal diode or selecting parts from a distributor's parametric selection table.

TABLE 9-1. Essential 1N4148 specifications

Parameter	Value
Forward current (continuous)	300 mA
Forward voltage (typical)	0.7V
Peak inverse voltage	75V
Switching time (typical)	4 nS

Note that the forward voltage will manifest as a voltage drop when the diode is in a circuit, and different diode types have different voltage drops. For a garden-variety red LED, the voltage drop is typically around 1.4V. Figure 9-8 shows how this would appear in a simple circuit.

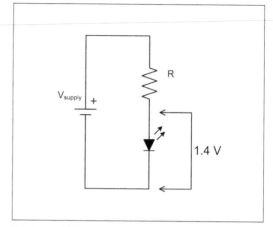

FIGURE 9-8. Voltage drop across an LED

The value of R in Figure 9-8 would be chosen to provide sufficient forward current to cause the LED to emit light, which can vary from manufacturer to manufacturer. Note that this is essentially a voltage divider, like the one described in Chapter 1. The same math would be used here to determine the value of R.

RECTIFIERS

A solid-state rectifier is useful for dealing with current flow of any significance, on the order of 500 mA or greater. They are typically used for AC rectification in power supplies, in protection circuits, and in voltage regulator circuits.

Figure 9-9 shows a 1N4004 type rectifier, which is a 1A device capable of handling up to 400V of reverse voltage. It is common in small linear power supplies. Other members of the 1N4000 series have PIV ratings from 50 to 1,000 volts, and all are rated at 1 A of forward current.

FIGURE 9-9. Typical small solid-state rectifier

Rectifiers come in packages other than axial lead types like the one shown in Figure 9-9. High-current rectifiers are available in so-called *stud-mount* packages, like the one shown in Figure 9-10.

FIGURE 9-10. A stud-mounted rectifier

Yet another variation on the rectifier theme is the bridge rectifier. A *bridge rectifier* is a full-wave rectifier; that is, whereas a single rectifier will convert an AC signal to a pulsed DC signal (as shown in Figure 9-6), a full-wave rectifier will convert both the positive and negative parts of the AC signal to DC output, as shown in Figure 9-11.

Another variation of a full-wave rectifier circuit uses only two rectifiers, but it requires a center-tapped secondary winding on the transformer to serve as a common return.

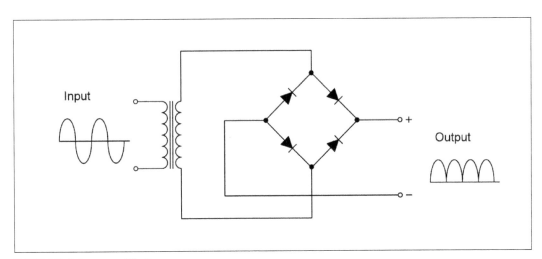

FIGURE 9-11. Full-wave rectification

The full-wave bridge does not require the center tap. Figure 9-12 shows a small bridge rectifier. Larger types, with a center hole for mounting to a chassis or heatsink, are also available.

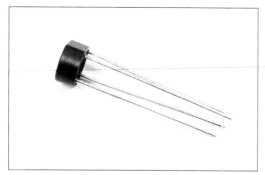

FIGURE 9-12. A small bridge rectifier

A small full-wave bridge, or even just four rectifiers, can be used to create a "don't care" DC input for a circuit, as shown in Figure 9-13. Instead of an AC source like a transformer, if the input is a DC source, the rectifiers in the bridge will ensure that the output will always be consistent with regard to positive and negative. Although this trick isn't commonly used, some devices do employ it. Some devices that use a wall transformer can accept either AC or DC input at the nominal voltage and work just fine because there is a full-wave bridge inside, and probably a voltage regulator and some filtering as well.

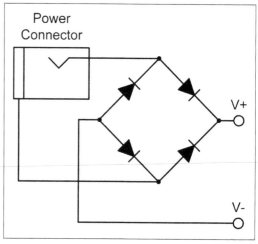

FIGURE 9-13. A "don't care" DC power input protection circuit

LIGHT-EMITTING DIODES

A light-emitting diode is a type of rectifier that has been manufactured in a way that enhances its ability to produce light when current flows through the silicon junction at the heart of the device. This involves a process called *doping*, wherein specific impurities are added to the otherwise pure semiconductor material used in the device. Although all semiconductors have added impurities (otherwise, they wouldn't work), an LED will generate light as a by-product of current flow.

LEDs come in a wide range of shapes, sizes, and output levels. Some types, designed for surface-mount applications, are very small, while others are large enough to be used for illumination purposes. LEDs also come in a variety of colors. The first LED components emitted light in the infrared. These were followed in the 1950s by devices that emitted visible red light. In the 1970s, yellow and green LEDs appeared. Blue LEDs didn't appear until 1994, when Japanese researcher Shuji Nakamura demonstrated a high-output blue device. These days, you can purchase LEDs that emit light from infrared to long-wave ultraviolet. Figure 9-14 shows a selection of various types of LEDs.

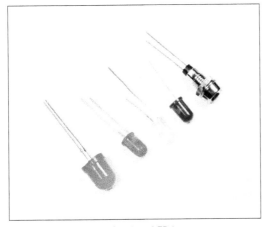

FIGURE 9-14. A sample of various LED types

Because the core of an LED is basically just a small piece of semiconductor, the outer casing can be molded into almost any reasonable shape. In addition to those shown in Figure 9-14, cylindrical and arrowhead shaped packages are also available. These "oddball" shapes sometimes find their way into consumer electronics.

An LED is primarily characterized by its output wavelength (color), forward voltage, and forward current. The *forward voltage* is the voltage at which the semiconductor junction in the LED will start to conduct, and the *forward current* is the current at which the LED will produce its maximum output. The lowest common forward voltage is 1.6V, and some low-current devices will glow with only 1 mA. Other devices have a forward voltage as high as 14V and a current of 70 mA.

ZENER DIODES

A Zener diode is a device that behaves in many respects like a typical diode. Physically, Zener diodes look identical to any other diode or rectifier. What sets a Zener apart is that, when the current is reversed, it will not allow the current to pass until a specific voltage level is reached. When this occurs, the Zener will conduct in the reverse direction.

This unique characteristics allow Zener diodes to be used as a form of voltage regulator, and they are often used to produce a reference voltage. A Zener with a rating of, say, 5.1V, can be used to shunt any voltage greater than 5.1V to ground. Anything less than 5.1V will not be affected.

EXOTIC DIODES

A solid-state silicon junction is sensitive to external energy, such as heat and light. This sensitivity can be enhanced to create a device that is useful as a light sensor. A PIN diode (the name refers to the P-I-N junction in the device, where the *I* stands for *Indium*) has a modified junction between its semiconductors. When used as a light detector, a PIN diode is connected in reverse such that no current flows until a photon strikes the junction. When this happens, the junction temporarily conducts and allows current to flow.

Solid-state lasers are a variation on the LED, but with some structural modifications to make it possible for the device to emit coherent light. Laser diodes are connected to a circuit much like an LED, and their output can be modulated to carry information. They are useful for fiber-optic communications systems, and infrared versions are used in military applications for target designation. They are also commonly used as pointers to annoy people at meetings and for teasing pets with a bright spot they can't catch. Figure 9-15 shows a module that can be used with something like an Arduino.

FIGURE 9-15. Small laser-diode module

Laser diodes are available in output powers ranging from around 5 mW up to and beyond 30W. The low-power types are commonly found in laser pointers, whereas higher power parts, in the 500 mW to 800 mW range, are used in CD and DVD drives. Devices with output powers above 1W are used in engravers, cutters, labeling systems for date codes, and some projection systems.

DIODE/RECTIFIER AXIAL PACKAGE TYPES

Axial packages for discrete diodes and rectifiers with axial leads are defined by a series of type numbers beginning with a DO prefix in the DO-204 family defined by JEDEC. The DO-35 package is commonly used for small-signal devices, and the DO-41 package is used with larger devices, such as the 1N4004 rectifier. Use the outline drawing in Figure 9-16 when looking at Table 9-2.

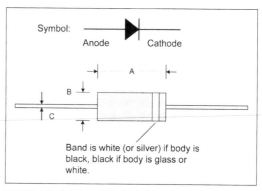

FIGURE 9-16. Axial-lead diode package dimensions

TIP Semiconductor schematic symbols typically use an arrowhead to indicate conventional current flow through the device. As discussed in Chapter 1, the conventional current flow is the opposite of how the electrons in a circuit move. In the case of diodes, this means that the terminal names (cathode and anode) have the meaning you might expect, in that electrons enter through the cathode, move across the semiconductor junction, and emerge at the anode. There is current flow when the cathode is negative with respect to the anode. However, thanks to Ben Franklin and general social inertia, current flow is described as moving from the anode to the cathode. We just have to live with it and remember to switch our thinking around as the need may arise.

TABLE 9-2. Standard JEDEC axial diode package dimensions[a]

JEDEC	A	B	C
DO-16	2.54	1.27	0.33
DO-26	10.41	6.60	0.99
DO-29	9.14	3.81	0.83
DO-34	3.04	1.90	0.55
DO-35	5.08	1.90	0.55
DO-41	5.20	2.71	0.86

[a] Dimensions are in millimeters; maximum values shown.

DIODE/RECTIFIER SURFACE-MOUNT PACKAGES

Diodes and rectifiers are available in surface-mount packages similar to those used for resistors and capacitors, and some employ the same package numbering system (see Chapter 8). Some have metalized ends for soldering. Larger packages, such as the DO-214, employ folded leads. Figure 9-17 shows three different surface-mount package types for diodes-type devices. The upper set of numbers is the minimum size, the lower set is the maximum, and numbers in parentheses are millimeters.

There are many other package types besides the ones shown in Figure 9-17. Note that, while not shown in the figure, the packages have markings to indicate which end is the cathode and which is the anode. As always, be sure to consult the datasheet for the specific dimensions and device characteristics.

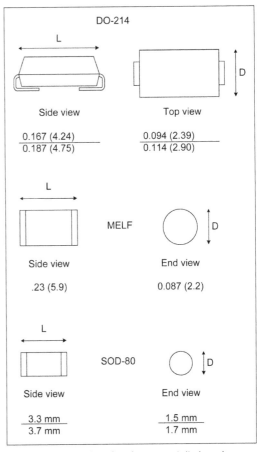

FIGURE 9-17. Examples of surface-mount diode and rectifier packages

LED PACKAGE TYPES

LEDs come in a variety of physical forms, from the common radial lead types like those shown in Figure 9-14, to minuscule surface-mount parts, jumbo-size devices used for lighting and electronic sign applications, and metal-enclosed laser diodes like the one shown in Figure 9-15.

In addition to round shapes in various diameters, LEDs are also available in rectangular shapes, such as the styles shown in Figure 9-18. The T1 standard lamp size is another popular form factor, with a diameter of 3 mm and an overall body length of 5 mm. Many 3 mm size LEDs are interchangeable with T1 type bulbs.

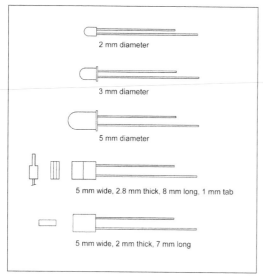

FIGURE 9-18. A small sample of available LED shapes

LEDs are available in surface-mount packages as well as through-hole types. The 0603 and 0805 sizes are popular (see Chapter 8 for a description of standard surface-mount package sizes), and kits containing a variety of colors are available for about $12 from distributors such as Digikey (see Appendix D). Figure 9-19 shows an outline drawing of a Dialight 598 series LED in an 0603 package.

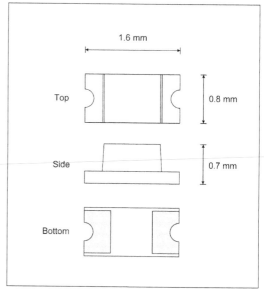

FIGURE 9-19. A Dialight 598 series 0603 package surface-mount LED

Other sizes are also available, from 0402 (1005 metric) to 1616. Side-looking packages allow surface-mount LEDs to be mounted such that they are visible at a 90-degree angle to the surface of the PCB.

Transistors

In 1947, the world's first transistor was created at AT&T's Bell Laboratories, and the world underwent some massive changes as a result. These early devices used germanium as the semiconductor, and they tended to be fragile and expensive. In 1954, the silicon transistor appeared, followed in 1960 by metal-oxide semiconductor (MOS) devices.

Transistors can be classified into two basic types: bipolar junction and field-effect. Bipolar junction transistors (BJTs) have unique operating characteristics quite unlike any other electronic component. Transistors can be used as amplifiers, regulators, or switches. In truth, the regulator application is closely related to high-power amplifiers, although most people probably don't think of a power supply that way. When used as switches, transistors formed the basis of the first solid-state logic circuits in computers, with later, and faster, devices finding applications in radio communications and radar systems. BJT devices are fabricated as either NPN or PNP. This refers to the type of semiconductors used to create the device and the order in which they are arranged internally. Appendix A discusses semiconductor junctions in more detail, and the texts listed in Appendix C have even more detail.

A field-effect transistor can be either an N-channel or P-channel device, depending on how it was made. FETs also come in a number of variations, from the original type to more modern MOSFET designs. Each type has its own unique set of characteristics, such as current handling capacity, voltage range, and switching speed, that make it suitable for a particular application.

Figure 9-20 shows the schematic symbols for both NPN- and PNP-type BJTs. Appendix B contains a comprehensive set of schematic symbols, including various types of transistors. Refer to Appendix A for an overview of solid-state concepts.

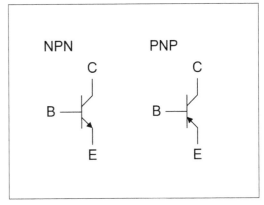

FIGURE 9-20. Standard schematic symbols for BJT transistors

The letters B, C, and E refer to the names of the terminals on a BJT device. These are base, collector, and emitter, respectively. In a BJT, current flows between the E and C terminals, and the B terminal acts as the control input.

SMALL-SIGNAL TRANSISTORS

Just as with small-signal diodes, a small-signal transistor is designed to work with low voltage and current levels. Small-signal transistors come in a variety of package styles, from surface-mount to multi-legged metal cans. Figure 9-21 shows a selection of package styles available.

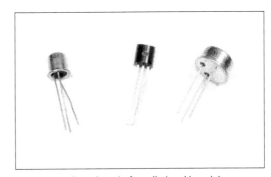

FIGURE 9-21. Assortment of small-signal transistors

ACTIVE COMPONENTS | 221

A selection of inexpensive small-signal transistors is handy to have around. Table 9-3 lists some types that are useful for switching, controlling low-current relays, and driving LEDs, and as amplifiers and oscillators. PN2222 (plastic TO-92 package) parts can be purchased for as little as 10 cents each from companies such as Alltronics (see Appendix D).

TABLE 9-3. Listing of low-cost small signal transistors

Number	Type	Vce	Ic	Pd	Ft	Packages
2N2222	NPN	40 V	800 mA	500 mW	300 MHz	TO-92, SOT-23, SOT223
2N2907	PNP	60 V	600 mA	400 mW	200 MHz	TO-92, SOT-23, SOT223
2N3904	NPN	40 V	200 mA	625 mW	300 MHz	TO-92, SOT-23, SOT223
2N3906	PNP	40 V	200 mA	300 mW	250 MHz	TO-92, SOT-23, SOT223

The parameters used in Table 9-3 are defined as follows:

Vce
  Maximum voltage across collector and emitter

Ic
  Maximum current at the collector

Pd
  Maximum device power dissipation

Ft
  Maximum usable frequency

POWER TRANSISTORS

As you might guess, a power transistor is designed to handle significant amounts of current, from several hundred milliamps to over 100 amps or more. Today, it is more common to see MOSFETs in the role of power transistors rather than BJT devices, mainly because the MOSFET can operate more efficiently at high power and voltage levels than the BJT type. Refer to "Field-Effect Transistors" on page 222 for more on FET devices.

Figure 9-22 shows some of the package options available. Notice that each type is designed to be mounted onto something, such as a panel or some other form of heatsink.

FIGURE 9-22. Assortment of power transistors

For most small projects, a power transistor is not necessary. But, should you ever need to control something like a high-current contactor, a heating element, or a hefty DC electric motor, a power transistor might be required.

FIELD-EFFECT TRANSISTORS

Whereas a BJT works by modulating the exchange of positive and negative charge carriers across a semiconductor junction, field-effect transistor devices behave in a manner similar to the operation of vacuum tubes. A FET controls the flow of electrons moving through its junction in proportion to the amount of voltage present on the gate pin of the device. In fact, there have been solid-state vacuum tube replacements made from high-voltage FET devices that plug in where a vacuum tube used to reside.

A MOSFET (which, as noted earlier, stands for metal-oxide semiconductor field-effect transistor) is a common type of FET. These devices have a low internal resistance when in the "on" state, and some types can handle a significant amount of current. They are often used in DC power switching circuits, and have also been put to use in the output stage of audio amplifiers. FET devices come in TO-3, TO-5, TO-92, TO-66, and TO-220 packages, as well as small-outline transistor (SOT) and other types of surface-mount packages.

CONVENTIONAL TRANSISTOR PACKAGE TYPES

Transistors and other devices are packaged in a variety of styles, some of which have fallen into disuse over time as surface-mount technology has become more prevalent. But for some transistor types, a through-hole package is still the only feasible way to go, particularly if the device needs to be mounted on a heatsink of some sort. Radial lead parts are usually mounted directly on a PCB in an upright position.

Table 9-4 lists some of the transistor package types that are available. Some of these are shown in Figure 9-23, but not all, since some are now either considered obsolete or are uncommon, but you might still encounter them in the wild at some point.

TABLE 9-4. JEDEC transistor package types

Case	Mounting	Package type
TO-3	Flange mounted	Metal can
TO-5	Radial leads	Metal can
TO-18	Radial leads	Metal can
TO-46	Radial leads	Short metal can
TO-52	Radial leads	Metal can
TO-66	Flange mounted	Metal can
TO-92	Radial leads	Molded plastic
TO-126	Flange mounted	Plastic/metal
TO-202	Flange mounted	Plastic/metal
TO-218	Flange mounted	Plastic/metal
TO-220	Flange mounted	Plastic/metal
TO-254	Flange mounted	Plastic/metal
TO-257	Flange mounted	Metal can
TO-264	Flange mounted	Plastic/metal
TO-267	Flange mounted	Metal can

Transistors that are flange mounted usually require a flat surface of some sort, either a metal plate, chassis panel, or a heatsink. It is possible to simply solder a TO-126, TO-202, and TO-220 (and similar types) directly to a PCB if it won't be dissipating very much power. Heatsinks are available for these types that attach to the mounting flange of the transistor, allowing it to stand free.

Figure 9-23 shows some examples of common transistor packages. Note that the TO-66 is essentially a smaller and slightly

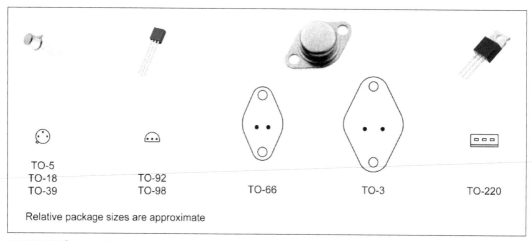

FIGURE 9-23. Common transistor package types

narrower version of the venerable TO-3 package.

Transistor pin definitions can vary from one type to another and, of course, a FET or a TRIAC (triode for alternating current; discussed shortly) will have different connection types than a BJT device. Figure 9-24 shows pin connections for some common package styles. Note that these are just a small sample of the possible variations. Just because something is in a TO-18 package, don't assume that it's a BJT with the same E-B-C pin order as shown in Figure 9-24. The pins might be in a different order, or it might be a UJT (unijunction transistor) or a FET.

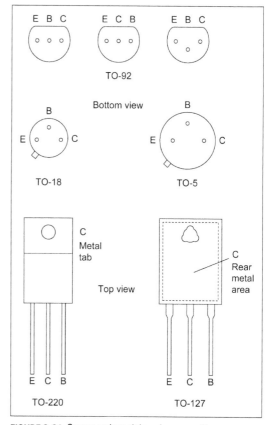

FIGURE 9-24. Common transistor pin connections

Refer to the datasheet, if possible, for the actual connections for a particular device. Often, a part purchased as a single item in a plastic retail display package will have the connection diagram on the package. If you have a part with no markings, or if it has a custom number assigned by the equipment manufactuer to prevent people from reverse-engineering their amazing technology, I would suggest discarding it. While it is possible to determine the basic characteristics of a BJT or a FET using test equipment, it's really not worth the effort in many cases. If a part costs only 50 cents, your time is a lot more valuable. Just chuck it in the trash and purchase the right part.

SURFACE-MOUNT TRANSISTOR PACKAGE TYPES

SMD transistor packages typically have three connection points or leads for single device packages. Some parts might contain more than one transistor and consequently more leads. The SOT-23 (small-outline transistor, type 23) package, shown in Figure 9-25 with a couple of other common transistor package, is common.

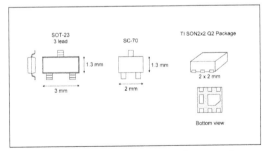

FIGURE 9-25. A selection of common surface-mount packages used for transistors

The SON2x2 package is used for a power MOSFET device, with the large area under the package being a heatsink connection.

Refer to the datasheet for the connections for a particular device. Most, but not all, SOT-23 and SOT-223 type packages use the arrangement shown in Figure 9-26. The SOT-223 is basically the surface-mount equivalent of the TO-220 shown earlier. The emitter pins are duplicated to accommodate the higher current capacity of the component. The SOT-223 package is also used with devices such as linear voltage regulators.

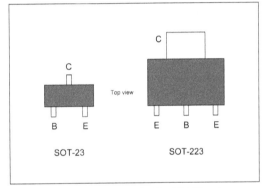

FIGURE 9-26. Pinout diagrams for the SOT-23 and SOT-223 packages

SCR and TRIAC Devices

SCRs (silicon-controlled rectifiers) and TRIACs (triodes for alternating current) are members of a class of solid-state active devices called *thyristors*. These are used exclusively for power control applications. Appendix B shows the schematic symbols used for SCR and TRIAC devices.

In terms of packaging, an SCR or a TRIAC looks a lot like a conventional transistor. The plastic TO-220 package is popular, and some devices with higher current capacity come in the TO-3 metal-can package. There are also small TO-5 and TO-92 versions, as well as some surface-mount types. Very high-power devices are available in stud-mounted packages, similar to the types used for rectifiers shown in Figure 9-10. The difference is that the SCR or TRIAC will have two terminal posts protruding from the body of the package instead of just one.

SILICON-CONTROLLED RECTIFIERS

A silicon-controlled rectifier is a four-layer semiconductor that behaves like a gated diode and passes current in only one direction when it is enabled. It doesn't conduct until a voltage is present on the gate terminal, after which it will continue to conduct until the current flow through the device drops close to zero (or some specific cut-off threshold). SCRs are available in different package styles, depending on the rated power capacity of a particular device. Low-current devices can be found in TO-92 packages, while higher current parts come in TO-220, TO-3, or stud-mount packages.

TRIACS

If you have a light dimmer in your house, chances are, you also have a TRIAC. Unlike an SCR, a TRIAC can pass current in both directions. Like the SCR, a TRIAC does not conduct until a voltage is applied to the gate terminal. Because of their ability to conduct in both directions, TRIACs are commonly used for AC power control by enabling current flow for only part of a positive or negative cycle. Appendix A shows an example of how a TRIAC is used to control AC current in a light dimmer using an RC phase shift circuit.

As with SCRs, TRIACs are available in a variety of packages. The TO-220 style is common, and high-power versions are available in TO-3 and stud-mounted packages.

Heatsinks

Solid-state components can get extremely hot during operation as heat builds up in the package of the device. Even though the current through a device might not exceed its rated maximum, or even come close to it, the device can be damaged if excessive heat is allowed to accumulate over time. Without some way to conduct the heat away from the part, it might eventually fail.

Power transistors are particularly susceptible to overheating, due to the amount of current they are capable of handling. One tried-and-true way to deal with the issue is to use a heatsink. Purpose-made heatsinks can be had for almost every power transistor package type, including the TO-202, TO-220, TO-66, and TO-3 styles. In many situations, the part is simply mounted to a panel, chassis, or other large metallic mass that can conduct heat away from the part. Insulating shims can be placed under the case of the transistor to isolate it, and nylon or phenolic fiber washers are typically used to isolate the mounting screws.

In other cases, there might be sufficient air flow to allow the part to mount directly to a freestanding heatsink, such as the TO-220 type shown in Figure 9-27. This type of heatsink is designed to stand vertically on a PCB with the transistor bolted to it. A thin layer of thermal conductive compound or a special thermal pad is sometimes placed under the body of the transistor between it and the heatsink to improve the thermal conductivity.

FIGURE 9-27. TO-220 vertical mount freestanding heatsink

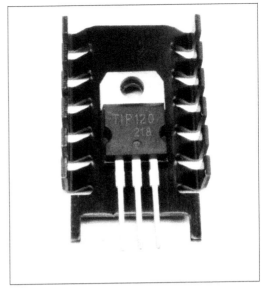

FIGURE 9-28. TO-220 on vertical heatsink

There are similar styles available for the TO-66 and TO-3 packages as well. Since the heatsink isn't physically connected to anything around it, there is no need to electrically insulate the body of the transistor, which simplifies mounting. Figure 9-28 shows how the transistor is aligned on the heatsink (although in this case the mounting hardware has been omitted for clarity).

Figure 9-29 shows a vertically mounted heatsink for a motor driver IC on a PCB. The part is similar to a TO-220, only it's much wider to accommodate more pins. It is used as a motor controller, so it has to be able to deal with some significant amounts of current.

FIGURE 9-29. Heatsink on a motor controller PCB

Small transistor packages can also have thermal issues. Small press-on heatsinks composed of stamped metal fins are available for TO-5 and TO-18 packages. Figure 9-30 shows two examples of a TO-5 press-on heatsinks, along with a TO-5 part for comparison. A TO-18 heatsink is similar, just slightly smaller. Heatsinks for the TO-5 and TO-18 packages are also available as extruded aluminum parts, tubes with flared tabs at one or both ends, and stamped metal assemblies that somewhat resemble hats.

FIGURE 9-30. TO-5 press-on finned heatsink

Heatsinks are also available for TO-92 type parts, although in some situations you can mount the part against a metal panel or even a large heatsink for another part by using a small bracket. The downside to this approach is that the leads sometimes must be extended with wires (and the soldered connections covered with heat-shrink tubing) in order to reach the PCB.

> A freestanding heatsink is a potential source of high voltage, such as in power supplies. Even in a low-voltage circuit, a heatsink might be at full V+ potential, and creating a short to ground can do some serious damage. Never touch a freestanding heatsink with your fingers or a grounded tool while the circuit is active.

Heavy-duty heatsinks are usually made from aluminum extrusions, which are then cut and drilled for a specific transistor package style. These are found in the output stages of high-power amplifiers, linear DC power supplies, and motor controllers, or anywhere else a lot of heat needs to be pulled away from a part and safely dissipated.

Note that, in order for a heatsink to be effective, it must itself have some way to dump the heat it has accumulated from a transistor or other part. You typically accomplish this by moving air over the heatsink with a fan of some type. In some instances, as on the heatsink of the CPU in a PC, the fan can be mounted directly onto the heatsink. In other designs, the heatsink fins protrude from the chassis of the device (or might be part of the chassis itself) and the flow of air in the local environment is sufficient. This type of arrangement is often seen with high-power audio amplifiers used in automobiles, VHF transceivers for mobile applications, and battery-powered AC inverters.

Integrated Circuits

A typical integrated circuit is a small square or rectangular piece of silicon crystal onto which active components and circuit traces are placed using photo-lithographic means. This piece of silicon is usually referred to a *chip*, although the entire integrated circuit (IC) is also called a chip. The surface features in modern ICs are extremely small, allowing extremely high circuit density. Modern microprocessor chips might contain millions of individual transistors.

The origin of the IC goes back over 60 years but didn't become commercially viable until around 1958 when Jack Kilby of Texas Instruments patented the first practical IC using germanium as the base semiconductor material. This was followed shortly thereafter by Robert Noyce's development of a silicon-based IC at Fairchild Semiconductor.

ICs can be broadly divided into two categories: linear and digital. The boundary between the two types is not always distinct, depending on what is in the circuitry of the IC. Linear devices are basically anything that isn't digital, such as operational amplifiers (op amps). Linear devices are designed to work with a continuously variable signal, voltage, or current, and theoretically can have an infinite number of operating states. A digital device operates on discrete binary values at specific voltages to perform certain logic functions and can have only a limited number of operating states.

An op amp IC is the modern equivalent of a circuit that was originally developed for use in analog computers. The operating principle of an op amp is straightforward: it is designed to use feedback to force its input to zero volts by changing the value of the output. By carefully selecting the components that form the feedback circuit, a small input voltage can cause a large change in the output. Refer to Appendix A for a brief overview of op amp circuit theory.

A digital IC can be anything from a simple logic gate (AND, OR, NAND, NOR, etc.) to a microprocessor or microcontroller. Digital circuits operate on bits: discrete values of either 1 or 0, true or false. Chapter 11 discusses logic ICs and microprocessors, so the main focus here will be on the packages available.

Still other types of ICs can be a mix of linear and digital, such as analog-to-digital converters and direct digital synthesis (DDS) devices. IC packaging has also been used for things like resistor arrays, transistor arrays, and diode arrays.

CONVENTIONAL IC PACKAGE TYPES

When integrated circuits were developed, one of the first challenges facing engineers was how to package them. Early packages came in styles that included flat, rectangular, ceramic bodies with leads protruding from two opposing sides. These were the original *flat-pack* styles. Later, the dual-inline package (DIP) styles were developed, again in ceramic. The plastic packages that are common today came later and have largely replaced the old ceramic types.

Figure 9-31 shows the basic dimensions of the two most common types of DIP packages. The size of a DIP package reflects, generally, the internal complexity of the IC inside of it and the number of external connections it provides. Small circuits such as a 555 timer or an op amp are usually packaged in an 8-pin style. Logic circuits such as the 74xx series devices typically come in 14- or 16-pin packages, with some larger devices using a 24- or 28-pin package. Memory devices are usually packaged in 24-, 28-, or 32-pin packages to provide for the memory addressing pins needed. Small (e.g., 8-bit) microprocessors usually have many interface pins for external memory and interface chips, so they are often packaged in a 40-pin DIP package. Larger microprocessors, such as some ARM-based parts, and the Intel and AMD CPUs (central processing units) found in PCs, can have over a hundred pins and come in pin-grid array (PGA), ball-grid array (BGA), and high-density surface-mount styles. Microcontrollers (microcontroller units, or MCUs), on the other hand, typically don't have pins to access external memory, so the pin count can be very low even though the internal logic may be quite complex. For example, the ATtiny85 MCU from Atmel or the PICmicro from Microchip Technology are both available in 8-pin DIP packages.

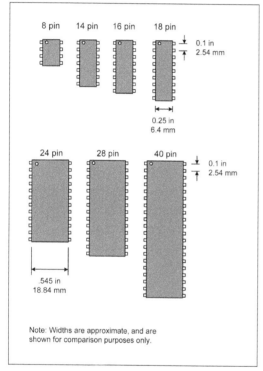

FIGURE 9-31. Through-hole DIP packages for integrated circuits

Figure 9-31 is intended to show the relative size differences between different types of DIP packages. It is not a precise reference. The width dimensions shown in Figure 9-31 are for the body width of the part, not the tip-to-tip width for the pins. There is some degree of variance possible in both the body width and the width-wise pin spacing, so the dimensions shown here must be considered to be approximate. The pin-to-pin spacing or pitch, at 0.1 inches or 2.54 mm, is a standard that just about every IC manufacturer follows. The definitive reference for DIP IC dimensions is the datasheet provided by the manufacturer. Fortunately, you can generally trust the layout templates in most PCB layout software packages to get DIP packages right, and a part that is slightly off-nominal will still fit on the PCB or a solderless breadboard. In fact, it is the PCB that defines the actual final pin-to-pin width dimension for a DIP IC, and the leads usually require some forming during assembly in order for the part to fit correctly into the pad holes on the board. There are special IC insertion tools made for just this purpose.

There are also 4-, 6-, and 64-pin DIP styles. The 4- and 6-pin styles are used for things like opto-isolators, but the 64-pin package is rare nowadays. Packages with high pin counts are typically fabricated in surface-mount styles.

SURFACE-MOUNT IC PACKAGE TYPES

Surface-mount packages for active devices come in a variety of sizes and styles. Some are large enough that they can be mounted on a PCB by hand with a good soldering station and some kind of magnification. Other types, such as BGA packages, with the connection points located under the body of the IC in a grid pattern, really cannot be hand-mounted and must be installed using specialized automated soldering equipment. Some packages are incredibly small, such as the C8051F30x series from Silicon Labs.

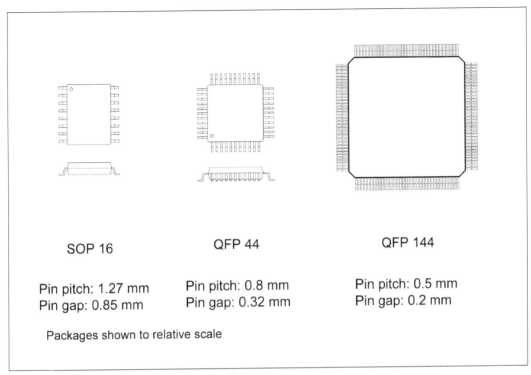

SOP 16

Pin pitch: 1.27 mm
Pin gap: 0.85 mm

QFP 44

Pin pitch: 0.8 mm
Pin gap: 0.32 mm

QFP 144

Pin pitch: 0.5 mm
Pin gap: 0.2 mm

Packages shown to relative scale

FIGURE 9-32. Examples of low-, medium-, and high-density SMT IC packages

These parts contain a complete 8051-type microcontroller in a surface-mount package measuring 3 mm × 3 mm.

Surface-mount IC packages vary in style, from some that aren't much different from their DIP equivalents to those with leads so closely spaced that a workbench microscope is needed to work with them. Figure 9-32 shows some of the common surface-mount IC package types that are available.

The small-outline IC (SOIC) package family comprises a number of related packages. These range from the common SO, SOM, and SOL types to the high-density QSOP package. Table 9-5 lists some of the more commonly encountered members of the SOIC family (note that numbers in parenthesis are mm).

TABLE 9-5. SMD IC packages in the SOIC family

Name	Description	Body width	Lead pitch
SO	Small Outline	0.156 (3.97)	0.050 (1.27)
SOM	Small Outline Medium	0.22 (5.6)	0.050 (1.27)
SOP	Small Outline Package	0.3 (7.62)	0.050 (1.27)
SOL	Small Outline Large	0.3 (7.62)	0.050 (1.27)
VSOP	Very Small Outline Package	0.3 (7.62)	0.025 (0.65)

Name	Description	Body width	Lead pitch
SSOP	Shrink Small Outline Package	0.208 (5.3)	0.025 (0.65)
QSOP	Quarter Small Outline Package	0.156 (3.97)	0.025 (0.65)

I should point out that this table is primarily for comparison purposes. The key features for each type are the body width and the lead pitch. The number of leads will depend on what is in the package, which can can range from 8 to 56. The types of leads (gull-wing, J-lead, flat) can vary from one manufacturer to another, even for packages with the same name, and the body dimensions can also vary slightly from what is listed here. As stated before, always check the datasheet for a part before committing it to a design.

Figure 9-33 shows the packages listed in Table 9-5 to scale.

HIGH-CURRENT AND VOLTAGE REGULATION ICS

There are some types of parts that might look like a transistor, but are actually an IC. Voltage regulators are a perfect example of this. Other examples are various power op amps in TO-3 and large plastic packages that look like a wide TO-220 with many more pins. Power op amps are found in DC power control circuits, and some types are specifically designed for use in audio amplifiers. In fact, almost all home audio equipment made today uses these types of monolithic power amplifier modules.

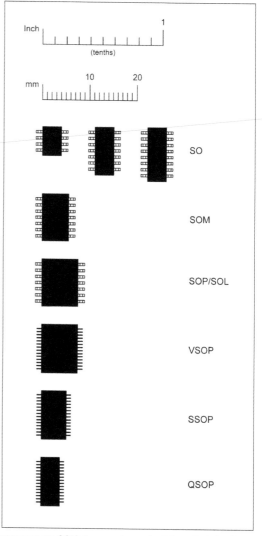

FIGURE 9-33. SOIC-type packages for integrated circuits

A linear voltage regulator IC is a handy thing when you need to produce a specific voltage level for a circuit, or some part of a circuit. These are basically a complete linear power supply in a TO-220 or SOT-223 package. The most basic types have terminals for input, output, and ground. Other types have the ability to adjust the output voltage via a control voltage (often set by a small potentiometer or a zener diode).

The 78XX and 79XX parts have been around for quite a while and are readily available. The "XX" is replaced with the output voltage of the device. The 78XX parts produce positive voltages, and the 79XX parts are used to produce negative voltages. These regulators are fixed output, with output voltages of 5, 8, 10, 12, 15, and 24 volts. Most are rated for around 1A of output current. I like to keep a small stock of 7805, 7905, 7812, 7912, and 7815 regulators on hand.

Low-current regulators are also available in TO-92 packages, such as the LM317. This is an adjustable device with an output voltage from 1.2 to 32V with 100 mA of output current. They are useful when working with analog-to-digital or digital-to-analog convertors that rely on a reference voltage (analog interfaces are discussed in Chapter 13).

Summary

The realm of active components encompasses a broad range of devices, all of which share the common feature of using an external source of current and having the ability to modify the way in which current moves through them in nonlinear ways. Some are externally controllable, whereas others rely on the intrinsic behavior of the solid-state material from which they are fabricated.

The basic P-N junction is the primary building block on which silicon-based solid-state discrete devices are built. If a device has just a single P-N junction, it is probably a diode or rectifier of some sort. Devices with multiple P-N junctions include BJTs, UJTs, FETs, SCRs, and TRIACs.

Integrated circuits incorporate multiple BJTs, FETs, and other components into a design that can be simple, such as an op amp or a 555 timer, to something as complex as the CPU in a computer with millions of individual components.

For experimenting and hacking, having a decent selection of discrete solid-state parts on hand is always a good thing. A bag of cheap 1N4148 diodes and some 1N4001, 1N4002, and 1N4003 parts come in handy when you are dealing with power supply circuits. LEDs can be purchased in the form of *kits* from a number of suppliers. Although these are basically just a box or a large bag containing smaller bags of parts, you can always sort them out and store them neatly yourself.

A basic selection of transistors (such as the ones listed in Table 9-3) can be used for anything under about 300 MHz. Unless you plan to work with RF, you generally won't need anything faster. If you would like a large selection to work with, inexpensive kits of transistors are available, with a good selection of various types in neatly labeled little plastic bags.

When it comes to working with surface-mount parts, you will of course need the correct tools (such as the soldering equipment shown in Chapter 3) and to be able to use the soldering techniques illustrated in Chapter 4. SMT can be extremely challenging, so unless there is a definite need for it, I would suggest avoiding it, at least initially. A prototype is expected to look clunky, and if you are hacking something for fun, there's really no need to try to make it production-ready. It doesn't really matter, for example, if a hacked RoboSapien robot toy looks like it's wearing a Buck Rogers backpack, so long as it works.

When you are working with solid-state components, static charge is your enemy, and it never sleeps. You should seriously consider purchasing an ESD wrist strap, at the very least. Anti-static mats aren't very expensive, and putting one on your workbench (or on the kitchen table) will make your ESD prevention measures even more effective. It is extremely annoying to have just one part of a particular type left, only to discover that when you picked it up from its storage container some stray static charge got to it. It's even worse to build something, test it, and then have it fail in use a week or a month later because one of the parts was partially damaged by ESD during construction.

Lastly, I can't stress enough how important it is to get and read the datasheet for a part. I have seen parts in TO-92 packages that look like transistors but are really something else entirely or, worse yet, might have a common part number but have a pin-out that is uncommon. ICs can also be a problem in terms of packaging. Although the 7400 and 4000 series logic parts (discussed in Chapter 11) are generally consistent, microcontrollers, microprocessors, and other complex logic parts can have completely arbitrary pin-out configurations. Without a datasheet to guide you, it is nearly impossible to work with these components.

CHAPTER 10

Relays

Relays may be an old technology, but they are still essential in electronics. A relay is basically just an electrically operated switch. Some are tiny and handle only small amounts of current at low voltages, whereas other types, called *contactors*, are huge (the size of a small refrigerator) and can safely deal with hundreds of amperes and thousands of volts. But regardless of size and power capacity, all relays and their close cousins the contactors use the same basic principle of operation.

This chapter describes various types ranging from low-current TTL-compatible reed relays to high-power types used to control AC. Techniques for controlling a relay from a low-voltage circuit are also covered, as well as some examples of how relays can be used in control and logic circuits.

Relay Background

Relays have been around since about 1830. In fact, they are probably one of the oldest types of electrical components (other than perhaps switches). Relay-based switching systems replaced human telephone switchboard operators in mid–20th century, and some early computers were built using relays, such as the Zuse Z3 (1941), the Atanasoff-Berry Computer (1942), and the IBM ASCC/Harvard Mark I (1944). Although they haven't changed much in terms of operation, they have evolved into a myriad of types over the past 180 years.

Regardless of the actual internal physical arrangement, all electromechanical relays operate on the principle of electromagnetism as the force driving a mechanism of some sort. The mechanism might be an armature, metal reeds, or a contactor bus bar type of arrangement.

ARMATURE RELAYS

An armature relay has an internal mechanism that transfers the motion induced by a solenoid coil (an electromagnet) into physical motion, usually by means of some sort of lever action. The motion of the armature can be used to control multiple sets of contacts, and in some designs, a small pushbutton is brought out through the shell of the relay to allow manual operation. Figure 10-1 is a simplified illustration of a basic SPST type of relay.

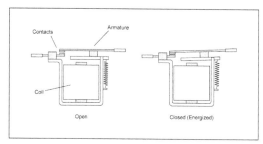

FIGURE 10-1. A basic electromagnetic relay

Armature type relays can have one or more sets of contacts (SPST, SPDT, DPST, DPDT, 3PDT, and so on), with the multiple contacts all mechanically connected to the same armature lever. Since they are basically electromechanical switches, the same general contact descriptions apply. But relays have the advantage of an electromagnet, which can exert considerable force. Relays with 10 or even 20 sets of contacts are sometimes found in applications such as elevator controls and old-style telephone switching systems. If you look ahead to Figure 10-8, you can see three sets of SPDT contacts through the clear plastic shell of the relay.

REED RELAYS

Some types of relays use a set of thin metal strips called *reeds* for the contacts. These are sealed into a small glass tube, and a coil wound around the tube provides the electromagnetic force that causes the reeds to bend and make contact. Figure 10-2 shows how this type of device works.

Any suitable magnetic field will cause the reeds to flex and make contact. In fact, the sensors used in security systems to detect an open door or window are often nothing more than the reed in its glass tube without a coil. A permanent magnet in the window or door frame closes the contacts when it's close to the sensor. Figure 10-3 shows a bare reed sensor module suitable for use with something like an Arduino or BeagleBone single-board computer.

CONTACTOR

In a high-current, high-voltage type of relay known as a contactor, a solenoid coil is used to pull in a bar or frame that holds the contacts, as shown in Figure 10-4. The result is

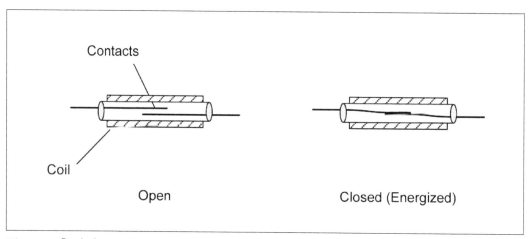

FIGURE 10-2. Reed relay operation

rather like a piston, with the bus bar making a shorting connection between the input and output terminals of the device.

Large contactors can be startling when they are energized, producing a loud bang as the bus bar is pulled in by the solenoid.

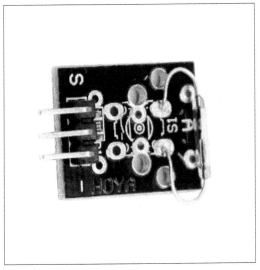

FIGURE 10-3. Reed contacts without a coil

Relay Packages

Some relays are minuscule, while others are huge. Relays come in printed circuit board form factors with both through-hole and surface-mount packages. They can be found in packages that use a socket, making them easily replaceable. Other types have lug terminals for use in industrial and automotive applications, and still others have large bus-bar connections for applications such as high-power motor controllers. It all depends on how much current they are designed to handle at a given voltage and how much voltage and current is necessary to drive the relay mechanism.

PCB RELAYS

Small relays for PCB applications are available in types that range from reed relays to compact armature devices that are capable of handling up to 120 VAC at 10A or so. The previous section showed what a reed relay looks like (in Figure 10-3). Figure 10-5 shows a compact relay designed to be mounted to a PCB.

The contact ratings for a relay of this type can be substantial, even through the relay coil itself requires only 70–85 mA at 5V DC to operate. The actual coil current depends on the resistance of the coil, and this can vary between models and manufacturers. Note, however, that even 70 mA is way beyond the output capability of most ICs, so some type of relay driver IC or a transistor is needed to operate the relay, as described in "Controlling Relays with Low-Voltage Logic" on page 243. You can find relays of this type controlling things like lighting, heater elements, small electric motors, and as *drivers* for larger heavy-duty relays and contactors.

Reed relays are also available in packages that look like a typical 14-pin IC, as well as packages that have bare wire leads for the contacts and the coil. A reed relay might have a coil with a high enough resistance to allow it to be driven directly from a microcontroller or logic IC, but a reed relay usually won't carry as much current through its contacts (the reeds) as an armature type of relay.

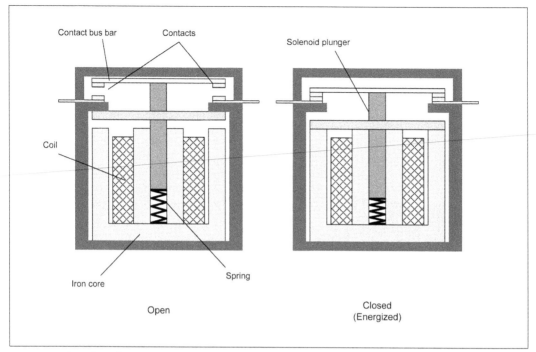

FIGURE 10-4. Operation of a heavy-duty contactor

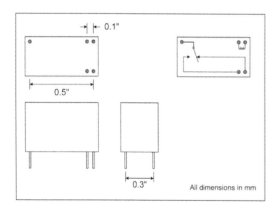

FIGURE 10-5. A PCB-mounted low-voltage relay

Both miniature armature and reed relays are available in through-hole and surface-mount packages. The surface-mount relay shown in Figure 10-6 is an example (all dimensions are in millimeters).

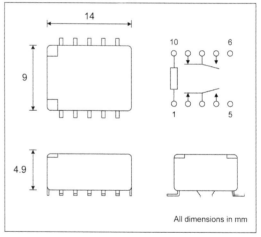

FIGURE 10-6. Surface-mount, low-voltage relay

This happens to be the package drawing for a Panasonic TQ series relay, but similar parts are available from Omron, NEC, and other manufacturers. The TQ2SA-5V has a 5V DC coil at 178 ohms with a nominal power consumption of 140 mW, and nominal contact ratings of 0.5 A at 125V AC.

LUG-TERMINAL RELAYS

Relays with terminals specifically designed to accept lug connectors, such as those used in automotive and industrial applications, are available in both vertical and horizontal mounting styles. Coil voltage ratings are available in both DC and AC values covering the range from 5V DC to 110V AC, or more. Figure 10-7 shows an example of an automotive lug-terminal relay with right-angle mounting brackets. The coil operates at 12V DC, and the SPDT contacts are rated for 12V DC at 40 A.

FIGURE 10-7. Right-angle-mounted lug-terminal armature relay

This type of relay uses the crimped lug connectors discussed in Chapter 7 and the crimping tool shown in Chapter 3. Although it is possible to solder to the terminals, this is generally not a good idea. Long ago, when the base of the relay assembly was made of Bakelite, it wasn't as big of a problem, because Bakelite can tolerate the heat of soldering. Modern relays are made using plastics with much lower melting temperatures, and it is possible to deform the base and cause the terminal to shift during soldering.

SOCKETED RELAYS

Many early relays used solder terminals for connections, so replacing one was an exercise in desoldering and resoldering the wiring. As this was a tedious and error-prone process, clever engineers devised a means of using sockets for relays, along the same lines as the sockets used for vacuum tubes. Some socketed relays use round sockets with eight contact positions, like the one shown in Figure 10-8, while others use a rectangular socket with the holes arranged in a grid pattern. Both types typically bring out the relay connections to screw terminals, and either spade or ring lugs are used to connect the wiring.

FIGURE 10-8. A typical octal (eight-pin) relay socket with screw terminals

These types of relays are mostly used in industrial applications that involve switching high voltages and large amounts of current. They are also available in an 11-pin form.

> **TIP** Although octal sockets might seem like a throwback to the days of vacuum tubes, they are still quite common and readily available. It is also possible to purchase just the octal plug and put your own electronics into it. This is useful for applications where you might want a sealed module (like, say, a sensor data collector for a remote environmental monitor) that can quickly and easily be replaced if necessary.

Selecting a Relay

Relays have two sets of primary specifications: coil and contacts. The coil will have a nominal operating voltage and resistance, although sometimes the manufacturer will give a power value instead of a resistance. If the current isn't specified, a quick application of Ohm's law will tell us how much current we can expect the relay coil to draw, and we can use the power specification to figure out the coil resistance (see Chapter 1).

The contacts should be rated to handle the load they will be controlling and then some. To be safe, it's a good idea to derate the contacts by 50%, meaning that if you want to control a 240V, 40 A contactor with a 24V AC coil that draws 36 mA with a *driver* relay, the smaller relay will need to have contacts that can handle 72 mA at 48V AC (twice the current and voltage actually required). A smaller relay is used in this case because the contactor uses AC for its coil, not DC. This makes it more challenging to control with just a solid-state driver circuit, although it is possible (and not uncommon). Using a small intermedite relay to handle the AC for the contactor's coil keeps things simple.

A small relay rated for 100 mA contact current should do fine for this application. It also implies that, for a situation like this, you could use a miniature PCB-mount relay, like the units shown in Figure 10-9 (this is a bank of four, with built-in drivers).

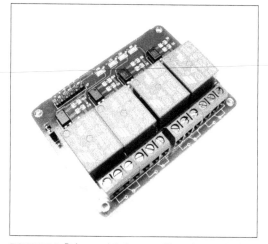

FIGURE 10-9. Relay module for use with a microcontroller

However, a small relay might require anywhere from 20 to 50 mA for its coil, which means it can't be controlled directly from a standard logic IC or microcontroller. In this case, either a driver IC or a transistor driver (as described in "Controlling Relays with Low-Voltage Logic" on page 243) will be needed. The board shown in Figure 10-9 doesn't have this problem, because the drivers are already on the PCB.

A cascade of relays is not an uncommon situation. Figure 10-10 shows how a sequence of relay driver, small relay, and contactor can be connected to control a high-current, high-power system. Conceptually, this can be extended as far as necessary, so that, in theory, a 5V logic signal could control hundreds of amperes of current.

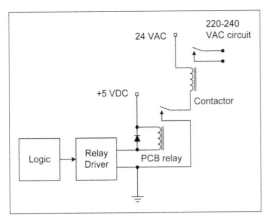

FIGURE 10-10. Using a small relay to control a larger relay

Relay Reliability Issues

Modern relays are fairly reliable, with claims for some small low-current types of over 10,000,000 mechanical cycles with no load. But relays are mechanical devices, and that implies wear and tear on the operating parts.

In a relay, a failure occurs when either the contacts can no longer pass current effectively or the relay's coil will no longer operate the contact mechanism. Two major sources of relay failure are *contact arcing* and *coil overheating*.

CONTACT ARCING

Of course, the more load (power) a relay carries, the more the contacts will wear due to arcing. At some point, a loaded relay might actually burn the contacts to the point where they no longer make good contact. In other words, the contacts start to become resistors, or even open circuits. This can be a major problem when a relay is controlling an inductive load, such as a motor or another relay.

You can reduce arcing using a resistor-capacitor (RC) *snubbing* circuit to help damp out the arc at the contacts. Figure 10-11 shows one way to do this. This technique works for both DC and AC circuits, and it can help extend the life of a relay's contacts considerably.

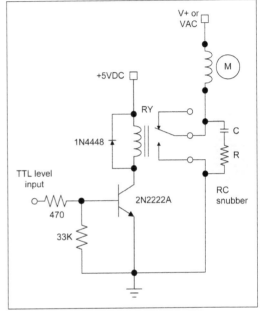

FIGURE 10-11. Relay contact snubber

> **TIP** Remember, if you need some assistance with the schematic symbols or electrical concepts presented in this section, refer to Appendix A or Appendix B.

The idea behind a snubber is that the combination of R and C will reduce the effect of spikes experienced by the relay contacts when power is removed from an inductive load, in this case a motor (M). When current flow through an inductive load stops, the magnetic field around the windings in the load will collapse, generating a large voltage spike. The RC circuit acts to stretch the spike so that by the time it reaches its maximum voltage, the relay contacts are far enough apart that an arc can't form between them. Appendix A includes the equations for calculating values for an RC circuit.

In some high-power contactors, the contacts are made from a silver alloy. Since there will always be some arcing in high-current/high-voltage applications, the contacts will start to oxidize. But with silver contacts, the result is silver oxide, which is itself a decent conductor.

COIL OVERHEATING

Coil overheating will cause the coil to deteriorate as the insulation becomes brittle and breaks down. This will eventually result in shorts in the windings, which will cause it to draw more current and get even hotter. It is not unheard of for a relay to burst into flames when the coil severely overheats, due to progressive shorts in the windings. The easiest way to avoid this situation is to ensure that the relay isn't being driven with more than its rated voltage. If possible, the circuit should be designed so that the default state of the relay is off, not on. Putting a fuse in series with the coil of a relay is another way to help prevent a catastrophic failure.

RELAY BOUNCE

Just like a mechanical switch, the contacts of a relay have a tendency to bounce when the relay closes. The resulting output from the relay looks something like Figure 10-12.

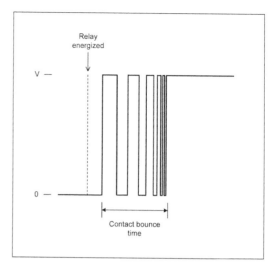

FIGURE 10-12. Relay contact bounce

If a relay is connected to another relay, or to something like a lamp, LED, or a motor, the contact bounce isn't a real big issue, except that it does prolong contact arcing and subsequent wear (each bounce is an arc, even if tiny). The arc suppression techniques mentioned earlier can help reduce the effects of contact bounce on the relay contacts.

If a relay is connected to a digital circuit of some type (as an input, for example), bounce can be a big problem. In a situation like this, the input will need to be *debounced*, either by logic hardware or by software. Relay bounce can also sometimes be heard if a relay is switching an audio signal. It's the "crunchy" blip or pop that occurs when the audio input is switched from one source to another using a set of relays.

Relay Applications

Relays are useful for routing signals, switching current, or as a form of logic for some applications. While is it possible to use a solid-state component to do switching and routing chores, the relay offers the advantage of low closed-circuit resistance, immunity to reverse current flow from inductive loads, and the ability to act as an isolated control transition between low- and high-voltage circuits.

CONTROLLING RELAYS WITH LOW-VOLTAGE LOGIC

A relay uses a coil to move the contacts, so it's an inductive load to whatever is driving that coil. An inductor will produce a current in the reverse direction when the energizing current ceases to flow and the magnetic field collapses, and the resulting voltage spike can be quite large. A relay coil can also draw a considerable amount of current, much more than most ICs can safely handle.

Some relays are made specifically for use with logic circuits, which means they will have a low coil-current (high resistance) and a built-in protection diode. However, the coil-current general-purpose relays with 5V DC coils can range anywhere from 20 to 80 mA, so this type of device should not be connected directly to something like a TTL logic chip or a microcontroller without some kind of interface circuit. The current through the coil will overwhelm the IC and probably damage it. Figure 10-13 shows one way to deal with this situation with an NPN transistor (a 2N2222A), a couple of resistors, and a diode.

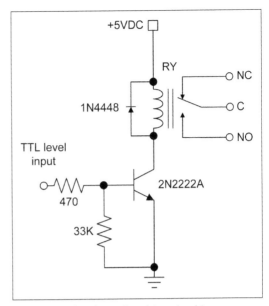

FIGURE 10-13. Simple one-transistor relay driver

This circuit is simple and effective, but relay driver ICs are also available that package from one to eight relay driver circuits into a single chip. The single-channel devices might be an option if you need to drive only one relay. Table 10-1 lists some of the available relay driver ICs.

TABLE 10-1. Relay driver ICs

Part number	Manufacturer	Internal logic	Drive current
CS1107	On Semiconductor	Single driver	350 mA
MAX4896	Maxim	8-channel driver	410 mA single, 200 mA all
SN75451B	Texas Instruments	Dual AND driver	300 mA

Part number	Manufacturer	Internal logic	Drive current
SN75452B	Texas Instruments	Dual NAND driver	300 mA
SN75453B	Texas Instruments	Dual OR driver	300 mA
SN75454B	Texas Instruments	Dual NOR driver	300 mA
TDE1747	STMicroelectronics	Single driver	1A
UDN2981A	Allegro	8-channel driver	500 mA max, 120 mA/channel

Note that the parts in Table 10-1 can be used to drive things other than a relay, such as lamps, valves, actuator solenoids, high-current LED displays, and so on. Also, many of the parts listed here will work with CMOS as well as TTL logic levels. Check the datasheets from the manufacturers for details.

SIGNAL SWITCHING

Figure 10-14 shows a device (an HP 3488A switch/controller) that is typically used to route signals between different types of measurement equipment and devices or circuits under test. The switching is done by banks of relays mounted on PCBs that plug into the rear of the unit. The 3488A can be programmed to perform switching actions at specific times, but it can also be controlled using a GPIB/IEEE-488 control interface to a PC or other control device providing the commands to route signals through the relay banks.

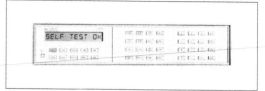

FIGURE 10-14. The HP 3488A switch/controller unit

Small relays can be used to switch standard video signals or select inputs for a measurement device. They were commonly used in the past to route telephone circuits, although that function has been largely replaced by solid-state components. For signal switching, reed relays are often used.

POWER SWITCHING

An example of using relays for power switching might be a situation where there is a need to control things like pumps, valves, and heaters for a marine specimen holding tank. Relays are a good choice for this, because a relay can operate a 120V AC pump as easily as a 12V DC pump. The main consideration would be the voltage and current ratings for the contacts.

Air conditioning units employ heavy-duty contactors to control various fans and compressors. The input to the contactors is often a low voltage like 24 VAC that is controlled by a thermostat. A conventional mechanical thermostat can be replaced with a microcontroller and a relay to create a custom programmable controller without the need to worry about creating a suitable circuit to

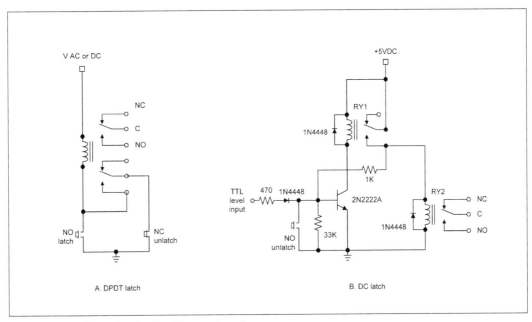

FIGURE 10-15. Latching relay circuits with manual reset

interface directly with the existing A/C control voltage.

RELAY LOGIC

As mentioned earlier, some of the first computer-like systems were based on relays. It was a natural choice, given that the telephone switching networks were starting to use relays for dynamic circuit routing. The rotary dial on old-style telephones wasn't an aesthetic design decision; it was like that because each pulse produced by the dial as it spun back to the start position drove a rotary relay at a switching office somewhere. The caller effectively modified the network wiring between her phone and whomever she was calling every time she dialed a number. Multiple banks of rotary relays allowed simultaneous calls to go through the system, and huge windowless buildings once held thousands or even tens of thousands of relays, all chattering away at the same time.

Creating a latching circuit is straightforward, as shown in Figure 10-15. In the simple design shown in circuit A, the relay is held closed by a second set of contacts. Once the relay is energized, the current flow through the contacts will keep it energized until the connection to ground is opened by the normally closed switch. Note that this circuit will work for both AC and DC, up to the maximum voltage that the switches and relay contacts can tolerate.

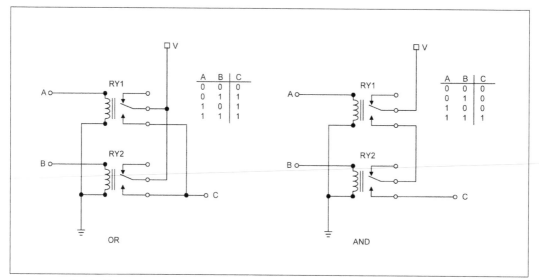

FIGURE 10-16. Simple OR and AND relay logic circuits

The circuit in part B of Figure 10-15 is intended for use with 5VDC only, and it's basically just a variation on the driver circuit shown in Figure 10-13. The main difference is that, when RY1 is energized, it will supply 5V back into the transistor via the 1,000 (1k) ohm resistor, thus keeping it in an on state. When the pushbutton switch is pressed, the base of the transistor will be grounded, current flow through RY1 will stop, and the relay will open, thus breaking the lock.

It's easy to construct OR and AND logic using relays (see Chapter 11 for more about logic devices), as shown in Figure 10-16. There are other ways to get the same results, using relays, but the idea here is to show how the standard truth tables for OR and AND can be satisfied with just a couple of SPDT relays. You can construct logic circuits for NOR, NAND, and XOR functions as well.

This isn't as esoteric as you might think, and relay logic appears in various guises in circuits found in industrial controls, home appliances, automobiles, and even in some avionics. Eventually, however, even these applications will become the domain of solid-state controls and switches. But for now, relays are still very much alive and well, and when you're dealing with the interfaces between systems operating at vastly different voltage and current levels, a relay may be the easiest and cheapest way to get the job done.

Summary

This chapter identified the three major types of relay mechanisms: armature, reed, and contactor solenoid. It also covered some of the available package types, including PCB, lug terminals, and sockets.

We also briefly examined some of the ways that relays can fail, and some of the techniques available to reduce the likelihood of failure. Lastly, we wrapped up with a look a relay latching and relay logic.

You should come away from this chapter with a sense of what types of relays are available and some of the ways they can be used.

While it is beyond the scope of this book to delve into the theory behind things like one-shot debounce timers and complex relay logic, Appendix C lists some excellent references if you want to explore the topic further.

CHAPTER 11

Logic

IN ELECTRONICS, THE TERM *logic* GENERALLY means a circuit of some type that implements a logical function. The logic circuit might employ some relays, as shown in Chapter 10. Or it might consist of multiple small- and medium-scale ICs containing logic components, such as gates or flip-flops. It can be as simple as a couple of transistors and some diodes, or as complex as a modern CPU with billions of circuit elements.

The study of digital logic can, and does, fill entire textbooks (some of which are listed in Appendix C), but the primary emphasis in this chapter will be on the actual physical components.

We'll start with a look at the building blocks of digital logic in the form of TTL and CMOS ICs. We'll also take a quick look at programmable logic, microprocessors, and microcontrollers. The chapter wraps up with an introduction to logic probes and other techniques for testing logic circuits, as well as some tips for working with digital logic components.

Refer to Chapter 9 for a description of the package types used with logic devices. Chapter 4 shows some techniques for soldering ICs to circuit boards, and Chapter 15 discusses printed circuit board layout considerations for logic devices.

There is no way that the various families of available logic devices can be adequately covered in just a single chapter. For example, a typical data book from a manufacturer of digital logic devices is a thick tome with anywhere from 400 to 800 pages. And that can be for just one logic family or class of devices.

There's no need to dread a future of wading through data books and online comparison tables, however. The key is to understand the basics of digital logic and the Boolean algebra on which it is based. Once you can define what you need the logic to do, or what it is already doing, then picking the appropriate type of logic device is largely a matter of practical considerations involving supply voltages, input and output voltage levels, and operating speed. Get the logic nailed down first; then worry about how to implement it with actual components.

Logic Basics

Digital logic is the application of formal logic theory, namely Boolean algebra, to electronic circuits. Boolean algebra was introduced in 1854 by George Boole in his book *An Investigation of the Laws of Thought*, and it defined a set of algebraic operations for the values 1 and 0, or true and false. In Boolean algebra, the primary operations are conjunction (AND), disjunction (OR), and negation.

The notation used for Boolean algebra uses ∧ for conjunction, v for disjunction, and ~ for negation. Thus we could write:

Z = (A ∧ B) v C

which states that the value of Z is equal to the result of A and B or C. This is the same as:

Z = (A AND B) OR C

The Boolean AND and OR operations always have at least two inputs and a single output. A digital circuit is the physical embodiment of a Boolean equation, with symbols for an AND gate, an OR gate, and negation, as shown in Figure 11-1.

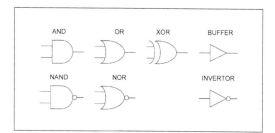

FIGURE 11-1. Common symbols for digital logic

The small "bubbles" used with digital logic symbols (as shown in Figure 11-1) indicate negation (or inversion, depending on how you want to think about it)—that is, if a logical true (1) encounters a bubble, it is negated and becomes a logical false (0), and vice versa. For example, Table 11-1 shows the truth table for an AND device (where A and B are the inputs).

TABLE 11-1. AND gate truth table

A	B	Output
0	0	0
0	1	0
1	0	0
1	1	1

The NAND (Not-AND) device, with a bubble on the output, produces the truth table shown in Table 11-2.

TABLE 11-2. NAND gate truth table

A	B	Output
0	0	1
0	1	1
1	0	1
1	1	0

The buffer simply passes its input to its output, and it's more of a circuit element than a logic operator. The invertor is a buffer with the output negated, such that a true input becomes a false output, and vice versa.

Don't be confused by the open circles often used to indicate terminal points in a circuit diagram. These do not indicate logical inversion. Only when the open circle is next to a digital component symbol does it mean that logical negation is applicable.

Just as the NAND gate is the negation of the AND gate, the NOR gate is the negation of the OR gate. The XOR (which stands for *exclusive OR*) is not a conventional Boolean operation, but it shows up in digital electronics quite often. Figure 11-3 shows each of the gate logic symbols from Figure 11-1 and its associated truth table. The buffer and inverter aren't shown, because their functions are easy to understand.

Recall the simple logic equation shown earlier:

Z = (A AND B) OR C

Now consider Figure 11-2, which shows what the example logic equation looks like as a circuit diagram.

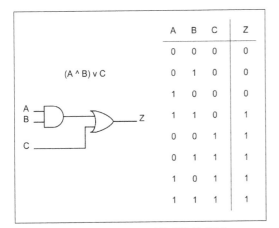

FIGURE 11-2. Circuit equivalent of (A AND B) OR C

The table to the right in Figure 11-2 is the truth table for this particular circuit. Truth tables show up quite often with digital logic designs.

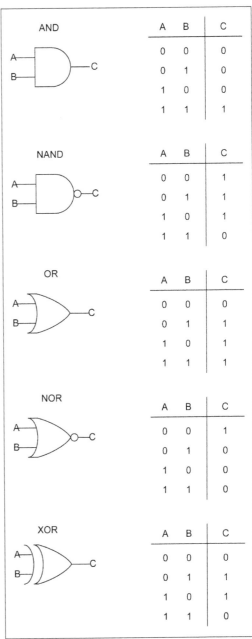

FIGURE 11-3. Truth tables for AND, NAND, OR, NOR, and XOR gates

Circuits that consist of the basic gates are called *combinatorial logic*, because they operate on combinations of inputs to produce a single output. Combinatorial logic has no state, except for that which exists instantaneously as a result of the input combination. When things like flip-flops are included in the mix, the result is referred to as *sequential logic*. A sequential logic circuit can have a "memory" of a previous state, and these types of circuits are found in things like state machines, counting circuits, pulse detectors, code comparators, and the CPU of a microprocessor or microcontroller.

An in-depth discussion of combinatorial and sequential logic circuits, while definitely an interesting topic, is beyond the scope of this chapter. Refer to Appendix C for sources of additional information.

Origin of Logic ICs

One of the drivers for the development of integrated circuits was digital logic. Early computers and logic systems utilized discrete components, initially with vacuum tubes and later with transistors. These were labor intensive to build, expensive to maintain, and expensive to operate, due to their power and cooling requirements. The development of the integrated circuit paved the way to smaller, faster, cheaper devices that used only a fraction of the power of the older designs.

An early form of logic gate, called *resistor-transistor logic* (RTL), was employed in the first commercial logic ICs that appeared in the early 1960s. The first computer CPU built from integrated circuits was the Apollo Guidance Computer (AGC), which incorporated RTL logic chips. These early RTL computers took human beings to the moon and set the stage for the electronics revolution that followed. Your smart phone, iPad, desktop PC, and the computer that runs your modern car can all trace their heritage back to something that had less computing power than a cheap MP3 player but was able to help guide a spacecraft from the earth to the moon, put a lander on the moon and then back into lunar orbit, and bring the crew safely back home again.

As time progressed, IC fabrication techniques continuously improved. This resulted in higher-density designs containing an increasingly larger number of components on the silicon chip. Early IC designs were called *small-scale integration* (SSI) and contained a small number of transistor components numbering in the tens of unique elements. The next step was *medium-scale integration* (MSI), where each IC might have a hundred or more unique circuit elements. *Large-scale integration* (LSI) is the term applied to IC devices with thousands of component elements on the silicon chip, and *very large-scale integration* (VLSI) is applied to devices with up to a million unique transistors and other circuit elements.

By way of comparison, early microprocessors had as few as 4,000 transistor elements. A modern CPU might have over 1 billion (1,000,000,000) unique transistor elements in its design.

Logic Families

In the world of digital electronics, types of devices are catagorized in terms of so-called *families*, referring to the underlying design and fabrication process technologies that are employed to build the logic devices.

The original logic families (which included RTL, DTL, and ECL, described momentarily) were derived from the logic circuits used in early computers. These early circuits were originally implemented using discrete components.

Today, there are four common families of IC logic devices in widespread use:

ECL
  Emitter-coupled logic

TTL
  Transistor-transistor logic

NMOS
  N-type metal oxide semiconductor

CMOS
  Complementary metal oxide semiconductor

For the most part, the two families you will encounter regularly are TTL and CMOS. ECL is still used for some specialized high-speed applications, but it is starting to become much less common. NMOS can still be found in some types of VLSI ICs, primarily CPUs and memory devices.

TTL is based on conventional BJT transistor technology, whereas CMOS employs FET metal-oxide-type transistors. Because the logic thresholds in CMOS devices are approximately proportional to power supply voltage, they can tolerate much wider voltage ranges than TTL. The bipolar devices used in TTL, on the other hand, have fixed logic thresholds.

CMOS devices can be implemented in silicon with very small dimensions, which has resulted in a rapid shrinking of CMOS chip size and a corresponding increase in circuit density per unit area. The reduced geometries, along with very small inherent capacitance in the on-chip wiring, has resulted in a dramatic increase in the performance of CMOS devices.

Logic Building Blocks: 4000 and 7400 ICs

The industry convention for naming monolithic IC logic devices is to use 4000 series numbers for CMOS parts and 7400 series numbers for TTL parts. Other logic devices have their own numbering schemes, some of which are standardized and others that are assigned by the chip manufacturer.

The members of the 7400 family of TTL monolithic IC devices and subsequent generations of parts in the 74Lxx, 74LSxx, 74ACTxx, 74HCxx, 74HCTxx, and 74ACTxx series are by far the most common basic units of digital logic. The older 4000B series CMOS devices are also still available and are useful in certain applications.

CLOSING THE TTL AND CMOS GAP

Initially, TTL and CMOS devices had very different voltage and speed characteristics and couldn't be used in the same design without some type of *level shifting* on the connections between devices. The reason for this is that TTL logic levels don't rise high enough to be recognized as a logical 1 by a CMOS device. Table 11-3 shows the difference between the logic levels of traditional TTL and CMOS logic devices.

TABLE 11-3. CMOS and TTL logic levels[a]

Technology	Logic 0	Logic 1
CMOS	0V to 1/3 VDD	2/3 VDD to VDD
TTL	0V to 0.8 V	2V to VCC

[a] VDD = supply voltage, VCC = 5 V ±10%

The compatibility issue was addressed with the introduction of the HC family of TTL-compatible devices. These parts have numbers with HC between the 74 and the TTL part number, and they are pin- and function-compatible with the original 7400 series devices. 74HC devices can be used with both 3.3V and 5V supplies.

The 74HCT family of devices was introduced to deal with the input-level incompatibility in 5V circuits. Internally, the logic is implemented as CMOS with TTL-compatible inputs. 74HCT devices work only with a 5V supply.

Table 11-4 details the CMOS and TTL families, by year of introduction.

TABLE 11-4. CMOS and TTL families by year of introduction

Family	Type	Supply V (typ)	V range	Year	Remarks
TTL		5	4.75–5.25	1964	Original
TTL	L	5	4.75–5.25	1964	Low power
TTL	H	5	4.75–5.25	1964	High speed
TTL	S	5	4.75–5.25	1969	Schottky high speed
CMOS	4000B	10V	3–18	1970	Buffered CMOS
CMOS	74C	5V	3–18	1970	Pin-compatible with TTL part
TTL	LS	5	4.75–5.25	1976	Low-power Schottky high speed
TTL	ALS	5	4.5–5.5	1976	Advanced low-power Schottky
TTL	F	5	4.75–5.25	1979	Fast
TTL	AS	5	4.5–5.5	1980	Advanced Schottky
CMOS	AC/ACT	3.3 or 5	2–6 or 4.5–5.5	1985	ACT has TTL-compatible levels
CMOS	HC/HCT	5	2–6 or 4.5–5.5	1982	HCT has TTL-compatible levels
TTL	G		1.65–3.6	2004	GHz capable logic

4000 SERIES CMOS LOGIC DEVICES

When you are working with CMOS, either as part of the new design or when integrating to an existing circuit, it helps to have a selection of parts on hand. Table 11-5 shows a list of general-purpose CMOS logic devices that are handy to have around.

> **TIP** While it's not absolutely necessary to keep a stock of parts on hand, it can save you time and money down the road when you really need something but can't find it locally or don't have the time to go run it down. Purchasing parts in bulk can also save a lot of money, so if you have some other people who might want to go in on an order, you can all realize the volume savings. Be sure to check out the resources listed in Appendix D for companies that sell overstock and surplus components.

TABLE 11-5. Basic list of 4000 Series CMOS devices

Part #	Description
4000	Dual three-input NOR gate and inverter
4001	Quad two-input NOR gate
4002	Dual four-input NOR gate OR gate
4008	Four-bit full adder
4010	Hex noninverting buffer
4011	Quad two-input NAND gate
4012	Dual four-input NAND gate
4013	Dual D-type flip-flop
4014	Eight-stage shift register
4015	Dual four-stage shift register
4016	Quad bilateral switch
4017	Decade counter/Johnson counter
4018	Presettable divide-by-N counter
4027	Dual J-K master-slave flip-flop
4049	Hex inverter
4050	Hex buffer/converter (noninverting)
4070	Quad XOR gate
4071	Quad two-input OR gate
4072	Dual four-input OR gate
4073	Triple three-input AND gate
4075	Triple three-input OR gate
4076	Quad D-type register with tristate outputs
4077	Quad two-input XNOR gate
4078	Eight-input NOR gate
4081	Quad two-input AND gate
4082	Dual four-input AND gate

> CMOS parts are very static sensitive, so always take the appropriate precautions. Working on a antistatic workbench pad or grounded workbench, wearing an antistatic wrist strap, and removing a part from its antistatic packaging only when it's absolultely necessary can help you avoid discovering that a small bolt of high-voltage static has punched a hole through the metal-oxide junction of a transistor inside a CMOS part and rendered it useless.

7400 SERIES TTL LOGIC DEVICES

If you plan on working with TTL-type logic on a regular basis, you might want to consider having a supply of 7400 series devices on hand. Table 11-6 lists general-purpose devices that cover the essential functions we've already discussed, as well as a few that can perform special functions, such as latching and buffering.

TABLE 11-6. Basic list of 7400 Series TTL and TTL-compatible devices

Part #	Description
7400	Quad two-input NAND gates
7402	Quad two-input NOR gates
7404	Hex inverters
7408	Quad two-input AND gates
7410	Triple three-input NAND gates
7411	Triple three-input AND gates
7420	Dual four-input NAND gates
7421	Dual four-input AND gates
7427	Triple three-input NOR gates
7430	Eight-input NAND gate
7432	Quad two-input OR gates
7442	BCD-to-decimal decoder (or three-line to eight-line decoder with enable)
7474A	Dual edge-triggered D flip-flop
7485	4-bit binary magnitude comparator
7486	Quad two-input exclusive-OR (XOR) gates
74109A	Dual edge-triggered J-K flip-flop
74125A	Quad bus-buffer gates with three-state outputs
74139	Dual two-line to four-line decoders/demultiplexers
74153	Dual four-line to one-line data selectors/multiplexers
74157	Quad two-line to one-line data selectors/multiplexers
74158	Quad two-line to one-line MUX with inverted outputs
74161A	Synchronous 4-bit binary counter
74164	8-bit serial to parallel shift register
74166	8-bit parallel to serial shift register
74174	Hex edge-triggered D flip-flops
74175	Quad edge-triggered D flip-flops
74240	Octal inverting three-state driver
74244	Octal noninverting three-state driver
74273	Octal edge-triggered D flip-flops
74374	Octal three-state edge-triggered D flip-flops

Most of the parts listed in Table 11-6 should be available as LS, ACT, and HCT types.

CMOS AND TTL APPLICATIONS

The disparity between CMOS and TTL has become less of an issue with each passing year. Some TTL devices are now basically CMOS logic with a 74xx part number. Microprocessors and microcontrollers are now built using low-voltage CMOS techniques and some types need low-voltage components to interface to them. The 74ACxx,

74ACTxx, 74HCxx, and 74HCTxx series of logic devices are essentially CMOS devices capable of operating at low voltages while still providing traditional TTL functions, and in the case of the ACT and HCT series, the ability to interface to conventional TTL devices operating at 5V.

So why would anyone want to buy a 4000 series logic device? A 4000 series gate has several unique features, such as high input impedance, low power consumption, and the ability to operate over a wide temperature range. For some applications, a 4000 series CMOS device can serve as an input buffer as well as a logic gate. For example, a level sensor for a water tank can be constructed with just a couple of 4000 chips, without any op amps. For more ideas of things to do with 4000 series logic, check out *The CMOS Cookbook* by Don Lancaster (listed in Appendix C).

What logic family to use largely comes down to what you want to interface it to. If you want to work with something that uses conventional TTL logic levels, the choice has largely been made for you. If you want to integrate a low-voltage microcontroller into a design, you might want to use CMOS parts for some of the "glue" logic in the circuit. There is no easy "if this, then do that" answer, unfortunately, and in the end it comes down to a series of design decisions based on operational parameters, project budgets, and interface requirements. And reading datasheets. Lots of datasheets.

Programmable Logic Devices

Programmable logic devices (PLDs) are ICs that contain unassigned logic elements and some means to configure the connections between them. Early PLDs used a form of fuseable link to determine how the internal logic elements would be arranged. The downside to this approach is that once a PLD was "programmed," it would forever be that way. If the programming was wrong, or if there was a glitch during the process, the only recourse was to throw the part away and start over. There was no going back and trying again.

You can still purchase one-time programmable (OTP) PLDs today, and they are often used in production systems where there is a concern that someone might *reverse-engineer* a PLD and extract its programming patterns. Devices that can be cleared and reprogrammed utilize flash memory, UV EPROM (erasable programmable read-only memory) components, or some other technique to hold the programming data for the device.

There are four main types of PLDs in use today, as shown in Table 11-7. The PAL and GAL devices tend to be small and contain a limited number of logic elements. The CPLD-type parts are more complex, each being roughly equivalent to several GAL-type devices in a single package. The FPGA-type devices can be extremely complex and have the internal logic necessary to implement sequential logic designs, such as microprocessors, memory managers, and complex state machines.

The fundamental logic elements of a PLD, such as a PAL or GAL, are relatively simple.

TABLE 11-7. PLD device types

Type	Definition	Remarks
PAL	Programmable array logic	OTP using fuseable links. Typically small in size with a small number of logic elements.
GAL	Generic array logic	Like a PAL but reprogrammable. Uses electrically erasable programming data storage (similar to an EEPROM, or electrically erasable programmable read-only memory).
CPLD	Complex programmable logic device	Equivalant of multiple GALs in one package. May contain thousands of logic elements. Typically programmed by loading the interconnect patterns into the device.
FPGA	Field-programmable gate array	Contains a large number (millions) of logic elements in an array or grid with programmable interconnections. Supports sequential as well as combinatorial logic.

PAL devices have logic elements arranged as an array of "fixed-OR, programmable-AND" functional blocks. Each block implements "sum-of-products" binary logic equations. To get a basic idea of how a PLD works, consider the circuit shown in Figure 11-4, which is just a part of the internal logic found in something like a PAL device.

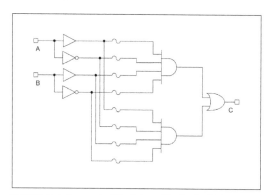

FIGURE 11-4. One part of the internal logic of a PAL or GAL device

Now, assume that we wanted to implement some logic, perhaps something like this:

$C = (A \wedge \sim B) \vee (\sim A \wedge \sim B)$

We can program this into the device by removing some of the fuseable links, which in turn allows only certain inputs to the AND logic elements, as shown in Figure 11-5.

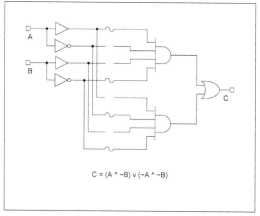

$C = (A \wedge \sim B) \vee (\sim A \wedge \sim B)$

FIGURE 11-5. PAL logic configured to implement a logic function

A PAL or GAL device can help to significantly reduce the parts count for a complex logic design. For example, one way to implement the equation shown previously with conventional logic gates would require two AND gates, two inventors, and an OR gate. That would be at least three conventional logic chips, with some unused gates left over. The PAL does it in one logic unit, with other logic units on the chip available for other functions.

There is, of course, much more to PLDs than what has been presented here. Some device manufacturers provide free programming tools for their parts. These support the use of hardware definition languages such as VHDL, Verilog, and Abel. There are also websites that provide free *IP cores* or predefined FPGA logic in VHDL or Verilog, for things like microprocessors and I/O controllers (and more). The ARM microcontrollers that are found in smartphones, tablets, embedded controllers, and digital cameras are sold by ARM not as silicon parts, but rather as IP cores that can be implemented in silicon by the customer.

If you are interested in exploring this end of digital electronics, I would suggest getting a book like Kleitz's *Digital Electronics: A Practical Approach* or Katz's *Comtemporary Logic Design* (both of which are listed in Appendix C). The websites of manufacturers such as Altera, Atmel, Lattice, Texas Instruments, and Xilinx offer lots of free information about programmable logic devices in general and their products in particular.

Microprocessors and Microcontrollers

A modern microprocessor or microcontroller is the result of years of refinement in IC fabrication processes. Internally, most modern microprocessors and microcontrollers use CMOS fabrication technology, for the reasons of circuit density and speed mentioned.

The term *microprocessor* usually refers to a device that comprises the central processing unit (CPU) of a computer but which relies on external components for memory and input/output (I/O) functions. A *microcontroller* is a device that has an internal CPU, memory, and I/O circuits all on one chip. Nothing else is needed for the microcontroller to be useful.

To help clarify, consider this: an Intel Pentium-4 is a microprocessor, and an Atmel AVR ATmega168 (as found on an Arduino board) is a microcontroller. The Intel part is a CPU, with memory management, internal instruction cache, and other features, but it needs external memory and I/O support (also known as the *support chipset*). The AVR chip, on the other hand, has an 8-bit CPU, 16K of internal memory, and a suite of integrated configurable I/O functions.

Microcontrollers are much easier to work with than microprocessors because, for many applications, all that is needed is the chip itself. In fact, most popular small, single-board computers are really nothing more than a microcontroller with some voltage regulation and USB interface logic.

Unless you have a specific need to create a custom design around a specific microprocessor or microcontroller, it makes more sense to buy a prebuilt module. In terms of low-cost boards, there are the Arduino units based on the AVR family of microcontrollers and numerous boards that use some form of the 32-bit Cortex-M3 ARM processor design. Figure 11-6 shows two different boards that use the STM32 processor from STMicroelectronics. Low-cost microcontroller boards are also available for microcontrollers such as the Texas Instruments MSP430 and Microchip's well-known PIC series.

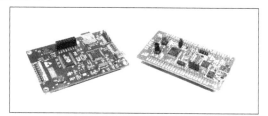

FIGURE 11-6. Two different STM32 ARM Cortex-M3 single-board computers

If you really need something with PC capabilities, many low-cost motherboards are available that use microprocessors from Intel, AMD, and Via. You can also find single-board PCs that are designed to connect to a common backplane (a so-called *passive backplane* that usually has just a set of edge connectors for PCBs to plug into), but these are generally intended for industrial applications and can be rather pricey.

PROGRAMMING A MICROCONTROLLER

There are a couple of ways to get a microcontroller to do what you want it to do, and both involve programming. The first is by creating a sequence of instruction codes that the microcontroller can interpret and execute directly. Normally, you do so using a tool called an *assembler*, and the technique is called *programming* in assembly language. In the early days of microprocessors and microcontrollers, this was the primary way to program them, because languages such as C didn't exist when these devices first appeared in the early 1970s.

Assembly language consists of a sequence of human-readable operation codes, or op codes, that an assembler converts into the binary values that the microcontroller will recognize and act upon. Op codes can have one or more associated values (called *operands*) for things such as a literal data value to load or an address in the program to jump to and resume execution.

Programs written in assembly language tend to be small, fast, memory-efficient, and difficult to read or modify. For example, a snippet of assembly language to read an incoming character from an input port might look something like Example 11-1.

Example 11-1. Example assembly language code snippet

```
INCHR:  LDA A   INPORT    ; get port status
        ASR A             ; shift status bit into carry
        BCC     INCHR     ; no input available
        LDA A   INPORT+1  ; read byte from input port
        AND A   #$7F      ; mask out 8th bit (parity)
        JMP     OUTCH     ; echo character now in A
```

Now imagine this expanded to hundreds, thousands, or even tens of thousands of lines. Assembly language can be difficult to write and difficult to read, and it doesn't readily lend itself well to modular programming techniques (although it can be done with some discipline). In other words, assembly language programming can be hard to do well, so it's no surprise then that this type of programming is now relatively rare. While in some cases it still makes sense to write low-level programs in assembly language, the advent of the C programming language provided for the second primary way to program a microcontroller.

C is an interesting language. It has been called "assembly language in disguise" by some, and one of its creators, Dennis Ritchie, once made the statement that "[C has] the power of assembly language and the convenience of...assembly language."*

The C language also has the advantage of portability. Assembly language programs will work with only one type of processor, but a C program can often be recompiled to work on many different types of processors. Most modern operating systems are written in C (or another portable language) with only small machine-specific parts written in assembly language for a particular microprocessor.

A good modern C compiler will generate code that, while perhaps not as "tight" and memory-efficient as an equivalent assembly language program written by a skilled programmer, is still respectable. The ability of a compiler to create efficient and compact code depends to a large degree on the microcontroller that will run the resulting program. Some microcontrollers, such as the original 8051 family, can be difficult to use with C because of the limited amount of internal RAM (256 bytes). Others, such as the AVR devices found in Arduino products, are easier to work with and include instructions that allow a C compiler to generate fairly efficient code.

TYPES OF MICROCONTROLLERS

These days, most microcontrollers come in 8-, 16-, or 32-bit types. Some, such as the Atmel AT89 series and the Cypress CY8C3xxxx family, are based on the venerable 8051 design created by Intel in the early 1980s. Others incorporate an ARM or MIPS 32-bit *core* in their design. Still others, such as the Atmel AVR series and Microchip's PIC processors are unique, and are available in 8-, 16-, and 32-bit variations. Table 11-8, Table 11-9, and Table 11-10 list some common examples of each type.

TABLE 11-8. Representative 8-bit microcontrollers

Name	Source	Comment
AT89	Atmel	8051 compatible
AVR	Atmel	Unique
CY83xxxx	Cypress	8051 compatible
68HC08	Freescale	Descended from 6800
68HC11	Freescale	Descended from 6800

---

* From Wikiquote (*http://bit.ly/dr-quote*), quoted in Cade Metz, "Dennis Ritchie: The Shoulders Steve Jobs Stood On," *Wired*, 13 October 2011.

Name	Source	Comment
PIC16	Microchip	Unique
PIC18	Microchip	Unique
LPC700	NXP	8051 compatible
LPC900	NXP	8051 compatible
eZ80	Zilog	Descended from Z80

TABLE 11-9. Representative 16-bit microcontrollers

Name	Source	Comment
PIC24	Microchip	Unique
MSP430	Texas Instruments	Unique

TABLE 11-10. Representative 32-bit microcontrollers

Name	Source	Comment
AT915AM	Atmel	ARM IP core
AVR32	Atmel	32-bit AVR
CY8C5xxxx	Cypress	ARM IP core
PIC32MX	Microchip	MIPS IP core
LPC1800	NXP	ARM IP core
STM32	STMicroelectronics	ARM IP core

SELECTING A MICROCONTROLLER

The type of microcontroller that is most suitable for a particular project depends primarily on processing speed, on-board memory, and the type of I/O functions required. To a lesser extent, the selection decision might also be influenced by ease of programming and the availability of free or open source development tools.

As a general rule of thumb, if all something needs to do is control a motor or two, or perhaps just collect data and pass it along, an 8-bit microcontroller might be more than sufficient. A device like the AVR ATmega168, found on Arduino boards, runs at around 16 MHz and has I/O functions such as analog input and PWM output. The software tools to program the device are open source and freely available.

If, on the other hand, you need to control a color LCD display while capturing image data with a digital camera, a 32-bit microcontroller running at 100 MHz or more would be more appropriate. It could also be more expensive.

Although many of the 32-bit microcontrollers can execute instructions fast enough to run Linux (the Raspberry Pi, for example), speed isn't everything, and it's important to bear in mind that many things happen slowly in the real world. If there's no need for speed, then don't pay for it. Meeting the I/O requirements for a project is probably more important than worrying about how many instructions per second a microcontroller can execute.

The availability of programming tools is another key consideration. Before settling on a particular microcontroller, check to see what kind of development tools are available. If you're planning to use something like a Raspberry Pi, Arduino, BeagleBone, or MSP430 Launchpad, see what the board supplier recommends. Also remember that a clone of one of these boards can usually use the same tools and techniques as an "official" board.

For microcontrollers like the AVR familiy, the MSP430, and the ARM-based chips, you can use Linux and the GCC *toolchain* to compile C or C++ code into binary code these processors can load and execute. There are also other open source compilers, linkers, and programming tools available for 8051-based devices. Microchip offers a free version of its development tools for the PIC microcontrollers.

Finally, unless you are familiar with JTAG and the interfaces used to program a microcontroller using that method, you might want to steer clear of some of the low-cost 32-bit ARM boards from Asia found on eBay and other places. There's nothing wrong with these boards, but they typically show up in a bag or box with no CD and no documentation. It's up to you to hunt down the details and fill in the blanks, and if you're new to all of this, that can be a daunting task.

Working with Logic Components

In many ways, working with logic circuits is easier than with analog systems, if for no other reason than that the digital logic is, for the most part, electrically simpler. That's not to say that it's carefree, however, as there are some caveats and cautions that are unique to the world of digital electronics.

PROBING AND MEASURING

Checking a logic circuit is relatively simple, because the various signals will be in only one of two states: on or off. Figure 11-7 shows a tool made specifically for this purpose.

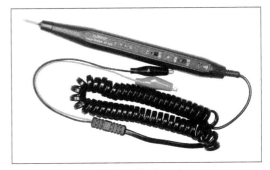

FIGURE 11-7. Hand-held digital logic probe

Using this tool is straightforward. You first connect it to the DC supply for the circuit using the red and black alligator clips. When you place the probe tip on the pin or lead of one of the circuit components, the lights on the probe will indicate if the signal state is high, low, or pulsing. Switches on the probe allow you to select for TTL or CMOS logic levels and also capture and display the last logic state, or specifically detect a signal that consists of pulses.

When using a DMM to measure signal voltages in a logic circuit, bear in mind that a typical DMM generally won't show anything reliably except a static (or slowly changing) voltage. If you attempt to measure a signal made up of fast pulses, you might see some small amount of DC voltage, and the AC scale on the meter might show that something is there, but neither measurement mode will be accurate.

To accurately measure a digital signal, you need an oscilloscope, and to observe the behavior of multiple digital signals simultaneously, you really need to use a logic analyzer. Chapter 17 discusses both types of test instruments.

TIPS, HINTS, AND CAUTIONS

The following are some general tips, hints, and cautions to keep in mind when buying logic components and building digital circuits.

Selecting Logic Devices

- Avoid selecting oddball or discontinued parts, unless you really don't care about building any more similar gadgets in the future. Fascinating logic devices have come and gone over the years, but once that super-cool combination gate/adder/latch thing you've found goes out of production, or the stock at the surplus vendor runs out, you won't be able to buy any more of them.

- Purchase only what you are comfortable working with, both in terms of logic functions and physical package types. You can always go back and revise your design later on.

- When building a circuit that uses devices from different logic families, always check to make sure that they are electrically compatible before actually acquiring them.

Physical Mounting and Handling

- When using through-hole parts, consider using a socket unless there are space and cost constraints. A socket makes it easy to change out a part if needed, and sockets are particularly useful with EPROM or EEPROM memory devices.

- Use good electrostatic discharge (ESD) prevention techniques. Work on a grounded mat, and use a grounded wrist strap.

- Never install or remove an IC while power is applied to the circuit.

Electrical Considerations

- When connecting multiple logic devices to a single device, take into account the current sink and current source ratings of the parts.

- Make sure that parts are interface compatible. Older TTL parts cannot be directly connected to older CMOS parts.

- Avoid connecting a logic IC input directly to Vcc (V+). Making the connection through a low-value resistor (between 220 and 480 ohms) is safer.

- Never directly connect the output of a logic IC to either ground or Vcc.

- Ground the inputs of unused gates and flip-flops. If left to float, the internal logic can change states or even go into oscillation, and this can induce spikes on the power supply lines.

- Use decoupling capacitors at each logic IC. A 0.01 µF part is typical. The decoupling cap should be connected between the Vcc and ground pins on the device, and it should be as close to the device as possible.

ELECTROSTATIC DISCHARGE CONTROL

Static charges are the mortal enemy of solid-state components. Devices based on CMOS technology are particularly susceptible, but any solid-state device can be damaged under the right conditions. ESD control includes safe practices for component storage and handling.

The first step is to obtain and wear a grounded wrist strap when working with static-sensitive parts. You can pick up a decent production-line-grade wrist strap from most electronics suppliers, or you can order one (or several) online. Even one of the cheap things sometimes found at computer supply outlets will work in a pinch, but don't expect it to last very long. Read and follow the instructions for connecting the strap to ground. Some straps have a built-in resistor to limit current, but some don't. Some have an alligator clip to connect to a metal ground, while others use a banana-type plug.

The second item is a grounded mat for the workbench. These mats are made of a conductive, high-resistance material that is intended to dissipate stray static charges. Like a wrist strap, a mat will have a lead that must be connected to a solid earth ground.

Lastly, if you plan to stock logic ICs, you really should have a supply of high-density anti-static foam sheets on hand. These are made of a black or dark gray material, usually about 1/4-inch thick. The idea is to push the leads of an IC into the foam. The foam contains carbon or some other high-resistance conductive material and prevents a potential difference between the pins of an IC. For this reason, you should pick up a piece of foam containing ICs with your wrist strap securely in place before removing any of the parts. This discharges the foam and any parts on it. Plucking a part from the foam without first making sure that there is no overall charge on it can put an IC in a position where some leads are grounded by your fingers while some are still in contact with the foam. The part is in grave peril at that moment, and it could get damaged by a static discharge through it from the foam to you.

Summary

A digital circuit is essentially the physical implementation of Boolean logic, along with some finite state machine theory and other concepts. It is also one of the fundamental technologies of the modern world, and without digital circuits, we wouldn't have PCs, the Internet, engine controllers for our cars, or programmable thermostats for our homes.

While it's not necesary to use solid-state components to build a digital circuit (see Chapter 10 for a couple of examples of relay logic), modern logic ICs are compact, consume very little power, and best of all, are cheap. In this chapter, we've looked at the two main families of digital logic ICs: TTL and CMOS. We've also examined some of the more recent hybrid devices that straddle the divide between TTL and CMOS by incorporating the ability to connect to either type.

CHAPTER 12

Discrete Control Interfaces

THIS CHAPTER COVERS THE BASICS OF USING a discrete signal on a single logic I/O port to sense and control things in the physical world. It also touches on topics such as buffers, logic-level translation, and current sink and sourcing considerations.

A *discrete* interface involves a single signal, typically binary in nature. This is probably the most common, and useful, type of interface encountered in digital electronics. It is also the simplest. It is either true or false, on or off. The microwave oven is powered on, or it isn't. The key is in the ignition, or it isn't. The infrared motion sensor is either active, or it isn't. And so on, and so forth. The opposite of discrete is *analog*, the realm of indefinite variable values. Chapter 13 covers analog interface concepts and components.

The term *discrete* comes from the realm of programmable logic controllers (PLCs) used in industrial control systems, and it has an advantage over a more general term such as *digital* in that it specifically implies a single signal or circuit intended for use as an interface to some external device. The term *digital* could mean anything from a single circuit carrying one bit of information between ICs to the multiple signals found in a parallel digital bus. Of course, the term *digital* can also mean an interface that responds to or generates binary signals for use with external devices, but I've elected to use the term *discrete* to make the distinction clear.

So, who is this chapter for? It's for anyone who wants to connect one thing to something else electronically with a simple on/off, true/false type of interface. It's for someone who wants to be able to sense when a door or window is open, or to be able to sense a true/false condition and enable or disable something in response. This chapter is also for those who want to extend, improve, enhance, or alter the behavior of an existing device or circuit. Not knowing exactly how the circuits at the other end work makes the task more, well, challenging, but not impossible.

The following terms are used extensively in this chapter (You can also find them in the Glossary, but we present them here for convenience):

DIO

In electronics and embedded computers, DIO typically refers to digital or discrete input/output.

Digital

Being of a numeric nature (i.e., comprising discrete numeric values, as opposed to continuously variable analog values). May refer to a measurement or signal that has only two possible values: 1 or 0, on or off. In electronics, digital devices are those components designed specifically to work with binary values.

Discrete

Something with two or more specific values, not a continuous range of values (i.e., analog). A term commonly used with programmable logic controller devices but can refer to any binary input or output signal.

Channel

A communications circuit (either wired or wireless) with specific endpoints. Can comprise a single signal or a group of signals.

Pin

A terminal point. May refer to an actual pin on an IC or one terminal position in a connector.

Port

Usually refers to a group of digital or discrete signals but may also refer to a single channel within a group.

The Discrete Interface

A discrete interface can be just a single connection, as in a single terminal on a PCB, or a single wire. It might also be a collection of terminals of one type or another, such as those found in the header strips on a small single-board computer like an Arduino or BeagleBone. What makes it discrete is that each terminal is individually controllable, rather than operating as a group, such as with a bus or the parallel port for a printer.

Some microcontrollers have terminals labeled as DIO, which, as mentioned earlier, usually means digital I/O. Multiple companies sell interface modules with multiple DIO lines, or channels. Sometimes these are arranged into groups of 8, 16, or 32 bits, which can be controlled individually or used in parallel.

For example, consider the small PCB shown in Figure 12-1. The main star here is a CY7C68013A microcontroller; the remaining parts on the PCB are there to provide the clock signal for the processor, regulate the supply voltage, and support a USB interface.

FIGURE 12-1. Small PCB with CY7C68013A microcontroller

The CY7C68013A is popular as a low-cost logic analyzer (see Chapter 17 for more on logic analyzers), and it is also used in numerous embedded devices. A board like this one can be purchased for around $10 on eBay.

Figure 12-2 shows a block diagram of the CY7C68013A IC, which is actually rather simple.

This particular version of the device has three DIO ports, labeled A, B, and D (other ports are not used in this version of the part, which is the 56-pin version). You can configure each pin on each port as an input, an output, or as an alternate function.

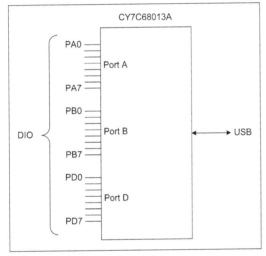

FIGURE 12-2. Block diagram of the CY7C68013A IC

The main thing we are concerned with here are the DIO pins on the device. As shown in Table 12-1, each DIO will accept an input voltage of up to 5.25V DC and output a minimum voltage of 2.4V DC. The CY7C68013A operates on 3.3V DC (typical). Outputs are rated at 4 mA of current, maximum.

TABLE 12-1. DIO pin voltage parameters for the CY7C68013A microcontroller

Parameter	Description	Min	Max
V_{IH}	DIO input HIGH	2	5.25V
V_{IL}	DIO input LOW	−0.5	0.8
V_{OH}	DIO output HIGH	2.4	–
V_{OL}	DIO output LOW	–	0.4

With this information, we can draw a couple of immediate conclusions:

- The DIO output functions of the CY7C68013A will most likely need low-voltage outboard components, or some type of voltage-level translation will need to be employed (see "Logic-Level Translation" on page 280).

- The DIO pins won't supply a lot of current, so directly driving something like a relay or an LED is not an option. A buffer or driver circuit of some type will be necessary (see "Buffering Discrete Outputs" on page 278 and Chapter 10 for more on relay interfaces).

With this example as our baseline, we can move on to examine some specific functions, and look at various ways to deal with discrete interfaces.

DISCRETE INTERFACE APPLICATIONS

In consumer electronics, a discrete interface might be used to sense when something is open or closed (like the tray on a DVD player), or it might be used to control an LED or operate a solenoid. If you peer inside a toy like a Robosapien (*http://www.wowwee.com*), you'll notice that the various actuators are discretely controlled. The little robot moves its arms and legs only at a fixed rate, so they are either in motion or they are not. They are discrete actions.

When viewing the world in terms of discrete interfaces, it becomes immediately obvious that this type of interface is, almost literally, everywhere. Previous chapters have shown discrete interfaces, but they weren't specifically called out as such. For example, the switches in Chapter 6, the relays in Chapter 10, and the logic in Chapter 11 are all used as discrete interface components. Here are four different catagories of applications that utilize discrete I/O functions:

User inputs
Pushbuttons on a front panel, or an old-style console game controller. The pushbuttons arranged along the sides of a display in a late-model luxury automobile or in an aircraft are discrete inputs to something, somewhere.

Limit switches
In a machine tool, various limit switches are used to detect when the machine has reached its physical limits and shut it down to prevent damage. Limit switches are common in many devices that incorporate controlled motion into their design.

Security systems
In a security system, almost all of the inputs to the local controller are discrete. Some systems have the ability to monitor temperature, but the main point is to monitor doors and windows. This is done with simple magnetic switches, such as the reed relays described in Chapter 10, hidden pushbutton switches, and sometimes snap-action type switches. If the system has infrared motion sensors, odds are that the output from those is a discrete on/off signal.

Power control
The ability to control the power to an external system, device, or mechanism is a primary application for a discrete signal. Lighting control, motor control, heater element power control, and launching a rocket are just some of the applications of discrete power control.

And there's more. Just look around and you'll see discrete I/O starts popping up everywhere. Sensing when a garage door is all the way down or all the way up is one example. A simple thermostat for controlling a heater or air conditioner is another example (the heater or A/C unit is either on or off, not somewhere in between). A machine on a production line that folds up a cardboard package for breakfast cereal is controlled by a set of discrete interfaces (it's probably a type of discrete sequential controller). The popular little toy rocket launcher found on some people's desks uses discrete actions to control the azimuth and elevation of the launcher, and when to emit a puff of air to fire the foam missile. It also incorporates discrete limit switches on the elevation and azimuth movements. There is nothing analog about the device. And the list goes on and on.

HACKING A DISCRETE INTERFACE

If you are designing your own discrete interface, you have full control over the operational parameters, such as voltage, timing, pulse width, and so on. However, if you want to interface with something like a CD player or the Robosapien toy robot, you will need to figure out how its discrete interfaces work and what the electrical characteristics are.

If you have schematics for a device available, you are most of the way there. Simply examine the schematic and you should be able to figure out the basic characteristics of the circuit (see Chapter 1 and Appendix A for basic electronics theory and Appendix B for an overview of schematics). The microcontroller example shown earlier illustrates the voltage and current parameters you would probably be most concerned with.

But what if there are no schematics available, or no datasheet for a part? The first step is to observe the interface while it is active: what, exactly, does it do? What is the highest voltage when it is active? How long is it active? How much current flows through it when it is active? Once you know these things, you can then move on to create an interface that will allow you to tap into, or even override, the existing discrete interface. Chapter 17 describes the types of test instruments that can be used to discover how an interface works, including a digital multimeter (DMM) and an oscilloscope.

But what if it isn't practical to poke at the interface while it's active? If the mystery interface is part of a battery-operated device, a reasonable first assumption is that the voltage will not exceed what the battery (or batteries) can provide. If the device operates from a plug-in transformer, like the ones described in Chapter 5, the same reasoning applies.

If the interface is an input, the main consideration is to not apply more voltage than what the device normally uses. It should be possible to look at the input with a DMM while it is active (i.e., the gadget is doing its normal functions) and determine the high and low levels, but in any case, a relay (see "Using Relays with Inputs" on page 274) or an opto-isolator (see "Optical Isolators" on page 274) can be used to provide a safe, voltage-indifferent interface.

Consider the mystery gadget shown in Figure 12-3. It has a sensor of some sort, a small DC motor, and a solenoid. It's shown here in schematic form because it doesn't matter for this example what the device actually does, only how it interfaces with its various component parts.

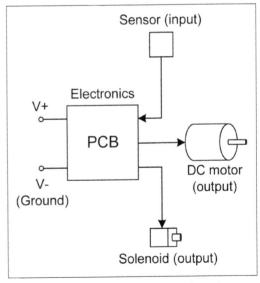

FIGURE 12-3. A block diagram for an example mystery device

Using Figure 12-3 as a reference, Figure 12-4 shows how to use a DMM to read the voltage on the sensor. If you suspect that it's a temperature sensor, you should be able to blow some warm air on it and watch the voltage reading change.

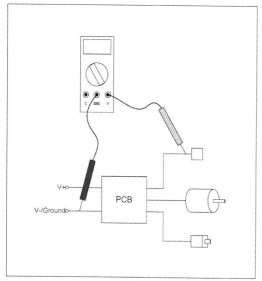

FIGURE 12-4. Measuring the voltage on an external sensors in the mystery device

If you notice that readings from the sensor wire don't seem to change, and it has both an AC and a DC voltage present, you might be dealing with some type of serial digital interface. Chapter 14 describes digital interfaces, including simple serial types that are often used with outboard sensors. This would be a good time to connect an oscilloscope (see Chapter 17) and take a look at the signal on the sensor wire.

To determine how much current the motor draws, you'll need to insert the DMM in series with the motor with it set to measure current, as shown in Figure 12-5. This isn't hard to do, but you will need to keep in mind that the current flows *through* the meter in this mode, so the motor won't work unless the meter is in the circuit. You will also need to cut the wire to the motor, strip the ends back about 1/4 inch to connect the meter, and then reconnect the wire ends when you are finished. Heatshrink tubing is perfect for insulting the reconnected wires (just remember to slide on a section of heatshrink before reconnecting and soldering the wires).

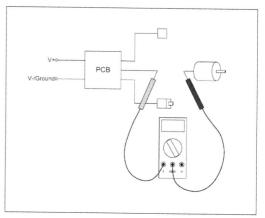

FIGURE 12-5. Measuring the current through a DC motor in a mystery device

By identifying the inputs and outputs of an unknown device, measuring the voltages and currents present when the device is active, and perhaps identifying some of the components, you can build up a profile. With this in hand, it will be much easier to interface your own circuit to the device with minimal guessing and hopefully avoid problems.

Discrete Inputs

A discrete input on one device is a discrete output on another. If you have full access to both ends of the interface, you can make

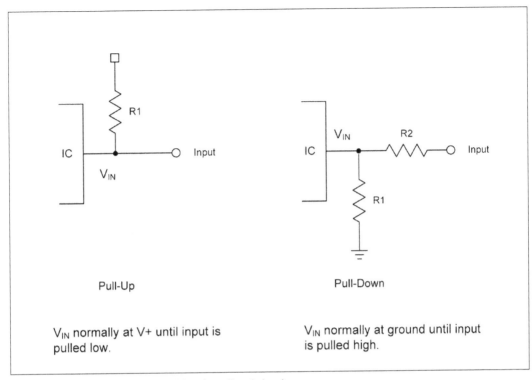

FIGURE 12-6. Pull-up and pull-down resistors for a discrete input

informed decisions about current, voltage, and timing. If one end is something of an unknown, it might be best to err on the side of caution and employ some type of isolation.

A discrete input usually doesn't require much in the way of current, just a voltage level sufficient for the circuit to sense it reliably. Depending on the impedance of the discrete input, it might be necessary to use either a pull-up or pull-down resistor to prevent build-up of stray voltage that could create erroneous input.

USING A PULL-UP OR A PULL-DOWN RESISTOR

Figure 12-6 shows both pull-up and pull-down circuits. In the pull-up circuit, R1 serves to hold the discrete input (V_{IN}) high until the external input is pulled low. In the pull-down circuit, V_{IN} is held low by R1, and R2 serves to limit the amount of current fed into the discrete input.

The value for R1 in the pull-up circuit might be around 22 k ohms, since all it has to do is provide a persistent voltage to the discrete input. It should be large enough so that, when the input goes to ground, the current through R1 is negligible. In the pull-down circuit, R1 can again be a high-value resistor, since it is just draining off any stray voltage to ground. R2 is a good idea to limit the amount of current fed into the discrete input, and it could be anywhere from 220 to 1,000 ohms, depending on the circuit voltage and the sensitivity of the discrete input. Also, bear in mind that in the pull-down circuit, R1 and R2 form a voltage divider and the discrete input might act as a current sink (see "Current Sinking and Sourcing" on page 278). So you won't see V_{IN} equal the external input voltage in many cases.

> It is not a good idea to apply more voltage to a discrete input than the supply voltage it normally uses internally. This can easily damage something. So if a discrete input is part of a circuit that uses 3.3V, don't apply more than 3.3V to the input, unless you know for a fact that it can handle a higher input voltage (many 3.3V microcontrollers can deal with 5V inputs, but not all). If you need to go from a high voltage to a lower one for the input, use a translator like the ones described later in this chapter.

USING ACTIVE INPUT BUFFERING

In some cases, it might be necessary to perform *level-shifting* in order to use a +5V TTL-level source with a 3.3V discrete input. Although many microcontroller devices will accept TTL-level inputs, some don't. If you are attempting to interface to an existing device without a schematic, it might be a good idea to consider using an active input buffer.

USING RELAYS WITH INPUTS

Yet another approach is to use a relay with a discrete input. Although it is the slowest form of input in terms of switching speed, it is also the safest. The contacts of a relay act as a switch connected to the discrete input, and when that is combined with the pull-up

and pull-down circuits shown in Figure 12-6, you can rest assured that the discrete input will receive the same voltage at which it is designed to operate. Figure 12-7 shows how this works with the pull-down circuit.

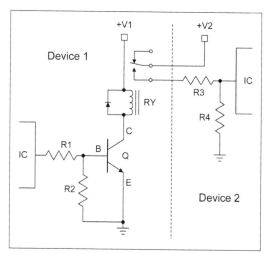

FIGURE 12-7. Using a relay as a safe discrete input

Once again, Figure 12-7 does not include resistor values, mainly because they will vary depending on the working voltages of the actual circuit. But, generally, R1 can be anywhere from 470 ohms to 2,200 ohms (2.2 k), and since R2 is there to ensure that the voltage across the base-emitter junction goes to zero when the input is removed, it can be something fairly large. A value between 33 k to 47 k ohms should work. R3 and R4 form a voltage divider, with R3 serving as a current limiter into the external discrete input, and R4 acting as a pull-down.

OPTICAL ISOLATORS

An optical isolator (also called an *opto-isolator* or *optocoupler*) is a device that uses an LED and a phototransistor of some type to couple a signal between two otherwise electrically incompatible circuits. For example, if you want to provide discrete signal feedback from a low-voltage circuit to a high-voltage circuit that doesn't share the same ground reference, you would want to use an optical isolator. Figure 12-8 shows a generic diagram of an optical isolator.

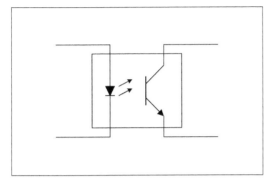

FIGURE 12-8. Generic optical isolator

When used with a discrete input, an opto-isolator can be used to pull down the voltage on the input when the LED is active, as shown in Figure 12-9.

Note that the circuit shown in Figure 12-9 will not invert the input. In other words, when the input goes low, the LED is active. When the LED is active, the transistor will conduct and pull the output low, as well.

Opto-isolators come in a variety of types and packages. There are simple phototransistor versions like the one shown here, as well as Darlington, AC input, and photo-triac types. Available packages range from four-pin plastic DIP to surface-mount types, and there are also tubular forms available with wire leads.

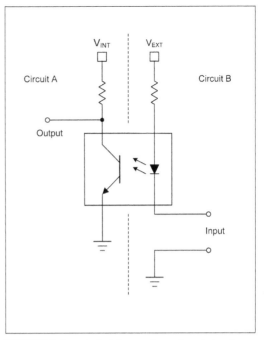

FIGURE 12-9. An opto-isolator used as a discrete input source

Many opto-isolator parts have numbers that begin with 4N or 6N, followed by a part number. One popular and common family of opto-isolators is the 4N25 family. These come in six-pin DIP packages as well as surface-mount types. Some types of opto-isolator and optocoupler devices are available in four-pin DIP packages. Figure 12-10 shows a 4N25 device in a somewhat unusual white DIP package.

FIGURE 12-10. A 4N25 opto-isolator device

Internally, the 4N25 looks like Figure 12-8, but with one additional connection. Figure 12-11 shows the internal schematic of the 4N25. A unique feature here is the connection to the base terminal of the transistor. In most cases, this would be left unconnected, but it is possible to alter the response behavior of the device by connecting the base terminal to a bias voltage.

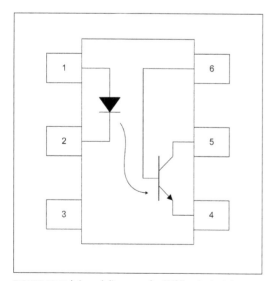

FIGURE 12-11. Internal diagram of a 4N25 opto-isolator

You can build your own quick-and-dirty optical isolator using just a couple of resistors—an LED and a phototransistor. Figure 12-12 shows the parts involved and the circuit diagram. The transistor is a Seimens BFH310, but just about any garden-variety type will work. The decision really becomes an issue only if you plan to push very short or high-speed signals through the isolator, and in that case, you probably shouldn't be trying to build your own, anyway.

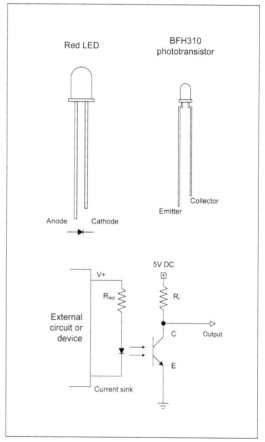

FIGURE 12-12. The parts needed for a home-grown opto-isolator

A light-tight package is essential for an opto-isolator. Phototransistors can sense stray light, so something needs to be placed around the LED and transistor. In

Figure 12-13, that something is a short peice of heatshrink tubing, and Figure 12-14 is the real thing, ready to use.

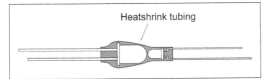

FIGURE 12-13. Inside a home-grown opto-isolator

FIGURE 12-14. The completed home-grown opto-isolator

Notice that the resistors shown in Figure 12-12 are not built into the finished isolator. Why? Because I have no idea where it might be used. If it's a 5V to 5V situation, a 180-ohm resistor will definitely light up the LED, but it doesn't need to be at full output to activate the phototransistor. So I'd probably go with something like a 220-ohm part, instead. The resistor used with the transistor should be capable of providing enough current for the transistor to work correctly, but no more. When the transistor sees light, it will pull the output line low (close to ground), and R_t will keep things from going up in smoke. Something on the order of 1,000 ohms will probably do the job in a 5V circuit.

Opto-isolators can be used for things other than just single-bit discrete signals. If the isolator is fast enough, it can be used to couple two circuits using a serial data channel (Chapter 14 discusses digital communications). Say, for example, you wanted to interface two microcontroller circuits, with one handling the I/O functions to a master system (perhaps using USB) and the other controlling various discrete I/O signals. Two opto-isolators are needed, one for each direction the serial data is moving between the microcontrollers. With this setup, if the control interface circuit is compromised in some way, the opto-isolators will prevent the discrete control circuit from also being damaged. The 6N26 high-speed opto-isolator, for example, can handle data rates of up to 1 Mbit/s.

Discrete Outputs

A discrete output that produces, say, 3.3V when it is active might not work directly with conventional 5V TTL logic (see Chapter 11). Your circuit might also draw more current than the original circuit was designed to supply, so it's possible to convert something in the original device into charcoal if you aren't careful.

Don't forget that an opto-isolator, like the one described in "Optical Isolators" on page 274, can also be used to couple the output of an external device into your circuit. The main consideration with this approach is to choose a current-limiting resistor for the LED that will allow for sufficient current to activate the LED without exceeding the discrete output's current limits. A driver transistor might be necessary, as described in "Simple One-Transistor Buffer" on page 278.

CURRENT SINKING AND SOURCING

A discrete interface can be either a current sink or a current source (and in some cases, both). The terms *sink* and *source* refer to how current moves into, or out of, the interface connection. Consider the diagram in Figure 12-15.

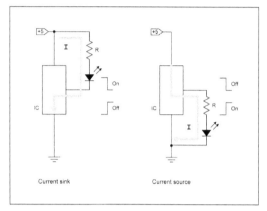

FIGURE 12-15. Current sink versus current source

In the case on the left, the IC is acting as a current sink. In other words, the current flow through the LED is passed into the IC and then on to ground. This is what you would expect to see with an open-collector type of device. On the right side, the IC is supplying the current necessary to activate the LED.

With a current sink, the connection into the IC (on the cathode side of the LED) will be high (at +5V in this case) until the IC closes the current path. When this occurs, the voltage on the cathode of the LED drops almost to zero. On the other hand, the voltage on the anode of the LED in the current source circuit will be zero (or very close to it) until the IC closes the path to the +5V supply.

This, then, is how you can determine if a discrete output is a sink or a source. Once you know that, you can determine the amount of current that can be safely handled by the discrete interface. Most ICs have sink and source limits published in their datasheets, but when in doubt, you should be safe if you limit the current to 10 mA.

BUFFERING DISCRETE OUTPUTS

Whereas a digital output is typically used with other digital circuits, a discrete output implies a connection to external devices in the physical world (things like motors, solenoids, relays, LEDs, and heater elements, for example). The device that the discrete output is connected to might not use the same DC supply voltage, or it might require more current than the discrete output can safely deliver. The solution to this is some kind of buffer to serve as an intermediary between the discrete output and the external device.

SIMPLE ONE-TRANSISTOR BUFFER

If you need to connect to a discrete output to control something that is beyond the sink or source capacity of a part in a circuit, you will need to use a buffer. One way to do this is to use a transistor. Figure 12-16 shows how a PNP transistor can be used with a sinking discrete output to drive a high-current load like a relay. The idea here is that the transistor will not "turn on" (i.e., saturate, or become fully conducting) until its base terminal is brought close to zero volts. The discrete output from the IC does this when it is enabled by pulling down the voltage across the resistor R2 through R1. The purpose of R1 is to prevent excessive current through the base of the transistor, and it should be as small as possible. R2 should be much larger than R1, as its sole purpose is to hold transistor Q in an off state.

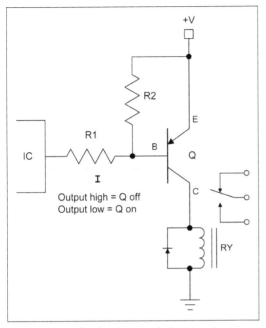

FIGURE 12-16. Buffer for current sink discrete output

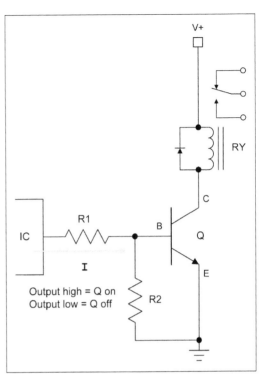

FIGURE 12-17. Buffer for current source discrete output

With a current source discrete output, an NPN transistor will serve as a buffer to allow the output to drive a high current load, as shown in Figure 12-17. As with the current sink buffer shown in Figure 12-16, the purpose of R1 is to limit the base current into transistor Q, and R2 holds Q in an off state until the IC generates a voltage at the discrete output.

Note that the relay in Figure 12-17 can be replaced with an LED or an opto-isolator like the 4N25 discussed previously, as shown in Figure 12-18. When using an LED or an opto-isolator, don't forget to put a resistor in series with the LED (R3) to limit the current through both it and the transistor, or smoke might result. In Figure 12-18, R3 could be something like 470 ohms if V+ is 5V.

DISCRETE CONTROL INTERFACES | 279

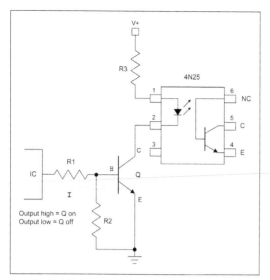

FIGURE 12-18. Using a 4N25 opto-isolator as a discrete output buffer

Logic-Level Translation

The widespread adoption of 3.3V logic has led to something of a dilemma when it comes to connecting things that use conventional 5V TTL logic levels to a discrete or digital input or output. While the circuits shown previously are suitable for interfacing to relays and motors, they are a bit of overkill if you just want to connect a TTL circuit to something like an Arduino (which uses 3.3V).

Fortunately, components are available that will handle the voltage translation for you. Some of them (such as the BSS138, NTB0101, or the TXB0108) will automatically sense the signal direction. Others (such as the SN74VLC245A) need external logic to change the direction of the signal, but if you need to go only one way, you're all set.

THE BSS138 FET

A field-effect transistor can be used as a bidirectional logic-level translator to connect a 3.3V discrete interface to a 5V TTL device, and one popular device for this purpose is the BSS138. The BSS138 is an N-channel MOSFET that comes in an SOT-23 surface-mount package, as shown in Figure 12-19.

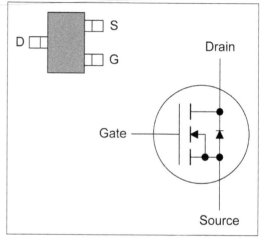

FIGURE 12-19. The Fairchild BSS138 FET device

AdaFruit and Sparkfun both sell a PCB with four BSS138 devices specifically for interfacing low-voltage logic to conventional TTL. In each case, these are small PCBs that contain four BSS138 devices. Connect the low-voltage side to the 3.3V logic and the high-voltage side to 5V logic, and you are good to go. The circuit itself is simple, as shown in Figure 12-20 (based on the circuit used in the Sparkfun BOB-12009 quad-level translator).

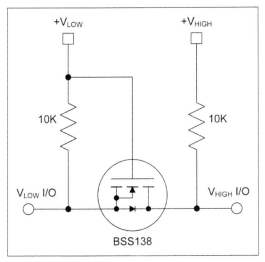

FIGURE 12-20. BSS138 logic-level shifter

Note that, unlike the relay circuits shown earlier, this circuit will not handle large amounts of current. The Fairchild version of the device is rated for 0.22A continuous current. It is good for interfacing logic signals, not driving heavy loads.

THE TXB0108

The TXB0108 is an octal bidirectional logic-level translator with auto-direction sensing. Internally, it consists of eight identical logic *cells* that perform the sensing and voltage-level transation functions. Figure 12-21 shows a block diagram of the internal architecture of the device.

The TXB0108 only comes in several different surface-mount package types, from a plastic small outline form to a 2.5 mm × 3 mm ball-grid array. The plastic small outline might be a challenge, but with a decent soldering station, it can be mounted successfully. Avoid the ball-grid array package, unless you plan to use a lot of these and an automated production system to build the circuit boards.

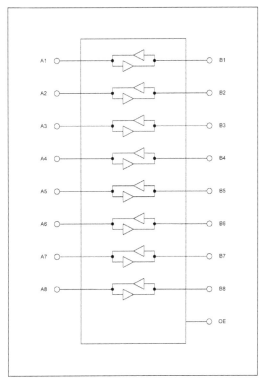

FIGURE 12-21. Internal block diagram of the TXB0108 octal logic-level translator

THE NTB0101

The NTB0101 is a one-bit (single-channel) logic-level translator with auto-direction sensing. It comes in an SOT891 surface-mount package, which has the six connection points tucked up under the device. That might be a problem if you don't have the equipment to deal with that type of surface-mount packaging. On the plus side, it is extremely small, with outside dimensions of only 1.05 × 1.05 mm.

Even though you might never use a part this small, it's still interesting to take a look

DISCRETE CONTROL INTERFACES | 281

inside. Figure 12-22 shows what the device looks like internally.

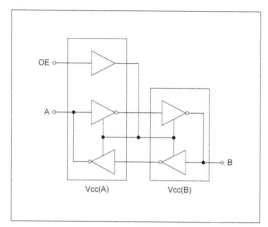

FIGURE 12-22. Internal schematic for a NTB0101 logic-level translator device

The OE (output enable) input controls the entire device, while the inverting buffers pass the signal from A to B (or B to A). Since two inversions in a row is the same as no inversion, the output will be the same polarity as the input.

Components

Table 12-2 lists the interface-level translation components covered in this section. There are, naturally, many more types available, but for the most part, these are fairly representative.

TABLE 12-2. Interface-level translation components

Part number	Manufacturer	Circuit type	Package
BSS138	Fairchild	N-Channel MOSFET	SOT-23 SMD
TXB0108	Texas Instruments	PMOS/ NMOS logic	SMD
NTB0101	NXP	Auto-sense Logic	SMD

Summary

Discrete inputs and outputs are where most circuits meet the real world, and the real world isn't always compatible with a particular circuit. This chapter presented a variety of ways to interface with discrete inputs and outputs. We've looked at circuits built using transistors and relays, optical isolators, and level translation ICs.

To reduce noise and the possibility of stray voltages, use pull-up or pull-down resistors. If you need to interface a circuit with limited current sinking or sourcing capability to something that draws a lot of current, then a relay, transistor, or FET interface would be a possibility. To interface newer 3.3V logic to older 5V TTL logic, a level translator is a compact and inexpensive way to get the job done.

As with almost every other aspect of modern electronics, there are a mind-numbing number of methods and parts from which to choose. It pays to do some research and see what's available, because it is likely that someone, somewhere, has already solved the problem and made a part or a small PCB module to do the job.

CHAPTER 13

Analog Interfaces

IN THE REAL WORLD, THINGS HAPPEN IN more or less continuous ways. When you walk, you don't take steps in discrete movements, but rather as a flowing series of motions (which has been described as "controlled falling"). When the valve for a garden hose is adjusted, it isn't set to some discrete amount of water flow in gallons or liters per minute. It just gets set to produce an output that "looks about right," and it would be very difficult to get it back to that same exact level on the following day. Digital electronics handle changes as discrete steps or values, not as a continuum; this difference in how numbers are handled leads to a need to translate from one domain to another.

Mathematics and computer science define two basic types of numbers: integer and real. An *integer* is simply a whole number: –5, 0, 1, 2, and so on. The real world is the domain of *real* numbers, and there are a lot of them out there. Between 0 and 1, for example, there is an infinite set of real numbers.

A real number can represent any value (a point) along a continuous number line from –infinity to +infinity. The set of real numbers can be further divided into rational and irrational numbers. The set of integers is also a subset of the set of real numbers.

Rational numbers can be expressed as the ratio of two whole numbers (such as 12/1, 6/4 or 2/40), which is why they are called *rational numbers*, not because they make sense. They can also be written in *decimal* notion (such 12.0, 1.5, and 0.05). Irrational numbers (such as pi or the square root of 2) are also in the domain of real numbers, but they cannot be expressed as fractions. The value of pi, for example, can be approximated with a fraction (22/7), but it is only an approximation.

An important concept to grasp here is that the set of real numbers is infinitely large. For example, consider the possible values between 0 and 1. In that range, you will find 0.0001, 0.45, 0.87, and 0.022, along with every possible value in between. No matter how finely it is divided, there are still more numbers. Such is the nature of the real world, which is why it is a challenge to convert the infinite range of real numbers into a form that a discrete machine like a computer can deal with and why you'll never be able to get the flow rate on the garden hose exactly the same tomorrow morning as it was yesterday.

Interfacing with an Analog World

Real numbers present a serious challenge for electronic sensors in general, and digital systems in particular. When a sensor measures some physical phenomenon (like, say, the temperature of an oven), it must first convert the physical manifestation of the heat into a voltage or current level that corresponds the physical phenomenon. Hence, the output of an analog sensor is just that: an analog, or close approximation, of the physical event in another form. In this case, it might be a voltage level that serves as an analogy for the original physical phenomenon (heat).

For a sensor, the challenge comes in the form of resolution. Consider the possible temperature values as an oven goes from 432 degrees to 433 degrees. It's not just 1 degree. Remember that there are an infinite number of real values between these two integers. Sensors are rated according to their measurement resolution, or how small of a degree of change can be reliably and repeatably detected. How much resolution is really necessary will depend on the system using the sensor. For a common kitchen oven, a resolution of +/− 1 degree is probably overkill (a cherry pie will cook just as well at 373.3 degrees as it will at 377 degrees, even if the cookbook does call for 375 degrees). But, for some applications, a more precise degree of measurement is essential.

FROM ANALOG TO DIGITAL AND BACK AGAIN

A signal from an analog sensor is a continuously variable voltage or current. While this is fine for an analog electronic circuit, it won't work with a digital system. The analog signal must somehow be converted into a form that digital electronics can work with, which means binary values. This involves the use of devices called analog-to-digital converters (ADCs). An ADC takes continuous samples of the analog input and generates a stream of digital values, one per sample.

But the act of conversion introduces a new set of problems. Because of the discrete numeric nature of a digital circuit, it is not possible to capture analog data and convert it into a digital form with a level of accuracy that will allow for a 100% faithful representation of the original signal. This effect, called *quantization*, arises as a consequence of obtaining or generating a sequence of measurements of a continuously variable signal at discrete points in time.

With analog inputs, any changes in the analog signal between sample events are lost forever, and the result is only as faithful to the original as the number of samples per unit of time permits. This is illustrated in Figure 13-1, which shows the difference between a signal sampled once every two seconds and the same signal sampled twice per second. Notice how the reconstructed result of the faster sampling rate is much closer to the original, but it is still not a 100% faithful reproduction.

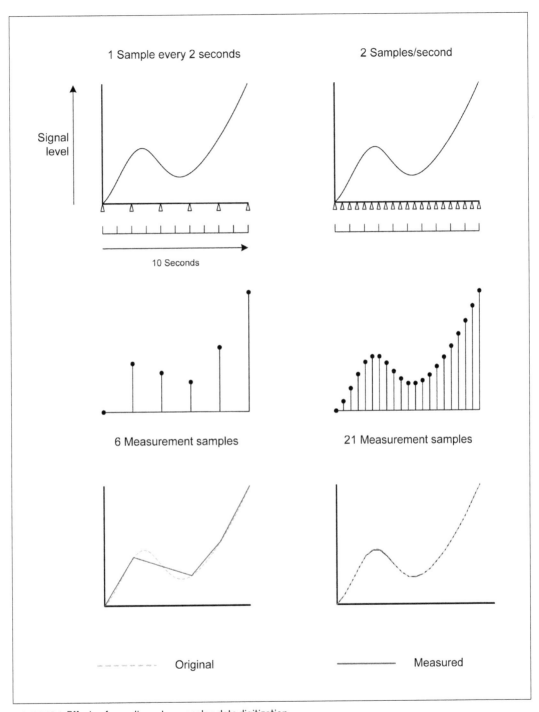

FIGURE 13-1. Effects of sampling rate on analog data digitization

Of course, not every situation needs a high level of fidelity in order to accomplish its objectives. In many cases, it is perfectly acceptable to take data samples at intervals of several seconds, or even minutes. This is particularly true when the measured input doesn't change very much within the sample period, such as might be the case with something like the temperature in a greenhouse, the water level in a holding tank, or the temperature inside a house. Other cases, such as the conversion of audio to digital form, require very high sampling rates in order to accurately capture the highest frequencies of interest and maintain a high-fidelity representation of the original input. The sound circuits in modern PCs use a sampling rate of around 44,100 samples per second. Because of something called the *Nyquist frequency* (or *Nyquist limit*), the highest input frequency that can be accurately captured is half the sampling rate, or in this case about 22 KHz. We won't delve too deeply into sampling theory here, but it's worth looking into if you plan to work with audio or even higher frequencies.

Analog data is typically converted to digital form with a resolution (or data size) ranging from 8 to 24 bits per sample. Resolutions of less than 8 bits or greater than 24 are not readily available, but are possible. With an 8-bit resolution, the data will range in value from 0 to 255 (or from –128 to +127 if negative values are used). Again, not every application requires a high degree of precision, and sometimes less is more than sufficient.

When you are converting from analog to digital (or vice versa), there is an inherent limitation in the conversion referred to as *quanti-*

zation error. In general terms, this is the error between the original signal and the digital values (or codes) resulting from the conversion. As shown in Figure 13-2, the lower the resolution of the ADC, the more pronounced the quantization error becomes. The graph covers only the first eight samples, and the input in this case is linear (it's a straight line). You can imagine it continuing on upward to the right, if that helps.

According to Figure 13-2 (which is only an approximation for purposes of discussion) a 9-bit ADC, with a range of 512 possible values (or codes), will generate a more accurate conversion than an 8-bit ADC with a range of only 256 possible values. The sampling events (the sample rate) are shown on the time axis as TS0, TS1, and so on. Note that it doesn't matter if the 8-bit converter is sampled at a fast rate; it cannot do any better than its fundamental 8-bit resolution, although it will be able to detect and convert fast changes in the input that are within its resolution.

Sample resolution can be expressed in terms of volts/step, or, in other words, the measurable voltage difference between each discrete digital value in the converter's resolution range. These are the codes mentioned earlier. Since ADCs generate binary codes instead of real-number values like 4.5 or 22.73, these codes must be translated into something that represents the original input voltage. This usually occurs in software, not in the ADC or the logic hardware. But in order to do the conversion, we need to know the scale. For example, if we have an 8-bit converter with a maximum full-scale input range of 0 to 10 volts, then each increment, or step, in the digital output code will be the equivalent of 0.039 volts. This can be expressed as:

- Resolution = $V_{max}/2^n-1$

Therefore, a 10-bit converter with a V_{max} of 10V can resolve 0.00978 volts/step, a 12-bit device can resolve 0.0024 volts/step, and 16-bit ADC can resolve 0.0001526 volts/step.

In Figure 13-1, the reason for the loss of fidelity between the sampling rates is a lack of samples to accurately track the changes in the analog signal in the slower example, not a lack of conversion resolution (the resolution isn't even mentioned, actually). In Figure 13-2, it is the lack of resolution that results in the loss of fidelity due to quantization error. The sampling rate and the sample resolution together determine how accurately an ADC can convert an analog signal to digital form.

Lastly, there's the issue of sensor error. If the analog input is noisy, or if the sensor produces an incorrect reading, it doesn't really matter what the sample rate or resolution might be; the result will still be erroneous. For this reason, many circuit designs take pains to ensure that the analog inputs are as free from extraneous noise as possible. It's common to use separate DC power inputs for the analog and digital sections of a circuit board so that the switching transients generated by the digital components don't bleed into the analog sections. Sensor inputs

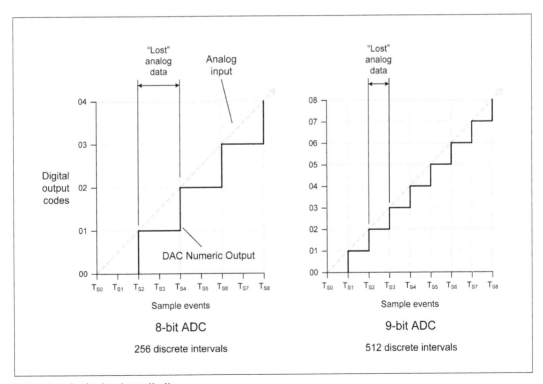

FIGURE 13-2. Analog input quantization

can be shielded (see Chapter 7), employ twisted-pair wiring (discussed in Chapters 7 and 14, and incorporate filtering of some type to suppress high-frequency transients and noise (see Appendix A for information about filters).

Converting digital data into analog data is another challenge for digital electronics. The process basically involves generating a voltage output that corresponds to a digital value, and the more bits used, the better the fidelity of the output. The device for the job is called a *digital-to-analog converter* (DAC). Just as with an ADC, the DAC has some inherent limitations with regards to resolution, and the devices also exhibit quantization. Figure 13-3 shows the relationship between the resolution and the output update rate (or sample rate).

For many applications, the output sample rate is not a critical parameter, and something on the order of once or twice a second between each output update will suffice.

This is assuming, of course, that whatever the DAC is intended to control does not need to change at a faster rate. But, as you can see from Figure 13-3, the faster the output sample conversion rate, the closer the resulting output will come to the original input. Using a DAC with higher resolution (12 bits instead of 8, for example) will also improve the quality of the output. But, due to the effects of quantization, it will never be exact. This is one of the primary complaints of audiophiles who claim that analog media like phonograph records are more true to the original sound than an MP3 digital file. They may well be right, but I can't hear the difference.

For some DAC devices, the output voltage range is established externally using a reference voltage. In other cases, the reference voltage is built into the DAC device itself. The output resolution is determined by the number of bits used to generate the output value and is just the output voltage range divided by the number of possible digital

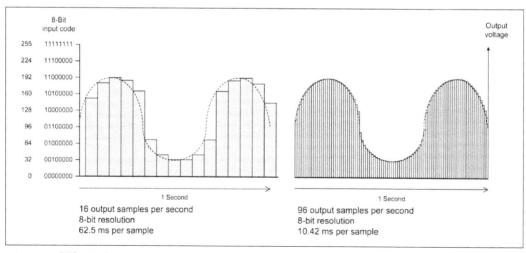

FIGURE 13-3. DAC output

input values. The actual accuracy of the output is a function of the linearity of the DAC, with some types being more linear than others. Linearity, in this case, can be thought of as how well the DAC will generate a straight line output given a continuously increasing range of numeric values to convert.

It is possible to "smooth" the output of a low-resolution DAC by using a passive filter, but if you really need high fidelity, a high-resolution (16- or 24-bit) DAC is the way to go. If, for example, a 10-bit DAC is used, it will generate 1,024 discrete voltage steps across its output range. A 16-bit DAC is able to output 65,536 discrete steps.

ANALOG-TO-DIGITAL CONVERTERS

Many microcontrollers come with a built-in ADC (or two, or three). These might be 8- or 10-bit devices, with sampling rates based on a divisor of the basic clock rate of the microcontroller. In other cases, an external ADC is necessary, such as when you're connecting to a PC or a microprocessor without a built-in ADC. An external ADC device can offer higher precision than a built-in function, and it can operate at a much higher sampling rate.

ADC devices come in the same packaging used with other types of ICs. Some types use a form of serial interface (I2C or SPI) rather than a parallel bus. This reduces the pin count on the device at the expense of conversion rate. Both through-hole and surface-mount packages are available. Figure 13-4 shows an inexpensive four-channel ADC module using an ADS1115 16-bit converter. This module can be used with any microcontroller or logic circuit that can support the I2C interface, and it can generate up to 860 samples per second. A similar module is available with a 12-bit converter for slightly less money.

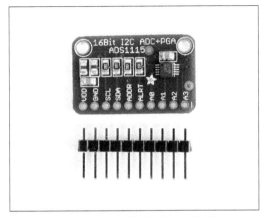

FIGURE 13-4. Four-channel, 16-bit ADC module (from Adafruit)

Another low-cost ADC is the MCP3008, an 8-channel, 10-bit device that uses an SPI interface and comes in a 16-pin DIP package. In fact, many good ADC ICs are available that require little in the way of control and data interface. They are easy to integrate into a circuit and simple to use. Table 13-1 lists some of the types available that utilize either the SPI or I2C interface.

TABLE 13-1. A sample selection of low-cost ADC ICs with SPI or I2C interfaces

Part #	Manufacturer	Bits	Channels	Interface
MCP3008	Microchip	10	8	SPI
AD7997	Analog Devices	10	8	I2C
TLV1548	Texas Instruments	10	8	SPI
MCP3201	Microchip	12	1	SPI

Part #	Manufacturer	Bits	Channels	Interface
AD7091	Analog Devices	12	4	SPI
MX7705	Maxim	16	2	SPI
ADS1115	Texas Instruments	16	4	I2C
MAX1270	Maxim	12	8	SPI

When considering an ADC, either as a built-in function in a microcontroller or as a standalone part, keep in mind these key points:

- Will the input voltage exceed the input range of the ADC? In some cases, this might damage the part, so it will need some type of voltage divider or limiter circuit to reduce the input level.
- Will the ADC sample rate be sufficient for your application? What is the highest frequency you expect the ADC to measure? Or, put another way, what is the least amount of time between significant changes in the input? If the input changes significantly (perhaps 1/10 of a volt) only over the course of several seconds, you probably don't need a fast ADC.
- Carefully read and follow the IC manufacturer's recommendations regarding PCB layout and power supply decoupling. This is particularly important when you are working with high-speed ADC devices, because they can be very sensitive to noise and voltage disturbances.

DIGITAL-TO-ANALOG CONVERTERS

Some microcontrollers include one or even two low-resolution DACs as part of their basic design, but many others don't. DACs come in 8-, 10-, 12-, and 16-bit resolutions (and other resolutions, as well), and most built-in DACs will be in the 10- or 12-bit category.

Figure 13-5 shows a module with a MCP4725 12-bit DAC, suitable for use with any microcontroller capable of supporting an I2C interface.

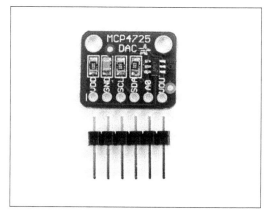

FIGURE 13-5. Single-channel 12-bit DAC module (from Adafruit)

Conversion speed is another consideration, but most modern DAC ICs are capable of operating up to 40 KHz or more.

With SPI or I2C devices, the two main parameters that effect conversion speed are the data transfer rate into the device over the serial link, and the settling time between discrete output levels.

TABLE 13-2. A sample selection of low-cost DAC ICs with SPI or I2C interfaces

Part no.	Manufacturer	Bits	Channels	Interface
MCP7406	Microchip	8	1	I2C
AD5316	Analog Devices	10	4	I2C
DAC104	Texas Instrument	10	4	SPI
MCP4725	Microchip	12	1	I2C
AD5696	Analog Devices	16	4	I2C

Here are a few points to keep in mind when using DACs:

- Don't exceed the output current capacity of the device. If you need to drive something with a high current draw (e.g., a lamp or motor), use a buffer or driver device. Several high-current linear driver ICs (i.e., op amps) are available for situations like this.

- If an output filter is needed, a simple resistive-capacitive filter might be sufficient (see Appendix A), but you might also want to consider an active filter of some kind. This has the added advantage of providing some degree of buffering to the DAC output.

Hacking Analog Signals

When confronted with an analog signal of unknown origin, you might be tempted to just toss an ADC on it and start measuring. But, before you do that, you need to make sure of a few key things.

When you are connecting to an unknown analog signal:

- If you're using an ADC that will work only with positive input voltages, verify that the signal won't go negative. In other words, measure the signal while the device is active and observe the behavior.

- Check the possible range of the analog signal. If you know the Vcc supply voltage for the external circuit, it's usually a safe bet that it won't exceed that.

- Check to make sure the analog output can tolerate additional impedance loading. In other words, will the behavior of the alien device change if you connect your ADC circuit to it? If so, you'll need to cobble up a high-impedance buffer to prevent unnecessary circuit loading.

If you need to isolate your circuit from the external circuit, you can use a transformer if the external signal is AC in nature. Small 1:1 audio and RF coupling transformers are readily available for applications like this. Another possibility is an active isolation circuit. Analog Devices offers an example in circuit note CN-0185 (*http://bit.ly/cn-0185*). Note that this approach is not simple, nor is it something suitable for a novice. If you are dealing with a DC voltage that won't change quickly, the circuit shown in Figure 13-6 might be suitable. Photosensitive resistors are described in Chapter 8.

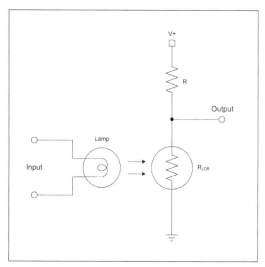

FIGURE 13-6. Simple low-speed signal isolator using a photoresistor

The circuit in Figure 13-6 employs a conventional miniature incandescent lamp, like a so-called *grain-of-wheat* type (because that's about how large it is). The entire assembly can be encased in a short section of heat-shrink tubing, just like the homemade opto-isolator shown in Chapter 12.

Why a light bulb? Because an incandescent lamp has a continuous output response to input voltage. An LED, on the other hand, will start to glow only when the voltage reaches a threshold and it starts to conduct. In other words, the lamp has a continuous response while the LED is discontinuous.

As the amount of light falling on the photoresistor (also called a *light-dependent resistor*, or LDR) increases, its resistance decreases. In this circuit, the output voltage will fall as the input voltage driving the lamp increases. If that's not what you want, you can swap the positions of R_{LDR} and R. Although simple, there are some points to consider with this circuit. First, is the input voltage suitable for the lamp? If it's too high, you'll need to add a voltage divider (discussed in Chapter 1) to adjust the input to the lamp. If it's too low, you might want to consider some kind of amplifier (which is beyond the scope of this chapter but is covered in several of the excellent texts listed in Appendix C). Second, the value of R will depend on the range of R_{LDR}, and it should be such that the maximum current through the LDR under maximum illumination does not exceed its rating or cause it to self-heat. Check the datasheet for the LDR to see its limits. Lastly, this circuit is not linear, and the dynamic range will depend on the value or R_{LDR}, R, and how the lamp is driven. Before you attempt to use it, it would be a very good idea to connect the lamp to a battery, a potentiometer, and a DMM and calibrate the output as a function of the input voltage.

Connecting a DAC to an external device is sometimes fraught with peril. If you have the schematic for the device you are trying to hack, it should be possible to figure the optimal means to make the connection. If not, you'll need to do some detective work to find the best way to gain control of the device.

Here are some final DAC hacking tips:

- Try to determine the current draw of the external circuit the original analog signal is controlling. You can do this by inserting a DMM in series into the circuit and looking at the current display while the circuit is active. If it's more than what your DAC is rated for, you might need to try to find another place to tap into the external device. You might be looking at the output of a high-current driver. Alternatively, you can use your own high-current driver to inject your DAC signal.

- You shouldn't try to "piggy-back" your signal on top of an existing analog control signal. The reason is that if your DAC is trying to output a high voltage level, and the external circuit is trying to pull the analog signal low, there could be some excessive current flow. You can build a simple DC mixer circuit with an op-amp and a few resistors, but it's still better to have just one analog voltage source active at a time. See *The Art of Electronics* (listed in Appendix C) for some circuit ideas.

Summary

This chapter provided a quick tour of the world of analog-to-digital conversion and back again. This is a field that has its own reference books, and what has been presented here is just the top of the waves when it comes to things like quantization, sampling rates, resolution, and other topics. However, the up side is that, for many applications that don't involve audio, video, or radar systems, you can just pay attention to the voltage range and the conversion resolution.

There are other types of ADC and DAC devices in addition to the ones listed here. The parts highlighted here were chosen because they have a simple interface that is easy to work with when you're using something like an Arduino, BeagleBone, Raspberry Pi, or similar microcontroller-based board. However, if you want to build your own control logic from the ground up, you might want to consider the ADC and DAC components that provide access via a parallel data bus. These devices offer high resolution and high-speed conversion rates but at the cost of more complicated control and data transfer circuitry.

To learn more about analog-to-digital and digital-to-analog conversion, refer to the excellent texts listed in Appendix C. You can also glean a lot of useful information from the Internet, at the websites of component manufacturers in particular.

CHAPTER 14

Data Communication Interfaces

CALL IT THE *Internet of Things*, OR *distributed systems*, or whatever the term of the day happens to be, but it is a fact that for a gadget to be as useful as it possibly can be, it needs to be able to exchange data with other things. After all, a gadget sitting all by itself in the corner with nothing to talk to is a lonely gadget.

In the past, communication of data between various devices and systems was accomplished through slow RS-232 or 20 mA current loop serial links, or bulky portable media such as magnetic tapes, removable disk packs, and floppy disks. Early in the history of computing, punched paper cards and rolls of punched paper tape were initially used for this purpose.

Nowadays, we have USB flash drives, SD memory cards, portable hard drives, and Bluetooth, USB, WiFi, and Ethernet communications. But the end result is the same: data from one device or system is passed to another for processing, storage, or display (or all three), just a lot faster today than 10,

20, or 40 years ago. Sharing data in real time is now a commonplace feature of things like networked refrigerators, coffee pots, thermostats, and home entertainment systems. There has been talk of allowing automobiles to communicate with one another on the road to help avoid accidents, or between cars and roadside wireless communications points to dispense traffic advisories and warnings in a way that doesn't require the driver to continuously listen to the radio or become visually distracted trying to read a flashing warning sign as it zips past.

In this chapter, we'll look at the SPI and I2C methods used for chip-to-chip communications. SPI and I2C are not the oldest forms of data communication, but gaining an understanding of these two techniques (synchronous and asynchronous) will help set the stage for what follows. Next, we'll cover the venerable RS-232 and RS-485 interfaces, the old-timers that are still going strong today. Then we'll look at USB, the ubiquitous interface, with some examples of the component hardware available to work with it. Ethernet comes next, in both wired and wireless forms. Lastly, we'll wrap up with a look at newer wireless technologies such as Bluetooth, ZigBee, and short-range VHF links.

As we go along, you might start to see some similarities between the various communications techniques. Some are synchronous (which requires a clock signal), and some are asynchronous (the timing is implicit in the stream of bits). Some use differential signaling (a +/− pair of wires), and some don't. Some use wires, and others operate at radio frequencies. But the one primary characteristic shared by all of the protocols we'll cover is that they are serial interface techniques.

We won't cover things like the parallel port on a PC (which has some interesting possibilities if you are willing to tilt your head to the side and look at it as something other than a way to control a printer), nor will we poke into the mysteries of the GPIB interface used in instrumentation applications. If you're curious, you can find out more about these topics from the books referenced in Appendix C.

If there are any technical terms or concepts presented here that might not be immediately clear, I'll make a point to give a reference to other places in the book that might help shed light on them, but in any case, don't forget about basic electronics theory in Appendix A, the Glossary, and the bibliography in Appendix C. Lastly, there are a lot of component parts listed in this chapter. If you want definitive information about them,

be sure to visit the manufacturer or vendor's website (either given here or in Appendix D) and download the datasheets and reference documents.

Also, you might notice that many of the images of PCB modules shown in this chapter are Arduino shields. It's not that I'm particularly partial to all things Ardunio (although I am rather fond of it); it's just that the advent of the Arduino has spurred something of a minor renaissance in hobbyist and experimenter electronics and microcontrollers. The result has been a flood of inexpensive Arduino clones, derivatives, and add-on modules using parts that would have otherwise gone buried in a consumer electronics device. If nothing else, these modules show just how easy it can be to use the parts listed in this chapter in your own creations.

> Although this chapter refers to specific part numbers and manufacturers by name, this does not in any way constitute an endorsement. I provide this information for your benefit by showing some examples of what is available at the time of this writing. For the latest information, and a broader view of what is available, see Appendix D. Appendix E lists all of the component parts mentioned in this book.

Basic Digital Communications Concepts

The primary concept behind all digital communications methods is that they pass data as binary values, either *serially* (one bit after another) or in *parallel* (with whole groups of bits moving in unison). Although the technology may have evolved over time, these

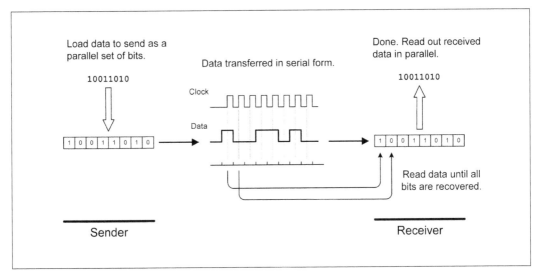

FIGURE 14-1. Serial data exchange

basic concepts apply to any form of digital communications.

A digital data stream can be sent over a wire or passed around in the form of radio waves. The radio signal gets converted back into a digital stream at the receiving end. Data that moves in parallel needs some way for the sender and receiver to coordinate who is talking and who is listening. And parallel data can be converted into a serial form and then reconstructed as parallel data at the other end. In the following sections, we'll cover the high points of these and other topics before moving on to examine particular examples.

SERIAL AND PARALLEL

Digital data communication interfaces can be divided into two broad catagories: serial and parallel. Serial data is where a digital value—a byte (8 bits), for example—is sent over a single channel (e.g., a wire) one bit at a time. At the receiving end, each bit is read and then reassembled once again as a byte. Figure 14-1 shows the process of serializing and reconstructing digital data.

Figure 14-1 shows what is known as a *synchronous serial interface*, meaning that the sending and receiving of data bits is coordinated by a clock signal sent from the sender to the receiver. The vertical dashed lines indicate when the receiver will look at the incoming signal to detect if it is either high (1) or low (0). This can occur at the start (rising edge) or end (falling edge) of each clock pulse. In this case, it's shown on the rising edge of the clock pulses. In the next section, we'll look at an asynchronous way to do this, which doesn't require a clock.

With a parallel data interface, an entire byte (or word, or even something larger) is transferred all at once from sender to receiver. As you might surmise, a parallel interface can be much faster than a serial type, since the parallel-to-serial (and back again) steps are eliminated. The downside is that a parallel interface needs a sufficient number of wires to carry all the individual bits. Figure 14-2 shows an example of a parallel interface.

For a parallel data transfer, only one control pulse (labeled "Data Strobe" in Figure 14-2) is absolutely necessary. When the receiver detects this pulse it will read in (or *latch*, in digital terminology) the data on the parallel lines into a data register. As with Figure 14-1, the dashed vertical represents the time when the data is actually sensed and loaded into the receiver's register.

High-speed parallel data interfaces have been used as interconnection channels between the processing modules of supercomputers, where the need for speed trumps the cost of the wiring and circuit complexity. The more pedestrian PC parallel interface used for things like printers and plotters operates on the same principles, just much more slowly. If you would like to find out more about parallel interfaces and some of the interesting ways the printer port on a PC can be hacked, take a look at some of the titles listed in Appendix C.

SYNCHRONOUS AND ASYNCHRONOUS

The terms *synchronous* and *asynchronous* refer to the way in which a data transfer is handled between sender and receiver. A synchronous interface relies on the use of a clock signal or transfer pulse for coordinating the data transfer timing, whereas an asynchronous interface does not. Figures 14-1 and 14-2 are examples of synchronous interfaces. Almost all parallel interfaces are synchronous, whereas serial interfaces aren't

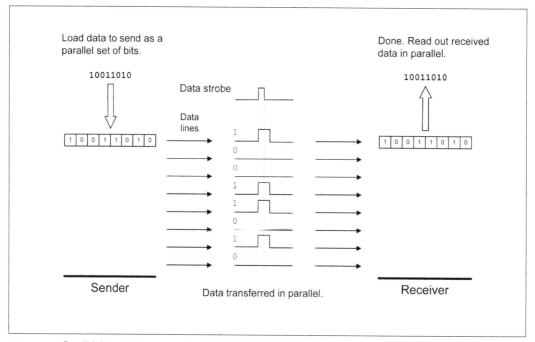

FIGURE 14-2. Parallel data exchange

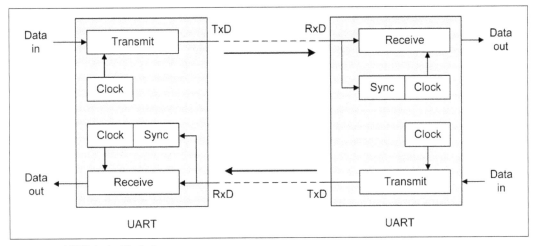

FIGURE 14-3. UART device block diagram

always synchronous but can instead be asynchronous.

An asynchronous data interface is almost always a serial interface. The asynchronous part comes from how a receiver for this type of interface detects incoming data and automatically synchronizes itself to the incoming stream of bits from the sender. Once the start of the incoming data stream is detected, the receiver will look for a specific number of bits, with each group of bits forming a byte (or character, in some cases) of data. Figure 14-3 shows a block diagram of an asynchronous serial interface that employs two UART (universal asynchronous teceiver-transmitter) devices to handle the parallel-to-serial conversion and then back to parallel again.

In Figure 14-3, you can see that the transmit part of a UART uses a definite clock rate to control how fast the data is pushed out to the receiver. The receiver, on the other hand, relies on a sync circuit to detect the incoming data from the transmitter and adjust an internal receive clock to match the data rate.

Most modern microcontrollers incorporate one or more UART functions into their design. If you want to use these functons for RS-232 or RS-485 interfaces, all you need to do is add some components to convert the signals to the appropriate voltage levels for the particular protocol. I use the term *function* here, because in modern microcontrollers, the UART is just part of the silicon chip, not a separate outboard component as in the past. There are standalone UART devices available, and "RS-232" on page 310 takes a closer look at how the data used with RS-232 is formatted and the data transfer speeds available. "RS-485" on page 317 covers the RS-485 interface.

SPI and I2C

This section covers the basics of the SPI and I2C short-range, chip-to-chip communications protocols. These are serial communications protocols that are easy to implement

DATA COMMUNICATION INTERFACES | 299

and easy to use. They also help to keep chip pin counts low by requiring only 2, 3, or 4 signals to implement, as opposed to 10 or more for a parallel interface. But convenience comes at a price (as always): for a given clock speed, a serial interface is not as fast in terms of bandwidth (the data equivalent of current, or amount of bits moved per second) as a parallel interface. Still, for many applications, the compactness and convenience outweigh the limited bandwidth.

Although other variations on short-range serial interfaces have been devised over the years, the two that still stand out and have passed the test of time are SPI and I2C, so these are the ones we will focus on here. Other interfaces, such as the Dallas/Maxim one-wire interface, have their place, and you can read up on this method, and others, in the application notes and manual available on the chip manufacturer's websites (see Appendix C).

SPI

The abbreviation SPI stands for *serial peripheral interface*. It is a full-duplex, four-wire synchronous serial interface designed for chip-to-chip communications, first defined by Motorola (now Freescale) around 1979. It has since become a de facto industry standard. You can find SPI interfaces on microcontrollers, I/O expansion ICs, SD memory cards, and sensor devices, just to name a few things. SPI is capable of very fast data transfers, with the only real limitations being the hardware's ability to generate and detect the clock signal reliably and move data without errors.

Half- and Full-Duplex

The terms *half-duplex* and *full-duplex* refer to the data transfer modes used with a pair of devices communicating over a channel of some sort. In a half-duplex system, each end of the connection has both a transmitter and a receiver, but they are never active simultaneously. Data moves in only one direction through the channel at any given time. USB and I2C interfaces are half-duplex, and RS-485 is typically implemented as a half-duplex interface. A full-duplex system has two separate channels with a transmitter at one end and a receiver at the other, and the channels move data in opposite directions simultaneously. The transmitter on a channel can send whenever there is data ready to transmit. SPI, RS-232, and Ethernet are examples of full-duplex digital data communication interfaces.

> Interestingly, it wasn't until relatively recently that anything like a standalone specification document was available for SPI. Prior to about 2000, information about the SPI protocol had to be gleaned from various microprocessor and microcontroller datasheets and user manuals. In 2000, Motorola released a semiformal SPI specification, the "SPI Block Guide." (*http://bit.ly/spi-blockguide*)

SPI devices communicate in a master/slave arrangement, where the master device always initiates the data exchange, rather like USB (the terms *master* and *slave* are historical at this point and refer to the control-response protocol implemented by SPI). SPI uses four signal lines: SCLK, MOSI, MISO, and SS, as defined in Table 14-1.

TABLE 14-1. SPI signal lines

Signal	Definition	Direction
SCLK	Serial clock	Master to slave
MOSI	Master out, slave in	Master to slave
MISO	Master in, slave out	Slave to master
SS	Slave select (active low)	Master to slave

In SPI terminology, a master device, typically a microcontroller or microprocessor, is connected to an SPI slave device. For every bit sent by the master on the MOSI line, the slave will return a bit at the same time on its MISO line. The result is that during each clock cycle (the SCLK line), a full-duplex data transfer occurs. Because SPI does not use device addressing, you must specifically select each slave device using the SS line. Figure 14-4 shows two different ways to arrange this.

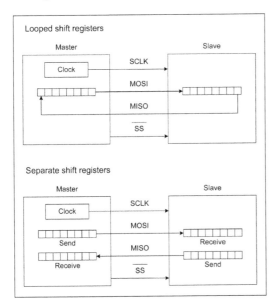

FIGURE 14-4. Master and slave shift-register operations for full-duplex communications.

Each slave device waits for a control input (the SS line) to go low. When this occurs, it will start to "clock in" data from the master device synchronously with the SCLK signals. Figure 14-5 shows a simplified timing diagram for SPI data transfers.

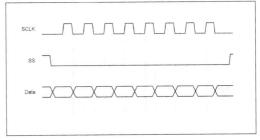

FIGURE 14-5. SPI data transfer timing diagram

In Figure 14-5, the data is changed (or toggled) on the falling edge of a clock pulse and read on the rising edge. Each of the odd-looking boxes on the data line represents a single bit of data, which can be either low (0) or high (1). When the SS line is high (inactive), a slave will cause its MISO pin to go into what is called a *high-Z* (or *high impedance*) state. This effectively removes it from the circuit until the SS line to that particular slave device is once again pulled low.

Figure 14-5 shows only one possible configuration for an SPI interface. The are four different modes that define the clock polarity and how the clock pulses will be toggled and sensed. Refer to the "SPI Block Guide" (*http://bit.ly/spi-blockguide*) for details about the clock polarity options. Note that the master and its slave devices must use the same clock and data modes in order to communicate, and most slave devices are hardwired when they are fabricated for one of the four possible modes. If a master is connected to

DATA COMMUNICATION INTERFACES | 301

multiple slaves with different clock modes, it will need to reconfigure itself for each slave as necessary.

You can connect multiple SPI devices to a single master by providing each one with its own SS line, as shown in Figure 14-6. There is no real limit to how many slaves a master can control; it's just a matter of having enough SS lines available. These are usually taken from the general-purpose DIO (digital or discrete I/O) lines of a microcontroller.

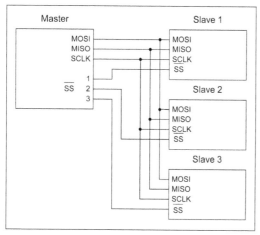

FIGURE 14-6. Multiple SPI slave devices with a single master

By now, you might be wondering how an SPI slave could have something to send back to a master if it is in the middle of receiving a command or data. The answer is that, unless the master is expecting something from the slave, it simply ignores whatever is sent back. This also applies to the slave when it is sending back a response to a command it received in a previous operation. It will ignore whatever the master sends and send back the response it has already prepared. Figure 14-7 shows how this works for a Maxim MAX7317 10-port I/O expander IC as a two-step process.

To read the state of one of the input ports on the MAX7317, the master first sends 16 bits off to the slave. The bits from D8 to D15 are an 8-bit command and address value. Bits D0 to D7 are data. When you are reading a port, only the command and address bits matter; the data bits are ignored (they are labeled as "Dummy data" in Figure 14-7). The master then raises the SS (or CS, chip select, as Maxim calls it) briefly and then sends 16 bits of dummy data while the 7317 returns 8 bits of dummy command and address data along with 8 bits of data containing the state of the port specified in the preceding command.

There is no limit on how many bits can be used to communicate with a slave device. Some devices use 8 bits, others use 16, and some might use more (such as SD flash memory cards). SCLK can be stopped and restarted in the middle of a transmission, if necessary. So long as SS is low, the interface is considered to be active (this is one of the advantages of a synchronous interface, by the way). Internal operations in a slave device typically occur when SS goes high (signalling the end of the transaction) and the slave device is deselected.

I2C

Like SPI, I2C (also written as I²C and pronounced "Eye-squared-cee") was designed to provide a short-range interface for connecting ICs, sensors, and other components at the circuit board level. It is not intended to be used to connect things to a PC, although it is possible to do that with the correct

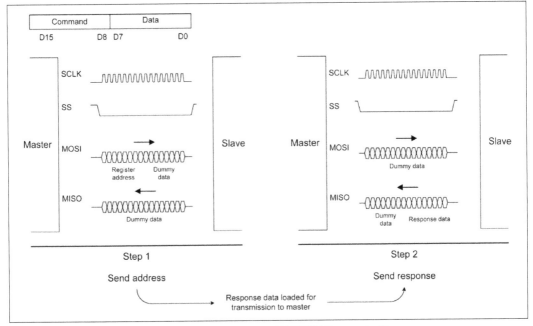

FIGURE 14-7. Example command and response sequence for a MAX7317 I/O expander IC

interface components. Some kits, such as the Velleman K8000 interface board, use this technique. Unlike SPI, the I2C interface is multi-master and uses an addressing scheme rather than chip select (SS) lines. Figure 14-8 shows multiple I2C devices connected in parallel.

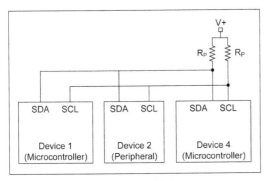

FIGURE 14-8. Multiple connected I2C devices

I2C uses only two bidirectional signal lines: the Serial Data Line (SDA) and the Serial Clock Line (SCL). There is no select line, and because it's a half-duplex interface, it requires only a single data line. These are *open-drain* lines, meaning that the drain of an internal FET is brought out to the SDA and SCL pins. It also means that external pull-up resistors are mandatory for an I2C interface. The pull-up voltage is typically from 3.3V to 5V, depending on the component's I2C interface specifications. A communication transaction on an I2C bus occurs when one of the devices, acting as the

> **TIP** Whereas SPI did not start out with an official specification, I2C was formally documented by Philips Semiconductors (now NXP) from the outset. The official I2C specification and user guide is available on NXP's website (*http://bit.ly/um10204*).

current master, places the two bus signals into a START condition. This serves as signal to other I2C devices that a master wants to communicate. When a start condition occurs, all other I2C devices will "listen" to the bus for incoming data.

After the START condition, the master sends the address of an I2C device. It also sends an indication of the type of action to be performed, either read or write. Once the rest of the I2C devices receive the address, they will compare it to their own. If there is no match, they simply wait in the listening state until the bus is released by a STOP condition. Otherwise, if the address matches one of the I2C devices on the bus, it will generate an acknowledge response to the master.

Upon receipt of the acknowledgment, the master will either start transmitting data or it will listen for the addressed slave to return data to it. This depends on whether the address was a write address or a read address. When reading data, the master responds to each byte from the slave with an acknowledgment. When the data transmission is complete, the master releases the I2C bus by setting it into the STOP condition.

I2C is relatively easy to work with, but that high-level simplicity hides the low-level complexity. For example, using I2C with an AVR microcontroller involves writing data to internal registers that control the two-wire interface (TWI) subsystem in the microcontroller. The TWI contains the logic necessary to set the START and STOP states, control the transmission speed (the bit rate), and perform address matching. It also handles the acknowledgments and checks for possible bus collisions (arbitration) if another device should happen to already be the master on the bus.

The steps necessary for an AVR microcontroller to carry out a complete I2C data transaction are the same as those outlined earlier. However, each brand of microcontroller with I2C support might do things in a slightly different way using slightly different logic, but the basic order of operations is defined as part of the I2C standard.

I2C supports either 7- or 10-bit addresses, depending on the devices used. In the original 7-bit design, shown in Figure 14-9, the least significant bit (LSB) indicates if the address will be used to read or write from the master device. The remaining 7 bits constitute the actual address of a specific I2C peripheral device on the bus.

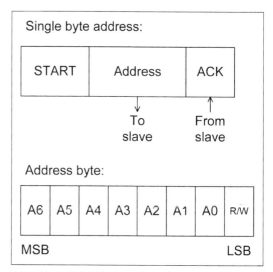

FIGURE 14-9. 7-bit I2C address format

A 10-bit address consist of 2 bytes, and the address is sent in two steps, as shown in

Figure 14-10. The most significant byte is sent first, then the least significant byte. Notice that, when the 10-bit form is used, the slave sends an acknowledge (ACK) for each of the 2 address bytes.

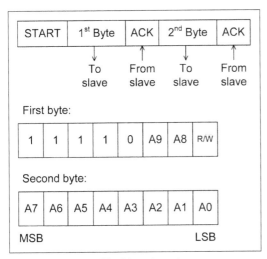

FIGURE 14-10. 10-bit I2C address format

An address is usually assigned to a part when it is created and put into production. For high-volume components, the usual method is to request an address assignment from NXP (formerly Philips). If you are connecting two microcontrollers, you might be able to assign any address you wish, but choose wisely.

Some address values, listed in Table 14-2, have been designated as reserved by NXP.

TABLE 14-2. I2C reserved 7-bit addresses

Address byte	R/W bit	Description
0000 0000	0 (write)	General call address
0000 0001	1 (read)	START byte
0000 001X	Don't care	CBUS address
0000 010X	Don't care	Reserved
0000 011X	Don't care	Reserved
0000 1XXX	Don't care	Hs-mode master code
1111 1XX1	1 (read)	Device ID
1111 0XXX	Don't care	10-bit slave addressing

Note that the address byte column in Table 14-2 shows the 8 bits in the address byte, and recall from Figure 14-9 that the least significant bit of the address is the read/write control bit. Also, when a bit position is marked as "don't care," it means that the bit value (whatever it may be) will be ignored by the I2C devices involved. For more detailed information about these reserved addresses, refer to the NXP I2C specification document (*http://bit.ly/um10204*). In most cases, the 10-bit address will be used the most.

There is no official list of I2C address assignments. NXP's position on this is that if it published all the known address assignments, people might decide to claim an unused address for a new product without going through NXP. This could, of course, lead to conflicts between an "official" address and a "rogue" address. Limor Fried (the founder of Adafruit) and her associates have started to collect and post the I2C addresses they run across (*http://bit.ly/i2c-table*).

The I2C specification defines four distinct speeds (or *bit rates*) for I2C interfaces, listed in Table 14-3.

TABLE 14-3. I2C bit rates

Designation	Name	Max rate (Kbits)
Sm	Standard-mode	100
Fm	Fast-mode	400
Fm+	Fast-mode Plus	1,000
Hs	High-speed mode	3,400

There is also an "Ultra Fast-mode" available for use with a unidirectional bus that supports a maximum rate of 5 Mbits/s, but it is not compatible with conventional I2C interfaces.

A BRIEF SURVEY OF SPI AND I2C PERIPHERAL DEVICES

Memory, discrete I/O ports, multi-axis accelerometers, color LCD displays, wireless communication modules, and many more things are all available with either SPI or I2C interfaces, or, in some cases, both. These two serial interface protocols are a major part of modern electronics, and many of the clever gadgets we take for granted today would be impossible without them.

This section is intended to give a glimpse of what is available, and provide some jumping-off points for your own investigations. The parts listed have been selected on the basis of functionality and availability, but this section lists only a small fraction of what is available. To see what else is available, check out the major electronics distributors listed in Appendix D. Be prepared to spend a few hours (or more) looking through the available parts.

Discrete I/O

If you need to output a set of parallel or discrete bits, but you are running out of digital I/O lines, using a serial-to-parallel device is one way to get there from here. Table 14-4 lists a few of these types of devices that use either I2C or SPI serial interfaces and provide from 8 to 16 discrete digital I/O ports.

TABLE 14-4. I2C and SPI discrete I/O port expansion chips

Part number	Manufacturer	Ports	Interface
PCF8574	Texas Instruments	8	I2C
MAX7317	Maxim	10	SPI
MCP23017	Microchip	16	I2C

Say, for example, you want to control a set of LEDs on a front panel using just two pins on the microcontroller. The PCF8574 from Texas Instruments provides 8 discrete I/O ports with an I2C serial interface. The Microchip MCP23017 is a device with 16 discrete ports that also uses an I2C interface. If SPI is preferred over I2C, the Maxim MAX7317 is one example of an I/O expander that provides 10 discrete I/O ports with an SPI interface.

ADC and DAC Devices

A multitude of ADC and DAC devices are available with either SPI or I2C interfaces, in a wide range of resolutions, conversion speeds, and number of channels. For a listing of ADC and DAC devices with serial interfaces, see Chapter 13.

Memory

Memory devices are probably the most common SPI and I2C peripherals. Available types include EEPROM (electrically erasable programmable read-only memory), serial SRAM (static RAM), and, of course, flash. The ubiquitous SD flash memory cards (and their smaller cousins, the microSD cards) use an SPI interface, and EEPROM memory is available with both SPI and I2C interfaces, as shown in Table 14-5. See Tables 14-6 and 14-7 for available serial SRAM devices and flash devices, respectively.

Displays

LCD display controller chips, such as the ILI9325C from ILI Technology Corporation, provide the control functions necessary to operate a touchscreen color LCD display with 256 × 320 display resolution. Modules based on this chip typically use an SPI interface, and the data rate is high enough to display near real-time video.

Standard two- and four-line LCD displays, like the one shown in Figure 14-11, are available with I2C and SPI interfaces. These are good for things that don't need to display a lot of information or color to be useful.

If you need a color display, something like Figure 14-12 might be what you want.

Table 14-8 lists some sources that sell displays like the ones shown in Figures 14-11 and 14-12.

FIGURE 14-11. Two-line LCD display module with SPI interface

FIGURE 14-12. Color LCD display module with SPI interface

TABLE 14-5. EEPROM memory

Part number	Manufacturer	Size (bits)	Organization	Interface
AT25010B	ATMEL	1Kb	128 × 8	SPI
AT24C01D	ATMEL	1Kb	128 × 8	I2C
AT25020B	ATMEL	2Kb	256 × 8	SPI
AT24C02D	ATMEL	2Kb	256 × 8	I2C
PCF85103C-2	NXP	2Kb	256 × 8	I2C
PCF8582C-2	NXP	2Kb	256 × 8	I2C
AT25040B	ATMEL	4Kb	512 × 8	SPI
AT24C04C	ATMEL	4Kb	512 × 8	I2C
PCF8594C-2	NXP	4Kb	512 × 8	I2C
AT25080B	ATMEL	8Kb	1,024 × 8	SPI
AT24C08D	ATMEL	8Kb	1,024 × 8	I2C
PCA24S08A	NXP	8Kb	1,024 × 8	I2C
PCF8598C-2	NXP	8Kb	1,024 × 8	I2C
AT25160B	ATMEL	16Kb	2,048 × 8	SPI
AT24C16D	ATMEL	16Kb	2,048 × 8	I2C
AT25320B	ATMEL	32Kb	4,096 × 8	SPI
AT24C32D	ATMEL	32Kb	4,096 × 8	I2C
AT25640B	ATMEL	64Kb	8,192 × 8	SPI
AT24C64B	ATMEL	64Kb	8,192 × 8	I2C
AT25128B	ATMEL	128Kb	16K × 8	SPI
AT24C128C	ATMEL	128Kb	16K × 8	I2C
AT25256B	ATMEL	256Kb	32K × 8	SPI
AT24C256C	ATMEL	256Kb	32K × 8	I2C
AT25512	ATMEL	512Kb	64K × 8	SPI
AT24C512C	ATMEL	512Kb	64K × 8	I2C
AT25M01	ATMEL	1Mb	128K × 8	SPI
AT24CM01	ATMEL	1Mb	125K × 8	I2C

TABLE 14-6. Serial SRAM memory

Part number	Manufacturer	Size (bits)	Organization	Interface
23A512	Microchip	512 Kb	64K × 8	SPI
23A1024	Microchip	1 Mb	128K × 8	SPI
N01S830HAT22I	ON Semiconductor	1 Mb	128K × 8	SPI
FM25H20	Cypress	2 Mb	256K × 8	SPI
PCF8570	NXP	2 Mb	256K × 8	I2C

TABLE 14-7. Flash memory

Part number	Manufacturer	Bits	Channels	Interface
M25P10	Micron	1 Mb	125K × 8	SPI
SST25VF010A	Microchip	1 Mb	128K × 8	SPI
SST25VF020B	Microchip	2 Mb	256K × 8	SPI
SST25VF040B	Microchip	4 Mb	512K × 8	SPI
SST25VF080B	Microchip	8 Mb	1Mb × 8	SPI
SST25VF016B	Microchip	16 Mb	2Mb × 8	SPI
M25P16	Micron	16 Mb	2Mb × 8	SPI
N25Q00AA11G	Micron	1 Gb	128Mb × 8	SPI

TABLE 14-8. SPI display devices

Product	Vendor/manufacturer	URL	Interface
1.8" color LCD display	Adafruit	http://www.adafruit.com	SPI
2.8" touchscreen color LCD display	Haoyu Electronics	http://www.hotmcu.com	SPI
3.2" touchscreen color LDC display	SainSmart	http://www.sainsmart.com	SPI

Other Peripherals

Table 14-9 lists some of the available peripheral devices with I2C and SPI interfaces (or, in some cases, both).

Given that there are so many types of components available with either an I2C or SPI interface, and sometimes both in the same device, the best way to find them is to look through catalogs from large electronics distributors and the manufacturer's websites. I've yet to find a definitive "all-in-one" listing of available I2C or SPI peripherals, but one might exist somewhere. However, given the highly volatile and constantly changing nature of the electronics industry, trying to

TABLE 14-9. Other I2C/SPI peripheral devices

Device	Manufacturer	Description	Interface
ADG714	Analog Devices	Eight-channel analog switch bank	SPI
ADXXRS450	Analog Devices	Single-axis MEMS angular rate sensor (gyroscope)	SPI
ADXL345	Analog Devices	Three-axis accelerometer	I2C/SPI
LIS3LV02DL	STI	Three-axis accelerometer	I2C/SPI
PCF8583	NXP	Clock and calendar with 240 bytes of RAM	I2C
SAA1064	NXP	Four-digit LED driver	I2C
TDA1551Q	NXP	2 × 22W audio power amplifier	I2C

compile and maintain a comprehensive listing might end up being an exhausting exercise.

RS-232

SPI and I2C employ techniques that predate most modern communications methods, such as synchronous clocked serial data and unique device addressing. But SPI and I2C are intended to be used within the confines of a PCB, not as an interface to external peripheral devices.

When you need an external device to communicate with another device like a PC, the simplest choice (and the most common until recently) has been RS-232, also known officially as EIA-232 due to a change of venue for the standards association that currently maintains it. But, since this interface has been known as RS-232 for the past several decades, I'll continue to refer to it that way here.

I think it's worthwhile to devote some time to RS-232, and RS-485 as well (covered in "RS-485" on page 317), because these interfaces are both still in common use and they are historically significant. Many microcontrollers and microprocessors have asynchronous serial interfaces as built-in features, needing only the appropriate signal level circuits to connect them to external devices. From a historical perspective, both SPI and early versions of RS-232 have common roots in early synchronous serial technology (it wasn't until UARTS became economically feasible that serial interfaces started dropping the clock signals). The RS-232 specification still defines a synchronous mode of operation, although no one uses it any longer (at least not that I'm aware of). The USB interface, along with some industrial interface standards, uses the concept of differential signalling employed by RS-485 and its predecessor, RS-422. If you understand RS-232 and RS-485 (in addition to SPI and I2C), you can apply that knowledge to other types of data communications, as well.

RS-232 is a voltage-based serial data interface. The difference between a logical zero and a logical one is determined by the voltage level present on the signal lines. Figure 14-13 shows how this works.

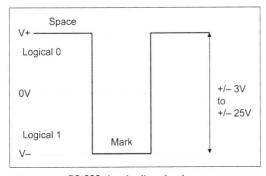

FIGURE 14-13. RS-232 signal voltage levels

Notice that RS-232 data signals employ negative logic; that is, a logical true or *mark* (1) is a negative voltage level, and a logical false or *space* (0) is a positive level. Also notice that RS-232 is bipolar, although some nonstandard implementations use zero volts as the mark level. Because most modern logic circuits don't have a negative voltage available, special driver ICs are used to generate the necessary voltage levels. We'll take a look at those later.

> TIP: Never mix "real" RS-232 with the TTL-level signals found on some microcontrollers. What comes out of the back of a PC can produce a negative 12 volts, if it's a full implementation of RS-232. This will almost certainly damage something like an Arduino or an 8051 microcontroller.

RS-232 is a full-duplex interface. It is also (usually) asynchronous, and all data clock synchronization is derived from the incoming data stream itself, not from an additional clock signal line in the interface (there is, of course, an exception to this: RS-232 can be implemented as a synchronous interface, but it is seldom used). RS-232 can be used in half-duplex mode, as well (see "Half- and Full-Duplex" on page 299).

Most RS-232 interfaces employ the ASCII (American Standard Code for Information Interchange) character encoding scheme, although the RS-232 standard itself does not specify a particular encoding technique. Data exchange over RS-232 takes the form of characters, and a character might be an actual ASCII character or just raw numeric data.

> TIP: The original ASCII defined characters with only 7 bits of data. Sending 8 bits for each character wasn't deemed necessary, since 128 possible characters can encode all the upper- and lowercase letters of the English alphabet and a host of punctuation and control characters as well. Sending only 7 bits per character also saved money, because it took less time, and time on a mainframe computer system was charged by the fraction of a second.

An RS-232 character consists of a start bit (a *mark*), followed by 5 to 9 data bits, an optional parity bit, and 1 or 2 stop bits (a space). Figure 14-14 shows the format for 8 data bits with no parity and 1 stop bit, and 7 data bits (true ASCII) with even parity and 1 stop bit. In both cases, the actual number of bits sent or received for each character or byte is 10 bits. Each unit of data, from the start bit to the stop bit (if used), is referred to as a *frame*.

DATA COMMUNICATION INTERFACES | 311

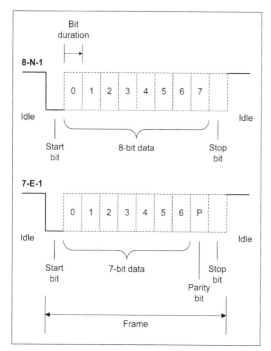

FIGURE 14-14. RS-232 data formats

For a specific format to work correctly, both ends of the communications channel must be configured identically from the outset. Attempting to connect a device configured as 8-N-1 to a device set up for 7-E-1 won't work, even though both ends are sending and receiving 10 bits of data per frame. It will, at best, result in erroneous data (garbage) at the 8-bit end and a lot of parity errors at the 7-bit end.

The speed (or *data transmission rate*) of an RS-232 interface can be defined in terms of either characters per second or bits per second. When referring to characters per second, we use the term *baud*. The *baud rate* is the number of distinct symbols moving through a communications channel per second, whereas the *bit rate* is the number of discrete bits moving through a communications channel per second. In simple digital communications schemes (such as SPI) that do not use start bits or stop bits and have no concept of frames, the bit rate and the baud rate are effectively the same.

Now for an example of how the term *baud* is misused. When someone refers to a "9,600 baud serial data channel," what he's really saying is that it is a 9,600-bits-per-second channel. Many times, the term *baud* is used incorrectly as a synonym for *bits per second*. This can be blamed mostly on modem manufacturers, which I suppose felt that saying their product was a 9,600-baud device (sounds fast, right?) sounded more impressive than the more technically correct 960-characters-per-second device (doesn't sound all that fast, does it?).

The distinction between bit rate and baud becomes important when you are dealing with a system that might use multiple bits to represent a single symbol and any associated parity and control bits, such as the character data shown in Figure 14-14. What this means is that a serial interface running at the so-called "9,600 baud" will not send or receive 1,200 character bytes per second (9,600/8), because at least 2 of the 10 bits in the frame are taken up by the start and stop bits, so only 80% of the frame contains actual data. The actual maximum symbol rate just happens to work out to 960 characters per second in this case. As a general rule, if you know the number of bits per data frame (which will be 10 for most cases involving RS-232), then dividing the bit rate rate by that number will give you the effective transfer rate in characters per second (CPS, or true baud rate).

RS-232 has some limitations to be aware of. For example, no more than one device can be connected to a single RS-232 port on a PC or other system. In other words, it's a point-to-point interface, as shown in Figure 14-15. It also has line length and speed limitations, because of the use of voltage swings to perform the signaling, and RS-232 tends to be susceptible to noise and interference from the surrounding environment.

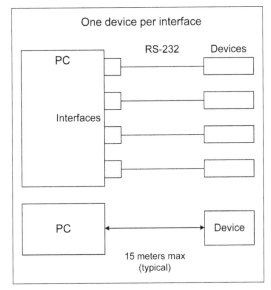

FIGURE 14-15. RS-232 connections

RS-232 SIGNALS

The RS-232 interface employs a number of signals for both data transfer and handshaking between two devices. Most of these signals are from the days when RS-232 was first defined and its intended application was to connect a terminal or mainframe to a modem. External modems are becoming scarce, but most RS-232 interfaces still retain the various lines shown in Figure 14-16. Note that these are for a DB-9 type connector. There are other signal lines that might be implemented with a DB-25 connector, but they are seldom used. If you're curious, you might want to find a copy of the EIA-232 specification and give it a read.

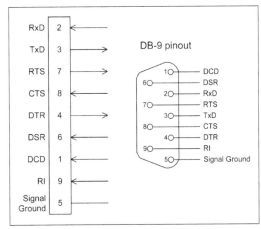

FIGURE 14-16. RS-232 signals

The basic RS-232 signals shown in Figure 14-16 and defined in Table 14-10 are really all that are necessary to implement a full RS-232 interface with hardware handshaking.

TABLE 14-10. RS-232 interface signal names

Signal	Definition
RxD	Received data
TxD	Transmitted data
RTS	Request to send
CTS	Clear to send
DTR	Data terminal ready
DSR	Data set ready
DCD	Data carrier detect
RI	Ring indicator

DATA COMMUNICATION INTERFACES

In many cases, all you really need are the RxD (receive) and TxD (transmit) data lines, and some devices are indeed wired this way. The CTS, RTS, DTR, and DSR signals are useful in cases where there is a need for strict data flow control, but at baud rates of 9,600 or less, where data is not moving continuously in large blocks, they can be eliminated. The DCD and RI signals are essentially useless these days, unless you want to connect to an old-style outboard modem. However, that doesn't mean the signals can just be left floating and unconnected. On a PC, for example, the control signals might need to be terminated correctly or the interface logic behind the RS-232 port might simply refuse to work.

DTE AND DCE

When working with RS-232 interfaces, you will no doubt encounter the abbreviations DTE and DCE, which translate to *data terminal equipment* and *data communications equipment*, respectively. Hailing from the days of mainframes and acoustic coupler modems, these terms were used to define the endpoints and link devices of a serial communications channel. The terms were originally introduced by IBM to describe communications devices and protocols for their mainframe products.

In the context of RS-232, DTE devices are found at the endpoints of a serial data communications channel. The *terminal* in DTE does not necessarily refer to a thing with a roll of paper and a keyboard (a teletype terminal, or TTY), or a CRT and a keyboard (an old-style computer terminal, or *glass TTY* as they were once known). It literally means "the end." Figure 14-17 shows this arrangement graphically.

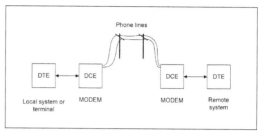

FIGURE 14-17. DTE and DCE

Another way to look at it is in terms of *data sink* and *data source*. A data sink receives data, and a data source emits it. Either end of the channel (the DTE devices) can be sinks or sources. The DCE devices provide the channel between the endpoints using some type of communications medium. For a system using modems as the DCE devices, this would typically be a telephone line, although VHF radio and microwave links have also been used in the role of communications medium.

Nowadays, modems are becoming something of an endangered species, although they are still used for data communications in some remote areas of the US and in places around the world that lack high-speed internet services. However, the wiring employed in RS-232 cables and connectors still reflects that legacy, which is why it's important to understand it in order to correctly connect things using RS-232.

The signals described in the previous section are named with reference to the DTE. In other words, on a DTE device, TxD is a data source, or output. On a DCE, it is a sink, or input, for the TxD of the DTE. This also applies to the RxD line. In effect, the DCE's data source and sink connections are functionally inverted with respect to the DTE's TxD and RxD lines, even though they have the same name. This might seem confusing, but the upshot of it all is that when you are connecting a DTE to a DCE, the interface is wired pin-to-pin between them (1 to 1, 2 to 2, and so on).

If you need to connect two devices that happen to be DTEs, you will need to use what is called a *cross-over cable*, or, if you don't need the handshake lines, a *null modem cable*. Figure 14-18 shows how the TxD and RxD lines would cross over for a DTE-to-DTE interface. It does not show how the handshake lines would need to be connected. For details about the use of handshaking, see "Handshaking" on page 315.

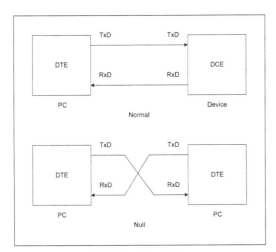

FIGURE 14-18. RS-232 null modem wiring

Most devices with a serial port are wired as DTE devices, although some can be configured as DCE. You can do this using jumpers, small PCB-mounted switches, a front-panel control, or even via software. The built-in serial port on a PC is typically implemented as a DTE.

HANDSHAKING

For devices with low-speed interfaces, or when the communications protocol is strictly a command/response format, you probably don't need the rest of RS-232's handshaking lines. In this case, you can use a ready-made commercial null-modem cable or a null-modem adapter like the one shown in Figure 14-19.

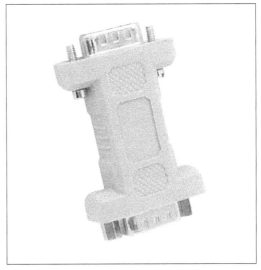

FIGURE 14-19. A DB-9 RS-232 null-modem adapter

RS-232's handshaking signals originated in response to its initial usage—namely, transferring data between a terminal and a remote host mainframe computer system, sometimes via a modem. The DTR, DSR, DCD, and RI signals mainly apply to devices such as modems. The RTS and CTS signals are used for flow control, regardless of what types of devices are involved.

DATA COMMUNICATION INTERFACES | 315

In the original version of the RS-232 specification, the use of the RTS and CTS lines is asymmetric—in the sense that the DTE asserts RTS to indicate that it has data ready to send to a DCE device, and in response, the DCE asserts CTS to indicate that it is in a state to accept data from the DTE. This request/accept protocol is used with half-duplex devices like RS-232 to RS-485 convertors, where the DTE's RTS signal is used to put the RS-485 device into transmit mode. "RS-485" on page 317 covers the half-duplex RS-485 interface. Note that this asymetric flow control scheme does not provide any way for the DTE to indicate that it is unable to accept data from a DCE, so the DTE needs to be able to accept and buffer whatever the DCE throws at it.

Of course, someone developed a nonstandard version of the handshake protocol, wherein the CTS signal is used to indicate that the DCE is ready to accept data and the RTS signal indicates that the DTE can accept data. This is known as *RTS/CTS handshaking*. In some operating systems, enabling an RTS/CTS handshake flow control is a configuration setting.

Some RS-232 interfaces are designed to operate without the handshaking lines, while with others, the handshaking logic can be disabled by software. But even if one device doesn't support handshaking signals, another device might, and it will need to be wired or configured to ignore them. Many people have puzzled over why an RS-232 port wasn't working, only to discover that they needed to correctly configure the handshaking lines for the port in order to get it to respond to an external device.

Figure 14-20 shows how a microcontroller can be connected to a DB-9 connector for an RS-232 port configured as DTE. This is sometimes referred to as a *handshake loopback*.

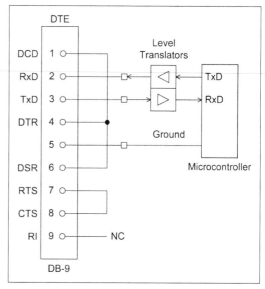

FIGURE 14-20. Connecting a microcontroller to a standard RS-232 port

Microcontrollers don't come with handshaking implemented for a serial port. If handshake signals are necessary, they must be taken from general-purpose DIO pins and controlled by software.

RS-232 COMPONENTS

With a microcontroller that already has a built-in UART, all you really need in order to connect it to something with an RS-232 interface are a set of level translators, also called *driver ICs*. These components convert 3.3V or 5V logic signals into the +/− voltages used for RS-232. Table 14-11 lists some of the interface ICs that are available. Some have driver/receiver pairs (a transceiver), and others have just drivers or receivers.

TABLE 14-11. RS-232 interface ICs (level translators)

Part number	Manufacturer	Transmitters	Receivers
LTC2801	Linear Technology	1	1
LTC2803	Linear Technology	2	2
MAX232	Maxim	2	2
MAX232	Texas Instruments	2	2
MC1488	ON Semiconductor	4	0
MC1489	ON Semiconductor	0	4
SN75188	Texas Instruments	4	0
SN75189	Texas Instruments	0	4

External UART IC components are available from multiple sources. These aren't generally suitable for use with microcontrollers because they require an address and data bus interface that might not be available with a microcontroller. They are commonly found along with microprocessors to provide RS-232 interface capability.

The MAX3100 or MAX3107 devices from Maxim are one type of UART IC, and Texas Instruments is a source for the venerable 16550 UART in the form of the PC16550D IC (Table 14-12 lists a few other available UART ICs). The 16550 is historically interesting, because many older (and even current) motherboards include an embedded version of this device in the IC chip set that supports the microprocessor. It might no longer be a separate outboard part in a PC, but it, or something very much like it, might still be there.

TABLE 14-12. RS-232 UART ICs

Part number	Manufacturer	Interface
MAX3100	Maxim	SPI
MAX3107	Maxim	I2C/SPI
PC16550D	Texas Instruments	Address/data bus
SC16IS740	NXP	I2C/SPI
TL16C752B	Texas Instruments	Address/data bus

RS-485

RS-485 (or EIA-485) is commonly found in instrument control interfaces and in industrial settings. RS-485 and its predecessor, RS-422, have a high level of noise immunity, and cable lengths can extend up to 1,200 meters in some cases. RS-485 is also significantly faster than RS-232. It can support data rates of around 35 Mb/s with a 10-meter cable, and 100 Kb/s at 1,200 meters.

So why should you care about RS-485? To be honest, you might not have a reason to, but as stated earlier, it does bear on the historical background of differential signaling schemes like USB. Outside of the historical curiosity aspect, should you ever want to work with things like a data acquisition module for an automated small-batch brewery, or a stepper motor controller module for that robot you've been wanting to build, then knowing about RS-485 might come in handy (search Jameco's website (*http://www.jameco.com*) for part number 184997).

RS-485 SIGNALS

RS-485 owes its speed and range capabilities to the use of differential signaling. Instead of using a dedicated wire to carry data in a particular direction, two electrically paired wires can transfer data in either direction, but not at the same time. The two wires in a differential interface are always the opposite of each other in polarity. The state of the lines relative to each other indicates a change from a logic value of 1 to a 0, or vice versa. Figure 14-21 shows a typical situation involving asynchronous serial data that incorporates a start bit and a stop bit. For comparison, it also shows the digital TTL input that corresponds to the RS-485 signals.

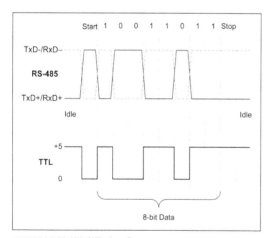

FIGURE 14-21. RS-485 signals

The TxD–/RxD– and TxD+/RxD+ designations in Figure 14-21 indicate that the + and – lines will connect a TxD output to an RxD input. The + and – indicate the logic true polarity of the signal (the – line is true (1) when low, and the + is true (1) when high). In many diagrams of RS-485 circuits, you will often see the lines labeled as just + and –. Note that the + and – signals always return to the initial starting state at the completion of the transfer of a byte of data.

LINE DRIVERS AND RECEIVERS

In an RS-485 interface, each connection point uses a pair of devices consisting of a differential transmitter and a differential receiver, as shown in Figure 14-22.

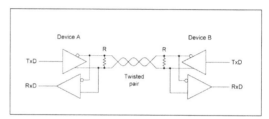

FIGURE 14-22. RS-485 differential transmitter and receiver

RS-485 might be implemented as a two-wire half-duplex interface, or it might be a bidirectional, four-wire, full-duplex interface, but for many applications, full-duplex operation is not required. Figure 14-23 shows a four-wire arrangement.

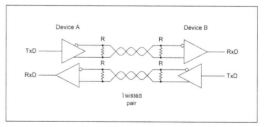

FIGURE 14-23. RS-485 full-duplex four-wire interface

RS-485 MULTI-DROP

RS-485 also allows you to connect more than one device (or node) to the serial bus in what is called a *multi-drop configuration*, as shown in Figure 14-24. To do this, you must be able to place the transmitter (output) section of an RS-485 driver into a Hi-Z (or high impedance) mode. This capability is also essential when RS-485 is connected in two-wire mode.

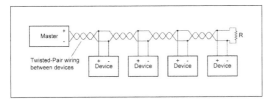

FIGURE 14-24. RS-485 multi-drop connections

The reason for this is that, if the transmitter were always actively connected to the interface, it could conflict with another transmitter somewhere along the line. In Figure 14-25, you can see how the drivers take turns being *talkers* (or data sources) when wired in half-duplex mode, depending on which direction the data is moving. There is really no need to disconnect the receivers, so they can listen in to the traffic on the interface at any time.

In a typical multi-drop configuration, one device is designated as the controller (or master) and all other devices on the RS-485 bus are subordinate to it, although it is possible to have multiple controllers on an RS-485 network. The default mode for the controller is transmit, and for the subordinate devices, it is receive. They trade places when the controller specifically requests data from one of the subordinate devices. When this occurs, it is called *turnaround*.

When using a half-duplex RS-485 interface, you must take into account the amount of time required to perform a turnaround of the interface. Even with an interface that can sense a turnaround electronically and automatically change its state from sender to receiver, there is still a small amount of time involved. Some RS-232 to RS-485 converters can use the RTS line from the RS-232 interface to perform the turnaround, as well.

RS-485 COMPONENTS

There are numerous transceiver ICs available for use with RS-485, as listed in Table 14-13. Many are available in through-hole DIP packages, but many are packaged in small-outline SMD styles, which are relatively easy to solder onto a PCB.

TABLE 14-13. RS-485 transceiver ICs

Part number	Manufacturer	Number of Transceivers
SN65HVD11DR	Texas Instruments	1
MAX13430	Maxim	1
MAX13442E	Maxim	1

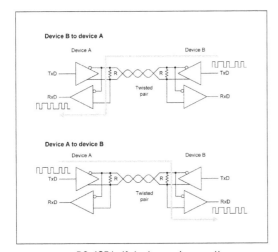

FIGURE 14-25. RS-485 half-duplex mode operation

Part number	Manufacturer	Number of Transceivers
SN75ALS1177N	Texas Instruments	2
SN65LBC173AD	Texas Instruments	4

Some of the parts listed in Table 14-13 are designed to operate in 3.3V circuit. Others can tolerate from 3.3 to 5V logic levels. If you elect to use an RS-485 transceiver IC, be sure to read the datasheet. Some have coupled logic for half-duplex operation. Other have TxD/RxD driver/receiver pairs that are not internally linked, so they can be used for full-duplex operation.

RS-232 vs. RS-485

Table 14-14 shows a comparison of some of the electrical characteristics of RS-232 and RS-485.

TABLE 14-14. Comparison of key features of RS-232 and RS-485

Characteristic	RS-232	RS-485
Differential	No	Yes
Max number of transmitters	1	32
Max number of receivers	1	32
Modes of operation	Half- and full-duplex	Half- or full-duplex
Network topology	Point-to-point	Multi-drop
Max distance	15 m	1,200 m
Max speed at 12 m	20 Kbs	35 Mbs
Max speed at 1,200 m	n/a	100 Kbs

In general, RS-232 is fine for applications that don't require high speed and with cable lengths shorter than 5 meters (about 15 feet) or so. If an external device utilizes very short (2- to 10-character) commands, returns equally short responses, and sometimes takes significantly longer to perform the commanded action than the time required to send the command to the device, an RS-232 interface at a speed of 9,600 baud is perfectly acceptable. In other cases, such as when you might want to have sensors or controllers distributed on a single communications bus, RS-485 would be the way to go for a serial data interface.

USB

USB was intended to be easy for end users to deal with. Consequently, when you are connecting a digital camera, tablet, or a BeagleBone board to a PC using a USB cable, all the details about the connection are hidden. Things like connection speed and capabilities are contained within both the external device and the driver software running on the user's computer. After the driver software is installed, the user need only plug in the device to start using it. But this ease of use for the user comes at the price of complexity for the person implementing the USB interface and the driver software.

Some microcontrollers come with built-in USB logic, and in other cases, you might need or want to add USB capability by using an outboard interface IC. It is also possible to convert RS-232 or TTL-level serial data into USB and back again using a single IC. In general, however, if the thing you are working with doesn't already have USB, then adding it should probably involve some careful thought. It might not be as simple as you think, because even if the hardware is straightforward, you will still need to create or buy the software to communicate with it. From a developer's or engineer's point of view, USB can be a challenge to work with.

> If you plan to work with USB, you should be prepared for something of a steep learning curve. USB is not a simple protocol, and there is a lot going on behind the scenes to make it all work as seamlesslty as it does—at least most of the time.
>
> This section is *not* a complete description of how to implement or use a USB interface. That can easily take an entire book, such as some of the titles listed in Appendix C. The intent here is give you some idea of what is involved and provide some suggestions for places to look for the additional details you might need.

USB TERMINOLOGY

As is true of almost every other specialized area of electronics technology, there are words and terms unique to USB technology. For convenience's sake, here are some of the terms used with USB that will appear in the rest of this section (they are also in the Glossary):

Data sink
A place where data is received.

Data source
A source of data (i.e., a sender).

Descriptor
A data structure within a device that allows it to identify itself to a host.

Device
A USB peripheral or function. Also used as peripheral device. See *function*.

Downstream
Looking out from the host to hubs or devices connected outward on a USB network.

Endpoint
An endpoint exists within a device, typically in the form of a FIFO buffer. They can be either data sinks (receiving) or data sources (sending).

Enumeration
When a USB device is initially connected to host, the host gets a connection notice and proceeds to determine the type and capabilities of the device.

Function
A function is a USB device, also referred to as a *USB peripheral* or just *device*. USB functions are *downstream* from the host.

Host
The host is the master on a USB network, and all other devices (or functions) respond to it.

Hub
A USB hub is used to expand the number of devices with which a USB host can communicate.

Interface
A set of endpoints in a device that act as either data sources or data sinks. An interface can have multiple endpoints acting as data sinks or data sources.

Peripheral
Another name for a device or function.

Pipe
  A logical connection between a host and the interface endpoints of a device.

Request
  Sent by a host to a device to request data or have the device perform an action.

Upstream
  Looking back toward the host from the perspective of the hubs and devices in a USB network.

USB CONNECTIONS

USB is a half-duplex asynchronous serial interface that uses a *master/slave* type bus with exactly one master and multiple slaves. The slaves are called *devices*, *peripherals*, or *functions*. The master is called a *host*, and only the host can initiate USB transfers. Peripheral devices always respond; they never initiate.

Electrically, USB is simple: just four wires. As shown in Figure 14-26, two of the wires carry data and use a differential signaling method similar to RS-485. The other two wires are used for DC power and ground. All four lines run through a shielded cable. Chapter 7 describes the connectors used for a host (an *A* type) and a device (*B*, mini, or micro).

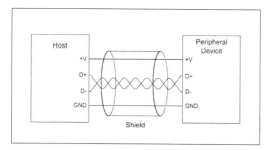

FIGURE 14-26. USB electrical wiring

A USB host can detect the devices connected to it and query each one to enumerate their type and capabilities. It can also detect when a device is connected or removed. Multiple USB devices are connected in a tree-like topology, with the controller at the base of the tree and the individual USB devices out at the branch tips. A *USB hub* is used to distribute signals between one controller port and multiple device ports. Figure 14-27 shows a USB network consisting of a host system with an internal hub, two external hubs, and eight USB devices.

USB CLASSES

The USB standard defines various classes of devices that utilize a USB interface. Each class has its own set of commands and responses, and each is intended for a specific set of applications. Table 14-15 lists some of the more common classes that you might encounter on a regular basis.

TABLE 14-15. Common USB device classes

USB class	Example(s)
Communications	Ethernet adapter, modem
HID (Human Interface Device)	Keyboard, mouse, etc.
Imaging	Webcam, scanner
IrDA	Infrared data link/control
Mass Storage	Disk drive, SSDD, flash memory stick
PID (Physical Interface Device)	Force feedback joystick

USB class	Example(s)
Printer	Laser printer
Smart Card	Smart card reader
Test & Measurement (USBTMC)	Test and measurement devices
Video	Webcam

You are probably already familiar with the HID and Mass Storage classes. These two classes include things like keyboards, mice, simple joysticks, outboard USB disk drives, and flash memory sticks (so-called *thumbdrives*). The HID class is relatively easy to implement, and most operating systems come with generic HID class drivers, so it is not uncommon to find devices implemented using the HID class that don't look anything at all like a keyboard or mouse. If a USB device uses a unique interface, it is up to the implementor to supply the low-level interface drivers needed by the operating system.

USB DATA RATES

The maximum data rate for a USB interfaces (measured in bits per second) can vary from 1.5 Mb/s to 4 Gb/s. The data transfer rates for USB are defined in terms of a revision level of the USB standard. In other words, devices that are compliant with USB 1.1 have a theoretical maximum data rate of 12 Mbits/s (megabits/second) in the full-speed mode, whereas USB 3.0-compliant devices have a maximum data transfer rate of 4 Gb/s (gigabits/second). Table 14-16 lists the specification levels and the associated maximum data transfer rates.

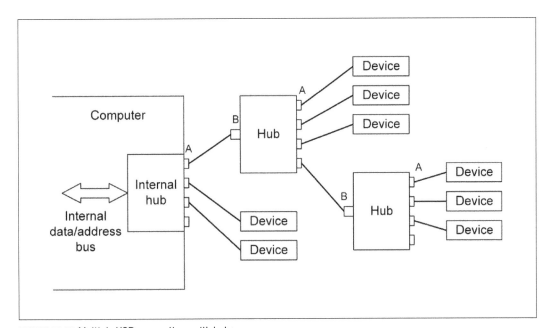

FIGURE 14-27. Multiple USB connections with hubs

DATA COMMUNICATION INTERFACES

TABLE 14-16. USB versions

Version	Release date	Maximum data rate(s)	Rate name	Comments/features added
1.0	1996	1.5 Mb/s	Low speed	Very limited adoption by industry
1.0	1996	12 Mb/s	Full speed	
1.1	1998	1.5 Mb/s	Low speed	Version most widely initially adopted
1.1	1998	12 Mb/s	Full speed	
2.0	2000	480 Mb/s	High speed	Mini and micro connectors, power management
3.0	2007	4 Gb/s	Super speed	Modified connectors, backward compatible

USB 3.0 is still rather new, and most USB devices that one might encounter in use will either be 1.1 or 2.0 compliant. You should also know that even though a device claims to be a USB 2.0 high-speed type, the odds of getting sustained rates of 480 Mb/s are slim. The time required for the microcontroller in the USB device to receive a command, decode it, perform whatever action is requested, and respond back to the host can be considerably slower than what you might expect from the data transfer rate alone. In addition, the ability of the host controller to manage communications can contribute to slower than theoretical maximum data rates. If the host is busy with other tasks, it might be unable to service the USB channel fast enough to sustain a high data throughput.

USB HUBS

A USB hub concentrates and passes data from downstream devices and other hubs to the upstream host. A root hub will pass the data directly into the host system. A PC (and most computers with USB capability) typically has a root hub built into its USB controller logic. Besides serving as the first-order router/distribution controller for a USB network, the root hub also detects when devices are being connected or disconnected. A root hub can be implemented either as a separate IC or as part of the chip set for the computer's microprocessor.

How USB devices and hubs are connected can also play a big part in how responsive the communications will be. A USB network using v1.1 hubs and devices is only as fast as the slowest device in the network, so just one 1.5 Mb/s device will drag the enter network down to that speed, even if some of the other devices can run at 12 Mb/s.

USB v2.0 hubs deal with this by separating the v1.1 low-/full-speed traffic from the v2.0 high-speed data. When purchasing new USB components, you should avoid v1.1 hubs and stick to v2.0 units. That way you can avoid having a high-speed v2.0 device run into a bottleneck due to v1.1 devices on your USB network, assuming that the host controller is itself USB v2.0 high-speed capable.

Hubs also participate in power management for peripheral devices. In a PC, the root hub can supply 500 mA of 5V power per A type connector. Of that, 100 mA is allocated to an external hub, leaving 400 mA for peripheral devices. A bus-powered hub is limited to 100 mA per type A port, so unless a downstream hub has its own power supply (i.e., it's a self-powered hub), the number of devices it can support at 100 mA each is limited to four. If it's self-powered, each of the hub's type A ports should be able to supply 500 mA.

DEVICE CONFIGURATION

When a device (perhaps an MP3 player or an Arduino, for example) is connected to a host, the first thing that happens is that the host becomes aware of the device due to a pull-up resistor on one of the data lines inside the device. Once a device is detected, the host requests a series of descriptors from the new device. These are data structures (tables) that contain information about the device's class, endpoints, and speed.

The sequence of actions that result when a device is connected, and which lead up to it being fully operational and ready for use, can be summarized as follows:

1. A device is connected to a USB host or hub and a data line is pulled high by the device.
2. Host issues a reset to the new device to place it into a known state with the default address of 0.
3. Host sends a request to endpoint 0 on device address 0 to get the maximum data packet size.
4. Host might send the reset command again and then issue a Set Address command that specifies a unique address for the device at address 0. If successful, the device assumes the new address.
5. Host sends query commands to the device at the assigned address to obtain information about the device, including the Device Descriptor, Configuration Descriptor, and String Descriptor.
6. Once the host has gathered sufficient information from the device, it will load the appropriate device driver.
7. The host now issues a Set Configuration command to the device, and it will now respond to commands specific to the type of device it happens to be.

If the data received from the device is consistent and complete, the host configures the device for operation. However, if the host is not satisfied with the data obtained from the device, it will ignore the device. When this happens, the host operating system will usually open a small error window stating that it encountered a problem with the new device.

USB ENDPOINTS AND PIPES

Each USB peripheral has a unique address on the USB network, assigned by the host. Each USB device has an interface that will respond to this address, and an interface can contain up to 16 endpoints. An endpoint can be either a data source or a data sink, which is the direction of the endpoint. Each endpoint is assigned a number from 0 to a maximum of 15. Every device has a default control endpoint 0, which is bidirectional. A

USB endpoint is somewhat like the notion of a port used for network communications.

Endpoints are part of a device, not something the host assigns. A device reports its endpoint numbers and characteristics when it is enumerated by the host during the initalization and configuration sequence. Once the host knows the endpoints, it uses both the device address and the endpoint number when communicating with a device. Figure 14-28 shows how endpoints exist in a device behind its assigned device address.

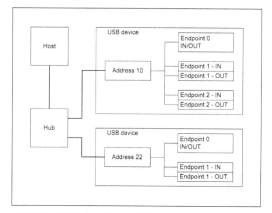

FIGURE 14-28. Endpoints in a USB device interface

A USB device sends and receives data using endpoints; the host client software transfers data through pipes. A *pipe* is a logical connection between the host and endpoint(s). Each instance of a pipe has a set of parameters associated with it, such as allocated bandwidth, transfer type (Control, Bulk, Iso, or Interrupt), data direction, and maximum packet and buffer sizes. Figure 14-29 illustrates the concept of a pipe.

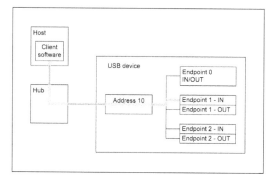

FIGURE 14-29. USB pipe

The USB specification defines two types of pipes: stream and message. A *stream* has no defined format, and it can be used to send or retrieve any type of data. A stream pipe can be used with bulk, isochronous, and interrupt transfer types, and it can be controlled by either the host or the device. *Message* pipes use a format defined in the USB specification, and they are always host-controlled. The direction of data flow is set by the request from the host. Message pipes support control transfers only.

DEVICE CONTROL

All USB devices must recognize and respond to the basic set of control commands (i.e., requests) used to enumerate devices and their capabilities. Specific requests sent by the host to a USB device will depend on the class of the device. For example, HID devices use a different set of requests than devices in the mass storage class.

Working through the possible requests and responses for each of the USB classes is way beyond the scope of this small section. I will say, however, that HID is relatively straightforward, which is why it is so often used for things like toy rocket launchers or dancing robots—and, of course, keyboards and mice. If you are working with something like a Raspberry Pi or a BeagleBone, you should have access to documentation on the USB port. If you don't have any documentation about the device you want to connect to with a USB cable, you will have to either search it out online, if it exists, or resort to reverse-engineering. See "USB Hacking" on page 327 for some tips and hints on reverse-engineering a USB interface.

USB INTERFACE COMPONENTS

Low-level components provide the interface functions necessary to implement either a host or a device, and some are capable of doing either function. Table 14-17 lists some of the IC components available for SPI to USB, I2C to USB, and RS-232 to USB.

TABLE 14-17. USB interface ICs

Part number	Manufacturer	Function	Interface
CP2102	Silicon Labs	USB to serial UART bridge	RS-232
CP2112	Silicon Labs	HID_USB to I2C bridge	I2C
FT232R	Future Technology Devices International (FTDI)	USB to serial UART bridge	RS-232
MAX3421E	Maxim	Peripheral/host controller	SPI
MAX3420E	Maxim	Peripheral controller	SPI

You should carefully study the datasheets and reference documentation provided by the various IC manufacturers. While electrically simple, some of these devices are internally complex and require you to have a good knowledge of USB to be able to use them correctly.

USB HACKING

A while back, someone came up with the idea of making a foam "rocket" launcher as a desk toy for bored office workers. You may have seen one, or been attacked by one (it doesn't hurt; it's just surprising and somewhat annoying). These toys use a USB interface to control the motion of the turret and an air puff mechanism that launches the foam missile across the room. Some recent versions even have a built-in video camera to assist with aiming. Now, what if you wanted to repurpose the turret mechanism for some other task? How does the USB interface work? What are the command codes? Without technical information from the manufacturer, you are basically starting from zero.

When you're faced with a USB interface and no documentation, sometimes the only way to find out what commands it accepts and how it responds is to reverse-engineer it. If you are adept with software development tools, you might want to take the approach described in an article by Pedram Amini (*http://bit.ly/py-launch*).

Another way to do this involves a USB communications analyzer that can watch the data traffic moving between a USB host and a device. As it turns out, Linux has this ability already built in, but of course that means that whatever you are trying to analyze must already have software for it running under Linux. It basically involves the kernel debug module, debugfs, and the usbmon facility. The Mengazi blog has a post (*http://biot.com/blog/usb-sniffing-on-linux*) with more information on the topic, and the Dlog blog outlines more approaches (*http://bit.ly/usb-traffic*).

There are a number of USB software diagnostic tools for different platforms available, some of them starting at free and then going up from there. Of course, expect the "free" software to have some limitations, but it might be sufficient to show you the commands and responses moving between your PC and the external device.

For an example of what can be done once the requests and responses for a device are known, check out Karl Ostmo's controller application for the Dream Cheeky USB missile launcher for Linux called pyRocket (*https://code.google.com/p/pyrocket/*).

Ethernet Network Communications

Ethernet is now over 30 years old. From its humble beginnings as a coaxial cable strung from computer to computer, it has evolved into a widespread form of computer networking that incorporates firewalls, routers, switches, bridges, and protocol translators into its architecture. In addition to the obvious locations like PCs, servers, and printers, Ethernet can be found in applications as diverse as industrial control systems, submarines, kitchen appliances, and traffic control systems.

ETHERNET BASICS

A complete description of Ethernet and networking is way beyond the scope of this book, but this section will focus on some key points to consider that might help make things clearer. For more information, refer to the texts listed in Appendix C.

But first, there's one very important thing to remember: Ethernet is the physical transport. The Internet protocol (IP) network—with its IP addresses, sockets, and subnets—is a logical construct that can use Ethernet to move data around. Sometimes, you might hear someone refer to these two concepts as if they were the same thing. They are not. Instead of Ethernet, the transport could just as easily be RS-232, RS-485, token ring, a radio link, lasers, or some other hardware communications scheme. The IP network uses the concept of IP addresses to route data between devices on the network, but it's doing this at a high level. At the transport level of physical wires and radio waves, data is moved from one point to another via unique hardware addresses, and that's all it does. It's up to the network software in the host system to process and manage the data packets. This section focuses on Ethernet and the IC components and modules available to work with it.

For the purposes of this discussion, the term *device* refers to the physical interface hardware and *host* refers to the system (PC, microcontroller, etc.) that is using the device to communicate. Hopefully, this will keep things consistent with the rest of this chapter. You might also encounter the abbreviation NIC, which stands for *network interface controller*. It appears quite often when you're working with PCs, and some folks might think it means "network interface card," but it's actually a general concept that applies to any device that can act as an Ethernet transceiver.

Modern (10BASE-T and up) Ethernet is a full-duplex, packet-based network with no masters and no slave devices. Any Ethernet device can communicate with any other device with which it can establish a connection in a peer-to-peer manner. But in order to establish that connection and exchange data packets, the two devices need to have fixed addresses.

Every device on an Ethernet network has a media access control (MAC) address, which is a unique identification number assigned to each device. Sometimes, this number is written into a device when it's manufactured, and other times, you need to assign it yourself. Without this identifier, an Ethernet device cannot send or receive packets of data through the network. Don't confuse the MAC address with the IP address of a host. As stated earlier, they belong to different levels of the network, although they are used together to correctly route data packets through the network.

Ethernet is a packet-based network scheme. An Ethernet packet is called a *frame* and consists of seven different data fields, as shown in Figure 14-30. Frames are specified in terms of octets, which are defined as 8 bits of data. The term *octet* was originally used because the size of a byte on early computer systems could vary, depending on the architecture of the computer. These days, the size of a byte has settled down to 8 bits, so byte and octet are synonymous.

Figure 14-30 is the original Ethernet frame, and it's still in widespread use. There are also variations, depending on the Ethernet specification in use. One variation in common use is known as Ethernet II. IPv6, which is very slowly replacing the original IPv4, also has its own Ethernet frame definition. Generally, you don't need to worry about it, because the NIC will identify the variations it can handle and process them accordingly.

When considering the speed of an Ethernet connection, you might encounter terms like *10BASE-T*. This translates to "10 MHz twisted pair." In earlier versions of Ethernet, the designations 10BASE2 and 10BASE5 also appeared, referring to coaxial cables interfaces. These are now obsolete, having been replaced by the common CAT-5 and CAT-6 cables with internal twisted-wire pairs and RJ-45 connectors.

There are three network speeds in common use today, as shown in Table 14-18.

TABLE 14-18. Ethernet network speeds

Network	Speed
10BASE-T	10 MHz
100BASE-T	100 MHz
1000BASE-T	1 GHz

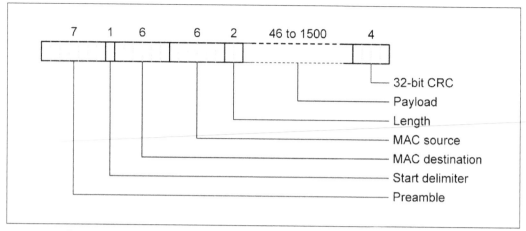

FIGURE 14-30. Ethernet frame contents

The theoretical maximum possible speed of an Ethernet connection is measured in bits per second, which is the same as the frequencies listed in Table 14-18. Because of the overhead associated with the frame headers and the additional layers of header data used by an IP network, buffer latencies, and various processing actions, the actual number of bytes per second that can be transferred between applications running on separate hosts isn't the same as simply dividing the speed by 8. Also keep in mind that, just because an Ethernet NIC is rated for 10 MHz, that doesn't mean that the device using it to communicate needs to shovel data into it that fast. The NIC will send a packet when the packet is ready, and if it takes a few hundred milliseconds for a microcontroller to write enough data to fill the frame payload, then a packet will be sent every few hundred milliseconds.

To ensure backward compatibility, the typical 100BASE-T interface can slow down to accommodate a 10BASE-T connection, and a 1000BASE-T interface can (or at least should) work with a 100BASE-T or a 10BASE-T connection. Because of this, you will often see network speeds given as 10/100, or 10/100/1000. Most new PCs, notebooks, and netbooks come with built-in 10/100 Ethernet ports. Some computer and tablet products don't have an Ethernet port, but these either rely solely on a wireless connection (discussed in "802.11" on page 338), or, if wireless is not an option, you can use a USB-to-Ethernet or Thunderbolt-to-Ethernet adapter.

Twisted-pair Ethernet is *point to point*, meaning that one host with one interface can connect to only one other Ethernet interface and the connections are one to one between the connectors. Because of this, you can't simply connect one host system to another (unless one of the NICs can sense which wires are the TxD pair and which are the RxD). Generally, in order to connect one host NIC to another, you need to use a crossover cable that swaps the positions of the TxD and RxD pairs inside the RJ-45 connector.

Hubs and switches are used to expand the network. A *hub* is a device that accepts Ethernet packets from a host and redistributes them to every other host that happens to be connected to the hub, as shown in Figure 14-31.

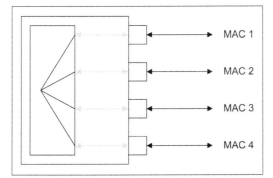

FIGURE 14-31. Ethernet hub

A network *switch* is essentially a "smart" hub, in that it will discover the MAC addresses of the devices connected to it and send only the data packets with the appropriate MAC address to the intended recipient. Some switches can be configured to distribute network traffic only from certain hosts to a specific host or group of hosts. Figure 14-32 shows a simplified diagram of an Ethernet network switch.

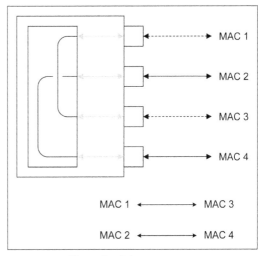

FIGURE 14-32. Ethernet switch

When a host receives a packet, the network interface hardware checks the MAC address in the packet to verify that it matches the interface's own MAC address. If it's a match, the interface will pass the data on to the driver software in the host, where the packets are reassembled into a form that the application software even further on up the chain can deal with. Figure 14-33 shows a simplified diagram of the entire network *stack*.

Once past the interface hardware, we're no longer working with Ethernet. As discussed earlier, Ethernet is just one way to move data around in a network; it's just the transport method. Past the network interface, it's just network data, regardless of how it got there.

From the perspective of the operating system on a computer, the network interface is a source or sink for data. The low-level software drivers that communicate with the hardware inspect the IP address of the data and pass the data to application software that has opened a port to the network interface.

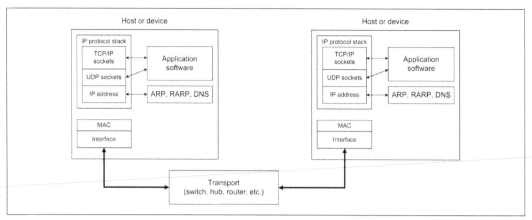

FIGURE 14-34. Ethernet and IP network components

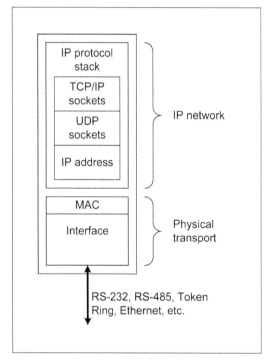

FIGURE 14-33. Ethernet and the IP network protocol stack

The notion of a port is similar to the endpoints used in an I2C network. They are logic constructs that allow an interface to send and receive data on behalf of multiple applications running in the system. Unlike I2C endpoints, a IP network port is bidirectional and full duplex. Figure 14-34 shows how the higher-level IP network functions in a host system use the low-level Ethernet transport to move data between systems.

The low-level software that does the receiving, reassembly, assembly, and sending is called a *protocol stack*. Some Ethernet controller ICs have a protocol stack already built into them, while others need a microcontroller to handle it. When a host system sends data to another host, it uses the functions in the protocol stack to assemble the outgoing packet. The IP address of the remote host is used to look up its MAC address and load the Ethernet frame payload with data. When the outbound packet is complete, it is passed into the Ethernet interface for transmission.

Figure 14-35 shows an example network composed of six hosts, a hub, and a switch, all connected to another network (or perhaps the Internet) via a router/firewall device.

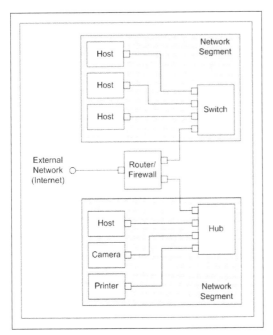

FIGURE 14-35. An example Ethernet network using a hub and a switch

ETHERNET ICS, MODULES, AND USB CONVERTORS

The microcontroller interface speed is an important factor to take into consideration when you are interfacing to an Ethernet controller IC or module. For an SPI interface, the speed of the serial interface ultimately determines just how fast data can move through the Ethernet IC. For example, a 10 MHz (10BASE-T) IC can theoretically move data at a maximum rate of 10 Mb/s, although in reality it can seldom achieve this because of latencies (slow downs) encountered in the network. A fast SPI interface can transfer data at a maximum rate of 3.4 Mb/s (assuming that the SPI master is capable of this), and that's much slower than the 10 Mb/s of "slow" Ethernet. Even running full out, the SPI master will never be able to push the Ethernet controller anywhere close

to its limit, and I2C is even slower. USB-to-Ethernet ICs and convertors fare somewhat better, since the maximum clock rate of USB V2.0 is 480 Mb/s, but the half-duplex nature of USB can get in the way and again slow things down.

FIGURE 14-36. Low-cost Ethernet interface module

Many Ethernet ICs use the media independant interface (MII) or reduced media independent interface (RMII) to connect to a microcontroller. This type of interface is designed to provide one part of a network interface and is usually found on 32-bit microcontrollers, such as the ARM Cortex-M3 and Atmel's 32-bit version of the AVR. MII or RMII allows a fast microcontroller to drive an Ethernet interface IC to near full network speed. For 8-bit microcontrollers, the only real option is an Ethernet controller or module with an SPI interface and slower network data transfer speeds.

For the most part, it's better to use a prebuilt module than to try to roll your own using just an Ethernet controller IC, but it's not that hard if you want to make your own PCB. There are a variety of Ethernet

TABLE 14-19. Ethernet ICs

Part number	Manufacturer	Function	Speed	Interface
AX88796C	ASIX	Ethernet controller	10/100	SPI
KSZ8851SNL	Micrel	Ethernet controller	10/100	SPI
LAN9512	Microchip	USB-Ethernet interface	10/100	USB
ENC28J60	Microchip	Ethernet controller	10	SPI
W5100	WIZnet	Ethernet controller	10/100	SPI

component parts to choose from, and Table 14-19 lists a few parts that have been spotted in low-cost modules for microcontroller applications.

Figure 14-36 shows one module that uses the W5100. This PCB also has an SD flash card socket, which makes sense, since it is an SPI module and the SD cards use an SPI interface.

USB-to-Ethernet convertors, such as the one shown in Figure 14-37, are common and readily available.

FIGURE 14-37. USB-to-Ethernet convertor

These convertors are useful when you want to add a second Ethernet port to a notebook or netbook computer that has only a single Ethernet port (which is most of them). They can also be used with a microcontroller that has a USB host port attached to it. Several companies make USB host port modules like the one shown in Figure 14-38.

FIGURE 14-38. Low-cost USB interface module

Wireless Communications

In order to communicate, devices need some way to transfer data. One way is to use wires, but wires tend to get in the way, they can add to the cost of a device or project, and they are a pain to run from device to device. The alternative is to go wireless.

BANDWIDTH AND MODULATION

The term *bandwidth*, as used in this section, refers to the actual range of frequencies employed by a band allocation in the radio spectrum, or by the amount of deviation exhibited by a transmitter at a particular frequency. When tuned to a specific frequency, a transmitter will also radiate some power at frequencies above and below the desired output frequency. Good transmitter design and tuning can eliminate much of this.

A radio carrier by itself doesn't really carry any information; it is just electromagnetic energy radiated into space at some specific frequency. Early radio systems (and even some in use today) used the technique of interrupting the carrier output to send information encoded in the form of Morse code or some other time-based scheme.

> **TIP** This is just a cursory overview of radio communications and is here mainly to set the stage for the upcoming descriptions of the wireless digital communications components to follow. If you want to learn more about radio technology, I suggest the books *Essentials of Short-Range Wireless*, *The ARRL Handbook*, and *The Radio Handbook*, all of which are listed in Appendix C. You'll also find a lot of good information available on the websites of wireless component manufacturers.

In amplitude modulation (AM) radio, the level (or amplitude) of the carrier wave is modulated by an input, typically audio. A receiver tuned to the same frequency as the transmitter need only extract the amplitude changes from the carrier to reconstruct the original audio modulation signal. AM is still used for commercial radio, and some early types of remote control and data communications used this technique as well, but it suffers from interference and slow data rates.

Frequency modulation (FM) was developed to overcome the problems of noise and interference experienced with AM radio. With AM, anything from a thunderstorm over the horizon to an automobile with bad spark-plug cables can interfere with the signal, but FM is largely immune to these types of interference. FM works by shifting the frequency of the carrier signal in proportion to the modulation input.

For example, in conventional commercial FM radio, a station operating at, say, 98.4 MHz uses a shift in its carrier frequency that corresponds to an audio input. In the US, FM radio stations are spaced 200 kHz apart, which is more than really necessary but avoids interference between stations. When an audio signal is used to modulate the carrier, it will shift in frequency depending on the polarity of the audio input. The *modulation level* (how far the carrier frequency will shift) depends on the level of the audio input but is constrained to keep the frequency within the 200 kHz channel. The upshot is that while the frequency of the FM station is nominally 98.4 MHz, the actual frequency at any given moment in time depends on the modulation level.

The modulation technique commonly used for digital communications is called *frequency-shift keying* (FSK). It is similar to FM except that information is encoded using discrete changes in the carrier frequency. A digital 1 or 0 causes the carrier to shift in frequency by a specific amount. A variation of

FSK, called Gaussian frequency-shift keying (GFSK), is used by the GSM mobile phone standard and in the devices described in this section. GFSK works well when you are dealing with binary 0 and 1 values, and the use of GFSK also helps to reduce the overall bandwidth of the signal at a given frequency. This, in turn, helps minimize interference with other devices operating on nearby frequencies.

THE ISM RADIO BANDS

A number of frequency ranges (or *bands*) have been designated industrial, scientific, and medical use (collectively referred to as the *ISM bands*). Internationally, these are defined by the ITU-R (International Telecommunication Union-Radio), with individual member countries allowing some or all of these frequencies within their own borders. Table 14-20 lists the ISM bands commonly used for consumer electronic devices in the United States.

TABLE 14-20. Common ISM radio frequency bands

Lower	Upper	Bandwidth	Center
26.957 MHz	27.283 MHz	326 KHz	27.120 MHz
433.05 Mhz	434.79 MHz	1.74 MHz	433.92 MHz
902.0 MHz	928.0 MHz	26 MHz	915.0 MHz
2.4 GHz	2.5 GHz	100 MHz	2.45 GHz
5.725 GHz	5.975 GHz	150 MHz	5.8 GHz

The 433 MHz and 915 MHz bands are region-specific. The 433 MHz band is assigned to Region 1, which includes Europe, Africa, Russia, the Middle East, and Mongolia. The 915 MHz band is assigned to Region 2 (the Americas). The 27 MHz, 2.45 GHz, and 5.8 GHz bands are worldwide. Other regions, as defined by the ITU, have their own frequency allocations, and some of the ISM bands are subject to local acceptance. The use of these frequencies generally doesn't require a license, but there are specific types of acceptable uses and the output power of the transmitter is restricted to limit the range and the potential for interference. In the US, this is governed by parts 15 and 18 of the FCC rules. Wikipedia has a good write-up on the ISM bands, including the ones not listed here.

It's interesting to note that one of the original driving reasons behind the creation of the 2.45 GHz band in the late 1940s was the development of medical diathermy equipment and the microwave oven. The ISM band at 27.120 MHz has been used for CB radio since about 1958, and because CB radios are licensed, the transmitter power they employ in the US is much higher than what would otherwise be allowed under Part 15 rules. Some types of radio-controlled toys still use 27.120 MHz, but they are prone to interference from nearby CB radios. It wasn't until the 1980s that devices such as cordless phones started to appear on the 915 MHz band, and later on the 2.45 and 5.8 GHz bands, as the necessary semiconductor technology evolved to the point where these applications became economically feasible.

ISM bands with higher frequencies offer better performance in terms of modulation capabilities, but the trade-off is reduced range. The 27 MHz band, for example, doesn't really have the bandwidth to support FM, but it does support AM and SSB just fine. These modulation methods suffer from external noise (lightning, electrical discharges, and other RF sources) but, on the other hand, it's not uncommon to hear a CB radio operator many miles away when conditions are right. The higher-frequency bands (915 MHz and above) can support FM and FSK modulation methods because of the increased bandwidth available, and these techniques are relatively immune to many types of interference that plague AM communications. Because of the higher frequencies used and low power output, they don't have the same range as the lower-frequency bands and are typically limited to line-of-sight or short-range applications.

If you have an application that needs range, but not particularly high data rates, you might want to consider the 433 or 915 MHz bands. If you're willing to put in the effort to implement an error-tolerant AM protocol (not a trivial project!), then 27 MHz might be an option if you need the range it can offer. This isn't as crazy as it might sound. If you have access to an old CB radio, or a shortwave receiver with the appropriate band, you can sometimes hear some odd bleeps and blips on the 27 MHz band, along with the usual chatter. Those odd little bits of sound might be anything from a remote-controlled toy to a remote monitoring device reporting to a base station.

There are more ISM band definitions than just those listed in Table 14-20, but for digital data communications, the 433 MHz, 915 MHz, 2.45 GHz, and 5.8 GHz bands are the ones that are most commonly used. Today, the 2.45 GHz ISM band, and to some extent the 5.8 GHz band, are used for things like wireless keyboards, mice, cordless phones, 802.11 (WiFi), Bluetooth, and 802.15 (which includes ZigBee). All these devices compete for the same limited bandwidth. So how do they avoid walking on one another? Well, most of the time the range and frequency seperation is sufficient to minimize collisions, but sometimes it isn't.

Each ISM band is divided into channels. If you've ever configured a wireless router, you've no doubt seen a reference to a channel in the configuration menus. In the 2.45 GHz band, the channels used for wireless LAN applications are separated by 5 MHz. Some devices, like the short-range devices discussed in "2.45 GHz Short-Range" on page 337, can squeeze even more channels into the 100 MHz bandwidth of the 2.45 GHz band by using narrow-band GFSK modulation. But even if two devices are on the same channel, most wireless devices are able to distinguish between the signals on the basis of signal strength and network addressing in the data packets.

2.45 GHZ SHORT-RANGE

If you want to define your own communications protocol for a short-range wireless data link, you can do that using a pair of low-cost transceivers operating at 2.45 GHz. A transceiver contains both a transmitter and a receiver. A popular IC component for this type of application is the Nordic Semiconductors nRF23L01+, a single-chip 2.45 GHz transceiver. Modules based on this IC can be purchased for around $2 each. Figure 14-39 shows one example of this type of product.

FIGURE 14-39. 2.45 GHz transceiver module

The nRF23L01+ supports data transmission rates of up to 2 Mb/s in the 2.45 GHz ISM band. This is a half-duplex device, meaning it can be either a transmitter or a receiver, but not both at the same time. It uses an SPI interface to communicate with a host microcontroller, and data is encoded and decoded using GFSK. It supports up to 125 channels within the 2.45 GHz band, and frequency hopping as well. You can download the datasheet for the nRF23L01+ (*http://bit.ly/nrf24l01*) from Nordic Semiconductor's website.

Note that the nRF23L01+ is just the RF transceiver. It doesn't impose any type of encoding/decoding scheme on the digital signal. Data to be transmitted is written to the device using the SPI interface, and data received (when in receive mode) is read out. The nRF23L01+ handles the task of asynchronous data reception and decoding internally. Defining the communications protocol and data format is up to the user.

802.11

A wireless local area network (WLAN) based on the 802.11 standard (or WiFi) is much more complex than just a small IC like the one shown in Figure 14-39. 802.11 components also tend to be a lot more expensive.

There are multiple variations on the original 802.11 standard. The *b* and *g* standards are the most common, with the 802.11*n* standard becoming more popular in recent years. The standards differ in terms of the ISM band used and the maximum optimal data rate, as shown in Table 14-21.

TABLE 14-21. 802.11 standards ISM band and data rate characteristics

Standard	Band (GHz)	Max data rate (Mb/s)
802.11	2.45	2
802.11*b*	2.45	11
802.11*a*	5.8	54
802.11*g*	2.45	54
802.11*n*	2.45/5.8	600

Each of the 802.11 standards employs a different type of signal modulation, but you don't usually need to be too concerned about that. So long as each end of an 802.11 type wireless link can support the appropriate modulation method, they should be able to communicate. This is why it's common to see an 802.11 device with a designation of 802.11*b/g*. This indicates that the device will autosense the modulation type and adapt accordingly.

The 802.11 standard (with no letter following it) was originally released in 1997, and by

1999 it had been superceded by 802.11*b*. Old 802.11 devices are now rare, although some 802.11*b* devices are still in use. 802.11*g* (2003) and, more recently, 802.11*n* (2009) have become the most commonly used standards of the 802.11 family.

Two new variants, 802.11*ac* and 802.11*ad*, have only recently been formalized, and devices capable of using these new standards are still in development. 802.11*ac* uses the 5.8 GHz band but supports data rates up to 6.9 Gb/s. 802.11*ad* is unique in that it operates on a 60 GHz band with a projected data rate of upward of 6.7 Gb/s.

As with wired Ethernet (described in "Ethernet Network Communications" on page 328), you might want to avoid the complexity of dealing with the low-level details of the interface and use something that will do the heavy lifting for you, such as the module shown in Figure 14-40. This particular part sells for about $85 USD from DFRobot (see Appendix D for its website URL).

The part of the unit that does the actual work is the rectangular module with the metal cover sitting in the middle of the PCB, in this case a WizFi210 from WIZnet. One of the things that makes this and other WiFi modules pricey (relatively speaking) is that chip manufacturers such as WIZnet have already gone through the process of certifying the operation of the module and obtaining an FCC ID number, which implies that it has been tested and verified to comply with FCC regulations. This is not an inexpensive undertaking, so purchasing a module that is already compliant and certified makes good sense.

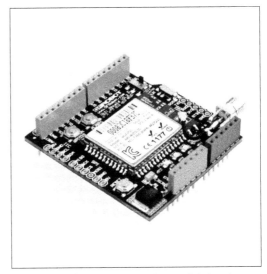

FIGURE 14-40. 802.11 *b/g* transceiver module (courtesy of DFRobot)

> Although there are multiple WiFi transceiver components available, it is a good idea to find one already mounted on a PCB with the appropriate antenna connector and associated circuitry. Laying a out a PCB for RF can be tricky, and it's something of a specialized activity. You can save yourself a lot of effort and aggravation by buying a product where someone else has already worked out the kinks for you.

A WiFi module intended for embedded applications, the Microchip MRF24WB0MA, is available for around $30 USD from distributors such as Digikey and Mouser (see Appendix D). It uses an SPI interface and is available already mounted on a PCB. Another possibility is the CC3000 802.11 *b/g* module from Texas Instruments, which is a popular part found in many low-cost WiFi modules. An evaluation board with the CC3000 installed (TI part number CC3000EM, available from Mouser and

TABLE 14-22. 802.11 b/g modules

Part number	Manufacturer	Internal controller	Interface
MRF24WB0MA	Microchip	Proprietary	SPI
SPWF01SA	STMicroelectronics	ARM Cortex-M3	SPI, I2C, UART
CC3000	Texas Instruments	Proprietary	SPI
WizFi210	WIZnet	Proprietary	SPI, I2C, UART

other distributors) runs about $38 USD, and bare modules (no supporting PCB) based on the CC3000 sell for between $17 to $28 USD in small quantities. The evaluation board is not much different from boards made for devices like the Arduino and BeagleBone, there's no need to do surface-mount soldering, and it's not that much more expensive, so it might be worth considering.

Table 14-22 lists the 802.11 modules mentioned in this section. The datasheets available for these parts describe the operation and provide reference designs if you really want to build your own PCB. If you do take this path, be sure to follow the manufacturer's recommendations for PCB layout.

All of these modules are complete plug-in 802.11 *b/g* functional blocks. All that's needed is a microcontroller or microprocessor (available through Allied, Digikey, Mouser, Newark, and other electronics distributors) to provide data and control.

BLUETOOTH®

This section discusses what is sometimes called *Classic Bluetooth*. With the release of version 4 of the specification from the Bluetooth SIG, a new specification came into being, called Bluetooth Low Energy, Bluetooth Smart, or Bluetooth LE. The new technology is a low-power, reduced-complexity protocol that is not backward compatible with Classic Bluetooth. "Bluetooth Low Energy (BLE)" on page 342 discusses Bluetooth Low Energy.

Bluetooth is a wireless technology standard designed for data exchange data over short distances. Bluetooth operates in the 2.45 GHz ISM band (see "The ISM Radio Bands" on page 336). It was invented in 1994 by the telecommunications company Ericsson, which originally envisioned it as an alternative to the then-prevalent RS-232 cables used to connect various devices such as PCs, PDAs (personal data assistants, like Palm Pilots), and printers. The Bluetooth trademark is protected by the Bluetooth Special Interest Group (SIG), and in order for a product to be called a Bluetooth device, it must be qualified to standards defined by the SIG.

Two key characteristics differentiate Bluetooth from other technologies operating in the 2.45 GHz band. The first is the protocol that it employs, which allows for the creation of ad hoc networks of devices with secure connections (pairing and bonding). The second is its ability to utilize a frequency-hopping spread-spectrum technique to improve communications reliability.

The frequency-hopping spread spectrum technique employed by Bluetooth involves dividing the transmitted data into packets, each of which is transmitted on one of 79 channels. Each channel has a bandwidth of 1 MHz. The newer 4.0 standard employs channels with a bandwidth of 2 MHz, which reduces the number of available channels to 40. Up to about 1,600 channel *hops* are performed per second, depending on the selected mode of operation.

To associate one device with another, Bluetooth uses a process called *bonding*. A bond is created as the result of a process known as *pairing*. When a bond is established between two Bluetooth devices, a special code called a *link key* is held by both devices and known to only those two devices. Once devices have been bonded, they can authenticate each other whenever they are in range, so the user doesn't need to repeat the process each time a link needs to be established.

Bluetooth employs a master/slave communications protocol, but there are no dedicated roles among devices. A master device can communicate with up to seven other Bluetooth devices in an ad hoc Bluetooth network. Any device can, by agreement with other devices, become the master until another device needs to take on that role and initiate a data exchange transaction.

Bluetooth devices are divided into three classes, each of which defines a maximum transmit power and a maximum operating range under optimal conditions, as shown in Table 14-23.

TABLE 14-23. Bluetooth classes

Class	Max power	Operating range
Class 1	100 mW	100 meters
Class 2	2.5 mW	10 meters
Class 3	1 mW	1 meter

The effective range of any link is limited by the device with the least range. A Class 2 device will work fine with a Class 1 device, but the resulting link between bonded devices will never be better than Class 2. Also be aware that, even though the range for a Class 1 device is given as 100 meters, in reality it can be much less, depending on the local environment.

The data rate of a Bluetooth device is defined by the revision level of the specification. Table 14-24 lists the defined data rates. In practice, the actual data rates will typically be lower due to processing overhead and other factors.

TABLE 14-24. Bluetooth data rates

Version	Data rate
1.2	1 Mb/s
2.0 + EDR	3 Mb/s
3.0 + HS	24 Mb/s
4.0	24 Mb/s

The best and least expensive way to put Bluetooth to work is to purchase a module. There are many to choose from, ranging in price from about $9 up to around $60 USD, depending on the data transfer speed and range capability. Table 14-25 lists some readily available modules that can easily be incorporated directly into a circuit. This list is by no means complete, so be sure to look around for a module that suits your needs (and your budget).

TABLE 14-25. Bluetooth modules

Part number	Manufacturer	Interface	Class
PAN1315A	Panasonic	UART	1 and 2
RN41	Microchip	UART	1
RN42	Microchip	UART/USB	2
RN52	Microchip	UART/USB	3
SPBT2632C1A	STMicrotechnology	UART	3
WT11i	Bluegiga	UART/USB	1
WT12	Bluegiga	UART	

Most Bluetooth modules come with the protocol already built in. This greatly simplifies things, since in most cases using a module just involves connecting it to your circuit. Once it's connected, you can then pair it with another Bluetooth device and start transferring data.

> **TIP** The *Bluetooth* wordmark and logos are registered trademarks owned by Bluetooth SIG, Inc. For more information about Bluetooth, see the SIG website (*http://www.bluetooth.com*).

BLUETOOTH LOW ENERGY (BLE)

Bluetooth Low Energy (BLE) is a low-power alternative to the previous versions of the Bluetooth specifications, now often referred to as Classic Bluetooth. Marketed as Bluetooth Smart, BLE is a wireless personal area network (PAN) technology defined by the Bluetooth SIG and intended for applications in healthcare, fitness, home entertainment, and security, among others.

The original Bluetooth specification was not specifically created with PAN applications in mind, but rather as a replacement for serial and parallel interface cables between a host computer and various peripheral devices. Consequently power requirements were not a major design consideration. This puts Classic Bluetooth at a significant disadvantage in ultra-low-power applications. Although Bluetooth has been incorporated into things like mice and keyboards, these products suffer from short battery life compared to other types of low-power wireless technologies. BLE is intended to provide significantly reduced power consumption and component part cost while still operating at a similar range and with enhanced data-transfer capability.

Version 3 Classic Bluetooth utilizes a synchronous protocol that is basically always on, by virtue of the periodic synchronization data exchanges that occur every 600 microseconds to maintain a communications link. Consequently, even an idle device continuously uses power. In addition, when a link between devices is broken, it can take upward of a second to reestablish the connection. This is too long for some PAN applications.

Like Classic Bluetooth, Bluetooth Low Energy uses a synchronous protocol but with a synchronization interval that can be varied dynamically from 5 ms up to several seconds. To conserve on energy expended to transmit data, a BLE packet can squeeze in more data than with a Classic Bluetooth packet, and the packet length is variable to accommodate different situations.

These features, combined with other power-saving techniques such as reduced transmit power, enhanced receiver sensitivity, maximized standby time, and reduced packet complexity, result in less power expended per bit. In other words, BLE can provide higher average data throughput rates while expending less energy to do it.

Bluetooth Low Energy uses a unique protocol and a simpler modulation scheme, and it is not backward compatible with the previous Bluetooth Classic protocol. However, the Bluetooth 4.0 specification permits devices to implement either the BLE or Classic protocols, or both, in a single device, and many transceiver chips now available can operate in *dual mode*. BLE operates in the same 2.4 GHz radio frequency band as Classic Bluetooth but utilizes a different set of channels. This allows dual-mode devices to share a single radio antenna.

In some applications, a BLE transceiver can operate for extended periods of time (on the order of months, or possibly even years) on a single coin cell battery such as a CR2032 (see Chapter 5). Many modern mobile and desktop operating systems such as iOS, Android, OS X, and Windows 8 provide native support for BLE. Other platforms are possible with the appropriate low-level software and transceiver hardware.

Transceiver chips for BLE are available from many of the same sources as Classic Bluetooth transceivers. Table 14-26 lists some of the IC components available.

TABLE 14-26. Bluetooth LE ICs

Part number	Source	Interface
CC2541 MCU/Controller 8051-based	Texas Instruments	I2C
CSR1010	CSR	UART/SPI
EM9301	EM Microelectronics	UART/SPI
nRF8001	Nordic Semiconductor	Proprietary

While there don't seem to be as many modules for BLE as for Classic, there are products available from several sources, such as the ones listed in Table 14-27.

TABLE 14-27. Bluetooth LE modules

Part number	Manufacturer	Interface
ABBTM-NVC-MDCS71	Abracon	UART/I2C/SPI
BLE112	BlueGiga	UART/USB
PAN1720	Panasonic	UART/SPI

If you don't want to go the do-it-yourself route, the SensorTag device from Texas Instruments is popular with folks using an iPad, iPhone, or Andoid device. It costs about $25 USD and includes a range of sensor input types, including temperature, humidity, pressure, accelerometer, gyroscope, and a magnetometer.

Adafruit sells a module for BLE, called the Bluefruit LE. This essentially an nRF8001 breakout that uses an SPI interface (which is what the nRF8001 uses) to communicate with something like an Arduino. Figure 14-41 shows what the board looks like. Adafruit provides a free, open source app to communicate with the Bluefruit, and there is a demo video on the Adafruit website.

FIGURE 14-41. Bluefruit LE breakout module from Adafruit

ZIGBEE

ZigBee is a wireless technology standard used to create short-range data communications links and ad hoc networks for applications that only require a low data rate, long battery life, and secure communication. The ZigBee specification is based on the IEEE standard 802.15.4 for low-rate wireless personal area networks (WPANs). The ZigBee specification adds a network layer and an application layer (along with other functions). For additional information about ZigBee and the ZigBee standard, visit the ZigBee Alliance (*http://www.zigbee.org*).

ZigBee devices can be used to create *mesh networks*, wherein multiple low-power, short-range devices cooperate to move data between intermediate devices. This allows widely separated ZigBee nodes to communicate over distances beyond the normal range of each individual device. Security is an integral part of the specification and is implemented through predefined shared encryption keys.

As with the other wireless technologies covered in this chapter, ZigBee operates in the ISM radio bands (see "The ISM Radio Bands" on page 336). In the US, it can use either the 915 MHz or the 2.45 GHz bands, although most commercially available ZigBee devices and modules use the 2.45 GHz band. In Europe, ZigBee devices operate on either the 868 MHz or the 2.45 GHz bands.

The data rate for a ZigBee link depends on the ISM band it is using, as shown in Table 14-28. A maximum data rate of 250 Kb/s isn't particularly fast, but for the monitoring and control applications that are the primary applications for ZigBee, it's usually more than fast enough.

TABLE 14-28. ZigBee data transfer rates by ISM band

ISM band	Data rate
868 Mhz	20 kb/s
915 MHz	40 kb/s
2.45 GHz	250 kb/s

The technology defined by the ZigBee specification is intended to be simpler and less expensive than other technologies, and in general terms, this is true. But it is also true that some Bluetooth RF modules are about the same price as a 2.45 GHz ZigBee module; so in the end, it really comes down to what you want to do and which technology best suits your needs.

Possible applications for ZigBee include wireless lighting and appliance control, distributed data acquisition, electrical service meters, robots, and traffic control systems. ZigBee RF modules with minimal interface and programming requirements are readily available (see Table 14-29).

TABLE 14-29. ZigBee modules

Part number	Manufacturer	ISM band	Interface
ATZB-24-BOR	Atmel	2.45 GHz	I2C, SPI, UART
XB24-AWI-001	Digi International	2.45 GHz	UART
MRF24J40	Microchip	2.45 GHz	SPI
CC2420	Texas Instruments	2.45 GHz	SPI
CC1120	Texas Instruments	868/915 MHz	SPI

When considering a ZigBee RF module, be sure to look for features like built-in protocol logic. This will greatly simplify things, because it makes the interface to the module a simple read or write data transaction, and there's no need to install software to deal with the ZigBee network and application layers.

Other Data Communications Methods

In addition to wires and radio waves, data can be carried over a light beam (a laser) and through the AC wiring in a home or office.

The remote controls used with television and stereo receivers (and just about everything else in the home entertainment section of the department store) use pulses of infrared light to send commands to various devices. Infrared has also been used to send documents to a printer from a handheld personal organizer, and some tablets include the ability to send commands or data using an infrared LED built into the tablet's case.

As with the short-range 2.45 GHz transceivers described in "2.45 GHz Short-Range" on page 337, you can devise your own communications protocol for an optical system. All you really need is an IR LED like the one shown in Figure 14-42 and a phototransistor receiver, like the module shown in Figure 14-43. You will also need a lens of some sort to help focus the light if you want to extend the range beyond just a few feet. A toy telescope or binoculars work well for this purpose.

FIGURE 14-42. Infrared LED module

If you think about the IR link as a type of wire, you can see that it would be easy to send an RS-232 type signal over the link. A second link, from the remote device back to the base location, would complete the

TxD/RxD loop, and presto, you would have a link that could theoretically work over long distances—assuming, of course, that the light source is powerful enough and the weather is clear. Systems like this have been implemented via solid-state IR lasers, and some people have used modified red and green laser pointers for this purpose.

> If you decide to work with lasers, be careful about where they are pointed. Setting up a visible laser link between two points at high elevations is not a good idea. If an aircraft flies through the beams, it can be bad news for you legally, and it can endanger the aircraft if the pilots are momentarily blinded by the laser reflections off the cockpit windows. A non-coherent (nonlaser) narrow-beam link is safer, but it will require some decent optics at each end and some serious tweaking to get it to work reliably. If you just want to set up a laser link between one structure and another a couple hundred meters away, then you're probably good to go—assuming it doesn't start to rain or blow dust in the middle of a big download, of course.

FIGURE 14-43. Infrared receiver module

Summary

The communications protocols covered here range from the straightforward simplicity of SPI, I2C, RS-232, and RS-485, to the complexity of USB and Ethernet. Fortunately, IC components and modules are available that hide much of that complexity and present an SPI or I2C interface. We also covered some of the complexity of 803.11 wireless networks, took a quick look at the underlying machinery of USB, and reviewed Bluetooth and ZigBee wireless technologies and some of the components available to implement them.

This chapter has covered a lot of ground and touched on a number of technologies, but the key theme is how the data is physically transferred. It might involve wires, radio waves, or light beams, but it all comes down to serial data communication using either synchronous or asynchronous methods. From what has been presented here, you should be able to make some good initial decisions about which type of data communication method is best for your application.

For wireless communications, a generic 2.45 GHz link is fine for short-range communication where security is not a major concern and you don't need to support a network architecture. If a secure short-range link is needed for intermittent data transfers, ZigBee would be a possible candidate.

Moving up in speed, Bluetooth offers both secure communications and fast data exchange rates of up to 24 Mb/s. For even faster data rates, with full networking capability, consider one of the 802.11 standards.

Outlining the communication technologies we've covered here, Table 14-30 is one way to organize a decision list based on communication speed.

TABLE 14-30. Communication types organized by speed

Connection	Speed	Data link type(s)
Wireless	Low	2.45 generic, ZigBee
Wireless	High	802.11b/g/n, Bluetooth, Bluetooth LE
Wired	Low	RS-232
Wired	High	Ethernet, RS-485, USB

Ultimately, your decision comes down to how much you're willing to spend, how much work you are willing to do to make a functional communications link, and how secure you need that link to be. Be sure to explore the resources available from the various manufacturers, distributors, and PCB module suppliers. Also, don't forget to check out the books listed in Appendix C for low-level, hands-on details.

CHAPTER 15

Printed Circuit Boards

THIS CHAPTER PRESENTS AN OVERVIEW OF printed circuit board (PCB) design and layout, with a focus on technique rather than specific tools. Special attention is given to topics such as PCB material, multilayer techniques, and surface-mounted components.

PCB History

Prior to the invention of printed circuit boards, assembling an electrical or electronic device was an arduous process. Insulated posts or metal tabs riveted to insulating strips provided the connection points for the various components, and insulated wires (most of the time) routed signals and voltages from one part of the circuit to another. All of it was assembled by hand. Figure 15-1 shows an example of the level of effort involved, with an image of the Atwater-Kent radio assembly line from some time in the early 20th century.

FIGURE 15-1. A vacuum tube radio assembly line at the Atwater-Kent factory circa 1925 (photograph from Library of Congress)

Starting in the late 1950s, the point-to-point wiring of electronic devices was gradually replaced by simple single- or double-sided printed circuit boards, as shown in Figure 15-2. The vacuum tubes are still there, but instead of wires running all over and resistors and capacitors suspended between terminal strips and solder posts, they are now mounted on a PCB. A circuit like this (which happens to be a plug-in module from a Hewlett-Packard 140A oscilloscope) was still largely assembled by hand, but the

349

process typically had fewer wiring defects and resulted in overall higher quality.

FIGURE 15-2. A plug-in module from a vacuum tube oscilloscope, circa 1963

With the widespread adoption of solid-state components, and later integrated circuits, the size of circuits shrank and is still shrinking today. A multi-layer PCB with both through-hole and surface-mount components is similar to the Texas Instruments MSP430 development board shown in Figure 15-3.

FIGURE 15-3. MSP430 microcontroller development board

It's tiny, but if it were built with 1980s technology, it would probably be two or even three times larger. Today, robots called *pick-and-place* machines grab surface-mount parts from a strip wound on a reel, place them on the board in the correct orientation at the correct location, and then go back and repeat the process for each surface-mounted part on the PCB. Some of these machines move incredibly fast, and they can populate a PCB with parts in a matter of a few minutes.

A look at the motherboard inside a modern PC will serve as an example of just how far PCB technology has progressed in the past 50 years. Today, many commercially produced PCBs cannot be assembled by hand, even if you wanted to. They are just too dense, and the parts are too small. Without automated PCB fabrication machines to make the base circuit boards, robots to place the parts on the PCB, and still more automation to solder and test the finished PCB, many of the electronic devices we have today could not exist.

But there is still a lot you can do with a simple PCB and the soldering techniques described in Chapter 4. This chapter takes a high-level look at how to go about designing and fabricating a PCB without ordering thousands at a time.

PCB Basics

A printed circuit board is built on a sheet of insulating material, called the *substrate*. Some types of PCBs have a layer of copper laminated to one or both sides. Unwanted copper is etched or routed away, leaving the circuit traces (the wiring) and the component mounting locations (pads). Other types start with a bare substrate board and add the copper to it to create the circuits on the board. Figure 15-4 shows what goes into a blank copper-clad, double-sided PCB.

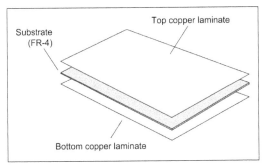

FIGURE 15-4. Basic double-sided blank PCB

Multi-layer PCBs are fabricated from stacks of thin substrate sheets, each with a copper circuit pattern on one side. After the circuit has been created on the substrate sheets, they are stacked and bonded to create a single PCB. A multi-layer PCB can have upward of 10 internal layers. A special type of pad, called a *via*, is used to connect traces between sides and layers, so that a circuit that begins on the top (component) side of the multi-layer stack might connect to a trace on the third layer, which in turn might connect to a trace on the solder (back) side of the PCB.

The substrate material is usually either a phenolic material (XXXP/FR-2) or a resin-impregnated woven fiberglass sheet (FR-4). The FR value indicates the flame resistance grade of the material.

PADS, VIAS, AND TRACES

A PCB layout consists of three basic types of shapes: pads, vias, and traces. *Pads* are where components are mounted to the PCB, which for through-hole parts (for example, a 1/8W resistor) will consist of a hole for a lead surrounded by an annular ring of copper. A *via* is typically smaller than a pad and, as noted earlier, is used to route a signal or power from one layer of a PCB to another (front to back for instance, or perhaps to an internal layer in a multi-layer board).

In double-sided or multi-layer boards, through-hole pads and vias are usually plated to create a thin tube of copper between the endpoints of the hole. For some homemade PCBs, this isn't possible (at least not easily), so a piece of bare wire can be soldered into the via on both sides. Through-hole components should also be soldered on both sides if there is no plating in the lead hole of the pad and there are traces on both sides.

Traces are the wiring of the circuit. Trace width determines the amount of current that can be safely passed through the trace without causing potentially damaging resistive heating. The thickness of the copper laminate also has an effect on the current carrying capacity.

SURFACE-MOUNT COMPONENTS

Surface-mount technology (SMT) devices are soldered directly to one side of a PCB (it is possible to have SMT parts on both sides of a PCB). There are no holes for the leads of SMT parts, just solid pads for the leads to attach to. Usually, solder paste is used (rather than regular stick solder) to make the connection, and the solder paste is melted through a process such as reflow soldering, infrared heating, or just a plain fine-tip soldering iron.

FABRICATION

PCB fabrication processing can take one of two forms: subtractive or additive. The *subtractive* process is the one most commonly used for small production runs, and it involves using an acid to remove unwanted copper from a PCB, leaving just the traces and the component connection pads. The *additive* process can take the form of full additive, starting with a bare substrate and building up the copper circuit pattern using a plating technique. A variation on this is the semi-additive process, where a substrate has a thin layer of copper applied, the unwanted portions are removed, and the final desired thickness is built up on the existing thin layer of traces and pads. Multi-layer PCBs are usually produced through the semi-additive process, because the plating used inside the connections between circuit boards (the vias) can be applied as part of the production process.

For an etching process (subtractive) the steps basically follow those shown in Figure 15-5. These are the same basic steps you would follow if etching a PCB in your own workshop, but the PCB fabrication houses have equipment to make the whole process a lot more precise and efficient.

After the board is etched, the various holes are drilled. Then pads and vias are plated to make a connection with both sides of the PCB (if it's a double-sided board, of course). The bare copper might also receive a thin plating of tin to make soldering easier. Finally, a commercial PCB fabricator can apply a solder mask to prevent the solder in a wave-solder machine from sticking to places where it shouldn't, and a silkscreen mask is applied to one or both sides. The silkscreen mask typically has component outlines, part numbers, and the name of the company or individual who designed the PCB.

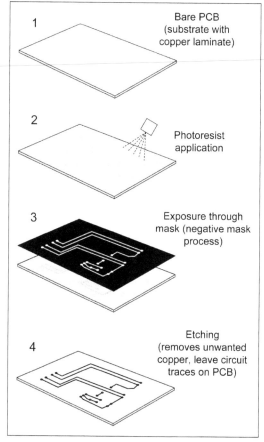

FIGURE 15-5. Basic steps for subtractive etching

Refer back to Figure 15-3. The material covering the board (it's red) is the solder mask. Over that is the silkscreen with some information about the board and a logo. The underlying traces are barely visible through the solder mask.

PCB Layout

PCB layout is the process of arranging components on a PCB within a given area and then deciding where to place (or route) the traces to connect them into a complete circuit. It is something of an art, and it's not uncommon to try multiple times until the component placement and trace routing work without conflicts or spacing that is too narrow. It is also not uncommon for a PCB layout to look nothing at all like the original schematic. Physically, it doesn't matter (generally) where components are placed, particularly with small PCBs. So you might find R1 at the lefthand side of a PCB, and R2 way over in the lower-right corner. So long as they are connected correctly, it doesn't really matter; what matters is that they are placed to take advantage of the available space on the board.

In the past, PCB layout was done using a sheet of clear plastic film to create the layout pattern with adhesive black press-on tape, pads, and vias. This was usually done at 4X size, and then scaled down with a special camera to create the final mask. The trace tapes came in difference widths, and pad shapes were available in 1X, 2X, 4X, and even 8X sizes. The designer could spend many days at a light table carefully applying the tape and picking up the individual pads and vias and placing them with the point of a razor knife. A photographically created grid on a transparent sheet was placed under the clear working sheet to provide placement reference. As you might suspect, the whole process was rather tedious. It also explains why old PCBs tend to have occasional wavy trace lines, rounded trace curves, and irregular trace spacing. Each one was done by hand.

Although it is possible to create a simple PCB layout on a clear piece of plastic with adhesive tapes and pad transfers or a special pen, it's easier to use a software tool made for that particular purpose. Some PCB kits employ a laser printer to transfer the layout pattern, while large PCB fabrication houses might use a photo plotter to create the positive or negative photo-resist exposure mask. A photo plotter uses a special head with a variable aperture that exposes an underlying photosensitive film as the plotter head moves across it. After exposure, the film is developed and the transfer mask is ready to use.

The following steps show one way to lay out a PCB using a tool like the PCB layout editor that is provided as part of the gEDA package. The layout editor can also be obtained as a standalone tool, and other software packages are available that also have PCB layout editors (Eagle, FreePCB, KiCad, and others). Refer to Appendix D for resource information and URL links. In general, however, the steps are basically the same for any PCB layout tool, or even if you are doing the layout the hard way with press-on tape and die-cut pads.

DETERMINE DIMENSIONS

First, decide on the dimensions of your PCB. If you have the parts available that will be used on the PCB, you can arrange them on a piece of graph paper to see how they will fit. Some CAE tools also provide package placement drawings as part of their component library, so you can model the topography of a

PCB using the package outlines supplied with the CAE tool (see "Place Components" on page 355).

To avoid situations where the lead of a component is blocked by another part, make sure to leave sufficient space for both parts and traces. If you are designing a single-side board, this is particularly important, and you might need to account for jumpers to route signals and voltages over one or more traces. For a double-sided board, leave enough room to place a via to take the trace to the other side of the PCB if you need to route under or around other parts.

ARRANGE PARTS

Second, decide where the mounting holes and pads for things like terminal blocks, I/O connectors, and wires will go. This establishes the physical mounting and interface geometry for your PCB. Figure 15-6 shows what a layout looks like after this step is complete.

When placing I/O and power connector pads, you will need to reference the placement from the edge of the PCB. Say, for example, you want to have a row of pin sockets with 0.1-inch spacing, like the ones found on Arduino, BeagleBoard, or MSP430 boards (see Figure 15-3). How close should the part be placed to the edge of the board? Generally, for something like this, it is a

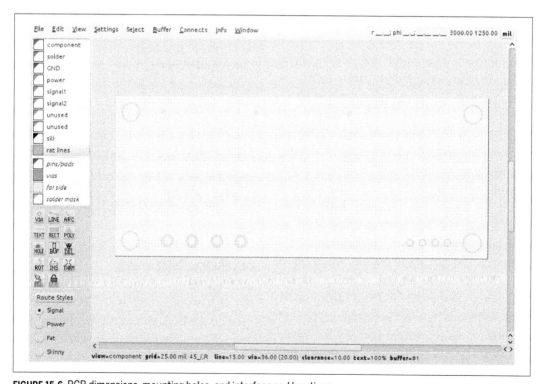

FIGURE 15-6. PCB dimensions, mounting holes, and interface pad locations

good idea to leave at least 0.15 inch to 0.2 inch of space between the pins and the PCB edge, as shown in Figure 15-7. It is generally not a good idea to route a trace between the row of connector pins and the edge of the board. It's just too easy for the trace to be damaged, and it precludes any trimming that might be needed to make the board fit in a tight location.

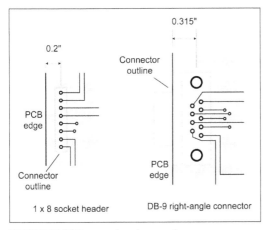

FIGURE 15-7. PCB connector placement

For things like right-angle DB connectors (DB-9, DB-25, etc.), the physical shape of the connector determines where it will be placed. The component library of a well-stocked PCB layout editor will have the general dimensions for common parts, but they should be checked, as each manufacturer can do things a bit differently. Figure 15-7 shows a DB-9 connector with a board edge-to-connector mounting-hole dimension of 0.315 inch. This number is obtained from the manufacturer's data sheet for the part.

PLACE COMPONENTS

Next comes component placement, which is shown in Figure 15-8. Here we have three silicon-controlled rectifier (SCR) devices and some Zener diodes, resistors, and capacitors. They are connected to the PCB with wires soldered though pads along one edge, as defined in "Arrange Parts" on page 354.

As with the connectors discussed in "Arrange Parts" on page 354, the *footprint* of each part is obtained from a library of component shapes that comes with the PCB layout tool. Select a part from the library and place it on the PCB layout area. You can easily move parts around before the traces are placed, so now is the time to get the best layout geometry for the available board space.

ROUTE TRACES ON THE SOLDER SIDE

Now comes the fun part: routing the traces. If you have a complex design, this is where you might want to turn it over to an auto-router and let it have a go at figuring out where to place the traces. The downside is that even the best auto-routers sometimes don't get it right, in which case you will have to go back in and rip up some, or all, of the auto-router's work and do it yourself. On the plus side, an auto-router will generally apply best practices and avoid some of the common mistakes that humans are prone to making.

This is a double-sided board, so we start with the solder (or back) side first. Figure 15-9 shows the traces for the solder side of the PCB. Note that some of them are rather hefty. These traces will carry a lot of current, so making them wider is equivalent to using a heavier gauge of wire.

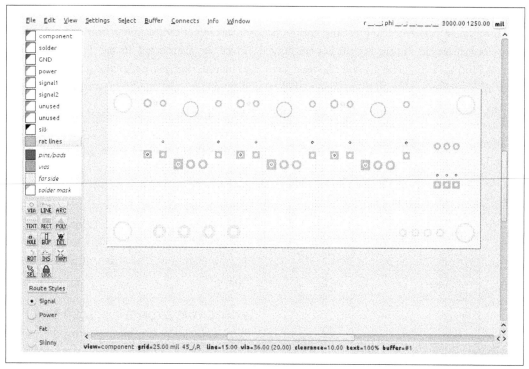

FIGURE 15-8. Component placement

ROUTE TRACES ON THE COMPONENT SIDE

Next comes the component (or top) side of the PCB. Here, you want to make sure not to create trace collisions or come too close to component pads. It's OK to go under a component like a resistor, so long as there is sufficient space between the traces and the component's pads. When working with DIP IC packages, you can also run a thin trace between the pads for the IC's leads. The leads of most surface-mount ICs are too closely spaced for this, so the typical solution is to route some of the device's leads to vias under the chip package and pass them through to the other side. This PCB layout doesn't need to do that, however. Figure 15-10 show the traces on the component side of the PCB.

CREATE THE SILKSCREEN

The last layout step is to create the silkscreen for the PCB. This is an optional step that results in the fabrication house creating a silkscreen and literally painting the component outlines, part numbers, and other information directly on the finished PCB. Figure 15-11 shows the PCB with both the top and bottom layers and the silkscreen visible. The layout tool created the part outlines for us using the component data in its library. The additional lettering is supplied by the layout designer.

GENERATE GERBER FILES

After finishing the layout, save it to disk and then create a set of files that can be used to fabricate it. That's assuming, of course, that

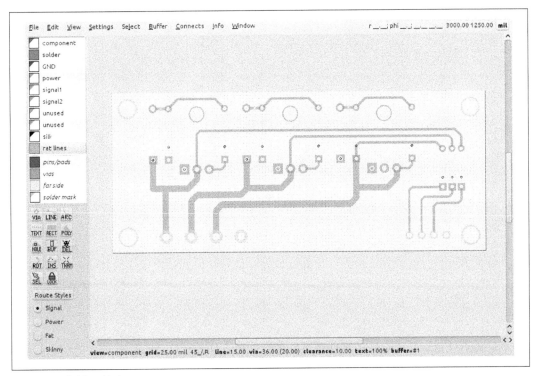

FIGURE 15-9. Solder-side traces

you have wisely elected to use a fabrication service (see "Fabricating a PCB" on page 357 for some thoughts on this).

The industry standard way of doing this is by means of so-called *Gerber files*, named after the company that developed automated photo-plotting equipment early in the history of PCBs. Each PCB layout consists of multiple Gerber files. These define the top pattern, the bottom pattern, any intermediate layers (for multi-layer boards), solder masks, a solder paste mask (for surface-mount components), top and bottom silkscreen templates, and a drill list that defines the sizes of the holes, how many of each, and where they are to be drilled. Figure 15-12 shows the gerbv tool displaying the Gerber files generated by the PCB layout editor.

Fabricating a PCB

If at all possible, consider using a low-cost PCB fabrication service instead of attempting to do it yourself. If you've never done it before, you might be surprised at just how difficult it can be to get everything right the first time. See Appendix D for a list of PCB fabrication service providers.

If you are lucky enough to have access to a PCB router, the Gerber files generated as part of the layout process can be used to direct the router. This is a quick way to make small PCBs, but it has some drawbacks. First, a routed double-sided PCB will not have plated-through vias, which means that each via will need to have a piece of wire soldered into it on both sides of the board and

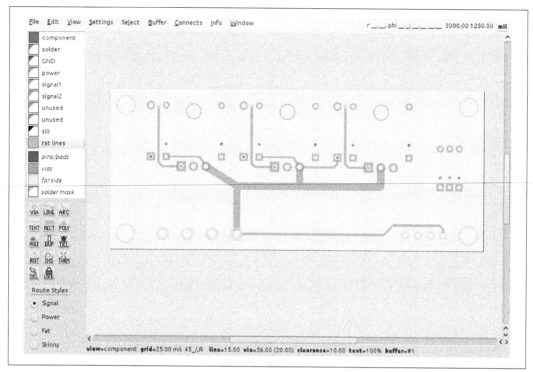

FIGURE 15-10. Component-side traces

then trimmed close to the PCB. Secondly, if the layout has a lot of open space between traces, then the router is going to spend a lot of time just cutting up and discarding copper. Lastly, the holes in the PCB must be manually drilled (usually, unless the router is a very fancy model).

If your PCB is a simple, single-sided design, doing it yourself using a good PCB kit might not be a bad way to go, but it has its own perils to avoid. Appendix D lists some sources for PCB kits that you might want to look into.

Etching a PCB by hand is an error-prone process that takes some time and patience to master. Each step needs to be timed correctly, the etchant needs to be mixed correctly, and the resulting board will still need to be drilled manually, just as with the routed PCB, and none of the vias will be plated. After investing in the equipment and supplies necessary (and your time, of course) to make good etched PCBs, you might find that it would have been cheaper to just pay someone to make the board for you.

If you do elect to have your PCB produced by a commercial fabrication house, then generally all that is needed is a set of Gerber files and some money. Be aware, however, that some PCB fabricators want you to use the tools they supply (usually free of charge) and might not be too eager to work with Gerber data files generated by a nonproprietary layout editor. If the reasoning for this is based on consistent data quality in a format that is

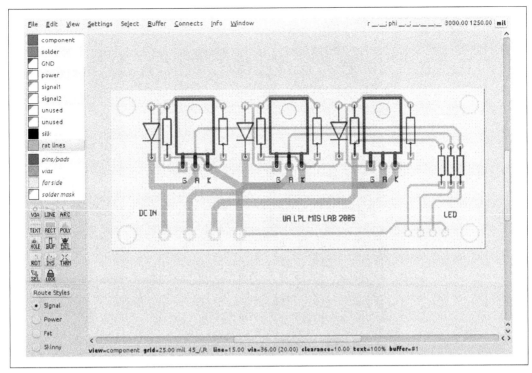

FIGURE 15-11. Silkscreen view

easy for the fabricator to process, and thereby hold down costs, then this approach could be justified. You will need to decide if you want to be tied to a particular layout tool, or if you want to pick your own tool.

PCB Guidelines

This section is by no means an exhaustive list of PCB layout guidelines, just a minimal set to help get you on the right path. As stated earlier, PCB layout is an art form, and it is also a science. When dealing with circuits that involve high-frequency RF or large switched currents, PCB layout designers must take into account things like capacitive and inductive coupling, trace inductance, stripline inductance and coupling, trace heating due to resistance, and other effects.

However, for most low-voltage, low-current circuits, these don't play a big role in the layout design, and just following some general common-sense guidelines can help you avoid some common mistakes and get a good end result.

If your PCB layout tool came with a user's manual, it would be a good idea to read it. It might contain some useful information, not only about the tool itself but also about PCB layout techniques in general. Another resource is, of course, the Internet, where you can find numerous tutorials on the subject.

LAYOUT GRID

Use the *snap-to-grid* feature of your layout tool. While not absolutely essential, it's there

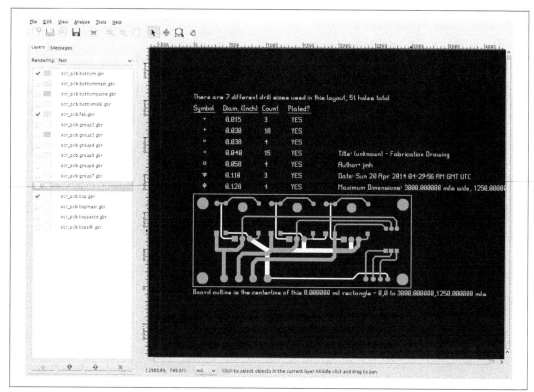

FIGURE 15-12. Reviewing the layout using a Gerber file viewer

to help position parts accurately and neatly on the PCB and make sure that features, such as mounting holes, end up where they are supposed to. The use of a grid also helps with small, dense PCB layouts, where every bit of space counts and trace spacing needs to be tightly controlled.

GRID SPACING

There are different opinions on grid spacing, with some sources suggesting 0.05 inch, others advocating 0.10 inch, and some suggesting .02 inch grid spacing. The grid selected for a layout tool is important, because it will determine how accurately parts can be positioned on the PCB layout. I typically use a 0.1-inch grid when working with a layout that involves conventional through-hole parts and components that have leads or pins spaced at 0.1 inch. For surface-mount components, I usually switch to a finer grid, on the order of 0.005 inch or 0.01 inch to accommodate the smaller parts. The downside to using a fine pitch grid is that it sometimes becomes invisible when the layout drawing is zoomed out.

LOCATION REFERENCE

Some tools define the lower-left corner as the (0, 0) origin point, or home location. Other tools use the upper-left corner, and still other tools let you define it yourself. In some cases, the selection of the origin for the layout can have a bearing on how the drill file will be interpreted. I typically use whatever the layout editor defines as the default, unless the PCB fabrication house has a specific requirement for origin location.

TRACE WIDTH FOR SIGNALS

Try to keep signal traces at least 10 mils (0.01 inch) wide, if at all possible. Although some fabricators support trace widths down to 7 mils, production quality can be tricky to maintain with very narrow traces.

TRACE WIDTH FOR POWER

Traces carrying power need to be wider, to provide a low resistance path for the current. The width is also dependent on the thickness of the copper cladding applied to the substrate, as a heavier copper clad is able to safely conduct higher current. Table 15-1 shows some very conservative rough guidelines for a PCB with 1 oz. copper cladding.

TABLE 15-1. Trace widths and power

Width	Power
0.010" (10 mils)	0.1 A
0.015" (15 mils)	0.3 A
0.020" (20 mils)	0.5 A
0.025" (25 mils)	0.7 A
0.050" (50 mils)	1.5 A
0.100" (100 mils)	3.0 A
0.150" (150 mils)	5.0 A

These numbers are only suggestions and are not set in stone. If you do the calculations, you will find that the calculated current values are higher, but that doesn't leave any margin. If your circuit will be operating in a high-temperature environment (e.g., alongside the engine of an automobile or in an enclosure that sits in the sun all day), using wider trace widths can help reduce possible failure due to thermal stress in the power traces. A Google search will turn up numerous trace width calculators.

TRACE SEPARATION

Keep signal traces separated by at least 10 mils, if possible. For traces that will be carrying significant current ($I > 100$ mA) that will be switched on and off, you might need even more separation to avoid inductive coupling into nearby traces. One way around this is to route traces carrying switched current near ground traces or ground areas on the PCB. The grounded traces and copper areas act as a shield for the potentially noisy traces.

VIA SIZE

Most PCB layout tools come with predefined via sizes. As with traces, use a larger via if it is carrying power from one layer to another. The minimum via size is usually specified by a PCB fabricator, based on what its in-house process can reliably produce. For most cases, you can use something like a 20 mils diameter hole, although 10 mils is also used. For example, the PCB tool used to create the layout illustrations in this chapter uses a default 20 mil via with an 8-mil thick annular ring, so the entire via diameter is 36 mils. For most situations, this is fine and most PCB fabricators should be able to handle it. For more information, check the documentation for your PCB layout tool and review the requirements for the PCB fabricator you plan to use (if any).

VIA SEPARATION

As with traces, vias should be separated from traces, pads, and other vias by at least 10 mils, if possible, although in some cases you can use a 7- or 8-mil separation for really tight layouts.

PAD SIZE

The size of the pads used for a particular part is usually specified as a fundamental part of the component layout definition in the tool's library. It doesn't hurt to check it, however, because whoever defined the part might have a different idea about pad size than what your design calls for. This is particularly true if you need to run traces between the pads of something like a DIP IC, or a TO-220 transistor. Most layout tools allow you to adjust the size of the pad's annular ring surrounding the hole, but you should not go below about 7 mils in thickness.

SHARP CORNERS

In general, you should avoid sharp (90 degree) corners, if for no other reason than they can be a problem during PCB fabrication (depending on the process). There is also a school of thought that claims that a sharp corner can radiate high-frequency electromagnetic energy and create interference in other parts of the circuit. Some studies have shown that this wasn't as big of an issue as was once thought. Still, in general, it's good practice to create a bend with a pair of 45-degree corners rather than a single 90-degree corner.

SILKSCREEN

If you are planning to have top (or also bottom) silkscreen designs on your PCB, be sure to avoid having the silkscreen go over any part of the board that will be exposed, such as pads or bare metal areas.

Summary

Creating a PCB is a process of translating a schematic into a physical object that uses current flow to perform a function. This chapter covered the basics of what a printed circuit board is made of and how you go about designing the pattern of wiring traces and mounting locations needed for the various components that will be soldered onto it. We also took a brief tour of the history of the PCB and showed how advances in electronics fabrication technology have led to the sophisticated, compact, and inexpensive (relatively speaking) devices that we use today.

This chapter is only the tip of the iceberg, so to speak. Entire reference books have been written on the subject of PCB design, and if you plan to do a lot of layout work, it might be a good idea to invest in one or two of the better ones. Another key point of this chapter is the necessity of practice. PCB layout is, in many ways, like learning to draw or paint. Your first attempts might look crude, but they will improve over time if you stick with it and gain the necessary skills and experience.

CHAPTER 16

Packaging

Packaging is a key aspect of electronics design, and some people make a lifelong career of it. For a consumer product, creating the package might involve the skills of an industrial designer or graphic artist, someone well versed in thermal analysis, an RF engineer to check for radio-frequency emissions, and a production engineer to make sure that the package can actually be built in a cost-effective manner. When a package is being designed for applications such as aerospace or marine applications, a whole new set of requirements comes into play. Physical volume, weight, electrical connections, mounting constraints, and even the type of paint used are just some examples. It can get quite involved.

However, for the most part, packaging for small- to medium-sized projects, particularly those that are prototypes or one-off devices, just involves applying a lot of common sense and asking some key questions. This chapter addresses those key questions and also examines issues such as plastic versus metal, sources for chassis components, and the potential of unconventional packages.

The Importance of Packaging

Every circuit or device that isn't going to spend its life on a workbench needs an enclosure of some sort to protect it. Even if it does end up living out its days in the back corner of a workbench, a package will prevent things falling into it or someone accidentally touching it while it's active.

Packaging can range from the simple to the complex. It all depends on where the device will reside and the type of environmental conditions it can be expected to encounter. If something will sit on a bench or a shelf, it might not need much more than an enclosure to keep out dust, stray bits of metal, and the occasional curious finger. If it's going to be outside, it will need to withstand the extremes of the local climate. A device that is intended to be wearable must contend with things like lint, dust, dirt, and sweat.

Types of Packaging

The material used for packaging generally falls into two basic categories: plastic and metal. Although wood is sometimes used for things like high-end designer PC enclosures, it's not really a good material when it comes to long-term durability and electrical shielding. The same goes for material like cardboard or foamcore. Not really a good idea.

PLASTIC

Plastic can be easily formed into just about any shape you might care to imagine, and once the final shape is fixed and set in a mold, you can make as many of the packages as you like, as quickly as the injection molding machine can push them out. Plastic is also easy to work with using a drill press and a rotary tool, and the harder types can even be machined with a lathe or vertical mill.

Early plastic enclosures used a material called *Bakelite*, invented in 1907 and named after Leo Baekeland. If you've even seen (or used) a vintage rotary-dial telephone, you've seen Bakelite. It was commonly used for everything from electrical insulators to radio and phonograph enclosures, and even in jewelry and children's toys. Today, Bakelite has largely been replaced by plastics such as ABS (acrylonitrile butadiene styrene), polystyrene, or PVC. Although modern plastics are easier to mold and less brittle, Bakelite is still used in applications where its unique properties are required.

Plastic does have some drawbacks when compared to metal enclosures. For one thing, many plastics don't tolerate UV very well, so it might not be a good idea to leave a plastic enclosure in a location where it gets direct sunlight. Plastic does not provide any shielding against unwanted electromagnetic radiation, so it needs to be coated on the inside with special conductive paint. Strength might also be a consideration, as a plastic part simply does not have the strength of a metal part of the same size and shape.

The mass production of a plastic part requires a mold, which is where most of the up-front expense lies. Creating a mold is a painstaking process, in both the design and the fabrication. When designing a mold form, the engineer must take into account things like unfillable gaps or corners, any volume change of the plastic as the temperature changes, and the design of any removable secondary mold parts to create complex internal shapes. Once the design is complete, a machinist creates the mold sections from metal. Only after the mold has been cleaned, polished, and checked for accuracy does it go out for production use. The mold design and fabrication process can cost many thousands of dollars, so it's not something to be taken lightly. If there is an error, the whole process might have to start over again.

Stock items such as plastic tubes, extrusions, and molded cases are also options for packaging with plastic. The advantage here is that the final package might not need much in the way of modifications to be immediately useful. Perhaps a few holes for indicators and connectors, and perhaps some internal parts like PCB card guides and a battery holder mounting arrangement (if a battery is used).

METAL

For creating electronics enclosures, metal can be cast into a mold, extruded into a tubular or rectangular shape, and formed into sheets, and a complex shape can be machined from a solid block of metal. Metal is extremely versatile, strong, and sometimes even lighter than an equivalent plastic enclosure, but it is not always very easy to work with.

Metal casting uses a mold created in damp sand into which hot molten metal is poured. As with a plastic mold, it is essential to get the mold right at the outset. Metal casting is typically used for heavy-duty enclosures intended for harsh environments. It doesn't show up much in consumer electronics, although it can be used for small items produced in large quantities. Some early portable MP3 players used cast metal cases, for example.

Extrusions are typically aluminum and come in a variety of shapes. The most common enclosure for elecronics is a rectangular extrusion with PCB guides formed into the inside of the extrusion. Extrusions can also be formed into a variety of shapes that can be used to create parts that will mount onto or inside something else, such as PCB card guides, brackets, or wiring channels.

Extrusions are available in various lengths with slots for holding panels and special adapters for connecting to other extrusions at 45- and 90-degree angles. Using these parts, you can build an enclosure using extrusions, some sheet metal, and the appropriate screws, nuts, and bolts to hold it all together.

Sheet metal fabrication has been used to create everything from the chassis in radio receivers to enclosures for computers. Almost all PC cases are made from sheet metal and fabricated using various bending, punching, and stamping techniques. Working with sheet metal requires a brake (for bending), a shear (for cutting), a drill press, a rivet tool, and perhaps a spot welder. For making large (greater than 1 inch) holes, or holes with a noncircular shape (such as D connectors), a hydraulic punch is essential.

Odds are that not many people will have all those tools lying about in their home shop. For this reason, custom sheet-metal fabrication, like mass-production plastic-injection molding, is typically used only when a large number of items need to be fabricated, or the cost and effort of building a one-off enclosure is outweighed by the need.

Machining from a solid block of metal can create a beautiful, lightweight chassis, but at the expense of machine shop time and a lot of wasted material. It is commonly used for low-volume things like parts of satellites and space probes, where the dimensions are unique to the particular application (space vehicles are largely hand-made things, and each is different). A machined enclosure might also be used as part of a prototype that will later be cast in metal or molded from plastic.

Stock Enclosures

Stock enclosures are available in a wide range of styles, size, and materials. Browsing through the available products from the enclosure and chassis manufacturers listed in Appendix D will give you some idea of the vast range of products available, and that's just a small sample.

PLASTIC ENCLOSURES

Companies such as Hammond Manufacturing make a variety of plastic enclosures for electronics, ranging from simple boxes to complete chassis with a carry handle.

Figure 16-1 shows one example, an ABS plastic box with a lid attached by two screws. It measures 1.38 inches long by 1.38 inches wide by .59 inches deep.

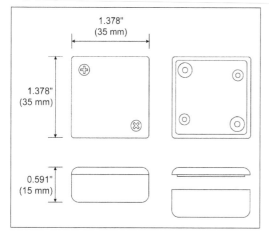

FIGURE 16-1. Small ABS plastic enclosure

Figure 16-2 shows what the enclosure looks like in real life (a ruler is included to provide a sense of scale).

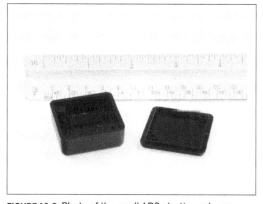

FIGURE 16-2. Photo of the small ABS plastic enclosure

What could possibly fit into such a tiny space? For starters, there is room inside for a PCB and even a battery. A simple display or some LEDs might be brought out through the cover (just drill or cut holes and let the display devices protrude through them). Attach a cloth or elastic loop to the back, and it could be something wearable. Glue one part of a square Velcro pad to the back and the device might be something you could put somewhere in a vehicle and take with you when you park it. A micro-USB connector could be brought out through the side of the enclosure. Here are some ideas for what could be made with a small enclosure like this:

- One part of a driveway light beam sensor (either the sensor or the IR LED sender).
- A really tiny CMOS camera.
- A fall sensor for an elderly person.
- Temperature and humidity display for use inside a vehicle.
- A koi pond water-level detector (to shut off the valve when the water level is high enough and turn it on when it is too low).
- Soil moisture detector (a red light might mean it's dry, and a green light could indicate sufficient soil moisture).
- An add-on to enhance an electronic toy (put an AVR microcontroller in the case and program it using the Arduino boot loader and software tools).
- A data logger to collect readings from a sensor and offload them later.

And the list goes on. But sometimes it's just not possible to cram everything into a small box, so a larger box is necessary. Figure 16-3 shows an enclosure commonly called a *project box*. This is an ABS box with a thin aluminum cover held in place with four self-tapping screws (see Chapter 2 for some caveats regarding self-tapping screws). It measures 1.5 inches high by 3.25 inches wide and 5.25 inches long (all measurements approximate).

FIGURE 16-3. Polystyrene "project box" with metal cover

It's common to get a cheap project box and then find out that the sides have warped inward so that the cover doesn't fit correctly. This is not uncommon with injection molded plastic parts, but if the box has stiffener ribs molded into it, the warping will be much less noticeable. Also, because of the warped plastic and loose production tolerances, the cover usually doesn't fit correctly to begin with. To make a cheap project box cover fit correctly and look presentable, you will probably need to trim the edges of the cover panel slightly on all sides. I use a sander like the one shown in Figure 16-4 to

do this. Ideally, if the cover fits correctly, it shouldn't bend or warp in the corners when the four screws are tightened. You will still probably need to pull out the sides of the box slightly when mounting the cover, because they will undoubtedly be warped to some degree.

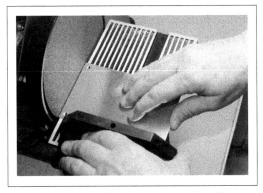

FIGURE 16-4. Reducing the overall size of a project box cover panel

Because of the hassles and aggravation, I don't recommend this type of enclosure. They tend to be poorly made and don't hold up well to rough handling. But then again, they are cheap, and sometimes it's about the only thing available on short notice. You can purchase them at many retail electronics outlets and supply houses.

> **TIP** The metal cover plates sold with plastic project boxes tend to be very thin. To get the best results when drilling holes, I recommend a drill press with sacrificial backing material, as described in Chapter 4. It is easy to warp the metal plate if too much pressure is applied during drilling without any back support.

Of course, plastic enclosures come in shapes other than rectangles and squares. Figure 16-5 shows some examples of the many different shapes that are available. Other shapes, not shown here, include game controllers, control pendants, and cases with a precut opening for a touchscreen display. It's easy to become lost for a couple of hours online looking at all of the types that are available.

FIGURE 16-6. A small, stock cast aluminum enclosure with gasket

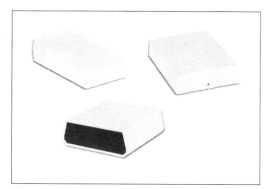

FIGURE 16-5. Plastic enclosures in various shapes

CAST ALUMINUM ENCLOSURES

Stock enclosures are available in cast aluminum for applications where the enclosure needs to be rugged and able to withstand harsh conditions. Figure 16-6 shows a typical small cast aluminum enclosure.

Some cast enclosures are available with rubber gaskets to protect the contents from rain, snow, and dirt. If, for example, you wanted to build a device to monitor something in a forest (rainfall or soil moisture content, for example), a sealed cast enclosure would be a good choice. Connectors can be mounted through holes in the enclosure and also sealed with rubber or neoprene gaskets. Some types are available with fins on the exterior for heat dissipation.

EXTRUDED ALUMINUM ENCLOSURES

Extruded aluminum is popular as an enclosure, and vendors offer various styles and sizes. Figure 16-7 shows a generic example with circuit boards that slide inside and end covers to close it all up.

FIGURE 16-7. Rectangular extruded aluminum enclosure

If you are old enough to remember what a modem is and have actually used one, odds are you've seen a commercial product built into an extruded enclosure. These types of enclosures were once popular and were used for everything from telecommunications devices to miniature test equipment. They are excellent enclosures for homemade test instruments, audio gear, and electronic music recording or sound processing gadgets, to name just a few possible applications.

Variations on this theme use solid metal end panels rather than plastic, and these are suitable for use in automobiles or other vehicles. The metal shell can be used as a heatsink to dissipate heat from something like a voltage regulator or a power transistor. Figure 16-7 has metal end panels.

SHEET METAL ENCLOSURES

Several enclosure manufacturers sell boxes and chassis made of formed sheet metal, either aluminum or steel, like the ones shown in Figure 16-8. One common type is the two-piece box, which is fabricated along the lines of the diagram shown on the left side of Figure 16-8. A two-piece box is made from two sheet-metal parts that have been bent or stamped to fit across each other, so that one piece forms the front, back, and bottom of the enclosure, while the other serves as the top and sides. In terms of ruggedness and quality, these can range from cheap sheet-metal stampings held together with self-tapping sheet-metal screws to products with press-in nuts in the cover mounting flanges, baked-on finishes, and ventilation slots cut into the top cover. As with just about everything else, it depends on how much you are willing to pay for it.

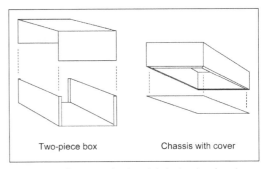

FIGURE 16-8. Common sheet-metal electronics chassis styles

Be sure to pay attention to the type of metal used in a sheet-metal enclosure. Steel is more durable, but it can be harder to work with than aluminum. A steel enclosure will also be heavier than an equivalent size in aluminum. Aluminum enclosures tend to be made from soft metal, and it can take some patience to get good results when you're cutting or drilling it. The upside is that you can use something like a rotary tool (discussed in Chapter 4) to make rectangular holes in an aluminum chassis or enclosure, or even a nibbler tool (also discussed in Chapter 4), but unless a steel enclosure is made with very thin gauge metal it can be a real chore to create noncircular holes in it.

Figure 16-9 shows a real-life example of a sheet-metal enclosure. It is made from steel and it's already painted, so all that's needed are the electronics to go inside and some rubber feet for it to sit on.

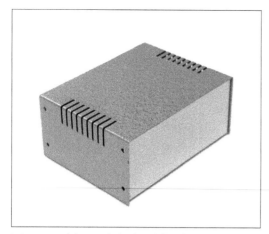

FIGURE 16-9. A two-part sheet-metal enclosure

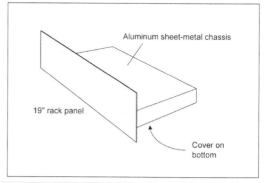

FIGURE 16-10. Rack mounted panel with sheet-metal chassis

The second common type of sheet-metal enclosure is shown on the right in Figure 16-8. This is essentially a box with an open side. Mounting flanges are formed along the open edges of the box, and a cover plate mounts over the open side with either self-tapping sheet-metal screws or machine screws (if press-in nuts have been used). It is commonly referred to as a *chassis*, rather than an enclosure, because generally people mount parts inside, perhaps with a transformer or other large items bolted to the top, and then mount the entire thing in some other, larger, assembly.

One way to use a flat chassis box like this is to bolt it to the back of a 19-inch rack panel at a 90-degree angle, as shown in Figure 16-10, and mount whatever controls, displays, and indicators are needed on the rack panel.

If you need something a bit larger and more rugged than a sheet-metal enclosure or chassis, a variety of enclosure kits are available. These kits allow you to build devices that look like commercial products, and these types of enclosures are often used in the production of commercial electronics devices.

Kits are more expensive than a simple enclosure, but low-cost products (such as the one shown in Figure 16-11) are available that require you to do the assembly work.

FIGURE 16-11. An enclosure kit for a bench or desktop electronic device

If you need a larger enclosure, you might want to consider some of the larger, heavy-duty enclosure kits that can be mounted into a 19-inch equipment rack. These types of kits aren't cheap (around $200 to $300 is typical). If you've ever looked at professional audio equipment in a music store, you've probably seen this type of chassis. Heavy-duty chassis like these are also found in television and radio broadcast studios, and the rack-mounted servers used in data centers are similar in design.

Technically, most of these would fall under the category of sheet-metal enclosures because they are formed from cut and stamped pieces of metal, usually steel. The main difference between the examples shown in Figures 16-9 and 16-11 is that one is very simple with a minimal number of parts, while the other has threaded holes for the screws, internal mounting points, braces, and other features.

Building or Recycling Enclosures

While there are many options to choose from when it comes to buying a ready-made enclosure, sometimes it's worthwhile (and fun) to build your own from raw materials or recycled items. Wandering through a secondhand store, browsing the electrical section in a hardware store, or even going through old toys that the kids have outgrown can help spark ideas and yield some interesting finds that you can use in your own project.

BUILDING PLASTIC AND WOOD ENCLOSURES

If you are comfortable with cutting sheet plastic, you can build your own enclosures. Ideally, you would want to have a drill press and a miniature table saw (such as the tools shown in Chapter 3). A rotary tool and a sander like the one shown in Figure 16-4 are also very useful. Figure 16-12 shows an exploded view of such an enclosure.

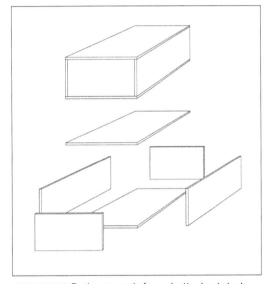

FIGURE 16-12. Enclosure made from plastic sheet stock

The box consists of six panels. Assuming that it is made from 1/8-inch to 1/4-inch sheet stock, you can assemble it using MEK to form plastic welds at the joints, or epoxy if the plastic isn't compatible with MEK (see Chapter 2 for information on adhesives). If you want to give the assembly a removable lid, you can use 1/4-inch by 1/4-inch pieces of extruded square cross-section posts cut from bar stock and attached to the inside corners, as shown in Figure 16-13. This makes the whole assembly much more rigid, and if you drill and tap the posts to accept a

#4 screw, they can be used to secure the lid on the enclosure.

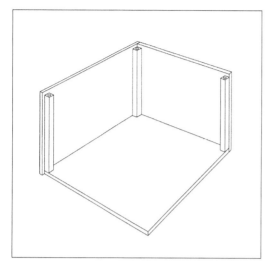

FIGURE 16-13. Using corner posts for reinforcement and cover attachment

Figure 16-14 shows the construction details. All dimensions are in inches unless otherwise stated.

Plastic can be worked with the same tools employed for metal. The primary differences involve the softness of the material and the tendency of plastic to melt and gum things up, so parameters like the tool speed on a mill or drill will need to be adjusted to compenstate. The lubricants used with aluminum or steel might not work well with plastic, particularly the solvent-based types. There are water-based lubricants available.

Cubes and rectangular boxes aren't the only options. You can make an enclosure with a sloped front panel or one that has a 90-degree bend to fit around the corner of something. The possibilities are limited only by your imagination and the materials you are using. Just bear in mind that, unless you really want to go to the trouble of making beveled edges, using a milling machine to make slots for things to slide into, and know how to sand and paint plastic, enclosures built with this technique have a tendency to

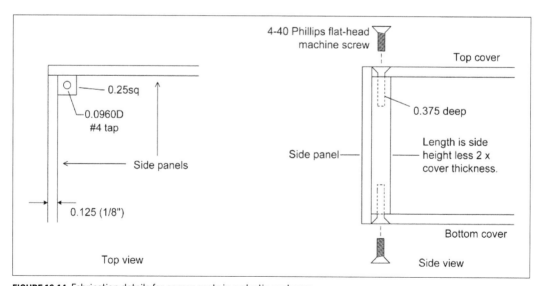

FIGURE 16-14. Fabrication details for corner posts in a plastic enclosure

look like hand-made items (fine-grit sandpaper and epoxy spray paint can help a lot with the aesthetics, however). Perhaps that's OK, depending on the application, but if you really need a polished look in a hurry, you might want to consider buying a ready-made enclosure. Consider your options before jumping into something like this, and if the trade-offs are worth the effort, you can end up with something that looks nice and will last a long time.

If you have woodworking skills, or know someone who does, and you have access to tools like a planer, jointer, and router table, then you can make enclosures from wood. Another option that is quite popular with the maker movement is to use a CNC (computer numeric control) router or a laser to cut pieces from wood panels. The wood can be thin plywood (birch is a common choice), or MDO (Medium Density Overlay) panels. A CNC router can produce clean cuts, but it's noisy and somewhat messy (lots of sawdust to deal with), and the cutting tool needs to be really sharp to avoid tearing the wood. A laser produces clean cuts with no rips or burrs, but it does tend to burn the edges of the wood pieces, leaving a dark outline around each part. But even with just conventional tools, you can still create good-looking things, like the device shown in Figure 16-15. This is a solar intensity measurement device I helped my daughter make for a science fair project. It might have the look of something high-tech, but it's really just some pine boards, a piece of plywood, and some paint.

FIGURE 16-15. A solar intensity measurement device made from wood

UNCONVENTIONAL ENCLOSURES

Just about anything that is the right size, easily sealed (more or less), and sturdy enough to take whatever physical abuse might get thrown at it has the potential to be an enclosure for electronics. An enclosure might be something that you might not think of as an enclosure for electronics, such as PVC or ABS pipe. You can get diameters ranging from 1/2 inch up to 8 inches at most hardware and home improvement stores, and a large number of attachments and adapters are available.

Figure 16-16 shows one way to use a section of PVC tubing as a sealed enclosure. This is an ideal low-cost housing for electronics used in an environment like a greenhouse, as an automated weather monitor, or as a remote sensor strapped to the side of a tree (for instance). It could also contain a level sensor (or two) and be mounted in a vehicle to determine when leveling jacks are extended to the correct height.

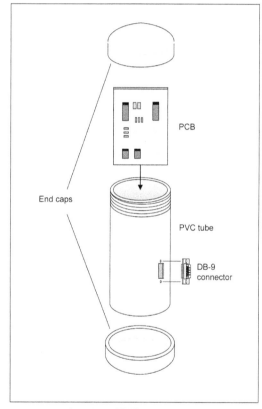

FIGURE 16-16. A section of PVC tubing used as an enclosure

Figure 16-16 shows a DB-9 connector, for a couple of reasons. First, DB-type connectors are available with glass-filled bodies and solid pins, which makes them suitable for applications like vacuum chambers and harsh environments. For many applications, a plastic-filled body would probably be acceptable, but the pins do need to be a solid machined type to block moisture entry into the tube. Secondly, the rectangular shape of the DB connector allows it to align vertically on the tube with a minimum amount of radius offset under the connector flange, which in turn reduces the amount of silicon rubber or epoxy resin needed to seal it to the tube. Remember that the connector can be wired to carry whatever signal and power lines are necessary, including USB, and a DB-25 could be used if more connections are needed. Once the mating connector is attached and the mounting screws are secure, the entire connector assembly can be encased in a generous application of clear silicon rubber. If it ever needs to be removed, it can be cut away with a sharp pocket knife.

Sometimes a useful little enclosure is an empty container that would otherwise be tossed in the trash or the recycle bin, like the little tins that mints come in. It turns out that the Altoids brand tins are popular for project enclosures. Searching "altoids tin projects" on Google returns around 127,000 results. Switch to images and take a look at what people have done with these ubiquitous little metal boxes. You'll find everything from ham radio modules to effects for guitar players, from MP3 players to set-top video players. You can even buy box lots of empty tins on eBay.

Steel food, coffee, and tobacco cans are another possibility, although you will most likely need to fabricate an end cover to replace the one that was removed when the can was opened. Metal cans can be pressed into service as antennas for wireless data links, as a housing for one end of a light-beam communication link, and as a chassis for mounting various sensors. The blind rivets described in Chapter 2 can be used to quickly and easily attach a mounting bracket.

The main thing to keep in mind when using a coffee or tobacco can in a situation where it can be exposed to rain or ambient moisture is that you will need to strip the old paint, prime the bare metal, and then repaint it both inside and out to prevent rust. That can be a lot of work if you don't happen to have a sand-blaster handy or a bench grinder with a wire brush wheel, so you might want to consider that.

Small utility boxes like the ones used for residential AC wiring can also be useful for some applications, and browsing the selection on the shelves at a hardware store or electrical supply outlet can be a good source of inspiration. Granted, most of the enclosures designed for use with electrical power wiring aren't all that visually appealing, because they're designed to be bolted to a flat surface or hidden behind a wall. But it's still possible to find something that will do the job, and it doesn't take too much effort to cut the mounting tabs off a molded box if you really don't need them.

REPURPOSING EXISTING ENCLOSURES

Occasionally, you might come across an intriguing enclosure with something already inside of it. Well, if you don't need whatever was originally packaged inside, you can pull it out and put in your own circuitry. Candidates for repurposing include old test equipment, junk PC components (particularly old outboard disk drive enclosures), and broken or discarded electronic toys like a spy voice recorder or a sing-along headset microphone. Lots of possibilities there. If you like harmless pranks, an old toy with something like an Arduino, or a wireless module like one of the Bluetooth LE types described in Chapter 14, stuffed inside can be a source of significant amusement.

While expending the time and effort to disassemble something and rebuild it to serve a new purpose might be considered to be *extreme hacking*, it is sometimes worth the effort if you really need a particular type of packaging for your own invention and simply cannot find it anywhere else. If you are careful, take your time, and don't mind expending the time on it, you can have a rugged, professional, and unique enclosure.

Designing Packaging for Electronics

One of the best ways to get ideas for packaging is to look and see what others have done to solve their packaging problems. For example, Figure 16-17 shows a personal pedometer built into a small plastic case. It has a flip-up cover, an LCD display and a reset button, a molded clip on the back for attaching it to your belt, and a removable fabric loop with a metal clip (not shown) for securing it to your clothing if you don't happen to have a belt.

FIGURE 16-17. A wearable personal pedometer

This is obviously a custom injection-molded item, and it appears to have been made with at least three different parts: case backshell, front cover, and panel insert. It wasn't disassembled because the device appears to have been glued together, and prying it apart could destroy it (not to mention irritate its owner). It uses some type of small battery (like the ones shown in Chapter 5), but the battery cannot be replaced. So when this thing dies, it's dead forever. That might be a good time to take it apart and see what's inside, if you're willing to wait that long.

The enclosure shown in Figure 16-1 could easily be used for a device like this. Plastic and metal belt clips can be purchased from multiple sources, and the ones seen in a quick Google search ranged from $3 to $6.

Figure 16-18 shows another interesting example of an innovative packaging solution. This is an old Fluke 1900A event counter and frequency meter.

FIGURE 16-18. Fluke two-part plastic instrument enclosure with carry handle

The Fluke Corporation made a series of electronic test equipment devices in the 1970s and 1980s featuring this two-part packaging scheme. It's a brilliant design in that it has only two main parts, the front panel with the attached PCB, and the shell. One screw in the rear holds it all together. The devices also included a carry handle that can be set to act as a stand or folded back out of the way, and only the front panel and internal PCB changes for each type of instrument.

Bud Industries manufactures a series of plastic enclosures that are similar to the Fluke design, including the IP-6130 and IP-6131 kits. Figure 16-19 shows an example (with nothing in it, of course). These are a bit more complicated than the Fluke package, because an enclosure kit must be capable of being adapted to a variety of applications. You can pick up an IP-6130 from an electronics distributor for around $26.

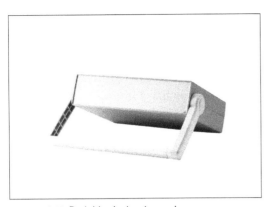

FIGURE 16-19. Portable electronics enclosure

DEVICE SIZE AND WEIGHT

If you're building something for your own personal use, considerations like size and weight might not be at the top of the list. However, if you think you might want to mass—produce your design, then these become major considerations.

Is the device wearable?

- Wearable devices need to be small and lightweight. Anything over about 1.5 × 1.5 inches (or around 38 × 38 millimeters) will probably be too big for a wrist, but things worn on a belt can be larger.
- Depending on the application, a wearable device should be light enough to be largely unnoticed. Something that has enough mass to bounce and tug with the wearer's every movement wouldn't be a good idea for an athletic wearable, but it might be acceptable for something used as a shop or construction worksite tool. Even then, something hanging from a belt that bounces around can be annoying.

Does the device need to be easily portable?

- If the device is something that a user might want to pick up and carry from one place to another, it should be light enough that it doesn't throw his back out when he lifts it.
- A portable device might need a handle (or two) if it's too large to conveniently pick up and carry one-handed. If handles are used, they must be sized to accommodate the weight of the device and comfortably fit a generic hand size.

Will the device reside in a static environment, like a workbench or a 19-inch rack?

- When you are building something for a workbench, the size should be just large enough to hold all the internal parts but not so large as to crowd out other things on the workbench.
- If the device will be mounted in a 19-inch rack, then the width is already constrained. The two other dimensions to consider are height and depth. Sometimes a small equipment rack won't accommodate anything with a depth of more than about 15 inches or so. Other racks, like those used for servers and networking equipment, can handle things up to 28 inches or more deep.

Will the device be integrated into a vehicle of some type?

- If your device is intended to be mounted in an automobile or a truck, then the weight isn't as big of a consideration as the size. It will need to be small enough to mount out of the way, perhaps under a seat, under the dash, or in the trunk.

- If your device will be used in something like a boat, it will need to be environmentally protected (see "Environmental Considerations" on page 378 for more on this). As with a device for a car or truck, the weight isn't as much a factor as the size.

ENVIRONMENTAL CONSIDERATIONS

Anything that will be exposed to the elements needs to be protected, and even things that are intended to be worn by a user should be protected if they might be exposed to sweat (sweat is very corrosive to electronics).

Will the device be exposed to rain, sea spray, or other forms of moisture?

- A plastic package with a gasket to seal the enclosure is acceptable for small items. The plastic should be thick enough to allow the cover to be tightly sealed without stripping screws or causing the enclosure to warp.
- A cast-metal enclosure is a better choice than plastic if your device will be mounted in an exposed location on a vehicle. Many cast-metal boxes are available with a gasket to seal the cover of the enclosure, like the one shown in Figure 16-6. Applying a thin layer of silicone grease to the gasket will help to prevent it from drying and rotting, and improves the water resistance of the seal.
- If the device will be mounted in an outside location, such as the control circuits for a remote weather station or some kind of data acquisition device, then a sealed cast-metal enclosure might be an appropriate choice. A section of PVC tubing (the so-called *Schedule 80* outdoor-rated type) can also be used. Assuming that a threaded end cap is used, the end cap will need to be sealed with Teflon thread tape or silicone grease to keep dirt and moisture out.

Will the device be exposed to direct sunlight?

- Direct exposure to sunlight can destroy a plastic enclosure over time. The UV from sunlight can break down the polymer chains in plastic, making it brittle and crumbly. Some plastics, like the PVC tubing known as *Schedule 80* can withstand direct sunlight for an extended period of time (on the order of years), but even it will eventually have problems.
- Try to shade an electronic enclosure from direct sunlight, if at all possible. Something like the white boxes used for official weather stations is one possibility. If nothing else, just putting your device inside a larger box, like an electrical utility box, can shade it from the direct glare of the sun.

Will the device be used in a dirty or dusty environment?

- For a dirty or dusty environment, the same general concepts for water resistance also apply. While dust isn't normally conductive, it does create thermal issues and it can get into controls like potentiometers and switches.

- If your device has a fan, odds are it won't be used in a moisture-prone environment, but it might be used in a dusty situation. Any fans or ventilation holes will need to be filtered to keep dust and dirt out, but still allow for air flow. Of course, this also means that someone will need to clean or change the filters occasionally.

THERMAL CONSIDERATIONS

Heat is one of the primary enemies of electronics. Heat degrades components, including PCBs, and it can lead to failures.

Will the device have a sealed enclosure?

- If a device is contained in a sealed enclosure, there is no way to air to circulate and remove excess heat, which means that the component parts will need to be rated to operate at the nominal internal temperate inside the enclosure.
- If at all possible, try to incorporate a heatsink for parts that generate heat. This includes things like voltage regulators, power transistors, and motor control ICs. A heatsink might be the wall of a cast-metal enclosure, or an add-on attachment for a TO-220 package (see Chapter 9).

Will the device be mounted near, or exposed to, a heat source?

- Try to avoid mounting your device in a location where it will be exposed to heat sources. For example, the space under the hood of a car or in the engine well of a boat can get very hot. The sun is also a major heat source, so things exposed to direct sunlight not only have to deal with direct UV, but also the IR (infrared, or heat) part of the solar spectrum. So try to mount things like a solar-panel motion controller gadget in a shady area, if possible, and run a cable out to the sensor that has to sit in the sunlight.

Sources

Table 16-1 lists the various sources for the commercially available enclosures discussed in this chapter. Although I've listed the manufacturer's name and URL, you can also find many of these at major distributors and from Amazon. I've not listed the prices I paid, because they can vary, depending on where you purchase a particular item. Shop around and compare to get the best price.

The enclosures listed in Table 16-1 are just examples of what is available; there are many styles, sizes, and types to choose from. Look around, compare prices, and if possible, go to local electronics parts suppliers and ask to look at their stock of enclosures. Some will be helpful and let you see samples of what they carry, and some might even have small enclosures sitting out on a display shelf. Others will want you to have a part number already in hand and won't let you look in the stockroom.

TABLE 16-1. Enclosure sources

Figure	Manufacturer/source	Part number(s)	URL
Figure 16-1	Hammond Manufacturing	1551NBK	http://www.hammfg.com
Figure 16-3	Parts Express	320-430	http://www.parts-express.com
Figure 16-5	Serpac	15-S, G10A, A31	http://iprojectbox.com
Figure 16-6	Retex	RI-400 series	http://www.retex.es
Figure 16-7	Hammond Manufacturing	1455N1201	http://www.hammfg.com
Figure 16-9	Hammond Manufacturing	1454R	http://www.hammfg.com
Figure 16-11	Hammond Manufacturing	1458G5	http://www.hammfg.com
Figure 16-19	Bud Industries	IP-6130	http://www.budind.com

Summary

This chapter has covered a wide range of electronics packaging types and techniques, from plastic to metal. One thing that I hope came through is that packaging is an exercise in trade-offs.

Here are the key trade-off considerations we've covered:

- Plastic is often cheaper than metal, but it doesn't have the structural strength of metal, and it isn't always suitable for large enclosures.
- Metal is stronger than plastic, and a metal enclosure can also serve as a heatsink, but it's more difficult to work with than plastic and can be more expensive.
- Designing packaging for harsh environments can be a challenge and usually ends up as a trade-off between environmental resistance and thermal considerations.

A completely sealed plastic enclosure can't readily dissipate heat, whereas a sealed metal enclosure can, but it will often be heavier and bulkier than the plastic equivalent.

The chapter also showed that off-the-shelf enclosures are a good choice for prototypes, and sometimes they can also be used for mass production. With mass production, there are up-front costs to be considered, mainly for mold machining and production tooling. Lastly, we took a brief look at salvaging an enclosure from an existing piece of equipment. Old test and computer equipment is a good place to look for enclosures, ranging from small to very large, and even low-cost consumer electronics can sometimes be repurposed. The major consideration here is the time and effort that will be expended versus the cost of buying something that's already close enough to get the job done.

CHAPTER 17

Test Equipment

THIS CHAPTER IS A QUICK TOUR OF WHAT IS available in the way of inexpensive test equipment, starting with the ubiquitous digital multimeter (DMM) and moving on to oscilloscopes, signal generators, and logic analyzers. The focus here is on low-cost tools that will help get the job done without costing a small fortune. In the world of test equipment, it's all too easy to spend a lot of money, with some types of equipment running upward of $30,000 each (or more). For the vast majority of situations you are likely to encounter when working with common electronics components and devices, that sort of precision and processing speed isn't necessary.

Now would probably be a good time to talk a bit about things like speed, accuracy, and bandwidth. We'll cover these topics in more detail later on in relation to each type of instrument, but the main point is that, for the vast majority of things you might want to build or modify, you don't need an oscilloscope with a bandwidth capable of displaying a 1 GHz signal, nor do you need a digital meter with 4 1/2 digits of resolution. You don't need an RF spectrum analyzer, or a high-precision pulse generator, or even a frequency counter. When you are working with things that interact with the real world in some fashion, the times involved are typically anywhere from 10 to 1,000 ms (0.01 to 1 second). Physical things usually don't move much faster than that. If you are working with a microcontroller, it will be running at a much faster rate, but you don't have visibility into the chip itself, just the inputs and outputs, and they are slow by comparison to the microcontroller's internal clock.

Basic Test Equipment

This section describes the two most fundamental types of test equipment: the digital multimeter and the oscilloscope. With just these two instruments, you can design, test, debug, and hack a wide range of electronic circuits. In fact, even an oscilloscope isn't absolutely necessary in many cases, but a DMM is essential.

DIGITAL MULTIMETERS

A modern DMM can be used to measure voltage, current, and resistance. Some models also include the ability to measure capacitance, inductance, and frequency, among other things. These devices are available in both portable (hand-held) and bench versions. There are also rack-mount variations for permanent installations.

DMMs are available in a range of prices depending on resolution, accuracy, and any additional features. Some basic models are available from around $10, although they might not offer much in the way of accuracy or functionality. Other models that feature increased accuracy, better measurement resolution, and enhanced features can range in price from about $30 to upwards of $300 or more.

If you have nothing else, the one thing that is essential is a decent DMM. Typically, the low-cost models are hand-held devices, designed for portability and battery operation. Figure 17-1 shows a basic DMM, and Figure 17-2 shows a fancier (and more expensive) model.

FIGURE 17-1. Basic low-cost DMM

FIGURE 17-2. DMM with enhanced features

The two main criteria for a DMM are resolution and accuracy. The resolution of a DMM is reflected in how many digits the device is capable of displaying. A meter with a 3 1/2-digit resolution will not show some very small values, but a meter with a 4 1/2-digit resolution will. The 1/2 refers to the most significant digit on the display, which is typically a 1. A meter that can display from 19,999 to 0.0001 is a 4 1/2-digit resolution device, and a 3 1/2-digit DMM can display from 0.001 to 1,999. Low-resolution meters tend to round measured values to fit in their display resolution range. So a voltage of 0.0006 V would show up as 0.001 V on a 3 1/2-digit meter, assuming that it rounds up.

USING A DMM

When poking around in a live circuit with a DMM, you can encounter a range of voltages. The DC supply will, of course, read at or near its design value. At other points in the circuit, the voltage will be less than the supply level. This is due to resistances and semiconductor junctions. Also, looking at an AC signal with a DMM set to the DC input will often show some small amount of voltage. This could be some DC present with the AC. If you think you should be seeing more than the DMM is showing, try switching to an AC input. If the signal is within the DMM's frequency range, you might see a higher reading for the AC component in the circuit.

> Always check the settings on a DMM before touching the probes to a live circuit. Bad things can happen if the probes are placed across a high voltage while the meter is set to measure resistance or current.

Although the negative probe (usually black) of the DMM can be connected to the local ground, with battery-powered meter, it doesn't have to be. You can measure the voltage drop across a resistor by touching the probes to either end of the part. This technique can be used to determine if any current is actually flowing through the resister: if there is no voltage drop (the voltage across it is zero), then no current is flowing.

> Never attempt to measure resistance in a live circuit. The DMM generates a small voltage when measuring a resistance, and this can wreak havoc with an active circuit. The DMM can also be severely damaged if you do this.

Measuring resistance with a DMM is straightforward, but be aware that measuring a part's resistance while it's in a circuit might give an erroneous reading, depending on what other parts are in the circuit.

To get a truly accurate measurement of small values of resistance, you should first touch the probe tips to each other and hold them tightly. If the DMM is capable of reading a small resistance, you should be able to measure the resistance of the probe leads. Subtract this from whatever reading the meter shows to get the actual resistance of the item between the probes. Note that some meters have the ability to automatically compensate for probe lead resistance.

Measuring current usually involves changing the connection of the positive (red) lead from the V input jack to the A jack. The meter in Figure 17-1 has two current input jacks: one for low-level readings and the other for up to 10A. Notice on the rotary dial that it can also measure both DC and AC current. The DMM shown in Figure 17-2 is essentially the same.

When using a DMM to measure current, it is important to bear in mind that the meter itself is part of the circuit. With a voltage measurement, it's more of a passive observer, but with current, it becomes directly involved. The reason is that a DMM uses an internal shunt resistance for current measurement. The *shunt* is a precision, low-ohm device (often just a piece of metal) that will exhibit a small voltage drop when current flows through it. The meter is actually reading the voltage drop across the shunt. To the external circuit, it appears as a conductor with a very low resistance.

Let's say, for example, that you wanted to know how much current is being used by a circuit designed to operate from a battery. Figure 17-3 shows one way to do this.

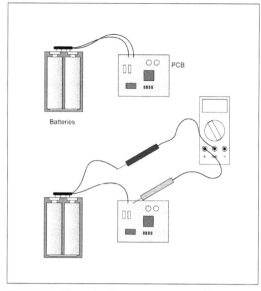

FIGURE 17-3. Measuring the current consumed by a circuit on a PCB

In order to measure the current used by the PCB, one of the leads from the battery pack will need to be disconnected. The DMM is then inserted between the battery and the PCB. An alternative approach would be to make a special cable with the appropriate connectors to allow it to be inserted between the battery pack and the PCB. It might even have a pair of banana-type plugs already connected to plug directly into the DMM.

> When the DMM is configured to measure current, *never* connect it directly across a power source. Most better models have an internal fuse, but that isn't always guaranteed to protect the meter from a fast voltage transient. And even if the fuse does sacrifice itself to save the meter, it can be a pain to replace the fuse on some DMMs.

OSCILLOSCOPES

An oscilloscope measures changes in voltage over time. That's it, but it doesn't really need to do anything else. By measuring an input signal over time, you can determine the voltage level, the frequency (if it's a periodic signal), and the rise and fall time of the start and end of a pulse. By using a current shunt (a type of low-value resistor), an oscilloscope can also measure current as a DC voltage across the shunt.

An oscilloscope is a versatile piece of test equipment that allows you to see what is going on inside a circuit. There are some models available that are about the same size as a smartphone. In fact, some are built into cases that look like smartphones, like the one shown in Figure 17-4. They typically won't work with high-frequency signals, but they're fine for looking at relatively slow events. The device shown in Figure 17-4 cost about $50 from a Chinese vendor through eBay.

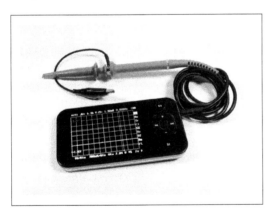

FIGURE 17-4. Miniature digital oscilloscope

Some oscilloscopes use a PC as the display. These devices plug into a standard USB port and serve as the *front end* to convert the external signals into a stream of data that a special application running on the PC can display as a waveform. If you elect to purchase a USB oscilloscope, make sure to read the specifications carefully, paying special attention to the sample rate. For example, one unit might sell for the amazing price of $34, but it might measure signals only up to 3 KHz. Another might sell for $70, but it might measure signals up to 20 MHz. Figure 17-5 shows one type of low-cost USB oscilloscope. Generally, these units are all variations on the same theme: small plastic enclosures with two BNC connectors for the probes and a USB connector (usually a type B). Some have additional indicators and controls, and the high-end units might have more memory to store the digitized waveforms.

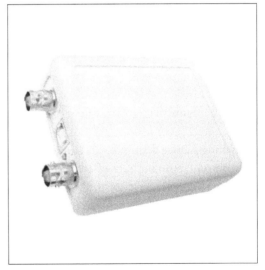

FIGURE 17-5. USB digital oscilloscope

You can use even the sound system in a PC as a low-speed oscilloscope by applying the signals from a circuit (with some appropriate buffering and protection) directly to the microphone or line inputs. Figure 17-6 shows one such application for Linux (xoscope).

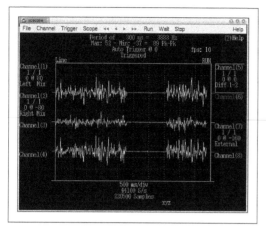

FIGURE 17-6. xoscope running on Linux (image from *http://xoscope.sourceforge.net/*)

While something like xoscope might seem like a neat idea, be aware that it usually won't measure DC voltages (it depends on the type of audio input used in a particular PC). Why? Because a capacitor on each audio input prevents DC from getting in. Also be aware that it won't measure signals with a frequency greater than 22 KHz, because the audio input will not respond to anything higher than that (it's limited by the maximum sampling rate of the audio input analog-to-digital convertor in the PC). Check out the official xoscope home page (*http://xoscope.sourceforge.net/*) for more details, and if you elect to install it, be sure to read the manpage.

It is also possible to use an Arduino as the hardware frontend for a digital oscilloscope running on Linux. Figure 17-7 shows the display window of the lxardoscope application. lxardoscope isn't particularly fast; in its basic configuration, its sampling rate limits it to around 1.5 KHz.

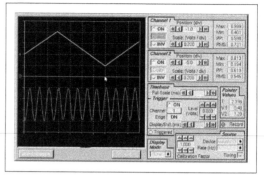

FIGURE 17-7. lxardoscope running on Linux (image from *http://sourceforge.net/projects/lxardoscope/*)

Be forewarned that this is something you'll need to compile yourself, and the Makefile that comes with it is set up for a 32-bit platform. To compile it to run on a 64-bit Linux machine, you will also need to install the `libforms2` package. It won't hurt to install the `libforms-dev` and `libforms-doc` packages as well. Replace the reference to the included `libforms.a` library in the Makefile with `-lforms` and add `-m64 -march=x86-64 -fPIC` to the `CC_OPTIONS` declaration.

Also note that you'll need to build a preamplifier circuit, so this falls under the catagory of *major project*. Still, it's an interesting example of how to integrate different subsystems into a functional whole, and I've included it for that reason. The archive from Sourceforge (*http://sourceforge.net/projects/lxardoscope/*) comes with documentation, schematics, and other interesting technical data. If nothing else, it's worth a look to see how someone else solved a particular problem.

If you would prefer something already built and tested, benchtop portable digital oscilloscopes capable of measuring waveforms up to 25 MHz are available for around $300. But if you really need a faster instrument, be prepared to pay upward of $1,000 for it, and some brands and models can be even more expensive. A web search for "digital oscilloscope" will return over 3.6 million results. That's a lot of shopping to do. Figure 17-8 shows an example of a compact portable digital storage oscilloscope.

HOW AN OSCILLOSCOPE WORKS

Oscilloscopes were invented early in the 20th century, when radio engineers realized that they really needed to be able to see what their circuits were doing. A voltmeter just didn't work for some situations. These early tools used a glass tube, similar to an old-style television picture tube, to display a dot that moved across the screen from left to right at a rate determined by a knob on the front of the instrument. Another knob was used to adjust the gain (also called *sensitivity*) of the input to keep the signal within the vertical limits of the display. Figure 17-9 shows a generic block diagram for an analog oscilloscope.

A basic old-style analog oscilloscope consists of a vertical amplifier/converter, a trigger/synchronization circuit, a horizontal sweep oscillator (or timer), and some type of display, typically a cathode-ray tube (CRT, similar to an old-style TV or computer monitor glass display).

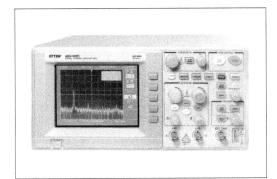

FIGURE 17-8. ATTEN ADS7202C digital storage oscilloscope

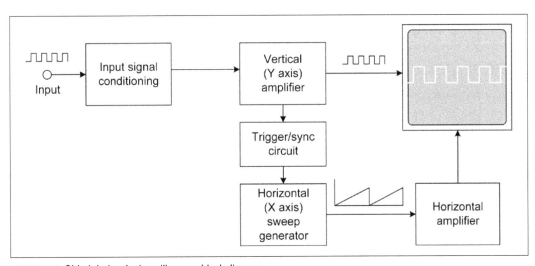

FIGURE 17-9. Old-style (analog) oscilloscope block diagram

The idea is to cause the display to respond to the input signal by moving the display point up or down as it sweeps across the face of the display. In older instruments with a glass CRT, the beam of electrons that creates the spot on the face of the tube is literally steered across the display while being deflected up or down by the vertical input signal. If you get the chance to work with an older CRT type oscilloscope, you should do so. It will help make some of the concepts masked by digital instruments much clearer.

Being able to see an input waveform is good, but if the horizontal timebase isn't in sync with the signal, it will appear to drift (or wander) across the display. The trigger circuit is used to lock the horizontal timing to the input and create a steady display. A trigger circuit works by sensing when the input signal has reached some threshold. When this occurs, the horizontal timebase is effectively reset to make the input signal appear to stand still in the display. Changing the horizontal sweep rate while the trigger is active has the effect of magnifying the input waveform in the horizontal direction. This allows you to zoom in on an interesting part of the waveform.

With a good trigger and a known hortizontal sweep rate, you can determine not only the peak-to-peak level of the input waveform, but also the frequency. Older instruments had a clear plastic plate over the face of the CRT with a grid machined or molded into it. Lamps along the side of the plate (hidden behind the bezel around the CRT) were used to illuminate the grid lines. Determining the frequency of the input was simply a matter of counting the number of vertical lines between repeating peaks in the waveform and multiplying by the time per division as determined by the horizontal rate.

The early oscilloscopes were all analog. Modern oscilloscopes are digital, meaning they convert the input signal into a stream of binary numbers (Chapter 13 describes analog-to-digital convertors). The binary data is used to generate a waveform on an LCD display. In a digital oscilloscope, the display processing is done virtually by the internal microprocessor after the signal is converted from analog to digital. Figure 17-10 shows a block diagram for a modern digital oscilloscope.

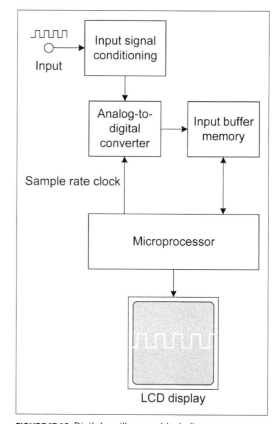

FIGURE 17-10. Digital oscilloscope block diagram

In modern digital oscilloscopes, measuring the frequency of the input signal is typically one of the functions done automatically by the input conversion and display logic in the instrument. A digital oscilloscope will also display the peak-to-peak level of the input signal, and many digital instruments have the ability to position *cursor bars* in the display to measure a particular part of a signal, such as the rise time of a pulse or the voltage level of a particular part of a waveform. Figure 17-11 shows the display generated by a digitial oscilloscope. This particular model doesn't have an LCD display, but it's otherwise fully solid-state and completely digital.

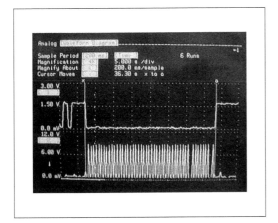

FIGURE 17-11. Digital oscilloscope display

USING AN OSCILLOSCOPE

As mentioned earlier, what you will be able to measure with an oscilloscope depends on how fast the instrument can respond to the input. This is largely a function of how fast the horizontial oscillator can run, but it also depends on the frequency response characteristics of the input amplifier section.

I mention this because if you use a low-speed instrument to look at a high-frequency signal, you might not see what you would expect. In an old-style analog oscilloscope, the input will generally fall off as the frequency exceeds the upper limit of the vertical amplifier circuit, and what's shown on the display will look like a solid bar or a fuzzy blur, if it shows much of anything at all. In a digital oscilloscope, the maximum input frequency is limited by the sampling circuit. A signal that exceeds half of the maximum sampling rate is subject to *aliasing*, which is a result of the Nyquist limit (a fundamental concept in sampling theory). An aliased signal will appear to be a different frequency than it really is, so other than demonstrating that there is really a signal present, it's generally useless.

It's really very simple to actually use an oscilloscope. Figure 17-12 shows a diagram of a generic modern digital oscilloscope.

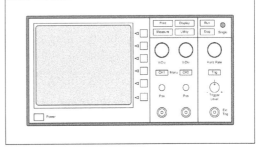

FIGURE 17-12. Generic digital oscilloscope front panel and controls

Here you can see that the instrument has two inputs, or channels. Each input channel corresponds to the vertical or y-axis input of an older analog oscilloscope, and each has a V/Div knob. This control determines the sensitivity of the input channel. Note that the control knobs operate in units of volts per division, where the division refers to the reference lines that appear on the display. You can see these in Figure 17-11. As you turn the V/Div knobs, the readouts on the display will change to indicate the active display range.

The buttons labeled *Pos* set the vertical position of the waveform on the display. You can adjust these to move a waveform display to where you want it and even have channel 1 overlap channel 2, which is sometimes useful when you're comparing two inputs.

The next major section of the front panel is the horizontal control. In an analog oscilloscope, this would be the *sweep frequency*: the rate at which the beam is driven across the face of the display. In a digital oscilloscope, it is essentially the amount of time from one side of the display to the other. To view a waveform with a frequency of, say, 1 MHz, you could set the horizontal control to around 0.00001 seconds (10 microseconds). That would be 1 microsecond per division if there are 10 divisions on the display. You should then see 10 cycles of the input waveform.

The last primary knob is the trigger level. As mentioned earlier, the horizontal rate of the display can be locked to the input signal. This results in a stable display that doesn't wander or drift over time. The trigger works by sensing when the input signal has exceeded a particular threshold, either positive or negative. Adjusting the trigger level allows you to select a part of the waveform to use to synchronize the display, and a digital oscilloscope will generally show what the trigger is currently set to by using a moving horizontal line on the display or a numeric readout in the display, or both.

The buttons along the side of the display are used for things like math functions, input mode selection, reference marker selection, instrument setup, and so on. The CH1, CH2, and Trig buttons might call up menus on the display, and the side buttons allow you to make selections from the menus. These button might also have dynamic functions that are available while the instrument is running and acquiring data.

A real digital oscilloscope will have more controls than this simple diagram, but they are all basically the same. Some just have more bells and whistles than others. If you acquire or have access to an oscilloscope, be sure to spend a little time with the user manual (if one is available, of course).

Here are a few tips and cautions for using an oscilloscope:

- Unless the oscilloscope is battery operated (as some modern digital oscilloscopes are), always use a grounded outlet. Leaving the chassis floating without a ground return can lead to situations where weird noise appears for no reason (it would most likely be a local AM radio station), or the chassis of the instrument can go *hot* and give a nasty shock, or even worse, cause a short and severely damage something (including you).

- Always ground the input probe. That's why it has a ground lead attached to it. With some oscilloscopes, the ground leads on the probes are connected to a common point inside the instrument, so you might get away with connecting one but not the other, but don't count on this.

- Never connect the ground lead on an oscilloscope probe to anything that isn't actually ground. Some instruments are designed to allow this, but some aren't, and it's unpleasant to discover that the ground lead really is ground when connected to a DC voltage in a circuit.

- With a digital instrument, be aware of potential aliasing. If you think you should be seeing a waveform at a particular frequency, but you are seeing something else, you might have exceeded the sampling limit for the instrument.

Advanced Test Equipment

At some point, you might find that you need something more than a DMM and an oscilloscope. When it comes to advanced test equipment, some of the most useful items are pulse generators, signal generators, and digital logic analyzers. These instruments can help wring out the kinks in a troublesome circuit and show what is going on at a specific moment in time.

But don't rush out and buy them just yet. In order to get the most from test equipment like this, you really need to have a good reason for using it and a good knowledge of *how* to use it.

PULSE AND SIGNAL GENERATORS

Pulse and signal generators are closely related, in that both can generate a repeating waveform. Typically, a signal generator will produce a sine wave, although one variation called a *function generator* will also output square, pulse, ramp, and triangle waveforms. A pulse generator isn't designed to output anything other than pulses (as you may have surmised from the name), but it is good at what it does. A quality pulse generator will output a single pulse, a burst of pulses, or a continuous train of pulses, all at a specific duty cycle. Many units can be configured to emit one or more pulses when a trigger signal is detected, and some models have the ability to dynamically vary the duty cycle in response to a control input (pulse-width modulation).

A new bench signal or pulse generator will run $300 and up, depending on the features and the brand. Since these are not items in high demand, they tend to be on the pricey side. Physically, digital signal, pulse, and function generators all look more or less the same. They have controls to set the output frequency, perhaps a knob to adjust the pulse duty cycle (if pulses are provided), and maybe a digital display to show the frequency or time of the output. A function generator will almost always have controls to select the type of output waveform.

Just as there are low-cost versions of oscilloscopes and DMMs, there are low-cost versions of signal and pulse generators. Figure 17-13 shows an Arduino-compatible DDS (direct digital synthesis) module that is capable of generating waveforms from 0 to 40 MHz using a AD9850 DDS IC. The AD9850 chip on the PCB can generate both sine and square waves, and it is relatively easy to interface to an Arduino (or some other single-board microcontroller). Toss in an LCD display, some connectors, and a nice case, and it would make a usable programmable signal generator.

FIGURE 17-13. An Arduino-compatible AD9850 DDS module

If you're interested in the AD9850, I would suggest downloading the datasheet from Analog Devices and studying it. It has a serial interface, but it uses a 40-bit internal register to hold a 32-bit frequency control word, a 5-bit phase modulation word, and a power-down function. That means it needs 40 bits of data each time the frequency or phase modulation is changed. This isn't hard to do in software, but it is beyond the scope of this book. You can usually find example code on the websites where the module shown in Figure 17-13 is sold, and an article on the Instructables website (*http://bit.ly/generate-sine*) describes how to build a cheap programmable sine wave generator using this module.

LOGIC ANALYZERS

The logic analyzer is a useful instrument for measuring and monitoring activity within digital circuits. Logic analyzers capture a set of digital inputs simultaneously and store the binary values in a short-term trace memory. The contents of the trace memory are

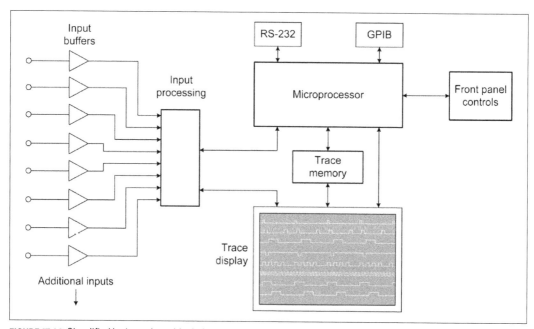

FIGURE 17-14. Simplified logic analyzer block diagram

then read out and displayed in the form of a timing diagram. Figure 17-14 shows a block diagram of a simple logic analyzer. This diagram would be applicable to a self-contained instrument, but there are other ways to achieve a similar result. For example, if the signals are changing relatively slowly, you can use the parallel printer port on a PC as a simple four-channel logic analyzer (the PC parallel port has four input lines available).

So, what does a logic analyzer show you? Figure 17-15 shows an example of the timing diagram generated by a logic analyzer. The idea here is that every input channel has its own line on the display. When a channel input goes high, the trace line goes up, and when it is low, the trace line returns to its zero position. A logic analyzer is particularly useful for visualizing how signals are changing state over time, both independently as a result of the action of other signals in the circuit. Logic analyzers often have 8, 16, or even 32 inputs.

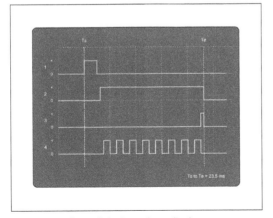

FIGURE 17-15. Example logic analyzer display

In Figure 17-15, you can see that a trigger (channel 1) causes the state of channel 2 to change to high, and at the same time, 8 pulses are generated (channel 4). After the eighth pulse, another signal (channel 3) occurs, which resets channel 2. What we have here might be a stepper motor controller moving the motor eight steps in response to an input trigger.

Also notice the Ts and Te vertical lines. Most logic analyzers provide some type of time reference markers that can be moved to different parts of the display. By using the reference markers, you can find out how long the trigger pulse on channel 1 was active, or the time between the falling edge of the trigger pulse and the start of the counter that generated the eight pulses. The digital oscilloscope display shown in Figure 17-11 is actually the oscilloscope function of a Hewlett-Packard 1631D logic analyzer, and you can see the X and O time reference markers on the display.

Some logic analyzers have the ability to display the data as a table of hexadecimal values, and still others have a built-in capability to observe and decode the signals used in a serial interface. Some can also be configured to start collecting and displaying data when a particular pattern appears on the inputs. What the instrument can do depends, to a large extent, on how much you are willing to pay for it.

Inexpensive USB logic analyzer modules are available that use the PC's display rather than provide one of their own. Figure 17-16 shows a diagram of such a device.

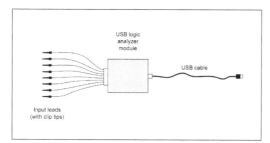

FIGURE 17-16. A simple USB logic analyzer device

Prices for USB logic analyzers vary, starting at about $150 and going up, depending on the built-in capabilities and number of input channels. The display used with these looks similar to the one shown in Figure 17-15. Keep in mind, however, that for $150 you'll get something that can sample data at a maximum of about 3 to 5 MHz. If you need to look at faster signals or need more than four or eight inputs, be prepared to pay considerably more.

Buying Used and Surplus Equipment

It is possible to set up a decent electronics lab with only used equipment, if you are careful about what you buy. Used digital meters, oscilloscopes, signal generators, frequency counters, and other items are available from local surplus outlets (if you happen to be lucky enough to have one near you), some electronics supply shops, and, of course, eBay. Ham radio operators occasionally hold swap meets, and you can often find bargains on older gear at one of these events. Flea markets are another possibility, but more often than not, what they are selling is simply junk that might be good for parts, but not much else. If you happen to have an electronic manufacturer or a university in your city that holds surplus sales, you might find something worthwhile there as well.

With used equipment, you also need to be prepared to do some minor repairs and calibration work yourself. Figure 17-17 shows a collection of older instruments that have accumulated over the years. All of it works, and every item has a service manual on file, and all of it needed some kind of minor calibration and repair work. Having documentation for your equipment is essential. Without the service manual, it is very difficult, at best, to perform the calibration steps necessary, or hunt down and replace a defective component should one fail.

FIGURE 17-17. Some older, but still useful, test equipment

Yes, the cart looks a little rough, to be sure, but it will get a new paint job at some point in the future. The paint doesn't make the equipment behave any better or worse, so it's not really a high priority. If you're curious, what you're looking at here, in top-to-bottom order, is a Sony/Tektronix 338 logic analyzer/protocol analyzer, a Tektronix 454 dual channel oscilloscope, a Fluke 8000A DMM, a Fluke 1900A multi-counter, and a HP 3478A DMM. The black box is a passive backplane PC with a GPIB/IEEE-488 interface.

One last comment about this equipment: it's slow. The oscilloscope is useful to only about 150 MHz, and the logic analyzer tops out at 20 MHz. But slow is OK. As stated earlier, most things, particularly things that interface with the real world, don't move very fast. Another big advantage of slow equipment is that it is cheap. Those who think they just have to have an oscilloscope capable of looking at GHz signals will pass by the older unit selling for a fraction of the cost of the fancy instrument, even when it can handle 90% of anything they might want to do.

Here are a few more caveats about used test equipment:

- Don't buy anything with vacuum tubes in it, no matter how cool it looks. Instrument-grade vacuum tubes are almost impossible to find these days, they use a lot of electricity, and they get hot. Really hot.
- Unless you are prepared to spend your money for something to use for parts, never buy an instrument that doesn't work, no matter what the seller says

about how easy it would be to fix it. Now, if someone gives it to you for free, that's a different story, but you will need to be prepared to spend some time working on it. With surplus equipment, it's often impossible to check the equipment before placing a bid on it, so you are effectively gambling. Never bid more than you are prepared to lose (like gambling in Las Vegas). As a wise man once said, there ain't no such thing as a free lunch.

- If ordering from an online seller, be sure to check the shipping cost. It's a rude shock to discover that the shipping charge for the great deal you found on an instrument is more than what you are paying for the instrument itself. Some of the less honorable sellers will attempt to make up for a low price by dishonestly jacking up the shipping costs.
- Get a service manual and, if possible, a user manual, for the equipment. There are many sources online for manuals, both free and for money. Sometimes the only way to get a manual is to pay for it, so that's another cost consideration. Other times you might get lucky and the instrument will come with manuals, but don't count on it. In Figure 17-17, you might notice the pouch on top of the 338. That's where the manuals, cables, and probes normally reside when not in use. In this case, I was lucky and the pouch contained everything it was supposed to have when I bought the instrument. That's more the exception than the rule.

- Buy only what you need. Something with lots of knobs and buttons might look like an awesome prop from a science-fiction movie, but you aren't making sci-fi movies, right? The dull, boring instrument with minimal controls will work just fine, because what's important is what it tells you.

Summary

The most essential instrument you will need for working with electronics is a DMM. A lot can be accomplished with just a good meter that can accurately measure voltage, current, and resistance. Features like data logging, range sensing, and the ability to measure capacitance and inductance are nice, but not essential. A 3 1/2-digit DMM is more than enough for most needs, although more precise instruments with 4 1/2-digit resolution are available if you really need one. Just be prepared to pay for the extra digit.

An oscilloscope should be the second thing to acquire for working with electronics. A decent oscilloscope can allow you to observe and troubleshoot both analog and digital circuits, and it can give you some idea of the peak voltage and frequency of the signal you are examining. A good two-channel digital oscilloscope can do a lot, but even a single-channel unit can be put to good use. One of the primary criteria for an oscilloscope is its frequency limit. Most inexpensive instruments don't go much beyond a megahertz or two, but then again, that might be sufficient for many projects. As the frequency response goes up, so does the price, with a multi-channel digital oscilloscope capable of gigahertz operation costing many thousands of dollars.

We also looked at some instruments that have very specific applications, including signal and pulse generators and logic analyzers. These are specialized for certain types of electronics, namely logic and signal processing circuits. Unless you have a specific need for them, they are not necessary. A good two-channel oscilloscope will easily display two logic states at the same time.

Throughout this chapter, one recurring theme has been that many electronics test instruments are available in low-cost forms, either as standalone devices or as add-on instruments for a PC. These include oscilloscopes, logic analyzers, signal generators, and even digital meters in the form of data acquisition modules with a USB interface. It all comes down to how much resolution and performance you really need. If you never plan to work with RF electronics. or have no interest in designing your own PC motherboard, there's really no need for fast, and expensive, instruments.

Finally, we took a quick look at the pros and cons of buying used test equipment. There is nothing wrong with older equipment, so long as it still works and produces accurate results. The main points are to use caution when making a purchase and make sure to obtain the relevant user and service manuals. With older used equipment, the price might be right, but it's the operation and maintenance that can present a challenge. It can also provide an invaluable learning experience, if you are willing to take advantage of it.

APPENDIX A

Essential Electronics and AC Circuits

This appendix presents brief overviews of specific topics in basic electronics theory, beyond the discussion presented in Chapter 1. It is intended for anyone who might be interested in exploring some of the theory behind electronics, or who perhaps might benefit from it for personal projects. Topics covered include voltage, current, power, series and parallel circuits, Thévenin circuit analysis, capacitance, and impedance. A terse overview of basic solid-state theory is included to introduce the concepts of semiconductors and their applications.

Appendix B contains a set of basic electronic schematic symbols. If you encounter something here that you don't recognize, be sure to look there. If you need more information than what is contained here, then you might want to look into the texts listed in Appendix C.

The main emphasis throughout this appendix is on the fundamental concepts, rather than the details. With a good grasp of the fundamentals, you'll find the more detailed concepts behind modern electronic components and circuits much easier to comprehend. So think of this appendix as a travel brochure to the land of electronics. What adventures you decide to have beyond this point are entirely up to you.

Units of Measurement

Table A-1 lists some of the most common characteristics that apply to electrical circuits. These will appear throughout the rest of this appendix as they are needed, and they are defined in Chapter 1, here in the appendix, and in the Glossary.

TABLE A-1. Standard units of measure used in electronics

Unit name	Symbol	Reference	Measurement
Ampere	A	I	Electric current
Coulomb	C	Q	Electric charge
Farad	F	C	Capacitance
Henry	H	L	Inductance

Unit name	Symbol	Reference	Measurement
Hertz	Hz	f	Frequency
Joule	J	E	Energy
Ohm	Ω	R	Resistance
Seconds	s	t	Time
Volt	V	E	Voltage
Watt	W	P	Power

In Table A-1, the Symbol column gives the nomenclature used with a definite value: 1V, 2.5A, 1 second, and so on. The Reference column gives the letter typically used in equations and schematic component references, as in C1, R10, E = IR, and so on.

In many cases, a fundamental unit of measure is impractical for normal usage, so it is specified in larger or smaller units through the use of a prefix. Table A-2 lists the most commonly encountered unit prefixes.

TABLE A-2. Standard value prefixes used with electronic units of measurement

Prefix	Symbol	Multiplier	Meaning
giga	G	1×10^9	one billion
mega	M	1×10^6	one million
kilo	k	1×10^3	one thousand
milli	m	1×10^{-3}	one thousandth
micro	μ	1×10^{-6}	one millionth
nano	n	1×10^{-9}	one billionth
pico	p	1×10^{-12}	one trillionth

Resistance is seldom given in milliohms or smaller, although these can occur when you're working with precision measurements of low values of resistance in the laboratory. You will usually see ohms given as kilo ohms, mega ohms, or just ohms, and values less than 1 ohm are usually given as a decimal value. Capacitance and inductance, on the other hand, are most often given in micro, milli, nano, or pico units.

Voltage, Current, and Power

The two key characteristics of electrical phenomena are voltage and current. Voltage can be viewed in several ways. The most common is to treat it as being analogous to pressure. Voltage can also be viewed as a type of potential energy. The third way is to view voltage as the measure of the electromotive force behind electron movement. Current is analogous to flow volume and is defined as the number of electrons moving past a particular point in a circuit in a specific interval of time.

The two concepts are closely related, and it is not always possible to speak of one without reference to the other. Without voltage, no current can flow, and without current flow, there is static electrical charge (a voltage), but no meaningful work can occur.

> **TIP** Refer to Chapter 1 for an overview of electric charge and electron movement. Many of the terms and concepts used here are defined there, so I won't duplicate that effort.

In a DC circuit, the relationships between voltage, current, and power are straightforward. There are no time, phase, or frequency aspects to worry about, as in the case with

AC circuits. "AC Concepts" on page 413 looks more closely at AC circuits, but in general, the discussion here about voltage, current, and power also applies to AC (with some frequency-dependent exceptions, which are covered in "Capacitance, Inductance, Reactance, and Impedance" on page 418). Current is discussed in "Current" on page 401, and power is covered in "Power" on page 402.

VOLTAGE

The unit of measure for voltage is the volt, abbreviated as V. In a DC circuit, voltage can be defined simplistically as the electric potential difference between two points.

In an electric circuit, energy is put into a system and either does some work, such as being converted to heat, or is expressed as electical or magnetic fields, as in the case of AC circuits and reactive components. The main point is that in order to create the potential gradient, something had to expend some energy in some fashion. That energy is returned when the potential becomes a voltage drop across a component and current flows. If the component is a resistor (or even just a wire), it will be expressed mainly as heat. If it is an inductor, the energy is stored in the magnetic field that forms around the coils of the inductor (or around a single wire). A capacitor stores energy in the form of an electric field between two plates. "Capacitance, Inductance, Reactance, and Impedance" on page 418 covers inductors and capacitors in more detail.

Another way to view voltage when current is flowing is as the force *pushing* current flow. The higher the voltage, the more readily it will force current flow through an impediment such as resistance (or an insulator, if the voltage is very high). By the same token, as the voltage or current is increased, more energy will be expended in the process, which is expressed as power, usually in the form of heat.

The amount of voltage between two points in an electric circuit is dependent on the voltage source, which itself applies some type of force to produce a current flow. A generator or dynamo uses electromagnetic force (EMF) to produce current flow at a given voltage. A power supply converts the voltage and current from a common AC power circuit (for example) into a specific voltage with a particular available current in either DC or AC form. A battery uses an electrochemical process to produce the electromotive force to generate a voltage potential.

Figure A-1 illustrates the three forms of voltage. A source of electromotive force produces a potential voltage. The EMF source might be a battery, a generator (dynamo), a bimetallic junction, or a solar cell. Whatever it is, it is a primary source of electrons with some amount of EMF giving them a potential voltage. This initial generation of EMF involves the conversion of some form of energy (chemical, electromechanical, thermal, or photonic) into electrical energy.

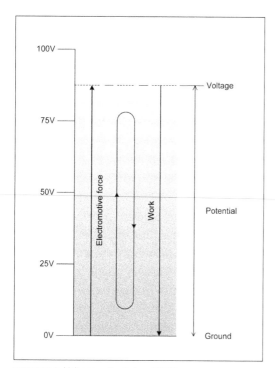

FIGURE A-1. Voltage potential and EMF

So long as the potential remains untapped, it will simply be a static potential. It is described in terms of volts, but not current or power (since it is static, no electrons are moving). When the potential is converted into current flow in a closed circuit, then the concepts of current and power come into the picture. The downward-pointing arrow in Figure A-1 indicates the drop in potential as current flows, some work is done, and the potential difference decreases back to zero. The voltage level shown is arbitrary and is solely for the purpose of illustration.

Figure A-2 shows a graphical representation of a DC voltage over time, which is just a straight horizontal line across the graph. This is what you would expect to see if you looked at a DC voltage with an oscilloscope. Again, the voltages were selected for illustration purposes.

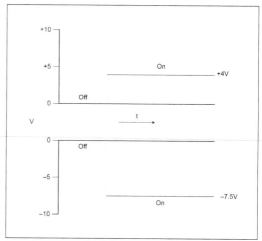

FIGURE A-2. DC voltage over time, a straight line

Notice that Figure A-2 shows both positive and negative voltages. The polarity of a DC voltage is largely dependent on what is used as the neutral (i.e., ground) reference; and in some cases, the positive terminal of a power source might be ground, in which case the voltages measured in the circuit would be negative. Some types of power supplies have both positive and negative outputs. This is often the case in PC power supplies, like the one described in "Power" on page 402.

It's important to bear in mind one essential fact about voltage: it exists only relative to two specific points. In other words, voltage is measured relative to something. When a voltage is measured relative to a common reference point (e.g., ground), it is a measure of the circuit voltage at that point. It doesn't really say anything about voltages across individual components, only how the circuit has affected the voltage at the measurement point (if there has been any effect at all).

Figure A-3 illustrates the measurement concept. Assuming that R1 and R2 are the same value, the voltage at point A will be some fraction of the supply voltage, which is shown as one-half of the supply voltage in this case. The actual value of the voltage at point A will depend on the values of R3, R4, and R5. Refer to Chapter 1 for an introduction to voltage dividers and to "Series Resistance Networks" on page 404, "Parallel Resistance Networks" on page 407, and "Thévenin's Theorem" on page 409 for a more detailed discussion of series and parallel resistance networks. The voltage at point B will always be whatever the supply voltage happens to be, which is 9V in this case.

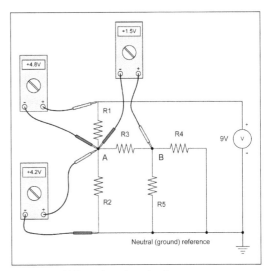

FIGURE A-4. Voltage drops in a circuit

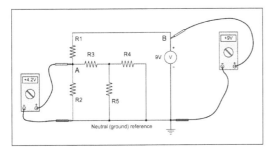

FIGURE A-3. Voltage measurement in a circuit

The voltage measured across a component is called the *voltage drop*. Figure A-4 employs the circuit from Figure A-3 to illustrate this distinction.

The voltage measured across R3 (between points A and B) in Figure A-4 is the voltage drop across the resistor, as is the voltage across R1. The measurement from point A to neutral across R2 is the voltage drop across R2.

CURRENT

Current flow is one of the two essential features of electricity (the other being voltage). Electrical current is the movement (or *flow*) of electrical charge through a conducting medium of some sort. Although it is typically used to refer to electrons as the charge carriers, it can also describe the flow of charge carried by ions within the electrolyte of a battery, for example. In principle, anything that carries electrical charge from one point to another is a form of current flow. It is analogous to the volume of fluid flowing through a pipe in, say, liters per second, or the number of steel ball bearings moving past a point in a tube in some specific interval of time (borrowing from Chapter 1).

Electrical charge without current flow is just static charge and does no work.

Current is measured in amperes, abbreviated as A, and often shortened to *amp*. The quick and easy definition of current states that one ampere of current can be defined as 1 coulomb of charge (6.24×10^{18} electrons) moving past a given point in an electric circuit per unit time, although the traditional formal definition involved the force generated between two conductors spaced some fixed distance apart in a vacuum. Recently, it has been proposed that the formal definition adopt the more straightforward definition, wherein 1 ampere is defined as 6.2415093×10^{18} elementary charges (either positive or negative) moving past a specific point in one second of time. Or, to put it another way:

1 ampere = 1 coulomb per second

POWER

Electrical current is used to perform work of some kind. This might be in an electric motor or a loudspeaker, wherein the current flow and associated electromagnetic phenomena are converted to mechanical motion. It could also be in the form of an incandescent lamp, a solid-state circuit (e.g., a radio or a computer), or an electric arc welder. In each case, the flow of electrons performs some type of work, with the result typically expressed as heat, radiant electromagnetic energy (RF), mechanical motion, or a combination of all three.

Power is a measure of energy expended as work is performed. Electrical power is simply the product of voltage and current, in that $P = EI$, where P is the power, in watts, E is the voltage, and I is the current, in amperes. From this, you can see that something like an electric arc welder operating at 25V (with the arc active) at 150A of current will produce 3,750W of power at the arc, all of which is concentrated into a small area about 1/8 inch to 1/4 inch in diameter. It's no wonder that it can melt metal.

Conversely, the power supply in a typical vacuum tube (valve) audio amplifier might produce 500V DC at around 250 mA (milliamperes). This results in 125W of available power. So, although the current is low, the high voltage is what gives the amplifier power to easily produce 25W or more of audio output to drive the loudspeaker.

Lastly, consider the switching-mode power supply (SMPS) in a modern PC. The input voltage and current are 8.5A at 115V RMS, which is 977.5W ("AC Concepts" on page 413 discusses RMS). Table A-3 shows the output specifications for a typical low-cost PC power supply.

TABLE A-3. Output specifications for a typical low-cost PC power supply

Output (V)	Current (A)	Power (W)
+3.3	20	66
+5	20	100
+12	18	216
+12	18	216
-12	0.8	9.6
+5	2.5	12.5

The supply is rated for 400W, so that implies that the specifications might be a bit misleading. If you add up all the output power, we get 620.1W. If you assume that one of the +12V outputs is simply a duplicate of the other and subtract 216W, then you get 404.1, which is close to the manufacturer's stated rating. Notice that there is a significant difference between the input power and the available output power. Some simple math (output/input × 100) shows that the supply is approximately 41% efficient. This isn't great, as a switching-mode power supply can do better than that, but it's acceptable for something around $40. The real downside here is that the wasted power (around 573 watts) is going to end up as heat, so the fan in this unit will be running quite a bit when it's under load.

Resistance

In DC circuits, the thing that mediates the relationships between voltage, current, and power is resistance, or resistivity, and its reciprocal, conductivity. In AC circuits, capacitance, inductance, and reactance also come into play, and those are discussed in "Capacitance, Inductance, Reactance, and Impedance" on page 418. However, in an AC circuit with no significant amount of inductance or capacitance, the concepts presented here for DC circuits will generally apply to root mean square (RMS) values for AC voltages (see "AC Concepts" on page 413 for the definition of RMS).

CONDUCTIVITY AND RESISTANCE

Electrical conductivity is a characteristic of all physical matter. The degree of conductivity ranges from zero (a perfect insulator) to infinity (a perfect conductor). The actual numeric value of conductivity can range from a very small number (for an insulator), to a very large value (for a good conductor).

The reciprocal of conductivity is resistance, which is what is normally used when we are considering the resistivity of a component. For example, a component with a resistance of 1 ohm has a conductivity of 1, whereas a component with a resistance of 1,000,000 ohms has a conductivity of 0.000001. A component with a resistance of 0 has an undefined (infinite) conductivity. From this, it follows that no matter how good a conventional conductor might be, it will never have infinite conductivity, and hence it will always have some amount of resistance. The only known exceptions are superconductors.

In electrical circuits, a resistor is a component that is designed to exhibit a specific resistivity. One way to consider this involves the concept of valence electrons presented in Chapter 1. Elements that can easily give and accept valence electrons tend to be good conductors, and consequently have a relatively low resistance to current flow. Those elements that have tightly bound electrons tend to be poor conductors, and hence good insulators. Resistors lie somewhere between insulators and good conductors and are made from various forms of carbon, high-resistance wire, and vapor-deposited films, among other materials. The material type and thickness are adjusted to produce the desired resistance value. Chapter 8 describes the various types of resistors that are available.

When current flows through a resistance, work is done to maintain the current flow. The higher the resistance (and the lower the conductivity), the higher the level of work necessary to maintain a given current flow. Varying the voltage across the resistive load will vary the amount of current *pushed* through the load. The relationship between voltage, current, and resistance is defined by Ohm's law.

OHM'S LAW

We can use the relationship defined by Ohm's law to help understand how voltage, current, and resistance interact, and subsequently, how power is related to these three characteristics as well.

Ohm's law is stated as:

$$E = IR$$

where E is the voltage in volts, I is the current in amperes, and R is resistance in ohms. The use of the symbols E and I is largely a matter of historical legacy. You can replace them with V and A if you wish, and some people do, but the old-school form is the one most widely recognized.

A load in an electrical circuit will have a specific resistance. If you know the resistance, it is possible to calculate the current at a given voltage. Once the current is known, you can calculate the power using the relationship given in "Power" on page 402:

$$P = EI$$

Ohm's law and the power equation are linear relationships, but in real applications, there are things like the power source current limit, load power handling capacity, and other factors to take into account that might limit the range of linearity. How a power source will behave when it reaches its current limit depends on the type of the source. In many cases, it will simply not generate any higher voltage once the current limit is reached. How a load will respond when its power dissipation limit is reached or exceeded depends on the nature of the load. It might melt, burst into flames, or simply become an open circuit and cease to conduct.

Now, let's have some fun with the power equation and Ohm's law. We can derive a useful variant of the power equation by simply replacing the E term with IR, which results in:

$$P = I^2 R$$

Substituting E/R for I in the original equation yields another useful variation:

$$P = \frac{E^2}{R}$$

So, if you know the resistance and the current, you don't really need to know the voltage drop in order to calculate the dissipated power. Or if all you know is the voltage drop and the resistance, you can still calculate the power using the second form.

SERIES RESISTANCE NETWORKS

Resistors can be arranged in series or parallel circuits to create new values and power handling capacity. This is useful when the value needed simply isn't available, but a lot of the wrong parts are on hand. It is also useful when a resistance needs to be able to handle a certain amount of power, but no such part is on hand (or might not even be readily available). A collection of resistors connected in series or parallel, or both, is commonly referred to as a *resistance network*.

When connected in series, the values of some number n of resistors (where $n > 1$) sum, as shown in Figure A-5.

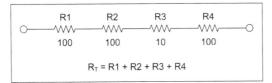

FIGURE A-5. Resistors in series

Note that in a series network, the amount of current flowing through each resistor is the same. You can calculate the voltage drop across each resistor, and hence the amount of power it will dissipate, by treating the series network as a multi-tap voltage divider, as shown in Figure A-6.

The first step is to sum the values of the resistors to get R_t and then use the supply voltage (10V) to determine the total current through the network. In this case, it works out to 0.03226A (32.26 mA):

$$I = \frac{E}{R_t}$$

$$I = \frac{10}{310}$$

$$I = 0.03226A$$

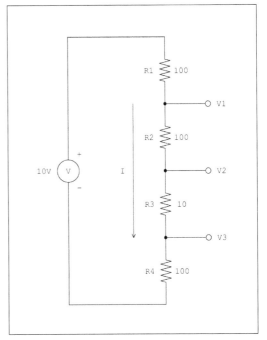

FIGURE A-6. Series resistance network as a multi-tap voltage divider

The next step is to sum the values of R2, R3, and R4, which gives 210 ohms. You can now treat the network as a two-resistor voltage divider and use Ohm's law to find the voltage for V_1. This turns out to be:

$$V_1 = 210 * 0.032253065 = 6.77V$$

Next, sum R1 and R2 to get 200 ohms, and R3 and R4 to get 110 ohms. Again apply Ohm's law to derive V_2:

$$V_2 = 110 * 0.032253065 = 3.54V$$

Finally, just use the value of R4 to get V_3, which gives you:

$$V_3 = 100 * 0.032253065 = 3.22V$$

Now that you have V_1, V_2, and V_3 (which have been rounded appropriately), you can work out the voltage drops across each resistor in the network, as shown in Figure A-7.

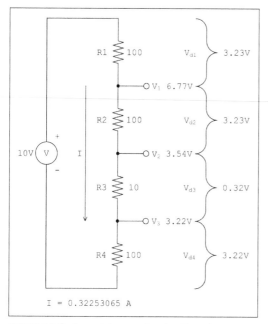

FIGURE A-7. Series resistance network voltage drops

As a sanity check, you can sum up the voltage drops and verify that they are equal to 10V:

$$V_t = V_{d1} + V_{d2} + V_{d3} + V_{d4}$$

$$V_t = 3.23 + 3.23 + 0.32 + 3.22$$

$$V_t = 10.0V$$

Finally, you can apply the voltage drops to determine what the power dissipation will be for each resistor, as shown in Table A-4. Remember that the network current is 0.03226A through each resistor, so the power is simply the voltage drop times the network current.

TABLE A-4. Calculation of power dissipation in a series network using voltage drops

R	V drop	Power (W)
R1	3.23	0.104
R2	3.23	0.104
R3	0.32	0.01
R4	3.22	0.104

The sum of the power values for Table A-4 is 0.322W.

The alternative approach, which is more straightforward but not as accurate, is to use the alternate I^2 form of the power equation. Since you know that the current in a series network is the same through all of the components, and you know the resistance values, calculating the power dissipation for each resistor is simplicity itself. Table A-5 list the results.

TABLE A-5. Calculation of power dissipation in a series network using $P = I^2R$

R	Power (W)
R1	0.104
R2	0.104
R3	0.01
R4	0.104

The sum of the power values for Table A-5 is 0.322W, which is the same as the preceding calculation.

PARALLEL RESISTANCE NETWORKS

A parallel configuration reduces the total resistance of the network, as illustrated in Figure A-8.

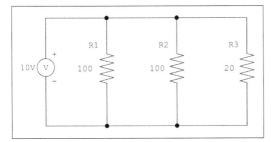

FIGURE A-8. Resistors in parallel

The total resistance of a parallel resistance network with more than two elements is given by:

$$R_t = \frac{1}{\frac{1}{R1} + \frac{1}{R2} + \ldots + \frac{1}{Rn}}$$

For two parallel resistors, you can use:

$$R_t = \frac{R1 \times R2}{R1 + R2}$$

If n resistors in parallel have the same value, the equivalent resistance is just the value of any one of the resistors divided by the number of resistors in parallel:

$$R_t = \frac{R}{n}$$

In order for this to work, the resistive elements *must* have the same value. If they are different values, you'll need to use one of the other equation forms to compute the total resistance of the parallel network.

In a parallel network, the voltage across each resistor is the same, but the current through each depends on the resistance of the component, and sums into the total current for the network:

$$I_t = I_1 + \ldots + I_n$$

Figure A-9 shows how each component in a parallel resistance network can have different amounts of power dissipation. In this case, there are three resistive loads, perhaps lights or maybe heating elements. Although the voltage across all three is the same, the amount of current flowing through each differs based on its resistance.

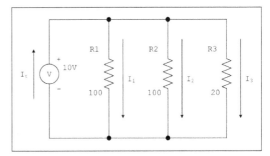

FIGURE A-9. Current flow in a parallel resistance network

To calculate the power, you can use the $P = I^2R$ form of the power equation, because the loads are in parallel and each can be dealt with separately from the other elements in the network. Using 10V as the supply voltage, you can determine the current for each using Ohm's law. Once the currents are known, $P = I^2R$ calculates the power dissipation for each resistor. Table A-6 shows the results of the math.

TABLE A-6. Calculation of power dissipation in a parallel network using $P = I^2R$

Load	Current (A)	Power (W)
R1	0.1	1
R2	0.1	1
R3	0.5	5

The total current demand on the power source will be 0.7A, with a total power dissipation of 7W. The I^2 form of the power equation is a good choice here, because in a parallel network, the voltage drop across each component is the same. What varies is the amount of current flowing through each load element.

The sanity check for Figure A-9 is simply to compute the parallel resistance of the network and derive the total power from that:

$$R_t = \frac{1}{\frac{1}{100} + \frac{1}{100} + \frac{1}{20}}$$

This gives an R_t of 14.285714286 ohms. Using $I = E/R$, you find that the total current would be 0.7A, and $P = EI$ gives 7W of power. This is the same as the I_t of 0.7A and the 7W obtained from Table A-6. Note that, while this is a valid answer, it does not tell you the power dissipation for the invidual resistors in the network, whereas the first form does.

Equivalent Circuits

When dealing with complex circuits, it is sometimes useful to represent the circuit in a simpler form. These representations are referred to as *equivalent circuits*, because they are a simplified version of what might otherwise be a very complex circuit.

The simplest DC circuit consists of a power source and a load element (e.g., the battery and light bulb from Chapter 1). The power source produces some amount of current at a given voltage. The current flows through the load and power is dissipated in some form, typically heat. Working with a simple circuit like this is much easier than trying to deal with the interactions within a complex circuit.

The use of equivalent circuits allows you to focus on the key characteristics of a circuit or system without becoming mired in the details. One way to think of an equivalent circuit is as a *black box* that has some particular characteristics (voltage, current, and resistance) but whose internal details are hidden from view, since they don't really matter from the perspective of whatever the black box happens to be connected to in the circuit.

VOLTAGE AND CURRENT SOURCES

In its most common form, an equivalent circuit is composed of passive, linear elements. In addition to the common symbols for resistors, capacitors, and inductors, a variety of symbols are employed with equivalent circuits to indicate voltage and current sources. You have encountered only a voltage source so far. Figure A-10 shows a more complete set.

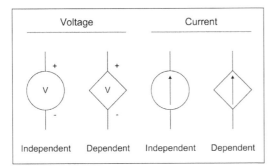

FIGURE A-10. Symbols used in equivalent circuit diagrams

Note that there are two types for both voltage and current: independent and dependent. An independent source is not affected by changes in the connected network, while a dependent source will change according to variables in the connected network. You might think of a dependent source as having a knob or lever that something else can use to alter its output. It is the dependent sources that allow an equivalent circuit to model an active circuit element like a transistor or operational amplifier.

LUMPED-PARAMETER ELEMENTS

Equivalent circuits are simplifications of more complex circuits and are composed of lumped-parameter elements, wherein something like a resistor in an equivalent circuit might represent many resistors (and other components) in a complex network (e.g., a Thévenin equivalent). This applies to capacitors and inductors as well (discussed in "Capacitance, Inductance, Reactance, and Impedance" on page 418). The main point of an equivalent circuit is to model the behavior of a more complex circuit by retaining the essential electrical characteristics of the original circuit in a simplified form that aids analysis.

THÉVENIN'S THEOREM

Thévenin's theorem states that any linear electrical network (complex circuit) that is composed of only voltage sources, current sources, and resistances can be represented by an equivalent voltage source and an equivalent resistance. When dealing with AC circuits, you can apply Thévenin's theorem to reactive impedances at some given frequency (see "Capacitance, Inductance, Reactance, and Impedance" on page 418 for more on reactive components).

Thévenin equivalent circuits are useful for analyzing power systems with variable loads. For example, a power source might have a complex internal circuit, but its Thévenin equivalent will have an equivalent behavior, which is much easier to work with. Another application involves the simplification of a set of circuits in a larger system so that each subcircuit can be represented by its equivalent.

Figure A-11 shows how a simple circuit with a voltage source and five resistors can be represented by an equivalent circuit with just a single voltage source and one resistor. From the perspective of nodes A and B, both the original and the equivalent would appear to be identical. Deriving a Thévenin equivalent is really nothing more than utilizing Ohm's law and the concepts of series and parallel resistances that were covered earlier.

When you are converting a circuit to its Thévenin equivalent, the first step is to deal with the voltage and current sources. Under

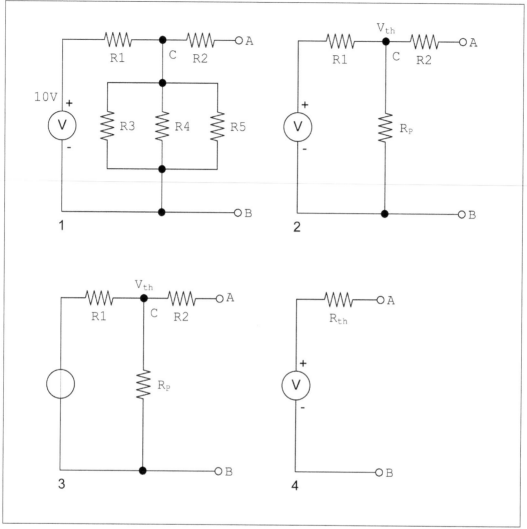

FIGURE A-11. Thévenin equivalent circuit example

normal conditions, a source will have some amount of internal resistance, usually rather low. You can elect to ignore this and treat the voltage and current sources as ideal sources. An ideal voltage source is replaced with a short circuit, and an ideal current source is replaced with an open circuit.

In order to define the equivalent circuit, you will need the equivalent voltage and resistance. For this example, let's pick some arbitrary values:

- $R1 = 100$ ohms
- $R2 = 100$ ohms
- $R3 = 200$ ohms
- $R4 = 100$ ohms
- $R5 = 200$ ohms

The first step is to determine the value of the parallel network composed of R3, R4, and R5:

$$R_p = \frac{1}{\frac{1}{R3} + \frac{1}{R4} + \frac{1}{R5}}$$

$$R_p = \frac{1}{\frac{1}{200} + \frac{1}{100} + \frac{1}{200}}$$

$$R_p = 50$$

The Thévenin-equivalent voltage (V_{th}) is the voltage at the output terminals A and B of the equivalent circuit. In this case, I've elected to treat the circuit as a voltage divider. This arrangement is shown in step 2 of Figure A-11. R2 is ignored on purpose. Since this is an open circuit calculation, there is no current flow through R2, and hence no voltage drop.

Now you can determine the open circuit voltage at point C (V_{th}) in the circuit:

$$V_{th} = V\frac{R_p}{R_p + R1}$$

$$V_{th} = V\frac{50}{50 + 100}$$

$$V_{th} = 3.33V$$

The Thévenin-equivalent resistance is the resistance measured across points A and B of the circuit. To calculate the equivalent resistance, replace independent voltage sources with short circuits, and independent current sources with open circuits. This is shown in step 3 of Figure A-11, where the sole voltage source has been replaced with a short circuit. This is equivalent to R1 and R_p in parallel, and R2 is now included in the calculation:

$$R_{th} = R2 + \frac{1}{\frac{1}{R_p} + \frac{1}{R1}}$$

$$R_{th} = 100 + \frac{1}{\frac{1}{50} + \frac{1}{100}}$$

$$R_{th} = 133$$

The result of our effort is shown in step 4 of Figure A-11, and Figure A-12 shows the final result.

Thévenin equivalent circuits are not perfect representations, and they have some limitations. The first involves linearity. Many circuits are linear only over a specific range, so the Thévenin equivalent is valid only within that range. Secondly, a Thévenin equivalent might not accurately model the power dissipation of the actual complex circuit. Still, even with these caveats, Thévenin's theorem is a handy tool to help you understand complex electrical circuits.

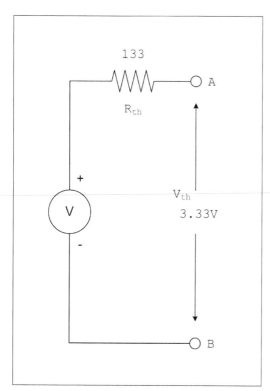

FIGURE A-12. Final Thévenin equivalent circuit result

EQUIVALENT CIRCUIT APPLICATIONS

Equivalent circuits are useful for more than just passive components. The linear behavior of active components such as transistors and op amps can be modeled with equivalent circuits and dependent sources. For example, Figure A-13 shows a simple model of an ideal op amp.

In Figure A-13, the idealized behavior is represented by the voltage output as a function of the voltage drop across the input resistance times the gain (G) of the device.

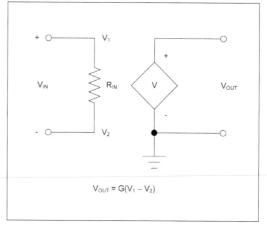

FIGURE A-13. An equivalent circuit example for an ideal op amp

In this form, this op amp equivalent can be dropped into an equivalent circuit model and treated like a resistor and a dependent voltage source. A simplified model like this is useful for a rough first-order analysis, and the level of detail can be increased as necessary to account for elements in the circuit such as a feedback path. "Operational Amplifiers" on page 445 covers op amps.

Equivalent circuits simplify the analysis of the voltage and current characteristics of a circuit. This, in turn, allows you to make better decisions regarding circuit power handling, battery current capacity, and voltage requirements.

There are other circuit analysis techniques in addition to Thévenin equivalent circuits, such as Kirchkoff's circuit laws and Norton's Theorem. Since the objective of this section is just to provide a high-level look at equivalent circuits, they aren't covered here. If you are inclined to learn more about equivalent circuits and circuit modeling, the texts listed in Appendix C describe these techniques in detail. Circuit simulation software such as PSpice (for Windows) also employs these techniques, and using a simulator is a lot less tedious than working out the equations by hand. Oregano and gpsim are examples of free analog circuit simulation software for Linux.

AC Concepts

Alternating current (AC), as the name implies, is a type of current flow that changes direction periodically. This is what makes it useful for power transmission, audio signals, and radio, but it also makes it more complicated. Whereas DC has only voltage as its primary characteristic, AC also has frequency, measured in hertz (Hz) or cycles per second, and phase, measured in degrees. In a sinusoidal waveform, such as the one shown in Figure A-14, a cycle is a complete waveform of 0 to 360 degrees, start to finish. AC also has a root mean square (RMS) value used to calculate power.

In most cases, the term *AC* refers to a sinusoidal waveform, as found with the mains current in many countries and shown in Figure A-14. Although you might think of something like the output of a signal generator producing a sine wave at, say, 1,000 Hz, as AC, that is more often referred to as a *signal*. The term *AC* is typically reserved for discussing power circuits, not signals, although they are the same thing and the fundamental concepts of AC circuits apply to both.

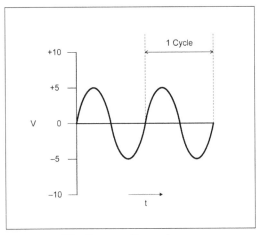

FIGURE A-14. AC voltage over time, a sine wave

The terminology associated with AC flow can sometimes be confusing and is dependent to a large extent on the context of usage. For example, when talking about the power wiring in a house, you would expect to hear *AC*, *AC voltage*, or *AC current*. These terms typically refer to the electrical power type in general, the voltage in the circuit, and the current flowing in the circuit, respectively. However, when referring to AC used to carry information (as is typically found in audio, radio, and instrumentation circuits), the common term is *AC signal* or just *signal*.

Alternating current has several unique primary features. As shown in Figure A-14, AC varies with time in a repeating cyclic fashion, the rate of which is its frequency. Frequency is the measure of the number of times the signal effectively changes direction in 1 second of time. Each time the voltage drops to zero, it is said to have reached a *zero-crossing point*.

Secondly, AC has both positive (high) and negative (low) peaks, but an RMS value is used for calculating power dissipation. Thirdly, AC has the characteristic of phase, measured in units of degrees from 0 to 360. Finally, with AC, the voltage and current can have different phases, which means that a peak in the voltage does not have to coincide with a peak in the current. This might seem counterintuitive at first, but it's a result of the reactance of a capacitor or inductor, which is covered in "Capacitance, Inductance, Reactance, and Impedance" on page 418.

WAVEFORMS

AC signals can occur with any one of a number of different types of waveforms, but the sine wave is the prototypical AC waveform. A sine wave is *pure*: that is, it comprises just one frequency. Other waveforms can be decomposed into a series of sine waves at various frequencies by means of Fourier analysis techniques (which we will not delve into here), but a pure sine wave cannot be decomposed any further. Figure A-15 shows a generic sine wave.

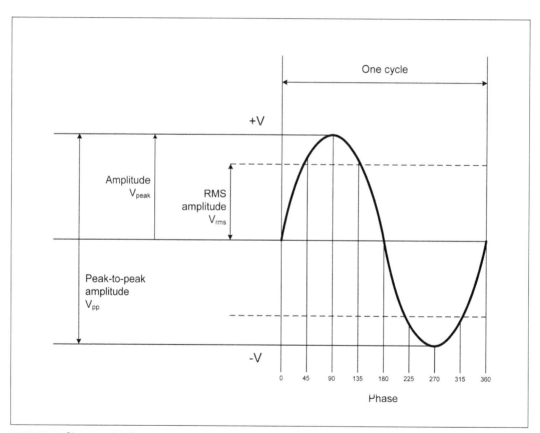

FIGURE A-15. Sine wave details

The sine wave gets its name from being defined mathematically by the sine function:

$$V_t = A\sin(2\pi f t + \theta)$$

where A is the amplitude, f is the frequency, t is time, and theta is the phase. Sometimes, you might see this form:

$$V_t = A\sin(\omega t + \theta)$$

where omega, the angular frequency, is actually just:

$$\omega = 2\pi f$$

The discussion of frequency, voltage, and power in "AC Frequency, Voltage, and Power" on page 416 will refer back to Figure A-15.

OTHER WAVEFORMS

Other periodic waveforms commonly encountered include square, triangle, pulse, and ramp, shown in Figure A-16.

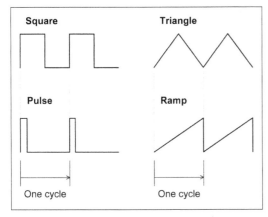

FIGURE A-16. Common types of electrical waveforms

Square and pulse waveforms appear mainly in digital logic circuits, because they easily represent the 1s and 0s of binary logic. The other waveform types appear in different contexts, such as power control circuits, timing circuits, music synthesis devices, and motion-control applications.

One of the most common, and most useful, of the nonsinusoidal waveforms is the *square wave* and its close relative the *pulse*. Although a square wave is usually drawn with a shape that implies instantaneous on and off times, in reality, square waves will include things like noninstantaneous rise and fall times, overshoot, and ringing, as shown in Figure A-17.

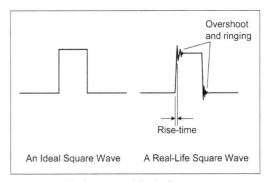

FIGURE A-17. Ideal versus real (typical) square wave

The overshoot and ringing occur because of the various inductance and capacitance effects in a circuit. Because even a wire has intrinsic inductance (as discussed in "Impedance and Reactance" on page 434), sending a pulse or square wave over more than a few feet of unshielded wire will result in a degraded signal at the receiving end.

When dealing with pulses and square waves, you'll often hear references to the duty cycle of the waveform. In fact, a square wave is actually a special case of a pulse with a 50% duty cycle. That is to say, it is *on* for one-half of a cycle and *off* for the other half. Figure A-18 shows some pulses at various duty cycles.

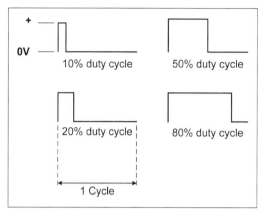

FIGURE A-18. Pulse duty cycle

AC FREQUENCY, VOLTAGE, AND POWER

Every AC waveform has a frequency. The inverse of a signal's frequency (*f*) is its period (*t*), which is the time interval between each repetition of the waveform:

$$f = \frac{1}{t}$$

$$t = \frac{1}{f}$$

This relationship applies to all periodic events, be it waveforms or the number of times a temperature sensor is queried in a given time interval. For example, if a video camera generates frames at a rate of 30 ms (milliseconds) per frame, it is operating at a frequency of 33.33 Hz. The time period of a 60 Hz signal is about 16.67 ms. A signal with a frequency of 10 KHz (10 kilohertz, or 10,000 Hz) has a time period of 100 μs (microseconds, sometimes written as u instead of μ).

Another essential characteristic is *amplitude*. There are three primary ways to describe the amplitude of an AC signal: peak amplitude, peak-to-peak amplitude, and RMS amplitude. Take a look at Figure A-15 again and notice that the peak value (the A term in the sine wave equations) refers to the maximum value on either side of the zero line. When we talk about the peak-to-peak value (often written as V_{pp}), we are referring to the range between the positive peak and negative peak.

Lastly, root mean square (RMS) amplitude is used to compute power (measured in watts, as with DC circuits) in an AC circuit. RMS is also known as the *quadratic mean*. For a sine wave $V_{rms} = .707 * V_{peak}$, and for other waveforms, it will be a different value. You can think of RMS as an average of the V_{peak}, and that is how it is used when computing power using:

$$P = \frac{V_{rms}^2}{R}$$

and, conversely:

$$P = I_{rms}^2 R$$

Now, here's something to consider: the AC power in your house is probably rated at something like 120V AC (volts AC), and in some parts of the world, it might be higher. That is its RMS value. The V_{peak} value is around 165 volts, and the V_{pp} is about 330 volts. The V_{pp} value isn't really something to get excited about, but it might be useful to know that the actual V_{peak} is 165 volts when you are selecting components for use with an AC power circuit. Just remember that the RMS value is used primarily to compute power.

AC PHASE

The last primary characteristic to consider is *phase*, as shown in Figure A-15. If you consider how AC is generated electromechanically, you can see that the phase angle corresponds to the rotation angle in an AC generator (also called a *dynamo* or *alternator*). The primary phenomena behind an AC generator, electromagnetism, is covered in "Inductors" on page 428. The main focus here is phase. Figure A-19 shows a simplified diagram of a single-phase generator.

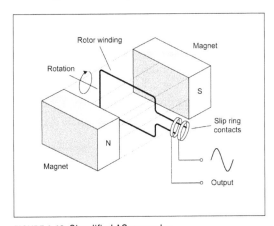

FIGURE A-19. Simplified AC generator

The output from the rotating coil between the magnets (the rotor) flows through a pair of slip rings. This allows for continuous contact with the rotor coil. As the rotor turns, the windings in the rotor intersect the magnetic field. This in turn causes current to flow. The direction of the motion of the windings relative to the magnetic field determines which direction the current will flow.

As the rotor in the generator turns, the output will vary at the same rate, with the result being a sine wave. The voltage of the output is a function of the angle of the rotor as it moves through the magnetic field produced by the stator assembly (and the number of windings in the rotor coil). Figure A-20 shows the phase angle relationship graphically. Note that the permanent magnets in this illustration (which are part of the stator and don't move) are shown only once. The rotor is tagged with a black box on one end to help keep track of it during rotation.

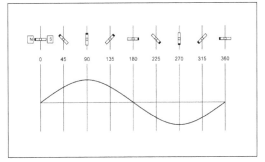

FIGURE A-20. Phase angle and AC generator rotation

Figures A-19 and A-20 describe a single-phase situation. In industrial and commercial AC power systems, it is common to find three phases (or in some cases six or even nine), because polyphase AC circuits tend be more efficient for power transmission and polyphase motors do not require the external starting circuit that a single-phase motor needs to produce a starting torque.

Phase plays a big role in AC circuits. An interesting thing to note about AC signals is that, when they are combined, the peaks and valleys of the waveforms add or subtract, resulting in a new waveform that is the algebraic sum of the originals. If two AC signals of exactly the same frequency but also exactly 180 degrees out of phase with respect to each other are combined, the sum is zero: they cancel out. You can hear the effect of *out-of-phase signals* when a pair of stereo speakers is miswired. If one is phase-reversed with respect to the other, the identical frequencies cancel, leaving only the differences. The result is often a thin, weak sound, or if the vocals are equally distributed on both the left and right channels, but the instruments are not, the vocalist might seem to fade into the background or vanish completely. This, by the way, is the basic principle behind noise-cancellation headphones.

In electronics, most of the AC will be of the single-phase type, and the main concerns in terms of phase involve phase lead or lag, phase shift, and phase angle detection. Although we have been talking about voltage phase up to this point, it is important to note that both the voltage and the current have phase, and they don't have to be coincident in time. In other words, some circuits can induce an angular difference between the voltage phase and the current phase.

Capacitance, Inductance, Reactance, and Impedance

This section deals with the passive components that are used with AC current: capacitors and inductors. We will look briefly at reactance and impedance but won't cover topics like phasor analysis or resonant circuits. These are very interesting topics, to be sure, but they aren't something that I can easily wedge into a summary like this and still do them justice. Out of necessity, this appendix is an abbreviated overview, and the main intent is to introduce some basic concepts and terminology. You are encouraged to seek out more detailed sources of information if you wish to learn more about the topics presented here.

The primary focus of this section is to look at the basic behaviors of capacitors and inductors, with a focus on understanding how they interact with changing current flow. In a DC circuit, a capacitor will not pass current after it has accumulated a charge, and an inductor will behave like a low-value resistor after it is energized. It is those moments when current flow starts or stops when the behavior of these components becomes apparent, and AC is continually changing. In an AC circuit, capacitors and inductors will exhibit *reactance*, which is the opposition to changes in voltage or current flow. A capacitor resists changes in voltage, while an inductor resists changes in current. This is due to how each type of component stores and releases energy. When combined with a resistance, the result is impedance.

CAPACITORS

In its most basic form, a capacitor is a passive component that consists of two parallel plates with a small gap between them, as shown in Figure A.21. The primary operating principle of a capacitor is the storage and release of electric charge. A capacitor does not permit DC to pass, but it does have the effect of allowing AC to pass.

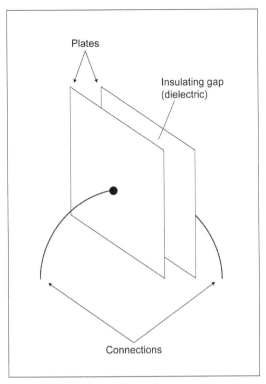

FIGURE A-21. A basic air-gap capacitor

As discussed in Chapter 8, most capacitors use a dialectic material (an insulator that can be polarized by an electrical charge) rather than an air gap. This allows the capacitor to be physically compact, and the characteristics of the dielectric can be tailored for specific requirements. Air can also act as a dialectric, but it can't be tailored for a specific application.

The fundamental unit of capacitance is the farad, abbreviated as F. One farad of capacitance produces a potential difference of 1 volt when charged by 1 coulomb. As defined in "Current" on page 401, a coulomb is equal to the amount of charge, in the form of 6.2415093×10^{18} electrons, produced by a current of 1 ampere flowing for 1 second.

In reality, the farad is an impractically large unit of capacitance, so it is typically specified in smaller units using the prefixes shown in Table A-7.

TABLE A-7. Standard value prefixes used with capacitance

Prefix	Symbol	Multiplier	Equivalent
milli	m	1×10^{-3}	1 mF or 1,000 µF
micro	µ	1×10^{-6}	1 µF or 1,000 nF or 1,000,000 pF
nano	n	1×10^{-9}	1 nF or 1,000 pF
pico	p	1×10^{-12}	1 pF

Although the farad is not a practical unit of capacitance for most applications, special types of capacitors are available with values measured in farads. These are often used as short-duration batteries for memory retention power.

When a voltage is applied to a capacitor, one of the plates will become charged in one polarity, while the other plate will take on the opposite polarity, as illustrated in Figure A-22. This is due to the electrostatic repulsion of like-charge particles (mentioned in Chapter 1), with the net result being that a capacitor will accumulate charges of opposite polarity on each of the plates. Note that Figure A-22 shows conventional current flow, not electron current flow.

ESSENTIAL ELECTRONICS AND AC CIRCUITS

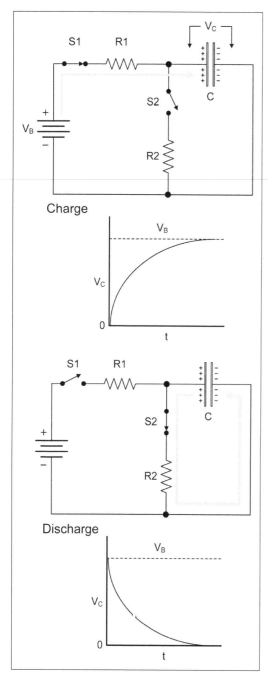

FIGURE A-22. Charge/discharge of a capacitor with DC

plates of the capacitor. The rate at which the charge accumulates is a function of both the capacitance of C and the value of R1 (discussed in "RC Circuits" on page 424). When switch S1 is opened and S2 is closed, the capacitor will discharge the energy it accumulated through resistor R2. The discharge rate is a function of the values of R2 and C.

Another point to take away here is that when charging, a capacitor will appear as an instantaneous short circuit to the power source. This is why resistors and large-value capacitors almost always appear together in power-supply filter circuits. It is generally not a good idea to connect a large capacitor directly to a power source, because even though the short-circuit condition is present for only a short period of time, it can lead to failure somewhere else in the circuit. For low-value capacitors, this effect is still present, but its impact on other surrounding components is negligible.

In an AC circuit, the capacitor will appear to pass the AC current as it charges and discharges with each cycle. This is shown in Figure A-23, where there is an AC source, a resistor, and a capacitor. Note that this is a greatly simplified representation of what really happens.

When switch S1 in Figure A-22 is closed and S2 is open, charge will accumulate on the

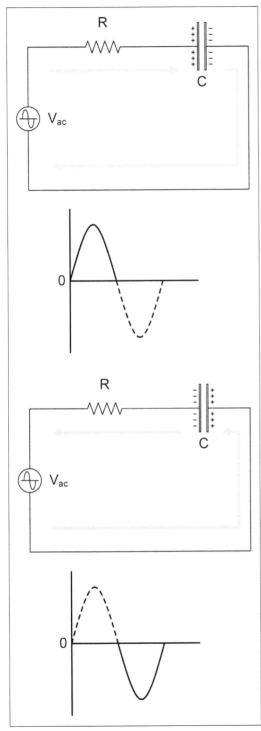

FIGURE A-23. Charge/discharge of a capacitor with AC

The ability of a capacitor to pass AC across what is effectively an open circuit prompted the 19th-century physicist James Clerk Maxwell to devise the notion of an electric displacement field. Not only did this help Maxwell visualize what was going on, but it led to the derivation of the electromagnetic wave equation that united electricity, magnetism, and optics.

Although calculations for electronics are done as if the components involved are ideal in their behavior, the reality is that there are no ideal components. The equivalent circuit for a typical capacitor, shown in Figure A-24, illustrates this.

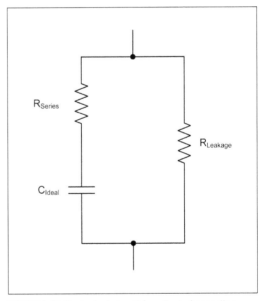

FIGURE A-24. Equivalent circuit for a typical capacitor

While the resistive aspects of a capacitor cannot be completely eliminated, they can be reduced. For the most part, however, the value of R_{Series} tends to be very low, while the value of $R_{Leakage}$ tends to be very high, so they can be safely ignored in most cases.

In addition to capacitive value and tolerance, capacitors are rated by working voltage, and exceeding the working voltage can result in the catastrophic failure of the part. In other words, it might explode. This becomes an important consideration when you are working with series and parallel networks of capacitors, as discussed in "Series and Parallel Capacitors" on page 422. Also, some capacitors are *polarized*, meaning that they have definite positive and negative connections. Electrolytic types are usually polarized, whereas ceramic types are usually nonpolarized. Incorrectly connecting a polarized capacitor will almost certainly damage it, sometimes causing a failure similar to an over-voltage condition.

In general, ceramic capacitors have working voltages into the hundreds of volts, whereas common electrolytic types range from around 10 to 100 volts. As a rule of thumb, the lower the value of the capacitor, the higher its working voltage can be.

Series and Parallel Capacitors

Just as with resistors, capacitors can be combined in series and parallel, although the math involved is a little different. Figure A-25 shows capacitors connected in series.

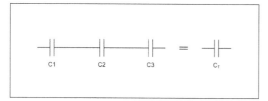

FIGURE A-25. Capacitors connected in series

When capacitors are connected in a series network, the total capacitance will be less than the lowest value component in the network (recall that for resistors the values are summed). The effect is equivalent to reducing the size of the plates of the capacitors in the network, thereby reducing the overall capacitance. This is given by:

$$C_t = \frac{1}{\frac{1}{C1} + \frac{1}{C2} + \ldots + \frac{1}{Cn}}$$

Capacitors in parallel, as shown in Figure A-26, will have a total value equal to the sum of the individual components in the network. This is equivalent to increasing the overall plate area and thereby increasing the capacitance.

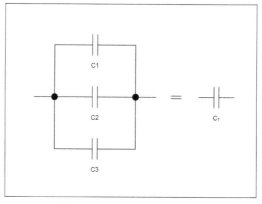

FIGURE A-26. Capacitors connected in parallel

$$C_t = C1 + C2 + \ldots + Cn$$

Notice that capacitors in series divide, whereas resistors sum. By the same token, capacitors in parallel sum, but resistors divide.

Capacitive Coupling

To DC, a capacitor is an open circuit, but it allows AC to pass due to the fact that the AC signal will alternately charge and discharge the plates of the capacitor. Figure A-23 illustrates this when a capacitor is connected to an AC source. Figure A-27 expands on that concept with a more complete circuit.

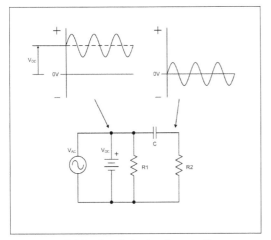

FIGURE A-27. DC blocking behavior of a capacitor

There are a couple of interesting things to notice in Figure A-27, where we have both an AC voltage source and a DC voltage source (a battery, for instance). The first is that AC and DC can exist simultaneously on the same wire (this is actually quite common in electronic circuits). Secondly, the AC signal will *ride* on top of the DC voltage, with the zero-crossing level of the AC signal at the maximum DC level. This is often referred as a *DC offset* or a *DC bias*, depending on the context.

Now notice in Figure A-27 that you can measure the composite AC-DC signal across R1, but the capacitor C will block the DC and allow only the AC to pass, so taking a measurement across R2 shows only the AC signal. We've neglected to consider the interaction between the resistors and the capacitor, which will affect how the circuit will respond at different frequencies. The next sections on RC circuits and reactance will cover these points.

Also note that when a DC offset or bias is present, the AC waveform no longer has true zero-crossing points when the phase is 0, 180, or 360 (0 again) degrees. In order to sense and utilize the zero-crossing points, the DC offset must be removed. A DC offset can also cause problems with bipolar circuits that are designed to *swing* between the V+ and V– supplies, such as op amps or audio amplifiers. In fact, a DC offset at the input of a direct-coupled audio amplifier can result in speakers with burned-out voice coils. Most speaker voice coils are not designed for continuous operation, but the coils will be continuously energized by the offset voltage. This, in turn, will cause the coils to overheat and eventually self-destruct.

For this reason, capacitive coupling is common in audio circuits, because the whole point of an audio circuit is to amplify, filter, or otherwise modify an AC signal (i.e., the audio). To deal with unwanted DC offset, capacitors are used at the inputs and between the processing or amplification stages to block unwanted DC and pass only the audio AC signal. The same technique is applied whenever a DC component in an input or output might cause a circuit to saturate, become unstable, or introduce distortion in the output.

Capacitive Phase Shift

When current flows into a capacitor, it takes some amount of time for the voltage to change. How long it will take depends on the value of the capacitor. It's akin to filling a tub with sand. The sand might be coming in at a fixed rate, but the tub will not fill instantly. At some point, the tub will be full of sand and the filling can stop. Assuming that the sand is supplied at the same fixed rate, a larger tub will take longer to fill. For any given amount of available current, a small-value capacitor will reach its maximum charge (i.e., it will be full) sooner than a capacitor with a large plate area. In both cases, when the charge is equal to the maximum supply potential, the current flow stops.

The main point here is that the current has to flow first, and the charge then catches up to it. The result is that the current leads the voltage, as shown in Figure A-28. Notice the 90-degree difference in the phases; this will come up again later in the discussion of reactance and impedance.

Conversely, when the AC voltage starts to head back to zero, the current again starts to flow, only now in the opposite direction. The current again leads the voltage and reaches its maximum negative value 90 degrees ahead of the voltage.

RC Circuits

As you've already seen, when current is applied to a capacitor, it will initially appear as a short, and current will continue to flow until the capacitor is charged. Placing a resistor in series with the capacitor, as shown in Figure A-29, limits the amount of current that can flow into it and consequently increases the time required to charge the capacitor.

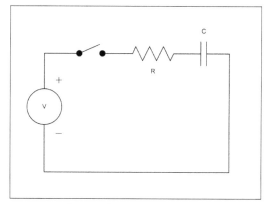

FIGURE A-29. Simple RC circuit

When the switch in Figure A-29 is closed, the capacitor will begin to charge. The amount of available current is limited by the resistor, R, so it will take longer for the capacitor to reach V_{max} than if R were not present, as shown in Figure A-30. The current starts at the maximum possible level (I_{max}) available through the resistor.

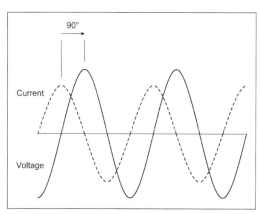

FIGURE A-28. Capacitive phase shift: current leads voltage

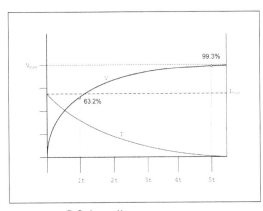

FIGURE A-30. R-C charge time

The period between the start of current flow and when the charge is close to 100% is divided into five intervals. The first interval, 1T, is at about 63.2% of the maximum charge on the capacitor. Each increase of 63.2% relative to the previous interval is defined by:

$$T = RC$$

R is in ohms, and C is in farads. This is called the *RC time constant*. T is also known as *tau* and is expressed in units of seconds. It is derived from the mathematical constant e, and some more math that defines the voltage to charge the capacitor versus time, as discussed regarding the universal timing equation in "RC Applications" on page 425. I_{max} is determined by the value of R and is the current into the capacitor when it initially appears as a short circuit. As the capacitor charges, the current will decrease until it reaches a value close to zero.

After each time interval, the capacitor will have charged another 63.2% from the previous tau value, as shown in Table A-8. In other words, with a supply of 1V, the capacitor will have 0.632V after 1 tau, 0.865V after 2 tau, and so on until at the fifth tau it is at 99.3% capacity with .993V.

TABLE A-8. RC tau time constant capacitor charging

Tau	Charge %
1	63.2
2	86.5
3	95.0
4	98.2
5	99.3

Looking again at Table A-8, you can see that the first tau point is achieved fairly rapidly, but after that, the charge gain for each successive tau interval slows down, because it's an exponential function. At 2 tau (2RC), the capacitor will have gained only an additional 23.3%, and between 2 tau and 3 tau, it will have gained only 8.5%.

Note that the 1T, 2T, etc., notation does not imply one second, two seconds, and so on. It is simply the point at which a 63.2% relative increase in charge occurs. The value of tau itself is time in seconds, and it can be anything.

RC Applications

The RC time constant is one of the most widely used concepts in electronics. For example, an RC circuit can be used to delay an action. Consider Figure A-31, which shows an RC circuit providing input to a 7408 TTL logic gate.

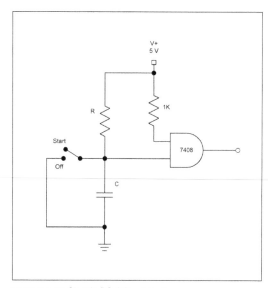

FIGURE A-31. Simple RC delay circuit

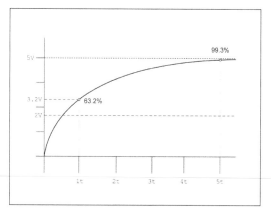

FIGURE A-32. RC time delay tau charge rate

The 1K resistor keeps one input of the gate pulled up, which prevents any false triggers from stray gate voltage or noise. When the switch is in the Off position, the junction of R and C is pulled to ground. When it is in the Start position, C can begin to charge through R. Since this is connected to a 7408 TTL gate, the key here is to determine the values for R and C such that it will reach the threshold of the gate in the desired amount of time. Let's pick 1 second as the target time.

We know that a TTL type gate will detect a high level input at about 2V, so when the voltage at the junction of R and C reaches 2V, the logic gate will detect a high input and its output will become high. The problem here is that with 5V supply, the 2V we want is less than the first tau of 63.2% by 1.2 volts, since 2V is actually 40% of 5V. Figure A-32 shows the situation.

We'll use the universal time constant equation, which allows us to determine the voltage across C at some point in time relative to a particular value for tau. Looking at Figure A-32, you can see that what we want is close to being half of 1 tau, so we'll pick tau of 2 seconds as our starting point, which should put the threshold level near the 1 second mark.

For the voltage across C at some time T, we can use a new equation, the universal time constant formula, which looks like this:

$$V_c = (V_s - V_i)\left(1 - \frac{1}{e^{\frac{t}{\tau}}}\right)$$

This equation describes the charging of a capacitor through a resistor at some specific point in time as an exponential function. Remember that RC simply determines the time between tau intervals, so to get the voltage at the in-between points, we need to compute the voltage along the charge curve at a particular point in time.

V_c is the voltage across C; V_s is the final voltage minus the starting voltage, V_i. In this case, it will be (5 − 0), or just the supply voltage of 5V. The time t is 1 second, and the value of e, Euler's constant, is approximately 2.7182818. The value of tau, as stated earlier, is 2. To determine the value of V_c after 1 second, plug in the values and turn the crank:

$$Vc = 5\left(1 - \frac{1}{e^{\frac{1}{2}}}\right)$$

This returns 1.967V for V_c after 1 second, which is probably close enough, given that common components have between 5 and 20% tolerance ratings. In reality, if the circuit needs a more precise time, it should probably be using a 555 timer IC (covered in Chapter 9) or a digital timer of some sort.

We have our tau of 2, so let's pick an R and see what kind of C we would need. Do this by solving the RC time constant equation for C, using 200 k for R:

$$C = \frac{2}{200,000}$$

This gives a value for C of 0.00001 farads, which is 10 μF. This would be an electrolytic component of some type. With an R of 200 k, you can expect to see a maximum of 25 μA into C with a 5V supply, and the value of R is just a number that was chosen as a starting point because I know that the larger R is, the smaller C will be for a given tau. We could also have picked a value for C and then solved for R.

In addition to serving as a time delay, the charging time of an RC circuit can shift the phase of an AC signal, an effect that is dependent on the impedance of the RC circuit, which is itself a product of the resistance and capacitance ("Impedance and Reactance" on page 434 discusses impedance). This behavior can be used for AC power control applications. Thyristor devices, such as silicon-controlled rectifiers (SCRs) and TRIACs (described in Chapter 9), are designed for use with AC power circuits. They are triggered into conduction at a particular voltage phase angle and continue to conduct until the phase returns to zero (or very close to it). This is how a standard household light dimmer works, as shown in Figure A-33.

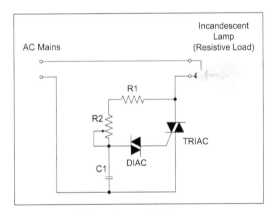

FIGURE A-33. AC lamp dimmer circuit

The dimmer works by enabling the TRIAC only for some portion of the AC waveform at a particular phase angle. The DIAC device acts as a trigger for the TRIAC, and it is controlled by the RC time constant of R1, R2, and C1, which form a phase delay network. By varying R2, the 1T point is shifted and the TRIAC is triggered at the phase angle corresponding to the delay. The TRIAC will remain in the *on* (conductive) state until the waveform returns to zero. The net result is that the circuit is selecting some portion of the AC phase at which the TRIAC becomes active. Figure A-34 shows how this works.

When the phase delay is long, the voltage across C_1 takes longer to reach the trigger level of the DIAC, less of the AC waveform is passed by the TRIAC, and the lamp will be dim. As the value of R_2 is decreased, the phase delay is reduced, C_1 will reach the trigger level more quickly in the cycle, and the TRIAC will start to conduct sooner so that more of the AC waveform is passed on to the load.

INDUCTORS

When current flows in a conductor, it induces a magnetic field around the conductor; and when a conductor moves through a magnetic field, a current is induced in the conductor. The interaction between electric current and magnetism has been observed in one form or another for over 400 years, but it wasn't formally spelled out until James Clerk Maxwell, building on the work of Michael Faraday and others, described the relationship between electric charge, magnetic poles, electric current, and magnetic fields in his *Treatise on Electricity and Magnetism* in 1873.

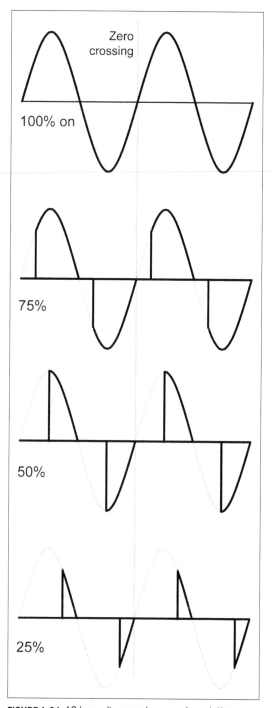

FIGURE A-34. AC lamp dimmer phase-angle variation

In its simplest form, an inductor is just a conductor, like a section of wire. As current flows through the wire, a magnetic field will form around it, as shown in Figure A-35. The strength of the field is related to the amount of the current flowing through the conductor. If the wire is formed into a coil, as shown in Figure A-35, the intensity of the magnetic field is increased proportionally to the number of turns in the coil.

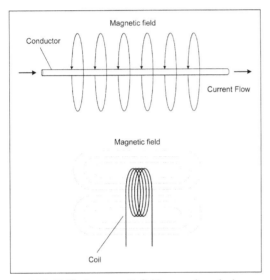

FIGURE A-35. Electromagnetic fields around conductors

Strictly speaking, an electromagnet, like the huge things used to pick up and move scrap metal in a scrapyard, is not an inductor. Once the current starts to flow, the magnetic field forms and will remain constant until the current flow ceases.

Inductance is the result of a changing magnetic field. During the times when the current is starting or stopping, an inductor is creating a changing magnetic field around itself, and this field interacts with the wire in the inductor as it expands and shrinks.

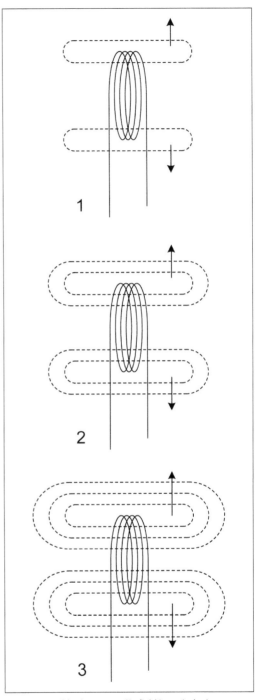

FIGURE A-36. Moving magnetic field in an inductor

ESSENTIAL ELECTRONICS AND AC CIRCUITS | 429

Since a magnetic field will induce current flow in a conductor as the conductor moves through the field, the same effect holds if the magnetic field moves through the conductor. In Figure A-36, the EM field grows in steps 1 through 3, and when the current flow through the coil returns to zero, the steps are reversed as 3 through 1. The current induced by the EM field will want to flow in the opposite direction as the current that is creating the EM field. In other words, an inductor will impede a change in current flow, as discussed in "Impedance and Reactance" on page 434.

Whereas a capacitor impedes (blocks) DC, an inductor impedes AC. But in a DC circuit, an inductor reacts only when the current flow is changing from off to on, or on to off. When the current flow in a DC circuit is steady, the inductor might present a slight resistance, but nothing else.

The unit of measurement of inductance is the *henry*. The simplistic definition of a henry can be stated as a rate of change of current of 1 ampere per second with a resulting electromotive force of 1 volt. The full definition is beyond the scope of this book, and it involves magnetic field strength and other things. If you're curious, refer to one of the texts listed in Appendix C. For our purpose here, the simple definition will suffice.

Like the farad, a henry is a large unit and not practical to work with, so most of the inductors you will encounter in electronics are rated in the millihenry (mh) range. As the value of an inductor increases, so does its effective impedance. The value of an inductor is a function of the number of turns of wire involved (one, in the case of a single wire), and what type of core material is used. Air does not contribute to inductance, but an iron core, for example, does.

Series and Parallel Inductors

For inductors connected in series, as shown in Figure A-37, the current through all the components of the network will be the same, but the voltage drop across each inductor might be different.

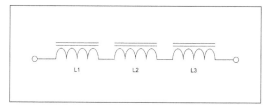

FIGURE A-37. Inductors in series

The total equivalent inductance is simply the sum of the individual inductive values:

$$L_{eq} = L1 + L2 + L3$$

Inductors in parallel have the same voltage potential across each component, but the current, depending on the inductance and the frequency of the signal across them, will vary. Figure A-38 shows a parallel inductor network with three components.

The equivalent inductance is calculated through the equation:

$$L_{eq} = \frac{1}{\frac{1}{L1} + \frac{1}{L2} + \frac{1}{L3}}$$

These simple equations work only when there is no inductive coupling between the components.

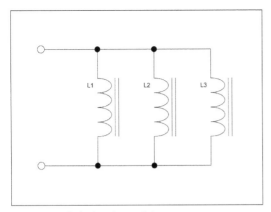

FIGURE A-38. Inductors in parallel

Inductive Phase Shift

When an inductor is energized, or when it is supplied with AC, the current flow is impeded but the voltage is not. The result is that the voltage leads the current, as shown in Figure A-39.

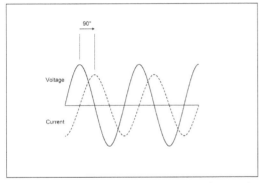

FIGURE A-39. Inductive phase shift: voltage leads current

To help understand why this is, recall that when current flows through an inductor, a magnetic field will be created (as shown in Figure A-36). As the magnetic field expands, it intersects the wires in the coil of the inductor. This, in turn, creates a reverse current flow (sometimes called *back EMF*) that will oppose the incoming current. The voltage remains unaffected, for the most part.

When the voltage reaches a steady state (which is at the minimum or maximum extents for the waveform of an AC signal), the current flow momentarily goes to zero and the coil's field ceases to expand (or contract), and no more reverse current flow is generated.

In the case of AC, when the voltage begins to drop in the second half of the waveform, the inductor again creates a reverse current flow as the magnetic field collapses and the field intersects the wires in the inductor's coil—only now the field is moving the opposite direction.

Inductive Kick-Back

The same effect that causes an inductor to induce lagging current when current starts to flow or with an AC signal is also responsible for the sharp voltage spike that is seen when a solenoid-type inductor (i.e., a relay) is de-energized. Consider the simple circuit shown in Figure A-40.

The graph on the right side of Figure A-40 shows what you could expect to see if you monitored the voltage drop across the inductor with an oscilloscope. The spike that appears when current ceases to flow in the inductor arises as the magnetic field in the inductor collapses. This occurs very quickly, and the spike is often very large (sometimes many hundreds of volts). As the field collapses, the effective direction of current flow will remain the same, but the voltage across the coil will be inverted, because it is now acting as a current source. Also, notice the the current flow in Figure A-40 is shown as electron flow, not conventional current flow.

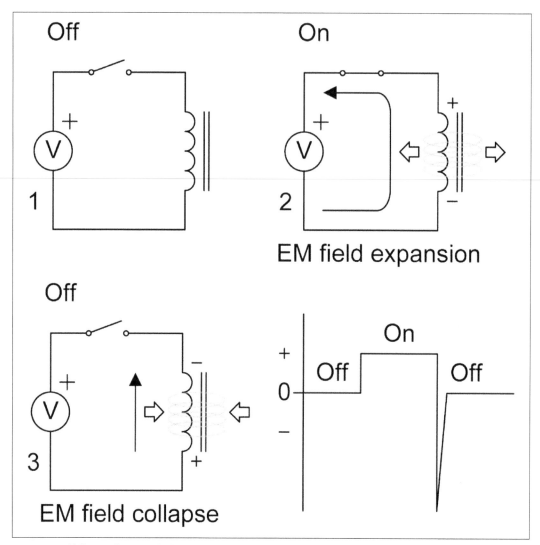

FIGURE A-40. Voltage spike created when an inductor is de-energized

The kick-back voltage from an inductor can create arcing across switch contacts, and it can easily destroy solid-state devices. Since in a DC circuit the spike will always be the opposite polarity of the voltage that energized the coil, a diode can be used to return the spike back to the coil and harmlessly dissipate its energy, as shown in Figure A-41.

When an active component, such as an IC or a transistor, is used to drive a relay, the kickback from the coil in the relay can destroy the solid-state part by suddenly presenting a voltage that is far beyond what the device might be rated for, as shown in Figure A-42.

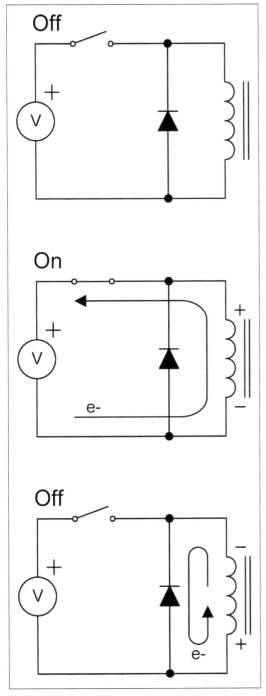

FIGURE A-41. Using a diode to snub the reverse spike from an inductor

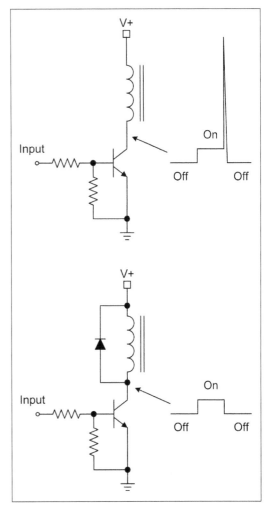

FIGURE A-42. Protecting a transistor with a kick-back snubber diode

Notice that, from the perspective of the transistor, the spike from the relay appears to be positive. This is because, when the field collapses, the coil becomes a momentary current source and the current flow is the same direction as when it was energized. The end of the relay coil connected to the transistor becomes positive, and the now negative end of the relay coil is connected to the positive power supply. It is equivalent to inserting a battery in place of the relay. With a diode installed across the coil of the relay, it effectively short-circuits the reverse voltage current flow from the coil due to field collapse, thereby saving the transistor from destruction.

IMPEDANCE AND REACTANCE

Capacitors resist changes in voltage, and inductors resist changes in current. In AC circuits, this opposition is typically called *impedance*, and the reactance of the component is a fundamental part of the impedance. This section provides a quick summary of reactance and impedance, but working through the math in detail is beyond the scope of this book. Refer to one of the texts on electronics in Appendix C for the low-level details. The *ARRL Handbook* and William Orr's *Radio Handbook*, for example, both cover these topics in some detail, with lots of practical applications for tuned circuits and antennas.

Reactance

The reactance is the opposition of a circuit element to changes in voltage or current, and it is a result of the circuit element's capacitance or inductance. So, although they are typically classified as passive components, capacitors and inductors also fall into the category of reactive components.

As shown earlier in the sections on capacitors and inductors, reactive components store and release energy. Capacitors store voltage in the form of an electric field, while inductors store current in the form of a magnetic field. Because work is required to store the energy in the components, they appear as loads when the current is changing. In an AC circuit, the current is constantly changing, so the load is persistent. This, in essence, is reactance. It is similar in some respects to resistance, but it is not the same thing.

It is reactance that gives rise to impedance. In an AC circuit, reactance is frequency dependent. Reactive components don't typically respond to DC (except for the appearance and cessation of current flow). An inductor, for example, will appear as a simple resistance to DC. A capacitor will appear as an open circuit. An ideal resistor, on the other hand, does not react to a change in current flow. It just presents a resistance regardless of whether or not the current flow is DC or AC.

The magnitude of the reactance of a capacitor is inversely proportional to frequency. In other words, the higher the frequency, the lower the reactance of the capacitor to the AC signal. The magnitude of the reactance of an inductor is also proportional to frequency, with the reactance increasing as the frequency increases. It is the opposite of a capacitor. Figure A-43 shows these relationships graphically. Note that reactance is denoted with the symbols X_L and X_C and specified in units of ohms.

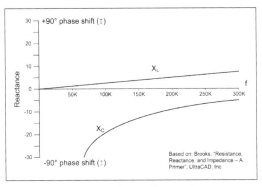

FIGURE A-43. Inductive and capacitive reactance versus frequency

These types of plots would usually employ a logarithmic scale to make them easier to comprehend, but it's informative to see them in their raw form.

The math necessary to find the reactance of a capacitor or inductor is straightforward. For capacitive reactance, you can do the following:

$$X_C = \frac{1}{\omega C}$$

And for inductive reactance:

$$X_L = \omega L$$

Note that C is in farads, L is in henries, and both X_C and X_L are in ohms. Also recall that:

$$\omega = 2\pi f$$

Impedance

Impedance is the sum of resistance and reactance. Ideal capacitors and inductors have only reactance, and they will cause a shift in the phase of the current relative to voltage. A resistor does have an associated phase shift, because it is not a reactive component.

In a purely reactive component, the phase shift is always 90 degrees. A capacitor causes the current phase to lag (–90 degrees) and an inductor causes the current phase to lead (+90 degrees). Impedance is what happens when a resistance enters the picture and modifies the current flow, although the term *impedance* is also used when we are discussing interaction of capacitors and inductors with an AC signal.

As shown earlier with RC circuits, a resistance can change the rate of charge on a capacitor and in the process alter the voltage-current phase relationship to some angle other than 90 degrees. This is the key principle behind the TRIAC circuit presented earlier.

The letter Z is used to denote impedance, and the relationship is given as:

$$Z = R + jX$$

Note that j is the square root of –1, an *imaginary* number. Conventional mathematics uses the letter i; but in engineering, the letter j is used to avoid confusion with I, the letter used for current.

One of the main applications of the concept of impedance arises when we consider a circuit as a whole, with all of its internal resistances and reactances. Recall that the unit of measure for reactance is ohms and that, for a particular frequency, a capacitor or inductor will have a specific reactance. It follows then that we could apply Thévenin's theorem to a circuit composed of resistances and reactances, and this is indeed the case. From this we can determine the impedance of the equivalent circuit. In the interests of keeping this appendix short and concise, we won't work through the math, but there are excellent examples in the text listed in Appendix C.

Passive Filtering

A combination of resistance, capacitance, and inductance can and often is used for passive filtering. *Passive filters*, as the name implies, have no signal amplifying elements such as transistors or op amps. Consequently, a passive filter has no signal gain. This means that the output level of a passive filter is always less than the input.

For low frequencies (from 0 to around 100 kHz), passive filters are generally constructed with simple RC networks, while higher frequency filters (above 100 kHz, or RF-type signals) are usually made from a combination of RLC components. This section looks at simple low-pass and high-pass RC filters as a way of showing how the concepts of reactance and impedance for RC circuits can be applied in practical ways. Although we won't get into LC and RLC filters, the concepts for these also follow from the material already presented, and in-depth discussions can be found in the texts listed in Appendix C.

Recall that a capacitor will pass high frequencies but block low frequencies (including DC). This implies that a capacitor in parallel across an AC signal will shunt high frequencies to ground, and a capacitor in series with the signal will pass high frequencies but block low frequencies. The inclusion of a resistor into the network allows the behavior to be adjusted for a specific range of frequencies.

Figure A-44 shows a simple RC low-pass filter. In this arrangement, the capacitor, C, will shunt high-frequency signals to ground, but allow low-frequency signals to pass through from input to output.

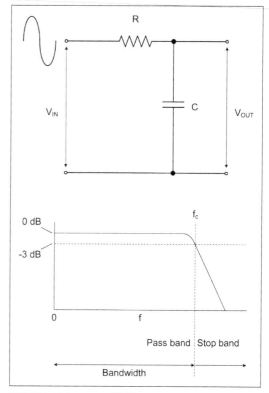

FIGURE A-44. Low-pass RC filter

The cutoff frequency, f_c, is defined to be the point where the output of the filter drops by −3 dB. Decibels are commonly used with AC signals, and it is a logarithmic scale. Decibels are used to express the ratio between two values, and in the case of Figure A-44, it refers to the decrease in the output of the filter at the cut-off frequency. The *cut-off frequency* is the point where the response begins the drop towards zero, and it is at this point where the capacitive reactance and the resistance are equal, and the output is 70.7% of the input.

The rate at which the filter will attenuate the signal is called the *roll-off* rate and is expressed in dB per decade, or dB per octave. A decade is a 10-fold change in frequency, and an octave is a 2-fold change in frequency. A passive low-pass or high-pass RC filter will typically exhibit a 20 dB/decade roll-off.

The cut-off frequency for a low-pass filter is calculated as follows:

$$f_c = \frac{1}{2\pi RC}$$

Assume you wanted to filter an input to an A/D converter (an ADC) to reject any noise that might be present. If the signal you want to sample will always be less than 100 Hz, you can rearrange the equation, pick a value for R, and solve for C at 100 Hz. Let's pick a value for R, say 10,000 ohms, and shuffle the preceding equation to solve for C:

$$C = \frac{1}{2\pi R f_c}$$

$$\frac{1}{2\pi * 10,000 * 100} = 0.000000159$$

Remember that C is in farads, so the capacitor needed is 0.159 µF. A 0.15 µF part is close enough, and it is a standard value. Since there is an inverse relationship between the value of R and the value of C, picking something like 10 k ohms for R is a reasonable place to start.

A high-pass filter takes advantage of the reactance of a capacitor to block signals below the cut-off frequency. Figure A-45 shows a high-pass RC filter. The equation for a passive high-pass filter is the same as for the low-pass filter.

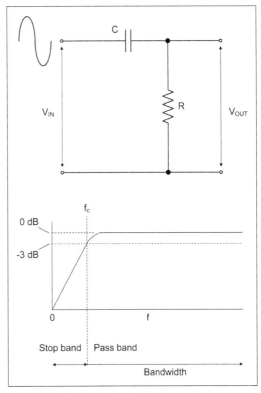

FIGURE A-45. High-pass RC filter

Both of the filters shown here are called *single-pole filters*, meaning there is only one network of R and C. *Multi-pole filters* use two of more RC filters to increase the roll-off rate of the filter response, but at the cost of increased signal attenuation and filter accuracy.

Solid-State Concepts

This section presents a high-level overview of solid-state devices. The brevity is out of necessity, since there is not enough room to cover the field with any significant level of detail. Ideally, you should be able to get a handle on the basic concepts and have a good idea where to turn next for more information.

As discussed in Chapter 1, copper, silver, and gold are good conductors, and things like plastics, ceramics, and sulfur are good insulators. Semiconductor materials such as silicon (Si), germanium (Ge), or gallium arsenide (GaAs) lie between insulators and conductors. In terms of conductivity, we can say that they are only partially conductive, hence the name *semiconductor*. But a semiconductor is not a resistor, because a semiconductor will exhibit nonlinear behavior under certain conditions, whereas a resistor will always behave in a linear fashion in response to current.

Silicon-based semiconductor components are fabricated on thin sheets cut from a silicon crystal grown in a special oven. Each circular sheet, called a *wafer*, can contain hundreds of individual parts. After the parts are created on the surface of the wafer, they are tested using automated machinery. Those parts that fail testing are tagged by their location on the wafer and will be discarded later when a diamond saw cuts the wafer up into individual chips. The chips are then mounted in a package of some type: small diodes in a package with two lead, transistors in a three-terminal package, and ICs in packages with anywhere from 8 to 144 leads or contact points, or more.

The manufacture of semiconductors is complex and utilizes highly specialized and expensive automated equipment. However, many millions of solid-state devices are produced each year, and the economy of scale, coupled with high levels of automation, allow prices to remain astonishingly low. The advances in technology have continued unabated over the years, with each new generation of semiconductor components being smaller, cheaper, and denser than the preceding generation. As Gordon Moore, Chairman Emeritus of Intel, was reported to have said in 1998, "If the automobile industry advanced as rapidly as the semiconductor industry, a Rolls-Royce would get half a million miles per gallon, and it would be cheaper to throw it away than to park it."

THE P-N JUNCTION

Modern solid-state devices are based on the physics of semiconductor materials that have been altered (or *doped*) to enhance their ability to conduct current in a particular manner. The doping takes the form of an infusion of so-called *impurities* (impure in the sense that they are not part of the original semiconductor). These atoms are the donors and acceptors of electrons that allow current to flow across the junction.

The basic arrangement of semiconductor material with different characteristics is called a *P-N junction*, and this is the basis for the ubiquitous diode. Adding a second junction results in a bipolar junction transistor (a BJT), and other arrangements are used to create field-effect transistors (FETs), SCRs, TRIACs, Zener diodes, LEDs, and integrated circuits. Figure A-46 shows a simplified diagram of a P-N junction.

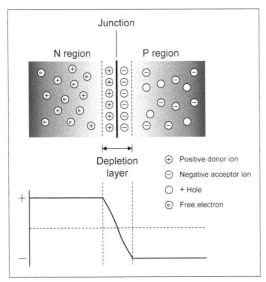

FIGURE A-46. Simplified P-N junction

Shortly after the junction is formed, some of the free electrons from the N side move to the P side. At the same time, positive *holes* (where an electron is missing) move through the semiconductor from the P side to the N side. When no more charges can move (or diffuse), the result is a region at the junction that is depleted of any free charge carriers, called the depletion layer, and a charge gradient appears across the junction.

DIODES AND RECTIFIERS

The interface between P- and N-type semiconductor material moves charge more easily in one direction than the other so that negative charges in the form of electrons can flow easily from the N to the P side of the junction, but not in the opposite direction. This is the basis for solid-state diodes. Figure A-47 shows how the symbol for a diode correlates with the P and N semiconductor material in the component. The semiconductor material is usually silicon crystal, but special-purpose diodes can employ germanium or gallium arsenide.

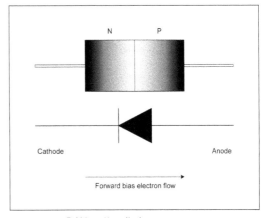

FIGURE A-47. P-N junction diode

Referring back to Figure A-46, you can see that the N-type material has a surplus of electrons (e–), also called *charge carriers*, while the P-type material has a surplus of positive charge carriers called *holes*. Note that there no charge carriers in the depletion layer.

For a simple P-N junction device, when a voltage is placed across a diode with the opposite polarity as the inherent potential across the depletion layer, the process of charge migration that initially occurred when the junction was formed will begin again, and as a result, current will flow across the junction. A P-N junction in this state is said to be *forward biased*. When an external voltage is applied to the junction with the same polarity as the junction's potential gradient, the depletion layer will simply continue to act as an insulator, blocking the movement of electrons and holes and thereby preventing current flow. This is called *reverse bias*.

The junction potential for silicon diode is approximately 0.7V. For Germanium diodes, it is around 0.3V, and Schottky diodes are about 0.2V. In order for a diode to be forward biased, the external voltage must exceed the junction potential. This is sometimes referred to as the *threshold voltage* of a diode. It conducts when the external voltage is greater than the threshold, but it does not conduct when it is less than the threshold level.

A common way to visualize what is going on in the P-N junction of a diode is called an *I-V curve*, as shown in Figure A-48. As you can see, the diode will not conduct until the voltage across it reaches the junction threshold, after which it conducts in a logarithmic fashion.

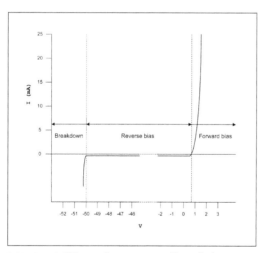

FIGURE A-48. I-V curve for a common silicon diode

One of the parameters used with diodes (and any solid-state part, for that matter) is the *peak-inverse-voltage* (PIV), which is the maximum voltage that the part can tolerate in the reverse bias state before the P-N junction breaks down. The graph in Figure A-48 shows a small amount of current flowing while the device is reverse biased. This is called the *reverse saturation current*, and it is typically very small, on the order of microamps.

In reality, the physical structure of a diode doesn't really look like Figure A-47. Semiconductors are fabricated over a series of steps that might involve photolithography, depositation of metals, oxidation, the introduction of doping impurities, and chemical etching. The entire operation takes place in a so-called *clean room*, where the air has been filtered to the point where there are almost no particles of dust or dirt that might contaminate the semiconductor and ruin the parts during fabrication.

Figure A-49 shows a more realistic view of a slice through a diode that has been created on a layer of semiconductor using a *fused junction* technique. This involves placing a tiny pellet of a P-type impurity (perhaps aluminum or indium) onto the surface of an N-type semiconductor crystal and then heating it until the pellet melts into the semiconductor. This leaves a P-type region where the pellet was placed, and the part of the pellet that doesn't fuse with the semiconductor becomes the anode terminal.

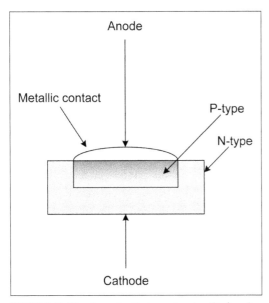

FIGURE A-49. Structure of a diode made using the fused junction process

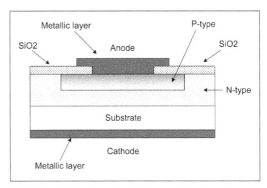

FIGURE A-50. Structure of a diode created using a planar diffusion process

Another technique, called *planar diffusion* or *epitaxial growth*, involves the diffusion of a P-type region into an N-type semiconductor that sits atop a conductive substrate material. A layer of silicon-dioxide (SiO2) is grown on top of the semiconductor, a section is etched away to expose the P-type material, and finally, a metallic layer is deposited into the space created in the SiO2 layer. This forms the anode of the diode. A metallic layer on the bottom of the substrate forms the cathode terminal of the diode. Figure A-50 shows a cross-section view of a planar diffusion diode.

SPECIAL-PURPOSE DIODES

In addition to small-signal diodes, other types of diodes are available for power rectification that are capable of handling significant amounts of the current. There are also Zener, avalanche, and Schottky types, as well as LEDs and photodiodes.

Zener Diode

A Zener diode is designed to operate in the reverse-bias mode without damage, at least until its maximum current rating is exceeded. When the reverse voltage across a Zener reaches a particular level, as determined at the time the device is fabricated, it will conduct. The I-V curve for a Zener looks similar to that for a standard diode, except that the Zener effect occurs at a much lower reverse voltage than the breakdown in a standard diode. Zener diodes are available in a wide range of voltages and are used to establish reference voltages and as signal limiters. Figure A-51 shows how a 5.6V Zener diode can be used to limit a variable voltage.

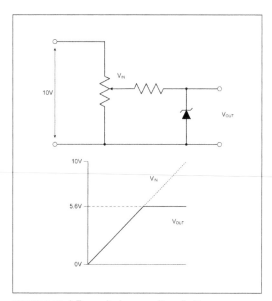

FIGURE A-51. A Zener diode as a voltage limiter

Avalanche Diode

An avalanche diode is similar to a Zener diode and, actually, the Zener and avalanche diodes exhibit both types of behaviors. But whereas the Zener diode is designed to go into reverse conduction at a specific voltage, the reverse voltage drop in an avalanche diode becomes low once it begins to conduct, and the transition to reverse current flow is very fast. This makes avalanche diodes ideal for protection circuits such as spike protection.

Photodiodes

A photodiode uses the energy of photons to move charge across a junction. A standard P-N junction is photosensitive, but a PIN junction offers improved response speed. A PIN junction comprises P- and N-type material with a thin layer of the element indium in between (hence the I in PIN).

When a photon strikes the junction of a diode, it creates an electron-hole pair. Since the junction has a voltage gradient, the electron is pulled into the N region and the cathode, while the positive hole is swept into the P region toward the anode. The result is a photocurrent.

A photodiode can be designed to operate in either a photovoltaic mode or in a photoconductive mode. In the photovoltaic mode, no bias voltage is applied to the diode, and the device will produce a voltage when it is illuminated due to the photovoltaic effect. In fact, solar cells are just very large photodiodes. In the photoconductive mode, the photodiode is typically reverse biased, and impinging photons will cause a reverse current flow that is proportional to the light level.

Photodiodes are found in a number of applications, ranging from IR remote-control receivers to camera light-level sensors, and from streetlight darkness sensors to the detector in an optical isolator.

LEDs

Light-emitting diodes (LEDs) are legion and can be found almost everywhere. They have largely pushed out the older incandescent technologies for miniature lamps and have started to make significant inroads into residential and commercial lighting applications. They are also cheap, and bags of LEDs can be had for the equivalent of pennies per device.

Internally, an LED is a diode, and it behaves just like any other diode. The big differences can be found in the materials used for the semiconductor and the fact that these materials allow the LED to produce light when current passes through it.

Figure A-52 shows the basics of how an LED produces light. When a bias is applied to the LED, electrons will flow from the N side toward the P side. At the same time, positive holes are moving from the P side toward the N side of the device. The charge carriers move into the junction with different potentials, with the electrons being at a higher potential. When an electron meets a hole, it essentially falls into the hole. In the process, it drops to a lower energy state and a photon is released. In silicon or germanium diodes, the same process occurs when there is current flow, but no light is emitted.

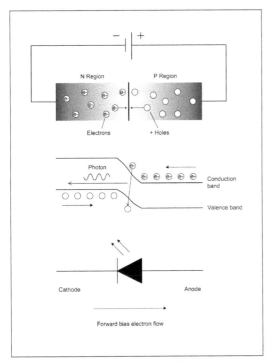

FIGURE A-52. Production of light in an LED

TRANSISTORS

Transistors are a fundamental building block of modern electronics. When they first appeared, they were a radical departure from the vacuum tubes (valves) they started to replace, and they were difficult for many people to understand. But the advantages of the transistor vastly outweighed the unconventional physics of their operation, and the vacuum tube soon started its march into oblivion. Compared to vacuum tubes, transistors are small and compact, use very little power, and don't generate massive amounts of heat. Without the transistor, it would not have been possible to create integrated circuits and achieve the levels of miniaturization that are common today.

This section is just a light overview of transistors. It is intended to serve as an introduction, not as a comprehensive tutorial. For a more detailed description of what is going on inside a transistor, and for descriptions of other solid-state devices such as UJTs, SCRs, and TRIACs, refer to the texts listed in Appendix C.

Bipolar Junction Transistors

Bipolar junction transistors (BJTs) come in two basic forms: PNP and NPN. The names arise from how these devices are fabricated from P- and N-type semiconductors. Figure A-53 shows a PNP device, and Figure A-54 shows an NPN device.

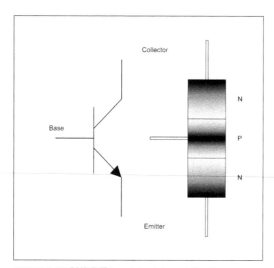

FIGURE A-53. PNP BJT transistor internal structure

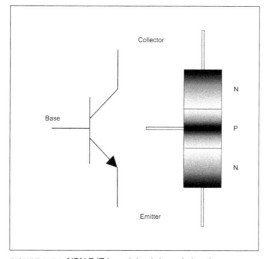

FIGURE A-54. NPN BJT transistor internal structure

For an NPN transistor, the collector must be more positive than the emitter. The base-emitter and base-collector junctions behave like diodes. In an NPN device, current can flow through base-emitter junction, but the base-collector junction will be reverse biased. A PNP is simply the opposite of an NPN.

A transistor works by modifying the current flow between the emitter and collector terminals by supplying a current to the base terminal. Thus, a small signal current can control a much larger current, and the amount of change in current between the emitter and the collector is a function of the gain (or h_{fe}) of the device.

Internally, the base and emitter terminals of an NPN device behave like a diode. The amount of current flowing through the collector, I_c, is approximately proportional to the amount of current flowing into the base through the base-emitter junction (I_b):

$$I_c = I_b h_{fe}$$

where h_{fe} is the current gain, or beta, of the transistor.

Transistors are nonlinear devices, so the relationship between the input and the output isn't a straight line. Depending on the h_{fe} and other characteristics, you might need to bias a transistor by applying a small constant DC offset to the base terminal. This will move the input signal into the linear region of the device's response curve. Without this bias, some part of the input signal might get cut off or distorted.

When a transistor is used as a switch, the idea is to push it into *saturation*: that is, into a state where the current is flowing freely between the collector and the emitter, and the voltage drop across the terminals is minimal. The trick to achieving saturation is supplying sufficient base current (I_b).

Field-Effect Transistors

A field-effect transistor (FET) differs from a BJT in that a voltage is used to control current flow through the device instead of a current. A FET has three terminals labeled *drain*, *source*, and *gate*. A small voltage change on the gate controls the current flow between the drain and source terminals. As with BJT devices, FETs come in both N and P types, and there are other variations such as MOSFETs, which as based on a metal-oxide semiconductor (the MOS part of MOSFET). Figure A-55 shows a cross-section view of the internal structure of a MOSFET device.

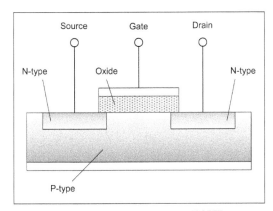

FIGURE A-55. Internal cross-section of a MOSFET

The primary current flow through a FET is from source to drain. The amount of current flow is determined by the voltage on the gate terminal. What is happening is that the gate voltage modulates the ability of charge carriers (holes and electrons) to move between the source and drain terminals. In this sense, a FET behaves much like a vacuum tube and, in fact, there have been solid-state replacements constructed for some types of vacuum tubes that utilize high-voltage FET devices.

FETs are also used extensively for signal and power switching. A FET that is in the *on* state (fully conductive) can present a very low resistance to current flow. Analog signal multiplexer chips, such as the Analog Devices ADG506A, use CMOS devices (a type of FET) to select input signals. Each channel has an *on* resistance of about 280 ohms.

Other applications for power control might utilize high-power MOSFETs to control DC power from a battery. By using a part with a low *on* resistance, the need for a relay can be eliminated. The IRF710 is an example of a power MOSFET in a TO-220 package. It has a maximum *on* resistance of 3.6 ohms and is capable of dissipating around 36W of power.

Operational Amplifiers

The operational amplifier (or *op amp*) is an old device. Early versions existed as vacuum-tube circuits developed for use in early analog computers. An op amp is a high-gain DC-coupled differential amplifier. The op amp schematic symbol uses + and − to denote the noninverting and inverting inputs, respectively, not the polarity of the input voltage.

As it is a differential amplifier, the output of an op amp is an amplified version of the difference between the inputs. In other words, if the noninverting (+) input is higher than the inverting (−) input, then the output of the op amp will be positive. If the inverting input is higher than the noninverting input, then the output will be negative.

Modern op amps are often powered with a bipolar supply, typically +/–15 VDC, although there are op amps available that will work with a single supply voltage. The use of a +/– DC supply allows the op amp to swing its output from – to + in response to the inputs.

Texas Instruments offers a downloadable version of the "Handbook of Operational Amplifier Applications" (*http://bit.ly/amp-handbk*) and Analog Devices offers a free "Op Amp Basics" guide (*http://bit.ly/op-amp-basics*).

BASIC OP AMP CIRCUITS

Feedback is used to set the overall gain of the op amp circuit. Without some kind of feedback, a high-gain op amp will respond dramatically to the inputs, with the output going full positive or full negative with just a minute amount of input difference. For this reason, op amps are seldom used without some kind of feedback network.

Op amp circuits can be either inverting or noninverting with regard to the original input. Figure A-56 shows a simple inverting op amp circuit (you can try it with a 741 device or something similar if you want to play with it).

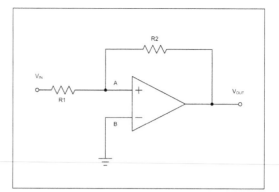

FIGURE A-56. Simple inverting op amp circuit

The basic principle behind the op amp circuit in Figure A-56 is straightforward: the op amp uses its output to attempt to force the difference between the inputs to zero volts. Since the output is coupled to the input via R2, the output voltage will appear at the junction of R1 and R2 (called a *summing junction*) and it will be summed with the input voltage through R1. The simple math to determine R1 and R2 looks like this:

$$Gain = \frac{-R2}{R1}$$

This is equivalent to:

$$Gain = \frac{V_{OUT}}{V_{IN}}$$

So the voltage gain of the circuit is the ratio of output to input, and it is set by the ratio of R2 to R1. The negative sign on R2 doesn't mean negative resistance; it simply means that this is an inverting amplifier.

If we wanted a DC amplifier with a gain of 4, perhaps to boost the signal level from a sensor or match the output of our circuit to something else, then we would select a value for R1 and then then solve for R2 with a gain value of 4. In Figure A-57 R1 has been set to 10K ohms, which means R2 will be 40K.

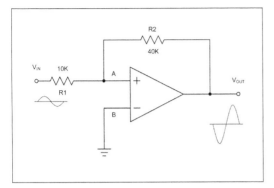

FIGURE A-57. Inverting op amp circuit with a gain of 4

Since the inverting input is always low (it's tied to ground), voltage at B will be zero. If A goes positive, then the output will swing negative so that the sum of the voltages at point A in the circuit will also be zero—hence the inversion. Another effect of this is that point A can be considered to be a virtual ground, since the op amp's output will always try to make it match the inverting input.

The output impedance of an op amp circuit like the ones shown here is typically a fraction of an ohm. The input impedance of an inverting amplifier is dependent on the value of R1. In some situations R1 might end up being small, and this can create impedance matching and input loading problems. The non-inverting form of the op amp circuit shown in Figure A-58 solves this problem.

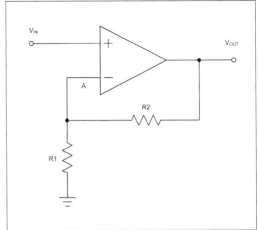

FIGURE A-58. Simple non-inverting op amp

Notice in Figure A-58 that the inverting input is supplied by a voltage divider comprised of R1 and R2. This adds a slight twist to the math for computing R1 and R2:

$$Gain = 1 + \frac{R2}{R1}$$

Because op amps are DC amplifiers by design, to work with AC signals and not have to deal with a DC bias hitching a ride on the input or output requires the use of coupling capacitors, which were covered in "Capacitive Coupling" on page 423.

SPECIALIZED OP AMPS

There are many varieties of op amps available. Some have very fast response times, and others are capable of handling large amounts of output current. Still others feature extremely low noise operation or low current consumption, or have been designed to minimize the amount of external circuitry required.

The LM386 is a good example of an op amp with a specialized output section that requires a minimal amount of external components. It is intended primarily as an audio amplifier. Figure A-59 shows the pin-out of the DIP package.

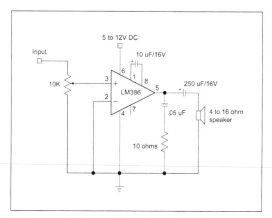

FIGURE A-60. Audio amplifier using an LM386

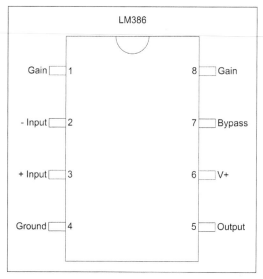

FIGURE A-59. The LM386 op amp IC

The circuit shown in Figure A-60 can drive an 8-ohm speaker directly, and because the output is coupled to the speaker with a capacitor, there need not be too much concern about stray DC on the input damaging the speaker. This little circuit is handy to have around if you need to trace an audio signal through a live circuit (in this case, an input coupling capacitor would be a necessity), check the output of a signal generator circuit, or integrate an audio output into something that didn't originally have one. Kits are readily available for LM386 amplifier modules in the $10 (USD) range.

Noise

Noise is the bane of electronic circuits, particularly those that deal with low-level signals. In general, noise is any unwanted spurious or periodic signal or voltage that might interfere with the intended operation of a circuit. It might be switching spikes from a nearby relay or power contactor, or it could be stray 60 Hz picked up from nearby appliances or power lines. It might even be a local AM radio station or a ham-radio operator with a high-power rig. This section describes some simple ways to deal with noise.

NOISE SOURCES

Noise can arise from many sources. Low-frequency noise, in the form of spikes and stray AC, is often the result of either power line spikes, inductive coupling, or a load switching on and off somewhere in the circuit. It can also occur when the ground return has a slight resistance, perhaps due to a poor connection or even corrosion. RF interference (RFI) can be a problem with wireless data links and other types of radio frequency devices, and under certain conditions, a strong radio signal can be detected and reproduced by an audio amplifier. If the transmitter is mistuned, and it's not your transmitter, there might not be much that can be done except to complain about it. However, in many cases, the solution is as simple as making sure that things are correctly shielded and grounded.

AC LINE NOISE

To deal with a noisy AC power line, such as the mains from a wall socket, you can use a filter, such as the one shown in Figure A-61. This is an off-the-shelf module that sells for about $20 (USD), and most major distributors carry items like this. If you disassemble old surplus equipment, you can often find one of these tucked away inside.

FIGURE A-62. Modular IEC power cable connector with built-in line filter

Both the filter module shown in Figure A-61 and the integrated module shown in Figure A-62 actually contain a surprising amount of circuitry inside. Figure A-63 shows what is tucked into a typical filter module like the one shown in Figure A-62.

FIGURE A-61. AC power line filter module

Some IEC-type power cord socket assemblies have a line filter built in, such as the unit shown in Figure A-62. These are available for about $2 each from distributors such as Jameco (see Appendix D).

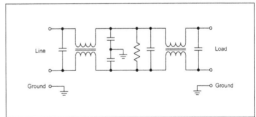

FIGURE A-63. AC line filter module schematic

This type of filter is designed to suppress unwanted signals at 100 KHz and above without interfering with the 60 Hz AC main. Because spikes typically have fast rise times equivalent to a high-frequency signal, the filter will help to suppress these. The arrangement of capacitors and inductors presents a significant insertion loss to both differential noise and common mode noise.

Although it is possible to build a filter from some capacitors, it's generally not worth the effort. For a few dollars, you can get something already built and tested.

AC RIPPLE

A small residual amount of AC on a DC power-supply line is referred to as *AC ripple*. In some cases, it might be small enough to not cause a problem, but in other cases it can introduce an annoying AC hum (in the case of audio), it can throw off a measurement from a sensor, or it can cause an actuator to misbehave. Cheap wall plug-in-type power supplies (covered in Chapter 5) can sometimes exhibit a significant amount of ripple. Of course, the ideal solution would be to use a better power supply, but sometimes you just have to work with whatever is on hand.

"Passive Filtering" on page 436 describes a basic single-pole low-pass RC filter. These are suitable for use as ripple filters, provided that the current through the filter is relatively low. The reason is the resistor. Referring to Figure A-44, if you use an R with a value of 33 ohms and a C with a value of 100 µF, you'll get a filter with a cut-off frequency of about 48 Hz. This is well below the typical 60 Hz AC encountered in the US. Ideally, you would want to keep the value of R as low as possible, to minimize the voltage drop through the filter and thus reduce the amount of power the resistor will need to dissapate.

SPURIOUS NOISE

Spurious noise and transient signals can occur when an unshielded signal line passes to close to something that is generating a significant electromagnetic field. Solenoids can cause problems, as can poorly shielded DC motors, transformers, and the wiring of circuits that might be carrying large amounts of current.

Logic devices can also introduce switching transients onto a DC power supply line, and for this reason, TTL-type logic typically employs small ceramic capacitors, with values of 0.01 and 0.1 µF commonly seen in circuits that operate at 50 MHz or less. These are also called *bypass capacitors*, and they are physically mounted close the IC and are connected between the Vcc (V+) pin and ground. Fairchild Semiconductor application note AN-363 (*http://bit.ly/an-363*) discusses power supplies, noise, and decoupling topics in logic circuits.

An arrangement of diodes called a *diode clamp* can be used to limit the voltage levels of transient spikes on the input of a circuit, as shown in Figure A-64. The idea here is that the forward voltage drops of D1 and D2, and D3 and D4, are summed, and since a typical silicon diode has a drop of about 0.7V, the diodes will not conduct until a spike exceeds 1.4V, either positive or negative.

Power Supply Mystery

Once, long ago, I worked as a test engineer for a company that produced switching-mode power supplies. One model of power supply was routinely failing final quality control (QC) tests, but the failure happened only when the supply was operating at or near maximum current load. We tried many things to attempt to isolate and resolve the problem. We investigated the transformer between the high-voltage and low-voltage sides of the supply (see Chapter 5 for an overview of what's inside a switching-mode power supply). We checked out the opto-coupler that provided the voltage level feedback to the high-voltage oscillator. We looked at every possibility, but nothing seemed to be wrong. The supplies simply failed and shut themselves down at high load levels when they should have continued to run.

After exhausting everything we could think of, someone noticed that a wire that connected one part of the low-voltage circuit to another section happened to be routed neatly under the transformer. Because the units operated at about 20 KHz, the transformer wasn't very large, and it was also unshielded. Just on a hunch, we moved the wire out from under the transformer and the unit behaved flawlessly. The wire, which happened to be the voltage sense line for the feedback circuit, was picking up the field from the transformer and inductively coupling the 20 KHz signal into the feedback circuit. Moving the wire solved the problem.

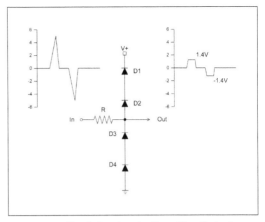

FIGURE A-64. Simple input protection with a diode clamp circuit

The R in the circuit shown in Figure A-64 is there to act as a current limiter, just in case the diodes are forward biased with enough current to damage them. The value of R should be high enough to prevent damage to the diode but low enough so as not to introduce a large voltage drop. If the input needs to have a greater range than +/- 1.4V, then Zener diodes might be a better choice. Finally, while the diode clamp circuit can prevent damage to components in the subsequent circuit, it still doesn't eliminate all of the spike. A capacitor can be used to smooth out the remains of the transient signal, but the ideal solution is to find the source of the transients and eliminate them.

RF NOISE AND RFI

If you suspect you might have an RFI problem, you should check to make sure that antenna cables are shielded, low-level signal lines are either shielded or employ twisted pair wiring (or both), and the shield grounds really are connected to ground. Mounting sensitive circuits in a grounded metal enclosure is also good.

When it comes to RFI, ground means a solid connection to the earth ground return. The negative terminal on a battery is not ground, and a power supply may, or may not, tie the negative terminal to ground. A wall power supply with only two outlet prongs and no ground pin will have a floating negative, whereas a bench power supply will typically have a terminal specifically for ground. If a device is operating such that it has no reference to real ground, it is effectively floating, and it will be susceptible to strong RFI and other noise. Connecting it to a real ground provides the return path needed to shunt the noise away.

You can also use an RC or RLC filter to block unwanted RF. An inductor will exhibit a higher reactance to high frequencies than to lower ones, and something as simple as a ferrite bead or a small choke (like the one shown in Figure 8-35) can block RFI on a DC signal line. An RC low-pass filter like the circuit shown in Figure A-44 can also be used to shunt unwanted high-frequency signals to ground.

Note that putting a filter on a signal line that is carrying something other than a slowly changing DC level or a low-frequency signal can cause unwanted side effects with the signal itself. A filter can distort the waveform of square waves and pulses, resulting in increased rise and fall times with an RC filter, and ringing with a choke or RLC filter. Basically, any signal with a fast rise time will be impacted by the filter. An AC signal (such as AC power or audio) will experience a phase shift, and in the case of an audio signal, some of the high-frequency harmonics might be supressed if the cut-off frequency for the filter is low enough.

Recommended Reading and Further Exploration

This appendix is just a high-level summary of the very deep field of electronics. The texts listed in Appendix C contain much more information, and I suggest that you look there to seek additional knowledge. I've listed some of the texts here that I feel are particularly relevant to the topics contained in this appendix.

The Art of Electronics by Paul Horowitz and Winfield Hill and *Electronics* by Allan Hambley both cover all of the topics presented in this appendix. *The Electrical Engineering Handbook* (Richard Dorf, Ed.) is a valuable reference that packs in a lot of detailed information, but it is mostly advanced material and can be a bit cryptic in places. Another text, written at a more introductory level, can help you make sense of what you might find is the *The Electrical Engineering Handbook*.

If you want to learn to work with op amps, Walter G. Jung's *IC Op-Amp Cookbook* is a good place to start. In addition to the theory (which is also covered in other texts), this book contains many interesting example circuits, all of which can be assembled on a solderless breadboard if you want to work with them.

The *ARRL Handbook*, an official publication of the American Radio Relay League, is my own go-to book when I want to find some information about RLC circuits and antennas. The *Radio Handbook* by William I. Orr is another favorite of mine. Both of these books are packed with hands-on how-to projects, and although most of the projects and circuits are oriented toward radio communications, they are treasure troves of information about inductors, tuned circuits, and high-frequeny electronics in general.

You can also find massive amounts of information free on the Internet. Some of what you find might not be entirely complete or totally accurate, but for the most part, enough eyes have been on it that the errors tend to get corrected. If I feel too lazy to pull down a book and elect instead to use Google to find something, I always make it a point to find more than one source and compare them.

APPENDIX B

Schematics

SCHEMATICS ARE THE LINGUA FRANCA OF electronics. Someone trained in electronics in China can look at a schematic created in Sweden and immediately understand what is being described by the symbols in the diagram. In its most basic form, a schematic shows the connections between the various components in an electrical or electronic device. More abstract forms can be used to describe functional relationships between components or subsystems or define an equivalent circuit.

The symbols used in electronic schematics have evolved over the years from early pictorial representations to the standardized symbols in use today. By the 1920s, most of the symbology used today was in regular use, with regional variations for some of the components. For example, even today, schematics created in places other than the US might use rectangles for resistors, along with other minor differences. The standard "IEC 60617 - Graphical Symbols for Diagrams" from the International Electrotechnical Commission (IEC) defines over 1,900 symbols, but obtaining a copy of the standard is rather pricey. This appendix describes a subset of the standard symbols in common use in the eletronics field.

An electronic schematic does not say anything about the physical arrangement of the components in a circuit. When you are reading a schematic and comparing it to an actual piece of hardware, it is not uncommon to find, say, R22 and R23 near each other on the schematic, but on opposite sides of the actual circuit board. A circuit's physical arrangement is a function of packaging and circuit board design. How the circuit moves electrons around between the components is the domain of the schematic diagram.

Wires and Current Paths

Figure B-1 shows a variety of common wiring symbols used in schematics. A wire, or current path, is shown as a solid line. When two wires connect, the junction is indicated by a black dot. If there is no connection, the wire lines simply cross. Schematics created before about 1980 used a *hump* (or *bump*) to indicate that one wire crossed another without connecting. This is seldom seen today, probably due to the greater complexity

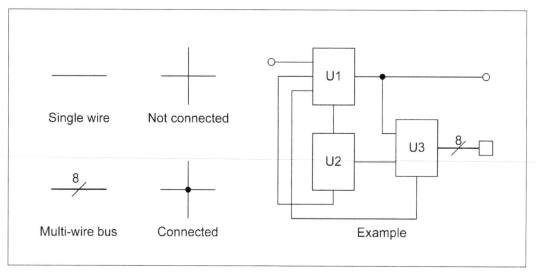

FIGURE B-1. Wiring symbology

of modern circuits with many wire paths in the diagram.

To indicate a group of wires run in parallel (i.e., a bus), a heavy solid line is often used with a slash and a number. The number indicates the number of independent paths in the bus. This is commonly seen in digital circuits where 8, 16, or more paths might connect two or more digital components in parallel.

> TIP: The wires that make up the connection paths in a circuit diagram are referred to as the *net*, or the *netlist*. Many schematic capture tools (some of which are discussed "Schematic Capture and EDA Tools" on page 457) can generate a netlist output for use as an input into other tools. When creating a schematic by hand or reading a schematic, you don't usually need to worry about the netlist, but it can be important when you're working with electronic design automation (EDA) tools.

Common Schematic Symbols

The symbols shown in this section are based on those found in various publications, such as the *ARRL Handbook* (published by the American Radio Relay League) and *The Art of Electronics* (Horowitz and Hill), and from many years of experience with creating and staring at schematics. They may, or may not, look like the symbols found in various schematic capture and EDA packages (see "Schematic Capture and EDA Tools" on page 457), but they are intentionally generic and should be close to what you will see in common usage.

> TIP: The small "bubbles" used with digital logic symbols (as shown in Figure B-11) indicate *inversion*; that is, if a logical true (1) encounters a bubble, it is inverted and becomes a logical false (0), and vice versa. See Chapter 9 for more on digital logic.

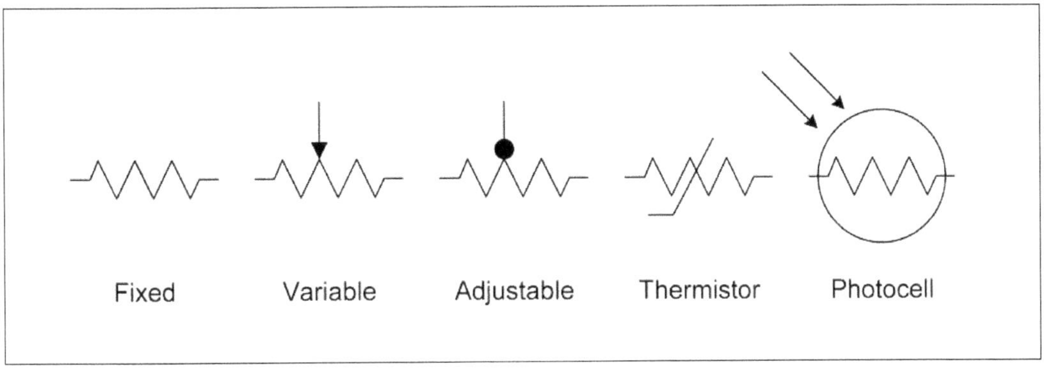

[Resistor symbols: Fixed, Variable, Adjustable, Thermistor, Photocell]

FIGURE B-2. Resistors

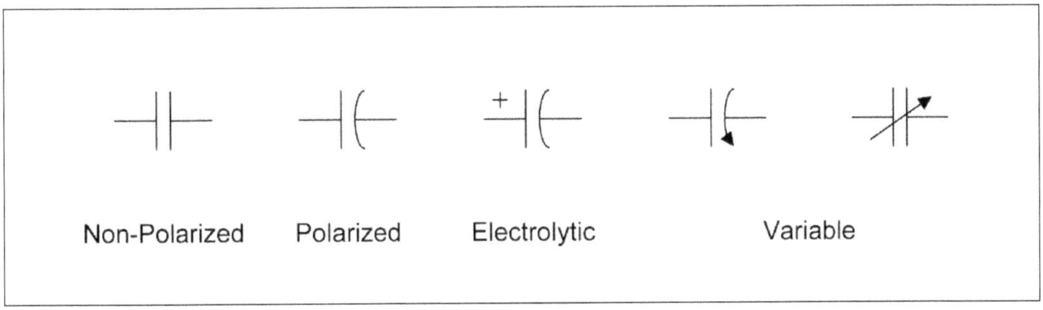

[Capacitor symbols: Non-Polarized, Polarized, Electrolytic, Variable]

FIGURE B-3. Capacitors

Schematic Capture and EDA Tools

There are multiple ways to create a schematic, ranging from a rough sketch on a napkin or the back of an envelope to a 17-inch × 22-inch piece of art generated with an EDA *(electronic design automation)* package and a special plotter or printer. For most people, the first, and least expensive, step is the notepad and a pencil.

Although just about any piece of paper will do, engineering notepads are readily available with 1/4-inch and 1/5-inch grids. They usually have a greenish tint, with heavy

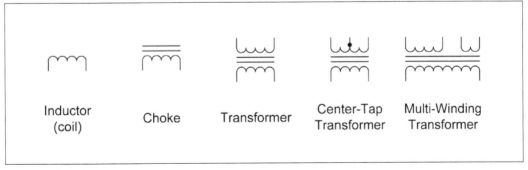

FIGURE B-4. Inductors

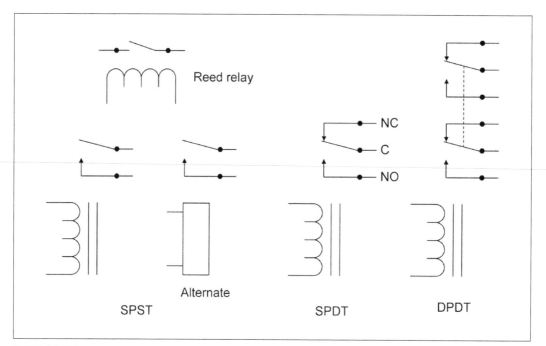

FIGURE B-5. Relays

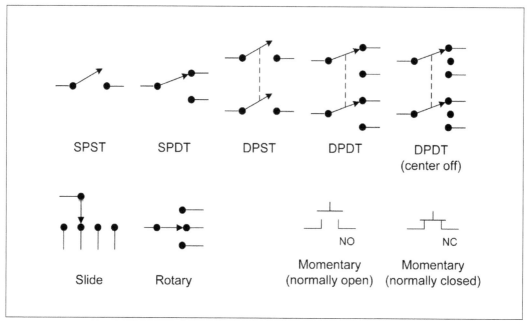

FIGURE B-6. Switches

458 | APPENDIX B

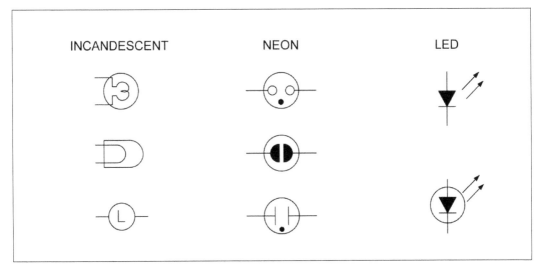

FIGURE B-7. Indictors and lamps

ruling on the back side that is faintly visible from the front. Grab a pencil (and perhaps a ruler), and you are ready to capture your design ideas on paper. While a nice schematic capture package can create pretty drawings, in reality, you can usually whip up a decent simple schematic with a pencil quicker than with a software package.

But when you need to be able to automatically generate a parts list, create a connection list (a netlist), export an EDIF file, or do a design rules check, then software is the way to go. A schematic created with an EDA tool also has the benefit of serving as the input to a PCB layout tool.

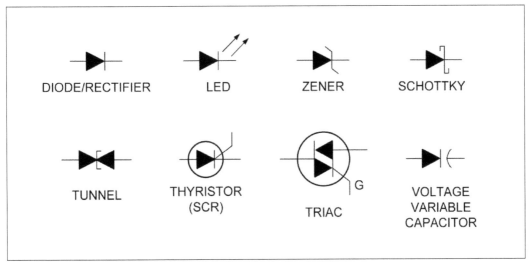

FIGURE B-8. Diodes and rectifiers

Software for electronics design comes in multiple forms. Some applications are intended for schematic capture only, others do PCB layouts, and some incorporate all (or most) of the tools necessary for end-to-end design of an electronic system. These full-up tool suites are known as *electronic design automation* systems, and they can handle everything from schematic capture to circuit layout. Some are also capable of 3D physical modeling and even thermal analysis.

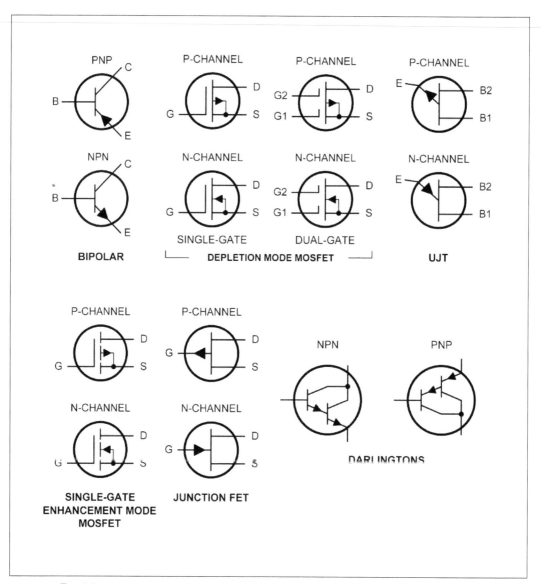

FIGURE B-9. Transistors

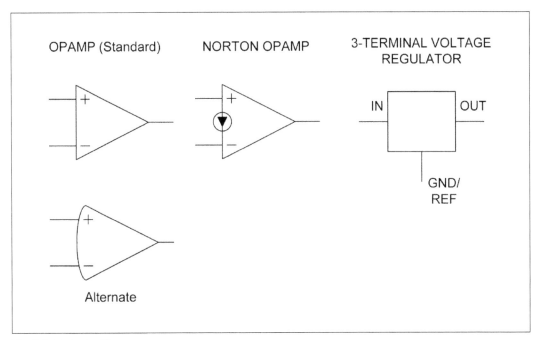

FIGURE B-10. Analog ICs

Several well-known, high-end EDA systems are available for electronics schematic capture and other design tasks. Products from Altium, Cadence (Orcad), and Mentor Graphics (PADS) are all excellent packages, but the price of these products can put them out of reach for an individual, a small makerspace, or even many small companies. Viable alternatives include open source programs (gEDA, KiCad, XCircuit), free limited-capability tools (EAGLE), and low-cost drawing tools (Visio).

Chapter 15 describes PCB layout concepts, which is a major part of the EAGLE, gEDA,

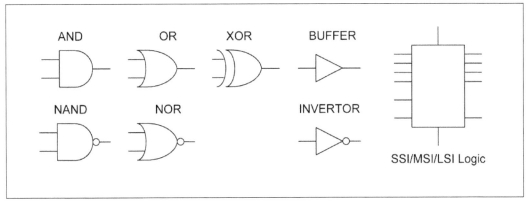

FIGURE B-11. Digital ICs

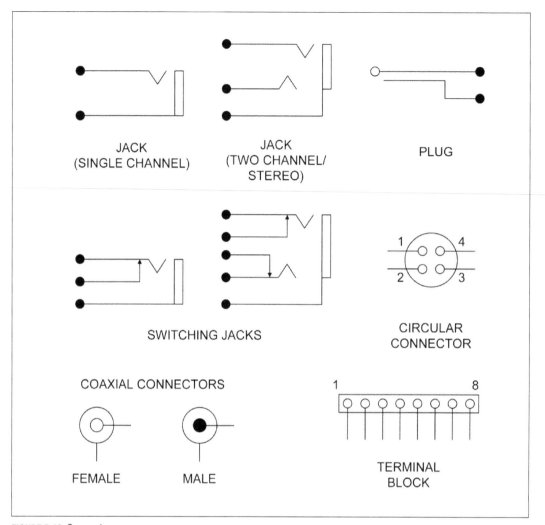

FIGURE B-12. Connectors

and KiCad tools. XCircuit is intended as a schematic capture only, and Visio is a general-purpose line-art graphics tool with the capability of creating schematics from primitive graphic elements or from predefined stencils. Each is described briefly in the following sections, along with a link to follow for more information or to download the software.

OPEN SOURCE AND COMMERCIAL TOOLS FOR SCHEMATIC CAPTURE

This section describes some of the most common and popular tools for schematic capture.

EAGLE

The EAGLE EDA tool from CadSoft (shown in Figure B-14) contains a schematic capture tool, a layout editor, autorouter, and library modules with symbols and layout templates for a variety of common parts. Interestingly, the entire tool package is installed on the target platform, but when EAGLE starts, it will ask if the user has a license. If not, then EAGLE can run as freeware mode with the following limited capabilities:

- Maximum PCB area is 4 × 3.2 inches (100 × 80 millimeters).
- A PCB can only have two layers (top and bottom).
- A schematic can have only a single sheet (1 page).

If you purchase a license, it will unlock more of the capabilities of EAGLE, but for many small designs, these limitations are not a major impediment (the entire design for a Arduino board will fit on a single sheet). The EAGLE EDA tool will run on Linux, Windows, and Mac OS X.

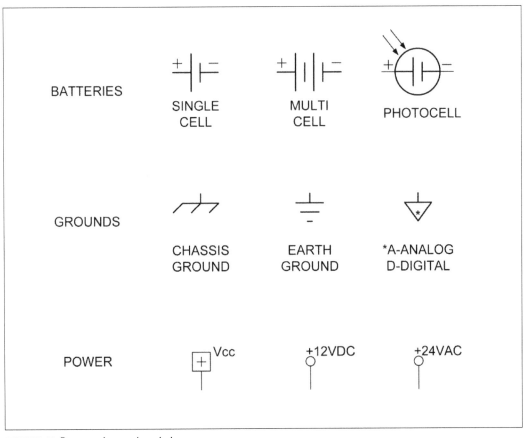

FIGURE B-13. Power and ground symbols

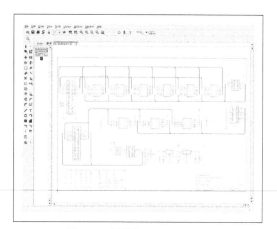

FIGURE B-14. The main EAGLE schematic capture interface

For more information about EAGLE and to download the free version, see CadSoft's website (*http://www.cadsoftusa.com*).

gEDA

Started in the 1990s, gEDA (shown in Figure B-15) has evolved into a complete suite of open source EDA tools. It includes schematic capture, behavior simulation, PCB layout, and BOM and netlist generation.

Schematics created using the schematic capture tool gschem can be printed to a PostScript file for printing or further conversion to other output formats (similar to how XCircuit works). gschem can also be used to create custom schematic symbols and block diagrams.

gEDA is a set of GNU/Linux or Unix-native programs, and there is no officially supported Windows version. The website does mention a way to build the gEDA tools for Windows using the GTK+ libraries, however. gEDA is available for most major Linux distributions as a set of installable packages.

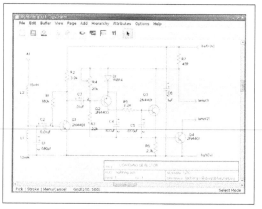

FIGURE B-15. The gEDA schematic capture tool, gschem

You can find more information about the gEDA projects and tools at the gEDA project home page (*http://www.geda-project.org/*).

KiCad

KiCad (shown in Figure B-16) is a GPL open source EDA software suite that consists of a schematic capture tool, a printed circuit board layout editor, a Gerber PCB file viewer, and a package footprint (physical form factor) tool. It also comes with a library of 3D models and can use an add-on rendering package to produce 3D CGI representations of what a finished PCB will look like.

KiCad runs on Windows, Linux, and Mac OS X, and is available for most major Linux distributions as an installable package.

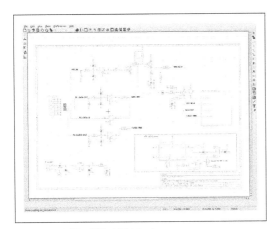

FIGURE B-16. The KiCad EDA tool

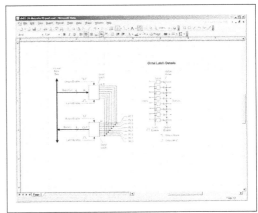

FIGURE B-17. Example Visio diagram

For more information, see the KiCad home page (*http://www.kicad-pcb.org/*).

Visio

For those of you who are saying, "Wait! What? Visio?" well, you might want to take a second look. It's actually a powerful graphics tool with a many more capabilities than are readily apparent on the surface, and it does more than create simple block diagrams and flow charts (see Figure B-17). Another useful feature of Visio is that it will integrate directly with Microsoft Word, and it can generate a huge variety of output file formats (DXF, JPG, PNG, SVG, TIFF, etc.).

If you do elect to use Visio, you will, of course, need to use Windows. I would also suggest the technical version of Visio, which comes with various useful stencils (templates) that would otherwise have to be created by hand. A suite of electronics symbols for Visio, based on the symbology found in the 2008 ARRL handbook, is available from RF CAFE (*http://bit.ly/rf-cafe*).

The technical edition of Visio also has a set of engineering stencils that include various schematic symbols.

XCircuit

The XCircuit application is intended for producing publication-quality schematics, although it can also be used as a generic line-art program for other applications. XCircuit is a not a full-up EDA tool, nor is it intended to be. But what it does, it does pretty well. Figure B-18 is a screenshot of XCircuit showing the contents of one of the libraries supplied with the tool.

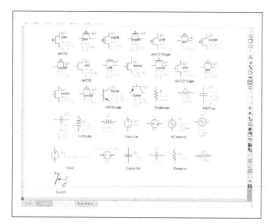

FIGURE B-18. The XCircuit schematic capture tool

The output of XCircuit is PostScript—nothing else. Schematics created with XCircuit can be imported into any word processor or graphics tool that can deal with PostScript. The XCircuit program itself is programmable using the Tcl language, so you can add functionality to meet your own specific needs.

For more information, see the XCircuit home page (*http://opencircuitdesign.com/xcircuit/*).

OTHER OPEN SOURCE SCHEMATIC CAPTURE TOOLS

You might also want to check out the following tools, which might appeal to you more than the ones highlighted previously. Each has its own unique set of strengths and weaknesses, which you'll want to research yourself:

- Oregano (*http://en.wikipedia.org/wiki/Oregano*)
- Open Schematic Capture (OSC) (*http://openschcapt.sourceforge.net*)
- Itecad (*http://www.itecad.it*)
- TinyCAD (*http://sourceforge.net/projects/tinycad/*)

APPENDIX C

Bibliography

THERE ARE MANY EXCELLENT BOOKS AVAILable that cover all aspects of electronics, ranging from hands-on projects to advanced theory. This is a short list of some of them. I own copies of all of these, and over the years some of them have become a bit worn. Nonetheless, I don't think I could pick a list of favorites, since I consider all of them to be valuable references and good sources of inspiration. If you want to acquire any of these for yourself, be sure to check and see if there is a more recent edition available, although you can save some money by opting for an older used book.

And, yes, there is indeed a shameless plug in here for my own book on instrumentation. Since I drew from it for writing this book, I felt it was proper to acknowledge it.

Electronics Theory

- Ed. Walter Kester. *Data Conversion Handbook*. Newnes. 2004. ISBN 978-0750678414.
- William Kleitz. *Digital Electronics: A practical approach*. Regents/Prentice Hall. 1993. ISBN 0-13-210287-0.
- Paul Horowitz and Winfield Hill. *The Art of Electronics, 2nd Edition*. Cambridge University Press. 1989. ISBN 978-0521370950.
- Richard Dorf (ed.). *The Electrical Engineering Handbook*. CRC Press LLC. 1997. ISBN 0-8493-8574-1.
- Allan Hambley. *Electronics, 2nd ed.* Prentice Hall, 1999. ISBN 978-0136919827.
- Randy Katz. *Comtemporary Logic Design, 2nd Edition*. Prentice Hall. 2004. ISBN 978-0201308570.
- Arthur Williams and Fred Taylor. *Electronic Filter Design Handbook, 4th Edition*. McGraw-Hill. 2006. ISBN 978-0071471718.

Engineering and Design Topics

- John Catsoulis. *Designing Embedded Hardware* O'Reilly. 2002. ISBN 978-0-596-00755-3.
- Walter G. Jung. *IC Op-Amp Cookbook*. Howard W. Sams. 1986. ISBN 0-672-22453-4.

- Don Lancaster. *The CMOS Cookbook*. Newnes. 1997. ISBN 978-0750699433.

Interfaces

- Jan Axelson. *Serial Port Complete*. Lakeview Research LLC. 2007. ISBN 978-1931448-06-2.
- Jan Axelson. *Parallel Port Complete*. Lakeview Research LLC. 2000. ISBN 978-0-9650819-1-7.
- Jan Axelson. *USB Complete*. Lakeview Research LLC. 2007. ISBN 978-1931448086.
- Benjamin Lunt. *USB: The Universal Serial Bus*. CreateSpace. 2012. ISBN 978-1468151985.
- Nick Hunn. *Essentials of Short-Range Wireless*. Cambridge University Press. 2010. ISBN 978-0521760690.
- Charles E. Spurgeon and Joann Zimmerman. *Ethernet: The Definitive Guide, 2nd Edition* O'Reilly Media, Inc. 2014. ISBN 978-1-4493-6184-6.

Construction Techniques and Projects

- ARRL. *ARRL Handbook*. American Radio Relay League. 2013. ISBN 978-1625950017.
- William I. Orr. *Radio Handbook*. Howard W Sams. 1979. ISBN 978-0672240348.
- John H. Moore, Christopher C. Davis, Michael A. Coplan, and Sandra C. Greer. *Building Scientific Apparatus*. Cambridge University Press, 2009. ISBN 978-0521878586.
- Erik Oberg (ed.). *Machinery's Handbook, 29th Edition*. Industrial Press. 2012. ISBN 78-0831129002.
- Christopher McCauley. *Shop Reference for Students & Apprentices*. Industrial Press. 2001. ISBN 978-0831130794.

Hands-On Learning

- Howard Berlin. *555 Timer Applications Source Book*. Howard W. Sams. 1979. ISBN 978-0672215384.
- Charles Platt. *Make: Electronics*. Maker Media, Inc. 2009. ISBN 978-0-596-15374-8.
- Forrest M. Mims III. *Engineer's Mini-Notebook Series, Volume 1 (MINI-1)*. Master Publishing. 2007.
- Forrest M. Mims III. *Engineer's Mini-Notebook Series, Volume 2 (MINI-2)*. Master Publishing. 2007.
- Forrest M. Mims III. *Engineer's Mini-Notebook Series, Volume 3 (MINI-3)*. Master Publishing. 2007.
- Forrest M. Mims III. *Engineer's Mini-Notebook Series, Volume 4 (MINI-4)*. Master Publishing. 2007.
- Mike Westerfield. *Building iPhone and iPad Electronic Projects*. O'Reilly Media, Inc. 2013. ISBN 978-1-449-36350-5.

Instrumentation

- J. M. Hughes. *Real-World Instrumention with Python*. O'Reilly Media, Inc. 2010. ISBN 978-0596809560.

Microcontrollers

- Massimo Banzi and Michael Shiloh. *Getting Started with Arduino*. O'Reilly Media, Inc. 2014. ISBN 978-1-4493-6333-8.
- Emily Gertz and Patrick Di Justo. *Environmental Monitoring with Arduino*. O'Reilly Media, Inc. 2012. ISBN 978-1-449-31056-1.
- Lucio Di Jasio. *Programming 16-Bit PIC Microcontrollers in C, Second Edition*. Newnes. 2011. ISBN 978-1856178709.
- Lucio Di Jasio. *Programming 32-bit Microcontrollers in C*. Newnes. 2008. ISBN 978-0750687096.
- Patrick Di Justo and Emily Gertz. *Atmospheric Monitoring with Arduino*. O'Reilly Media, Inc. 2013. ISBN 978-1-449-33814-5.
- Jonathan Oxer and Hugh Blemings. *Practical Arduino*. Apress. 2009. ISBN 978-1-4302-2477-8.
- Matt Richardson. *Getting Started with Raspberry Pi*. Maker Media, Inc. 2012. ISBN 978-1449344214.

Printed Circuit Boards

- Jan Axelson. *Making Printed Circuit Boards*. Tab Books, Inc. 1993. ISBN 978-0830639519.
- Matthew Scarpino. *Designing Circuit Boards with EAGLE*. Prentice Hall. 2014. ISBN 978-0133819991.

APPENDIX D

Resources

This appendix contains a collection of URLs for electronics distributors, sources for mechanical components, and vendors of surplus components of various types. It also includes a brief discussion of buying electronics components and other items from vendors on eBay, with some guidance and caveats.

This is not a comprehensive list, just the companies I am familiar with. This is not an endorsement for any particular company or product line.

Note that many of the sources listed here carry more than just what the list heading might imply, and some sources appear in more than one category to reflect the diversity of things they offer. Major distributors, for example, carry everything from passive components to hand tools and test equipment. Other companies, such as Adafruit and SparkFun, are more focused on the maker and hobbyist markets, but they too carry a wide range of products.

Software

OPEN SOURCE SCHEMATIC CAPTURE TOOLS

Itecad	*http://www.itecad.it/*
Oregano	*https://github.com/marc-lorber/oregano*
Open Schematic Capture (OSC)	*http://openschcapt.sourceforge.net/*
TinyCAD	*http://bit.ly/tinyCAD*
XCircuit	*http://opencircuitdesign.com/xcircuit/*

CAE SOFTWARE TOOLS

DesignSpark	Free, not open source	*http://bit.ly/designspark*
Eagle	Free, not open source	*http://www.cadsoftusa.com*
Fritzing	Free CAE tool	*http://fritzing.org/home/*
gEDA	Open source CAE tools	*http://www.geda-project.org/*
KiCad	Open source CAE tool	*http://www.kicad-pcb.org/*

PCB LAYOUT TOOLS

FreePCB	Windows-only PCB layout	*http://www.freepcb.com/*
FreeRouting	Web-based PCB autorouter	*http://www.freerouting.net/*
PCB	Linux open source layout	*http://sourceforge.net/projects/pcb/*

Hardware, Components, and Tools

ELECTRONIC COMPONENT MANUFACTURERS

Allegro	*http://www.allegromicro.com/*
Analog Devices	*http://www.analog.com*
ASIX	*http://www.asix.com.tw/*
Atmel	*http://www.atmel.com/*
Bluegiga	*http://www.bluegiga.com/*
Cypress	*http://www.cypress.com/*
Digi International	*http://www.digi.com/*
Fairchild	*http://www.fairchildsemi.com/*
Freescale	*http://www.freescale.com/*
FTDI	*http://www.ftdichip.com/*
Linear Technology	*http://www.linear.com/*
Maxim	*http://www.maximintegrated.com*
Micrel	*http://www.micrel.com/*
Microchip	*http://www.microchip.com/*
NXP	*http://www.nxp.com*

ON Semiconductor	*http://www.onsemi.com/*
Panasonic	*http://www.panasonic.com/*
Silicon Labs	*http://www.silabs.com/*
STMicrotechnology	*http://www.st.com/*
Texas Instruments	*http://www.ti.com/*
WIZnet	*http://www.wiznet.co.kr*
Zilog	*http://www.zilog.com/*

ELECTRONICS DISTRIBUTORS (USA)

Allied	*http://www.alliedelec.com/*
Digikey	*http://www.digikey.com/*
Jameco	*http://www.jameco.com*
Mouser	*http://www.mouser.com/*
Newark	*http://www.newark.com/*
State	*http://www.potentiometer.com/*

DISCOUNT AND SURPLUS ELECTRONICS

All Electronics	*http://www.allelectronics.com/*
Alltronics	*http://www.alltronics.com/*
American Science and Surplus	*http://www.sciplus.com/*
BG Micro	*http://www.bgmicro.com/*
Electronic Surplus	*http://www.electronicsurplus.com*
Electronic Goldmine	*http://www.goldmine-elec-products.com/*

MECHANICAL PARTS AND HARDWARE (SCREWS, NUTS, BOLTS)

All Electronics	*http://www.allelectronics.com/*
Alltronics	*http://www.alltronics.com/*
AmazonSupply	*http://www.amazonsupply.com/*
Bolt Depot	*http://www.boltdepot.com/*
Fastenal	*http://www.fastenal.com/*

McMaster-Carr	http://www.mcmaster.com
Micro Fasteners	http://www.microfasteners.com/
SDP/SI	http://www.sdp-si.com
W.M. Berg Co.	http://www.wmberg.com/

MICROCONTROLLERS, KITS, PARTS, AND SUPPLIES

Adafruit	http://www.adafruit.com/
circuits@home	http://www.circuitsathome.com/
CuteDigi	http://www.cutedigi.com/
DFRobot	http://www.dfrobot.com/
Electronic Goldmine	http://www.goldmine-elec-products.com/
Evil Mad Scientist Laboratories	http://shop.evilmadscientist.com/
JPM Supply	http://www.jpmsupply.com/
MCM Electronics	http://www.mcmelectronics.com/
Nightfire	http://www.vakits.com/
Parts Express	http://www.parts-express.com/
SainSmart	http://www.sainsmart.com
SparkFun	http://www.sparkfun.com/
Vetco Electronics	http://www.vetco.net
Velleman	http://www.vellemanusa.com

ELECTRONIC ENCLOSURES AND CHASSIS

Bud Industries	http://www.budind.com/
Context Engineering	http://contextengineering.com/index.html
ELMA	http://www.elma.com/
Hammond Manufacturing	http://www.hammondmfg.com/index.htm
iProjectBox	http://www.iprojectbox.com/
LMB/Heeger	http://www.lmbheeger.com/
METCASE/OKW Enclosures	http://www.metcaseusa.com/
Polycase	http://www.polycase.com/

Serpac	http://www.serpac.com/
TEKO	http://www.tekoenclosures.com/en/home

TOOLS

Adafruit	http://www.adafruit.com/
Apex Tool Group	http://www.apexhandtools.com
CBK Products	http://www.ckbproducts.com
Circuit Specialists	http://www.circuitspecialists.com
Electronic Goldmine	http://www.goldmine-elec-products.com/
Harbor Freight	http://www.harborfreight.com/
Maker Shed	http://www.makershed.com
MCM Electronics	http://www.mcmelectronics.com/
SparkFun	http://www.sparkfun.com/
Stanley	http://www.stanleysupplyservices.com/
Velleman	http://www.vellemanusa.com

TEST EQUIPMENT

Adafruit	http://www.adafruit.com/
Electronic Goldmine	http://www.goldmine-elec-products.com/
MCM Electronics	http://www.mcmelectronics.com/
SparkFun	http://www.sparkfun.com/
Surplus Shed	http://www.surplusshed.com
Velleman	http://www.vellemanusa.com

Printed Circuit Board Supplies and Fabricators

Most major electronics distributors sell things like etchant and single- and double-side clad PCB blanks with photo-resist applied. If you aren't comfortable with the chemicals and the procedures, consider using a commercial prototype PCB house.

PROTOTYPE AND FAST-TURNAROUND FABRICATORS

Advanced Circuits	http://www.4pcb.com/
ExpressPCB	http://www.expresspcb.com/
Gold Phoenix PCB Co.	http://www.goldphoenixpcb.com/
Sunstone Circuits	http://www.sunstone.com/
Sierra Circuits	https://www.protoexpress.com/

PCB KIT SOURCES

AngloEast	http://www.angloeast.net/	Laser printer-based decal transfer technique
Jameco Electronics	http://www.jameco.com/	Conventional acid etch and supplies
Think & Tinker, Ltd.	http://www.thinktink.com/	Various supplies for making PCBs
Vetco Electronics	http://www.vetco.net	Conventional acid etch kit

eBay

It is possible to find some amazing deals on eBay, but it pays to be careful about what you buy and from whom you buy it. Overall, eBay does a pretty good job of making sure that sellers aren't scammers in disguise, and PayPal makes payments safe and painless. I have also had good results purchasing items from various vendors in China through both eBay and Amazon. I've been able to find amazingly cheap things and I've never had any hassle with a Chinese seller. In fact, they go out of their way to make sure you get what you ordered and you like what you received. As a bonus, the shipping is usually free for small items (although it can take a while to get to you).

Shipping cost is probably the main drawback to buying anything on eBay. Most sellers are honest about the shipping, but there have been a few times when someone tacked on a "shipping and handling" charge that far exceeded what the shipping label stated for the cost. After experiencing this a few times, I started to pay close attention to the shipping costs stated in the item posting, and now I simply look elsewhere if it seems to be out of line. But sometimes the cost is just what it is. For heavy or bulky items, the shipping can be steep, particularly if the seller decides to use a pricey service rather than the cheapest method. If what you are bidding on or buying is something you just have to have, then that is what it will cost unless you can convince the seller to use the standard service instead of the express delivery option. It pays sometimes to look around locally before jumping on what seems like a good deal only to be shocked when the total with shipping comes due.

Lastly, beware of sellers who state "all sales final, no returns" in the item listing. This is a big red flag. If sellers aren't willing to stand behind what they're selling, you should probably think twice about giving them your hard-earned money.

Other Sources

If you have a local electronics supply outlet, I suggest visiting it to see what it has. In addition to parts and supplies, some have a section set aside for surplus test equipment. Occasionally, you can get a good deal or even haggle to get a better price. In the retail realm, there is Radio Shack, and most stores have a fairly decent selection of components, kits, and tools.

If there is a used tool store in your area, it might be worth your time to go and check it out. Where I live, we have several, and I've found some really great deals on interesting stuff, most of which only needed some cleaning and adjustment to make it work like new (it doesn't look like new, but that's not really important, at least not to me).

Some large manufacturing companies and universities have surplus property sales from time to time. These can be great places to find everything from old computer gear to office furniture. The downside is that they are sometimes overrun with people thinking they are going to get a great deal on something good, and willing to pay far more than what it is really worth. The truth is that the stuff at these sales is usually there because it is too worn out, too broken down, or too obsolete to be of any practical use any more. You can get great deals if you're looking for parts to salvage, or if you have the time and skills necessary to fix broken things, but otherwise, it might not be worth the effort.

Lastly, if you are fortunate enough to live in an area with an electronics surplus outlet, then by all means, go and check it out. These places can be a fascinating experience, and generally they are willing to negotiate the price, particularly for older items. When I was a teen, one of my first surplus acquisitions was an almost complete set of 19-inch rack-mount chassis for a sounding rocket launch control system. I never launched any rockets with it, but it was awesome anyway, and I learned an immense amount just taking it all apart and figuring out some of the circuits by following the traces on the PCBs. After all these years, I think I still have the red *Launch* button around somewhere in a box.

Datasheets, Application Notes, and Manuals

Almost every electronic component has a datasheet available for it. These are essential documents that describe the functional, electrical, and physical characteristis of a device. Even resistors have datasheets.

Application notes are intended to provide inspiration and guidance, and also to illuminate dark corners that a datasheet might not cover explicitly. Many successful product designs started from circuits described in an application note.

User manuals, sometimes running upward of 500 pages, are readily available for complex parts like microcontrollers and microprocessors, and manufacturers usually make them available for free. With a little effort, you can also find manuals for old test equipment and service manuals for consumer electronics. There are individuals and companies that specialize in locating and selling manuals for all types of electronic devices.

> **TIP** Never pay for a component datasheet or application note, if you can avoid it. The only time you might encounter a situation where you can't get the technical information any other way is when dealing with an old part that has been out of production for many years. Otherwise, you are paying for something you can get for free by downloading it yourself.

DATASHEETS

The history of datasheets goes way back to the dawn of the electronic age, when manufacturers realized that, if they wanted people to use their parts in a new design, they needed to be able to communicate essential characteristics to engineers so they could use the part correctly. As electronic components have become more complex, this has become an essential activity in the electronics component industry, and whole divisions of a large semiconductor company might be dedicated to doing nothing but testing, characterizing, and generally experimenting with the component products. For those with an inclination to poke and tinker, and a good background in math and semiconductor theory, this could be a dream job.

It's always a good idea to have the datasheets for the parts you are working with. There's no need to guess; the information is readily available in PDF form from the websites of companies like Atmel, Fairchild, Maxim, NXP, Silicon Labs, Texas Instruments, and many more. See Chapter 9 for a walk-through example of deciphering a datasheet.

APPLICATION NOTES

Semiconductor manufacturers have a long history of providing documents called *application notes* (or *app notes*) that describe how their products work, how they can be used in various applications, and how to interpret datasheet parameters. In the past, before the Internet and PDF files, it was common to see a row of bookshelves in an electronics engineering lab filled with printed paperback volumes of datasheets and application notes. The application notes were particularly popular, and it wasn't unusual to see books with some heavy wear and tear.

Today, you can download and print these documents for yourself as you need them. Personally, I like to keep printed app notes in three-ring binders so I can quickly find them when I need them, tag the pages, or even write in my own notes (using a pencil, of course!).

MANUALS

User manuals (also called *user's guides*) for components such as microcontrollers and microprocessors are readily available for free as PDF downloads from the manufacturer. These documents are typically written in a terse, cut-to-the-chase style, and they focus on the specific functional and operational characteristics of a particular device or family of devices. They aren't tutorials, by any stretch (although some manufacturers do have tutorials available for download as well). Still, when you need to know what a special function register does, or how to use the PWM (pulse width modulation) function of a device, the technical data in a user manual is invaluable.

If you need a manual for something like an old signal generator, or a consumer device like a video camera or a radio, you will need to do some hunting. If a user or service manual for something once existed, chances are, someone has a copy of it that she will sell for the right price. You have to decide if it's the right price, of course, but most of the time the cost isn't unreasonable. eBay is a good place to start, as many vendors offer copies of old technical documents, either as originals or as duplicates.

Caveat emptor: if you are buying a copy of an old manual, the odds are good that it is an unauthorized copy. In most cases, the original manufacturer might not really care, since the product has long been out of production and it gives them some free market exposure. That's assuming that they are still in business. But then again, they might care, and you might be in violation of their copyright. If in doubt, drop the manufacturer a quick note and ask how they feel about it.

You should also be prepared to inspect the document the minute it arrives, because some of the people who sell old manuals aren't always as careful when copying things as they should be. If you get something that isn't readable or is missing pages, return it, and check the listing on eBay before bidding or buying to make sure you can return it. As noted earlier, don't deal with someone who flatly states "all sales are final, no returns."

APPENDIX E

Components Lists

This appendix lists almost all of the IC components and modules mentioned in this book. While this collection is by no means comprehensive, it does contain enough representative parts from each category to provide a solid starting point for a new design.

The tables are organized into six categories as follows:

Control
    8-, 16-, and 32-bit microcontrollers

Memory
    SRAM, Flash, and EEPROM devices

Logic
    4000 series CMOS and 7400 series TTL

Interface
    ADCs, DACs, I/O expanders, relay drivers, and various assorted functions

Translation
    Discrete I/O level translator ICs

Data Communications Components
    RS-232, RS-485, USB, Ethernet, Bluetooth, and ZigBee

Some parts, such as the 4000 series CMOS logic and 7400 series TTL logic, are available from multiple sources. See Appendix D for web URLs for the manufacturers listed here, and be sure to check the various distributors for availability and possible alternate parts.

Controllers

8-BIT MICROCONTROLLERS

Part number	Manufacturer	Comment
AT89	Atmel	8051 compatible
AVR	Atmel	Unique
CY83xxxx	Cypress	8051 compatible
68HC08	Freescale	Descended from 6800
68HC11	Freescale	Descended from 6800

Part number	Manufacturer	Comment
PIC16	Microchip	Unique
PIC18	Microchip	Unique
LPC700	NXP	8051 compatible
LPC900	NXP	8051 compatible
eZ80	Zilog	Descended from Z80

16-BIT MICROCONTROLLERS

Part number	Manufacturer	Comment
PIC24	Microchip	Unique
MSP430	Texas Instruments	Unique

32-BIT MICROCONTROLLERS

Part number	Manufacturer	Comment
AT915AM	Atmel	ARM IP core
AVR32	Atmel	32-bit AVR
CY8C5xxxx	Cypress	ARM IP core
PIC32MX	Microchip	MIPS IP core
LPC1800	NXP	ARM IP core
STM32	STMicroelectronics	ARM IP core

Memory

SERIAL INTERFACE EEPROM ICS

Part number	Manufacturer	Size (bits)	Organization	Interface
AT25010B	ATMEL	1Kb	128 × 8	SPI
AT24C01D	ATMEL	1Kb	128 × 8	I2C
AT25020B	ATMEL	2Kb	256 × 8	SPI
AT24C02D	ATMEL	2Kb	256 × 8	I2C
PCF85103C-2	NXP	2Kb	256 × 8	I2C
PCF8582C-2	NXP	2Kb	256 × 8	I2C
AT25040B	ATMEL	4Kb	512 × 8	SPI

Part number	Manufacturer	Size (bits)	Organization	Interface
AT24C04C	ATMEL	4Kb	512 × 8	I2C
PCF8594C-2	NXP	4Kb	512 × 8	I2C
AT25080B	ATMEL	8Kb	1,024 × 8	SPI
AT24C08D	ATMEL	8Kb	1,024 × 8	I2C
PCA24S08A	NXP	8Kb	1,024 × 8	I2C
PCF8598C-2	NXP	8Kb	1,024 × 8	I2C
AT25160B	ATMEL	16Kb	2,048 × 8	SPI
AT24C16D	ATMEL	16Kb	2,048 × 8	I2C
AT25320B	ATMEL	32Kb	4,096 × 8	SPI
AT24C32D	ATMEL	32Kb	4,096 × 8	I2C
AT25640B	ATMEL	64Kb	8,192 × 8	SPI
AT24C64B	ATMEL	64Kb	8,192 × 8	I2C
AT25128B	ATMEL	128Kb	16K × 8	SPI
AT24C128C	ATMEL	128Kb	16K × 8	I2C
AT25256B	ATMEL	256Kb	32K × 8	SPI
AT24C256C	ATMEL	256Kb	32K × 8	I2C
AT25512	ATMEL	512Kb	64K × 8	SPI
AT24C512C	ATMEL	512Kb	64K × 8	I2C
AT25M01	ATMEL	1Mb	128K × 8	SPI
AT24CM01	ATMEL	1Mb	125K × 8	I2C

SERIAL INTERFACE SRAM ICS

Part number	Manufacturer	Size (bits)	Organization	Interface
23A512	Microchip	512 Kb	64K × 8	SPI
23A1024	Microchip	1 Mb	128K × 8	SPI
N01S830HAT22I	ON Semiconductor	1 Mb	128K × 8	SPI
FM25H20	Cypress	2 Mb	256K × 8	SPI
PCF8570	NXP	2 Mb	256K × 8	I2C

SERIAL INTERFACE FLASH ICS

Part number	Manufacturer	Bits	Channels	Interface
M25P10	Micron	1 Mb	125K × 8	SPI
SST25VF010A	Microchip	1 Mb	128K × 8	SPI
SST25VF020B	Microchip	2 Mb	256K × 8	SPI
SST25VF040B	Microchip	4 Mb	512K × 8	SPI
SST25VF080B	Microchip	8 Mb	1Mb × 8	SPI
SST25VF016B	Microchip	16 Mb	2Mb × 8	SPI
M25P16	Micron	16 Mb	2Mb × 8	SPI
N25Q00AA11G	Micron	1 Gb	128M × 8	SPI

Logic

4000 SERIES CMOS LOGIC

Part number	Description
4000	Dual three-input NOR gate and inverter
4001	Quad two-input NOR gate
4002	Dual four-input NOR gate OR gate
4008	Four-bit full adder
4010	Hex noninverting buffer
4011	Quad two-input NAND gate
4012	Dual four-input NAND gate
4013	Dual D-type flip-flop
4014	Eight-stage shift register
4015	Dual four-stage shift register
4016	Quad bilateral switch
4017	Decade counter/Johnson counter
4018	Presettable divide-by-N counter
4027	Dual J-K master-slave flip-flop
4049	Hex inverter
4050	Hex buffer/converter (noninverting)

Part number	Description
4070	Quad XOR gate
4071	Quad two-input OR gate
4072	Dual four-input OR gate
4073	Triple three-input AND gate
4075	Triple three-input OR gate
4076	Quad D-type register with tristate outputs
4077	Quad two-input XNOR gate
4078	Eight-input NOR gate
4081	Quad two-input AND gate
4082	Dual four-input AND gate

7400 SERIES TTL LOGIC

Part number	Description
7400	Quad two-input NAND gates
7402	Quad two-input NOR gates
7404	Hex inverters
7408	Quad two-input AND gates
7410	Triple three-input NAND gates
7411	Triple three-input AND gates
7420	Dual four-input NAND gates
7421	Dual four-input AND gates
7427	Triple three-input NOR gates
7430	Eight-input NAND gate
7432	Quad two-input OR gates
7442	BCD-to-decimal decoder (or three-line to eight-line decoder with enable)
7474A	Dual edge-triggered D flip-flop
7485	Four-bit binary magnitude comparator
7486	Quad two-input exclusive-OR (XOR) gates
74109A	Dual edge-triggered J-K flip-flop

Part number	Description
74125A	Quad bus-buffer gates with three-state outputs
74139	Dual two-line to four-line decoders/demultiplexers
74153	Dual four-line to one-line data selectors/multiplexers
74157	Quad two-line to one-line data selectors/multiplexers
74158	Quad two-line to one-line MUX with inverted outputs
74161A	Synchronous four-bit binary counter
74164	Eight-bit serial-to-parallel shift register
74166	Eight-bit parallel-to-serial shift register
74174	Hex edge-triggered D flip-flops
74175	Quad edge-triggered D flip-flops
74240	Octal inverting three-state driver
74244	Octal noninverting three-state driver
74273	Octal edge-triggered D flip-flops
74374	Octal three-state edge-triggered D flip-flops

Interface Components

SERIAL INTERFACE ADC DEVICES

Part number	Manufacturer	Bits	Channels	Interface
MCP3008	Microchip	10	8	SPI
AD7997	Analog Devices	10	8	I2C
TLV1548	Texas Instruments	10	8	SPI
MCP3201	Microchip	12	1	SPI
AD7091	Analog Devices	12	4	SPI
MX7705	Maxim	16	2	SPI
ADS1115	Texas Instruments	16	4	I2C
MAX1270	Maxim	12	8	SPI

SERIAL INTERFACE DAC DEVICES

Part number	Manufacturer	Bits	Channels	Interface
MCP7406	Microchip	8	1	I2C
AD5316	Analog Devices	10	4	I2C
DAC104	Texas Instrument	10	4	SPI
MCP4725	Microchip	12	1	I2C
AD5696	Analog Devices	16	4	I2C

SERIAL INTERFACE DISCRETE I/O EXPANSION ICS

Part number	Manufacturer	Ports	Interface
PCF8574	Texas Instruments	8	I2C
MAX7317	Maxim	10	SPI
MCP23017	Microchip	16	I2C

RELAY DRIVERS

Part number	Manufacturer	Internal logic	Drive current
CS1107	ON Semiconductor	Single driver	350 mA
MAX4896	Maxim	Eight-channel driver	410 mA single, 200 mA all
SN75451B	Texas Instruments	Dual AND driver	300 mA
SN75452B	Texas Instruments	Dual NAND driver	300 mA
SN75453B	Texas Instruments	Dual OR driver	300 mA
SN75454B	Texas Instruments	Dual NOR driver	300 mA
TDE1747	STMicroelectronics	Single driver	1A
UDN2981A	Allegro	Eight-channel driver	500 mA max, 120 mA/channel

SPI DISPLAY MODULES

Product	Vendor/manufacturer	URL	Interface
1.8-inch color LCD display	Adafruit	*http://www.adafruit.com*	SPI
2.8-inch touchscreen color LCD display	Haoyu Electronics	*http://www.hotmcu.com*	SPI
3.2-inch touchscreen color LCD display	SainMart	*http://www.sainsmart.com*	SPI

VARIOUS SPI/I2C PERIPHERAL DEVICES

Part number	Manufacturer	Description	Interface
ADG714	Analog Devices	Eight-channel analog switch bank	SPI
ADXXRS450	Analog Devices	Single-axis MEMS angular rate sensor (gyroscope)	SPI
ADXL345	Analog Devices	Three-axis accelerometer	I2C/SPI
LIS3LV02DL	STI	Three-axis accelerometer	I2C/SPI
PCF8583	NXP	Clock and calendar with 240 bytes of RAM	I2C
SAA1064	NXP	Four-digit LED driver	I2C
TDA1551Q	NXP	Two × 22W audio power amplifier	I2C

Translation

INTERFACE-LEVEL TRANSLATORS

Part number	Manufacturer	Circuit type	Package
BSS138	Fairchild	N-channel MOSFET	SOT-23 SMD
TXB0108	Texas Instruments	PMOS/NMOS logic	SMD
NTB0101	NXP	Auto-sense logic	SMD

Data Communications Components

RS-232 INTERFACE ICS

Part number	Manufacturer	Transmitters	Receivers
LTC2801	Linear Technology	1	1
LTC2803	Linear Technology	2	2
MAX232	Maxim	2	2
MAX232	Texas Instruments	2	2
MC1488	ON Semiconductor	4	4
MC1489	ON Semiconductor	0	4
SN75188	Texas Instruments	4	0
SN75189	Texas Instruments	0	4

RS-232 UART ICS

Part number	Manufacturer	Interface
MAX3100	Maxim	SPI
MAX3107	Maxim	I2C/SPI
PC16550D	Texas Instruments	Address/data bus
SC16IS740	NXP	I2C/SPI
TL16C752B	Texas Instruments	Address/data bus

RS-485 TRANSCEIVER ICS

Part number	Manufacturer	Transceivers
SN65HVD11DR	Texas Instruments	1
MAX13430	Maxim	1
MAX13442E	Maxim	1
SN75ALS1177N	Texas Instruments	2
SN65LBC173AD	Texas Instruments	4

USB INTERFACE ICS

Part number	Manufacturer	Function	Interface
CP2102	Silicon Labs	USB-to-serial UART bridge	RS-232
CP2112	Silicon Labs	HID USB-to-I2C bridge	I2C
FT232R	FTDI	USB-to-serial UART bridge	RS-232
MAX3421E	Maxim	Peripheral/host controller	SPI
MAX3420E	Maxim	Peripheral controller	SPI

ETHERNET INTERFACE ICS

Part number	Manufacturer	Function	Speed	Interface
AX88796C	ASIX	Ethernet controller	10/100	SPI
KSZ8851SNL	Micrel	Ethernet controller	10/100	SPI
LAN9512	Microchip	USB-Ethernet interface	10/100	USB
ENC28J60	Microchip	Ethernet controller	10	SPI
W5100	WIZnet	Ethernet controller	10/100	SPI

802.11B/G ETHERNET INTERFACE MODULES

Part number	Manufacturer	Internal controller	Interface
MRF24WB0MA	Microchip	Proprietary	SPI
SPWF01SA	STMicroelectronics	ARM Cortex-M3	SPI, I2C, UART
CC3000	Texas Instruments	Proprietary	SPI
WizFi210	WIZnet	Proprietary	SPI, I2C, UART

BLUETOOTH MODULES

Part number		Manufacturer	Interface
Class		BLE112-A-V1	Bluegiga
USB		2	RN41
Microchip		UART/USB	1
RN42		Microchip	UART/USB
2		RN52	Microchip
UART/USB		3	PAN1315A
Panasonic		UART	1 and 2
SPBT2632C1A		STMicrotechnology	UART

BLUETOOTH LOW-ENERGY COMPONENTS

Part number	Source	Interface
MCU/controller	CC2541	Texas Instruments
I2C	8051-based	CSR1010
CSR	UART/SPI	Proprietary
EM9301	EM Microelectronics	UART/SPI
Proprietary	nRF8001	Nordic Semiconductor

BLUETOOTH LOW-ENERGY MODULES

Part number	Manufacturer	Interface
ABBTM-NVC-MDCS71	Abracon	UART/I2C/SPI
BLE112	BlueGiga	UART/USB
PAN1720	Panasonic	UART/SPI

ZIGBEE MODULES

Part number	Manufacturer	ISM band	Interface
ATZB-24-B0R	Atmel	2.45 GHz	I2C, SPI, UART
XB24-AWI-001	Digi International	2.45 GHz	UART
MRF24J40	Microchip	2.45 GHz	SPI
CC2420	Texas Instruments	2.45 GHz	SPI
CC1120	Texas Instruments	868/915 MHz	SPI

Glossary

The terms and abbreviations defined in this glossary are representative of what you will typically encounter when working with electronics. This collection does not include specialized terms such as *charge carrier*, *bandgap*, or *stripline*, as these don't often come up unless you are working with semiconductor design or high-frequency microwave circuits, and those topics are well outside the scope of this book.

In some cases, when a term might be open to more than one interpretation, abbreviations in parentheses indicate the primary subject reference. For example, *annular ring* refers to a feature found in a PCB layout, not to the foam ring found in a high-end loudspeaker within a stereo system. The definition includes *PCB* in parentheses to indicate this distinction. If the abbreviation *SMD* appears, this means the definition refers to a *surface-mount device*. In other cases, the abbreviation commonly used with a particular term being defined is shown in parentheses, such as in the definition of *liquid crystal display (LCD)*.

A/D converter
See *analog-to-digital converter*.

AC
See *alternating current*.

accelerometer
A sensor (or transducer) for detecting and measuring acceleration.

ADC
See *analog-to-digital converter*.

additive process
A PCB fabrication process wherein the circuit pattern is produced by the addition of metal to a specially treated substrate. It is the opposite of the etching method used in a *subtractive process*.

Ah
See *ampere-hour*.

alternating current (AC)
A signal or power source that switches polarity periodically over time. See also *sine wave*.

alternator
An electromechanical device that converts rotary mechanical power into AC electrical power.

AM
See *amplitude modulation*.

American Wire Gauge (AWG)
A standard measure of wire diameter.

ammeter
An instrument used to measure current.

amp
Shorthand way of saying *ampere*. May also refer to an *amplifier*.

ampacity
The current carrying capacity of a conductor specified in amperes.

ampere (A)
The unit of measurement for electrical current, abbreviated as A. The symbol I is used for current in equations.

ampere-hour (Ah)
Unit of measurement for cell or battery capacity. It is based on the current that can be delivered for one hour.

amplifier
A circuit or device that produces an output that is a replica of the input with an applied increase in signal level (amplitude).

amplitude modulation (AM)
A modulation method in which the carrier amplitude changes in accordance with the input signal amplitude.

analog
A continuously variable value (usually voltage or current) that represents something changing in the physical world.

analog-to-digital converter
A circuit or device used to convert an analog input (typically a voltage) to a binary digital output (a number).

AND
A logic function that will return true if and only if all its inputs are true.

annular ring
(PCB) A ring of conductive material surrounding a hole drilled in a PCB used for pads, vias, mounting holes, and so on.

ANSI
American National Standards Institute.

apparent power
In a reactive circuit, the apparent power is the product of voltage times current and is measured in volt-amps.

armature
The rotating part of a motor or generator.

ASCII
American Standard Codes for Information Interchange.

ASIC
Application-specific integrated circuit.

auto-router
(PCB) A computer program that automatically determines the routing of traces in a PCB design based on component placement.

AWG
See American Wire Gauge.

battery
A DC power source consisting of one or more cells, typically electrochemical in nature.

BCD
See *binary-coded decimal*.

bias
A fixed voltage present at a node in a circuit, either intentionally or unintentionally. An intentional bias is used in many transistor

circuits, whereas an unintentional bias can occur due to stray induced current.

bidirectional
Describes a device that can accommodate signals traveling in either direction, or a communications circuit that supports data moving in both directions though a single channel.

bill of materials (BOM)
An inventory of the components used in an electronic circuit design. Also known as a *parts list*. Many schematic capture tools will generate a BOM automatically on demand.

binary
In digital electronics, the application of the base-2 number system, resulting in values consisting of 1s and 0s. Also refers to a system, signal, or device that has only two possible states.

binary-coded decimal (BCD)
A binary representation of a decimal number in which each decimal digit (0–9) is encoded in binary using four bits per decimal digit. Thus, the BCD value 0100 1001 is 49 in decimal.

bipolar input
An input that can accommodate a voltage ranging from negative to positive.

bipolar junction transistor (BJT)
A solid-state device in which the current flow between the collector and emitter terminals is controlled by the amount of current that flows through the base terminal. A transistor has two P-N junctions and is configured as either NPN or PNP.

bipolar output
An output that can produce a voltage ranging from negative to positive.

bit
A single binary digit. In a logic circuit, it can have a value of either 0 or 1.

bit banging
A software technique that employs a general-purpose port of a microcontroller to emulate a serial interface by writing out bits one at a time.

BJT
See *bipolar junction transistor*.

BOM
See *bill of materials*.

Boolean
Named after George Boole. In Boolean logic, a variable can have only one of two values: true or false.

break-before-make
A switch or relay that is designed to break (open) one set of contacts before engaging (closing) another set. This prevents the momentary connection of the old and new signal paths.

bridge
See *diode bridge*. For measurements, see *Wheatstone bridge*.

brownout
A condition where the voltage supplied to a system falls below the nominal operating range, but remains above 0V.

bus
Data path that connects to a number of devices. A typical example is the bus of a computer's circuit board or backplane. Memory, processor, and I/O devices might all share the bus to send data from one to another. A bus acts as a shared highway and is in lieu of the many devoted connections it would take

to hook every device to every other device. Often misspelled "buss."

byte
A group of eight binary bits used to represent digital data with a value from 0 to 255, or –128 to 127.

cable
A group of individual conductors insulated from one another. Also refers to large-gauge, single conductor wires.

capacitance (C)
The ability of two conductors separated by an insulator to store an electrical charge. Capacitance is measured in farads (F).

capacitive reactance (XC)
The opposition of a capacitor to alternating current.

capacitor
A passive electronic component that consists of two conductive plates separated by an insulating dielectric. See also *capacitance*.

CDIP
See *ceramic dual inline package*.

central processing unit (CPU)
The part of a computer system that performs arithmetic, logic, and control functions.

ceramic dual inline package (DIP)
A DIP IC package with a ceramic body.

ceramic pin grid array (CPGA)
An IC packet with a grid of pins and a ceramic body.

chamfer
A corner that has been rounded to eliminate a sharp angle.

chip
A reference to the small sections cut from a large wafer of semiconductor material (typically silicon), each containing various solid-state and passive components, that are encapsulated in a package of some type.

circuit
A collection of circuit devices and components connected in such a way so as to perform a specific function.

circuit breaker
A circuit protection device that automatically disconnects a circuit from a power source when an overload or short circuit occurs.

cm
Centimeter: 1/100 of a meter, aproximately 0.39 inches.

CMOS
See *complementary metal-oxide semiconductor*.

coil
Multiple turns of wire used to introduce inductance into a circuit, generate a magnetic field, or sense a changing external magnetic field.

comparator
A solid-state device that compares the values of two analog inputs and generates an output wherein the value depends on which input is higher than the other.

complementary metal-oxide semiconductor (CMOS)
A semiconductor fabrication technology in which p- and n-channel MOS-type transistors are used in tandem.

component side
The side of the PCB on which most of the components will be mounted. Also referred to as the *top* or *silkscreen side*.

conductance (G)
The reciprocal of *resistance*, measured in Siemens, it is the ability of a conductor to allow the flow of electrons.

conductor
A wire, PCB trace, or mechanical component that provides a path for current in a circuit.

conductor
(PCB) A conductive copper path on the surface on a PCB substrate that includes traces, pads, and vias. Can also refer to large areas used for ground planes or heatsinks.

conductor width
The physical width of a conductor on the printed circuit board. See also *trace width*.

contact bounce
Occurs when the contacts of a switch or relay make and break repeatedly (i.e, bounce) for a brief time after contact closure. See also *debounce*.

coulomb
Named for the French physicist Charles Coulomb, a coulomb is the base unit of electrical charge and is equal to 6.25×10^{18} electrons.

CPGA
See *ceramic pin grid array*.

CRC
See *cyclic redundancy check*.

CTS
RS-232 signal: Clear To Send.

current
This term can refer to either the flow of electrical charge in a circuit, or the rate of flow of electrical charge past a particular point measured in amperes.

cycle
One complete waveform (0 to 360 degrees) of alternating voltage or current.

cyclic redundancy check (CRC)
A check value calculated from the data that is capable of catching most errors. The CRC calculated by a sender is compared to a CRC calculated by a receiver, and if the two values match, then no error has occurred.

D/A converter
See *digital-to-analog converter*.

DAC
See *digital-to-analog converter*.

data acquisition system
A system for acquiring data in analog or discrete forms, or both, and either storing the data for later recall or transmitting it to another system for processing.

data sink
In serial communications, a sink defines a place where data is received.

data source
A source of data in serial communications (i.e., a sender).

DC
See *direct current*.

DCD
RS-232 signal: Data Carrier Detect.

DCE
Data Communications Equipment (used with RS-232/EIA-232 communications equipment)

debounce
The removal of the results of the phenomena that occurs when a switch or a relay makes and breaks contact several times when the contacts are closed, literally bouncing briefly.

descriptor

(USB) A data structure within a device that allows it to identify itself to a host.

device

(USB) A USB peripheral or function. Also used as *peripheral device*. See *function*.

DIAC

A *diode for alternating current* is a type of diode that conducts current in both directions. A DIAC will pass an AC signal after the voltage level has reached the breakover voltage, V_{BO}, either positive or negative. DIACs are commonly used as triggers for TRIAC devices.

dielectric

An insulator that is capable of concentrating and holding electric fields without conducting. A primary part of a capacitor used to separate two metallic plates.

differential signaling

A method for connecting two components, subsystems, or devices that employs two wires. One wire is always the inverse of the other, and a receiver need only look at the difference between the two and ignores any voltage common to both.

digital

Can refer to a measurement or signal that has only two possible values: 1 or 0, on or off. In electronics, digital devices are those components designed specifically to work with binary values.

digital multimeter (DMM)

A device for measuring voltage, current, resistance, and perhaps other parameters, that employs an internal ADC and a numeric display.

digital-to-analog converter (DAC)

A circuit that converts a binary signal to an equivalent analog form. A DAC cannot generate a truly continuously variable output, due to the nature of the binary input, but rather generates output in discrete steps. Sometimes referred to as a *D/A* or a *DAC*.

DIO

In electronics and embedded computers, DIO typically refers to *digital* or *discrete input/output*.

diode

A two-terminal semiconductor device with a single junction that allows current to pass in only one direction. Diodes can also be constructed using a vacuum tube, point-contact, metal-semiconductor junction (Schottky) or selenium-wafer technologies.

diode bridge

An arrangement of four diodes (or rectifiers) that always produces the same output polarity regardless of the input polarity. Typically employed for full-wave rectification or as input power protection.

DIP

See *dual inline package*.

direct current (DC)

Electrical current that flows in only one direction when active.

discrete

Something with two or more specific values, not a continuous range of values (i.e., analog). A term commonly used with programmable logic controller (PLC) device, but can refer to any binary input or output signal.

DMM

See *digital multimeter*.

double-sided board
A PCB wherein both sides of the substrate have a layer of copper laminate applied to produce traces.

downstream
(USB) Looking out from the host to hubs or devices connected outwards on a USB network.

DPDT
Double-pole/double-throw. A switch or relay with two conductors (poles) and two possible positions for each (throw).

DPST
Double-pole/single-throw. A switch or relay with two conductors (poles) and only one possible position for each (throw).

drain
One of the three terminals of a FET device. A voltage on the gate is used to control the current flow between the source and drain.

DRAM
Dynamic random-access memory (RAM). A form of random-access memory that uses a continuous clock signal to refresh stored data. If the clock signal stops, the data is lost. See also *SRAM*.

DSR
RS-232 signal: Data Set Ready.

DTE
Data Terminal Equipment (used with RS-232/EIA-232 communications equipment). The device that serves as the input/output terminal for communications channel.

DTR
RS-232 signal: Data Terminal Ready

dual inline package (DIP)
A type of IC package with two rows of pins. The spacing between the rows (the width) and the distance between the pins (the pitch) are defined by industry standards. DIP IC packages come in pin counts of 4, 6, 8, 14, 16, 18, 20, 22, 24, 28, 32, 40, and (rarely) 64.

duty cycle
The ratio of pulse width to period, indicating the percentage of time that a pulse is present during a cycle. A square wave a special case of a pulse with a duty cycle of 50%.

EDA
See *electronic design automation*.

EEPROM
Electrically erasable programmable read-only memory. A type of read-only memory that can be erased with an electrical pulse or specific voltage level. See also *EPROM, PROM,* and *ROM*.

EIA
Electronic Industries Alliance.

EIA-JEDEC
Electronic Industries Association/Joint Electron Device Engineering Council.

electric power (P)
The rate at which energy is consumed in a circuit or load and dissipated as heat or mechanical motion. Electrical power is typically measured in watts (W) but can also be expressed in Joules or some other unit of measurement.

electricity
The physical phenomena arising from electrons moving from a higher potential to a lower potential level through a conducting medium.

electrolyte
A chemical solution used as an electrically conductive medium in battery cells and some types of capacitors.

electrolytic capacitor
A type of capacitor in which the dielectric is formed through the reaction of an electrolyte and the electrodes.

electromagnet
A device that produces a magnetic field as the result of current flow through a coil of wire.

electromotive force (EMF)
The voltage developed by any source of electrical energy such as a battery, generator, or solar (photovoltaic) cell. Force, in this case, does not mean mechanical force, but rather potential.

electronic design automation (EDA)
Refers to the use of software tools to help automate the process of designing, simulating, and fabricating electronic circuits and systems.

electrostatic discharge (ESD)
Usually refers to the sudden brief release of stored static electricity. An ESD might measure many thousands of volts, and although very brief, there is sufficient energy to damage or destroy sensitive semiconductor devices.

EMF
See *electromotive force*.

EMI
Electromagnetic interference. Unwanted noise arising from electromagnetic radiation that interferes with the correct operation of other devices.

endpoint
(USB) An endpoint exists within a device, typically in the form of a FIFO buffer. Endpoints can be either data sinks (receiving) or data sources (sending).

enumeration
(USB) When a USB device is initially connected to host, the host gets a connection notice and proceeds to determine the type and capabilities of the device.

EPROM
Erasable programmable read-only memory. A form of read-only memory that can be erased using high-intensity UV light through a clear window built into the device package.

ESD
See *electrostatic discharge*.

etch
The use of an acid to remove unwanted copper from a PCB substrate, leaving just the conductors of the circuit pattern. See *subtractive process*.

F
See *Farad*.

Farad (F)
The unit of measurement for capacitance. Commonly used in smaller units, such as microfarad, or nanofarad.

FET
See *field-effect transistor*.

fiducial
On a PCB, an etched feature or drilled hole used for alignment during assembly operations. Fiducials are particularly important when working with multi-layer PCBs.

field-effect transistor (FET)
A type of transistor in which the voltage applied to a control terminal (the gate) creates a field within the device that modulates conduction between a source and drain terminal. See also *JFET* and *MOSFET*.

filter
A circuit designed to pass a specific range of frequencies while suppressing all others. A filter can be either passive or active.

firmware
Software that is loaded into the nonvolatile memory of a device that defines the functionality of the device.

floating
A signal line is said to be *floating* if it is not connected to a ground, ground-referenced signal source, or Vcc.

flyback diode
Refers to the diode commonly used to reduce voltage spikes seen across inductive loads when power is removed and the magnetic field collapses.

FM
See *frequency modulation*.

footprint
In PCB layouts, the physical area used by a component, including the pad pattern. Can also refer to the area covered by any component, such as a transformer or an entire PCB.

forward bias
The voltage potential applied to a solid-state P-N junction that will result in current flow through the junction when the junction has become saturated.

FPGA
Field-programmable gate array. A family of general-purpose logic devices that can be configured dynamically via a programming interface to perform complex logic functions. It is often used for prototyping production ICs or ASIC devices.

frequency
The number of complete cycles per second of a periodic waveform. Measured in hertz (Hz).

frequency fodulation (FM)
A modulation method in which the carrier frequency varies in proportion to changes in input signal amplitude.

full-duplex
A communications channel providing simultaneous data transfer in both directions. See also *half-duplex*.

function
(USB) A USB device, also referred to as a *USB peripheral* or just *device*. USB functions are *downstream* from the host.

fuse
A protective device designed to mechanically fail and interrupt current flow through a circuit when the current exceeds the rated value of the device.

G
Gram(s).

gain
The amount of amplification accomplished by a circuit. For example, a gain of three means the output is scaled up to three times the amplitude of the input.

gate

A logic circuit for performing a specific logic function. See *AND, OR, NAND, OR,* and *NOT*.

gauss

A unit of measurement of the intensity of a magnetic field.

generator

A device that converts rotational mechanical energy to DC electrical energy. See also *alternator*.

Gerber file

(PCB) A type of data file used to control a photo plotter or similar device, usually as part of a set of related data files. Named after the Gerber Scientific Co., the manufacturer of the original vector photo plotter. Gerber files define the top, bottom, and inner layers; the silkscreen; the solder masks; and a drill list used to make a PCB.

GHz

Gigahertz.

gigabit

1 billion bits per second, abbreviated Gb/s.

glitch

Generally used to refer to an undesirable behavior, a momentary pulse, or an unexpected input or output.

ground

Can refer to the reference point in an electrical circuit from which other voltages are measured, a common return path for electric current back to a power source, or a direct physical connection to the earth. Generally assumed to be at zero potential with respect to the earth.

ground plane

(PCB) A conductor layer, or portion of a conductor layer, used as a common reference point for circuit returns, shielding, or heat sinking.

H

The unit of inductance. See *henry*.

half-duplex

Data transmission over a single communications channel in either direction, but not simultaneously. Each end of the channel takes turns being either a receiver or a transmitter.

heat sink

A mechanical device or a physical surface attached to a component to aid in the dissipation of heat by conducting thermal energy away from the device and dissipating it.

henry (H)

The unit of inductance named for Joseph Henry, an American physicist. See also *inductance*.

hertz (Hz)

The unit of measurement for the repetition of cyclic waveforms, in cycles per second. Named for Heinrich Hertz, a German physicist.

hi-Z

Short for *high-impedance*. A condition where the output of a device or circuit is not being driven and which presents little or no load to other circuit components.

host

(USB) The master on a USB network. All other devices (or functions) respond to it.

hub

(USB) Used to expand the number of devices with which a USB host can communicate.

HVAC
An industry term for *heating, ventilation, and air conditioning*.

IEEE
Institute of Electrical and Electronics Engineers.

IERC
International Electronic Research Corp.

IGFET
See *metal oxide field-effect transistor*.

impedance
Represented by the letter Z, a measure of the opposition to current flow and is measured in ohms. In a DC circuit, impedance and resistance are the same, whereas in an AC circuit, it is dependent on frequency and the capacitive and inductive elements in the circuit. See also *reactance*.

impedance (Z)
The opposition to current flow arising from the combined effects of resistance and reactance in a circuit, and measured in ohms. See also *reactance*.

inductance (L)
Can refer to opposition to a change in current flow due to a magnetic field, or the related ability to store and release energy in a magnetic field. See also *reactance*.

inductive kickback
A rapid release of energy in the form of an inverse voltage across an inductor when the initial current flow ceases and the magnetic field collapses.

inductive reactance (XL)
The opposition that an inductor presents to an alternating current in the form of a counter electromotive force (cemf) and expressed in ohms.

inductor
An electrical device, typically consisting of some type of coil, that is designed to provide a specific amount of inductance measured in henrys (H). A single length of wire or a PCB trace can also be an inductor due to the magnetic field that naturally occurs around a conductor.

input
The source of a signal or data into a circuit.

insulation
An insulating material used with wire and various components to prevent short circuits and reduce shock hazard.

insulator
Any nonconductive material that exhibits a very high resistance to current flow.

integrated circuit (IC)
A solid-state device fabricated on the surface of a silicon wafer or similar semiconductor material that incorporates multiple microscopic components and associated conductive paths. See also *chip*.

interface
(USB) A set of endpoints in a device that act as either data sources or data sinks. An interface can have multiple endpoints acting as data sinks or data sources.

internal resistance
The resistance present within a power source, or within a switch or other component.

inverter
A type of logic device that acts as a NOT function where the output is always the opposite of the input.

IR
Infrared light that falls below the visible light spectrum. IR light has a lower frequency and longer wavelength than visible light. Radiated heat is also a form of IR.

ISA
Industry-Standard Architecture, a bus interconnection method used in older PCs prior to the introduction of PCI.

ISO
International Standards Organization.

ITU
International Telecommunication Union, a part of the UN that deals with all of the telecommunications.

JEDEC
Joint Electron Device Engineering Council.

JFET
See *junction field-effect transistor*.

joule
A measure of energy or work, abbreviated J. For electrical power, 1 joule is 1 watt of power applied for 1 second (a watt-second). As a unit of electrical energy, it is defined as 1 coulomb of electrical charge raised to a potential of 1 volt. Named for James Joule, the British physicist who established the mechanical theory of heat and discovered the first law of thermodynamics.

junction field-effect transistor (JFET)
A type of FET made by diffusing a gate region into a channel region. Voltage applied to the gate controls current flow in the channel by either depleting or enhancing the charge carriers in the channel. See also *FET* and *MOSFET*.

k
Kilo. A metric unit representing 1,000, and usually written in lowercase. For example, 4 kg means 4 kilograms.

kb
Kilobit(s).

keep-out zone
(PCB) The area on or near a component that the circuit board layout design cannot use due to thermal management and mounting constraints.

kg
Kilogram(s).

kHz
Kilohertz.

kilo (k)
The metric prefix for units of measurement equal to thousands (1,000 or 10^{-3}).

kilowatt-hour (kWh)
Unit of energy consumption equal to 1,000 watt-hours, or the energy consumed at the rate to 1 kilowatt (kW) for a period of 1 hour.

km
Kilometer(s).

kW
Kilowatt(s).

large-scale integration (LSI)
Refers to an IC fabrication technology used to create ICs with very complex circuits and high active component counts. Applies to devices with thousands of component elements on the silicone chip. See also *very large-scale integration*.

kWh
See *kilowatt-hour*.

latency
> The time interval between input and output events (stimulus and response). See also *propagation delay*.

LCC
> (SMD) Abbreviation of *leadless chip carrier*, a type of IC package, usually ceramic, that has no leads (pins), but instead uses metal pads at its outer edge to make contact with the printed circuit board.

LCD
> See *liquid-crystal display*.

lead
> The electrical contacts for a component, typically wires or pins extending from the body of the part. Can also refer to the wires used to connect a meter or other instrument to a circuit.

LED
> See *light-emitting diode*.

light-emitting diode (LED)
> A semiconductor diode that emits light when forward-biased.

linear
> Describes a circuit or component where the output is directly proportional to the input as a straight-line relationship.

liquid-crystal display (LCD)
> A display type that employs electrically controlled changes in the reflectivity of a liquid crystal material.

lithium batteries
> A battery with a chemistry based on the element lithium, typically used for low-power, high-reliability, long-life applications such as memory retention. See also *lithium-ion batteries*.

lithium-ion batteries
> A battery with a chemistry based on the element lithium, typically used as power sources for portable equipment, and with larger sizes, as power for electric vehicles. They are usually rechargeable. See also *lithium batteries*.

load
> A component, device, or apparatus that uses the electrical energy of a circuit to perform some form of work.

LSB
> Least-significant bit. Binary numbers are usually written with the MSB in the left-most position, and the LSB (2^0 place) in the furthest-right position.

LSI
> See *large-scale integration*.

mA
> Milliampere, or milliamp. 1/1,000 of an ampere.

mask
> (PCB) A chemical and heat-resistant material applied to a PCB by a silkscreen process that prevents etching, plating, or soldering in specific areas on a PCB.

mega (M)
> Prefix for units of measurement equal to millions (1,000,000 or 10^6).

MESFET
> See *metal-semiconductor field-effect transistor*.

metal-oxide field-effect transistor (MOSFET)
> A transistor type that uses an electric field across an oxide layer to control conduction. Also called *insulated gate field-effect transistor* (IGFET).

metal-oxide varistor (MOV)
A discrete electronic component that can be used to divert excessive voltage to ground. Also known as a *surge suppressor*.

metal-semiconductor field-effect transistor (MESFET)
A transistor that uses a metal-semiconductor (Schottky) junction to create the conductive channel, rather than using a P-N junction as in a JFET or a metal-oxide semiconductor layer found in MOSFET devices.

MHz
Megahertz. One million hertz, or cycles per second.

micro (u)
Prefix for units of measurement equal to millionths ($1/1,000,000$ or 10^{-6}).

microcontroller
A computer on a single integrated circuit (IC) with a core processor, memory, and programmable input/output functions.

Microprocessor
A complete computer CPU contained on a single integrated-circuit(IC) chip. Does not include memory or programmable I/O functions. See also *microcontroller*.

milli (m)
Prefix for units of measurement equal to thousandths ($1/1,000$ or 10^{-3}).

MISO
SPI Signal: Master In, Slave Out.

mm
Millimeter(s).

MOSFET
See *metal-oxide field-effect transistor*.

MOSI
SPI Signal: Master Out, Slave In.

MOV
See *metal-oxide varistor*.

MQFP
(SMD) Metric quad flat pack.

ms
Millisecond(s).

multi-layer board
A type of PCB that consists of two or more thin PCBs, each with traces, ground planes, and perhaps power planes, that are assembled into a single board. Plated pads and vias make the connections between the layers.

multimeter
An instrument that can measure voltage, current, or resistance in various ranges. Can refer to an analog meter (VOM) or a digital meter (DMM).

Murphy's Law
"Anything that can go wrong, will go wrong." Attributed to Capt. Edward Murphy around 1948 during testing at what is now Edwards Air Force Base.

mV
Millivolt (mV). $1/1,000$ of a volt.

MW
Megawatt(s).

mW
Milliwatt(s).

n-channel metal-oxide semiconductor
A type of transistor in which n-type dopants are used in the gate region (*channel*). A positive voltage on the gate allows current flow.

nA
Nanoampere(s).

NAND
NOT-AND: A form of gate or logic circuit that gives the inverted output of AND logic. Returns true if any inputs are false.

NC
Not connected or *normally closed* (when used with switch contacts).

negative
An object (if referring to a static charge) or the terminal of a voltage source that has an excess of electrons, or the polarity of a point in a circuit with respect to some other point.

NiCd
See *nickel-cadmium battery*.

nickel metal hydride
A rechargeable-battery technology based on nickel-hydrogen chemistry that employs positive electrodes of nickel oxyhydroxide (NiOOH).

nickel-cadmium battery
A rechargeable-battery technology that employs nickel oxyhydroxide (NiOOH) and metallic cadmium electrodes. Now largely obsolete.

NiMH
See *nickel metal hydride*.

nMOS
See *n-channel metal-oxide semiconductor*.

NO
Normally open, used with switch and relay contacts.

nonlinear
A circuit or component whose output versus input is not a straight-line relationship.

nonvolatile (NV)
Refers to a form of memory that retains stored data when power is removed.

NOR
NOT-OR: A form of gate or logic circuit that gives the inverted output of OR logic (i.e., the output will be true only if all inputs are false).

ns
Nanosecond(s), one one-billionth of a second (10^{-9}).

Ohm
The unit of measurement for resistance named for German physicist Georg Simon Ohm.

Ohm's law
A basic circuit law that defines the relationships between current, voltage, and resistance, usually written as $E = IR$.

ohmmeter
An instrument for measuring resistance.

OLED
See *organic light-emitting diode*.

op amp
See *operational amplifier*.

open-collector
Refers to an IC pin connected to the collector terminal of an internal transistor. This is a current sink, and when the transistor is active, the pin will be pulled to ground.

open-drain
Refers to an IC pin connected to the drain terminal of an internal FET. This is a current sink, and when the transistor is active, the pin will be pulled to ground.

operational amplifier (op amp)
A type of amplification circuit or component particularly well suited to use with IC technology. The ideal op amp is an amplifier with infinite input impedance, infinite

open-loop gain, zero output impedance, infinite bandwidth, and zero noise. With the appropriate feedback, an op amp can perform a wide range of functions.

OR
A form of gate or logic circuit that will output a true value if any input is true. Multiple inputs may be true at the same time.

organic light-emitting diode (OLED)
An LED made with organic materials that emits light when a voltage is applied.

oscillator
A circuit or device that generates AC signals with a specific frequency.

oscilloscope
An instrument that displays a plot of voltage over time, either on a CRT (an analog instrument) or an LCD type display (a digital instrument). Also called a *scope*.

p-channel metal-oxide semiconductor (pMOS)
A semiconductor in which p-type dopants are used in the gate region. A negative voltage on the gate enables current flow.

pad
(PCB) For wires and through-hole parts, a pad comprises a hole large enough to accommodate the component lead or wire, surrounded by an annular ring of copper. For surface-mount parts, a pad is an area of bare conductor where the component lead will be soldered. Also referred to as a *land*.

parallel interface
Transfers whole sets of bits (e.g., bytes or words) at once using a set of wires.

PCB
See *printed circuit board*.

PDIP
See *plastic dual inline package*.

peripheral
(USB) Another name for a *device* or *function*.

permeability
A measure of the ability of a magnetic material to conduct magnetic lines of force.

pF
Picofarad, 10^{-12} of a Farad.

phase
A time relationship between two electrical quantities such as current and voltage, or between two signals.

photo-resist
(PCB) A thin film of light-sensitive material applied directly to the bare copper laminate of a PCB. Once exposed to light and processed chemically, it will resist the etching acid, leaving only the desired circuit pattern.

pipe
(USB) A logical connection between a host and the interface endpoints of a device.

pitch
(General) Refers to the spacing between pins or leads of components and connectors.

pitch
(PCB) A term used to describe the nominal distance between the centers of adjacent features or traces on any layer of a printed circuit board. For example, a part with a 0.1-inch pitch will have leads spaced 0.1 inch center-to-center.

plastic dual inline package (PDIP)
A DIP with a molded plastic body.

plating
(PCB) A process that applies a thin layer of copper to inside of the hole drilled for a pad

or via in a PCB substrate. A plated hole serves as a conducting path between sides or layers of a PCB.

PLCC
(SMD) Plastic leaded chip carrier. A square surface-mount chip package in plastic with leads (pins) on all four sides that bend under the package. A PLCC can be used with a socket.

pMOS
See *p-channel metal-oxide semiconductor*.

polarity
The positive or negative voltage (or charge) at a given point in a circuit, the orientation of a part (such as a polarized capacitor), or at the output of a device such as a power supply or battery.

positive
An object (if referring to a static charge) on the terminal of a voltage source that has a deficit of electrons, or the polarity of a point in a circuit in respect to some other point.

potentiometer
A three-terminal resistor with an adjustable center connection that behaves like a variable voltage divider.

power supply
A device or circuit used to convert one form of electrical current to another. A common type converts AC current to a DC output. Other types can convert one DC voltage level to another.

primary cell
A device that uses up its chemical ingredients to produce electrical energy and therefore can not be recharged.

printed circuit board (PCB)
A method for mounting electronic components that integrates the physical support in the form of a thin sheet of material with the necessary circuit wiring. See also *substrate*.

PROM
Programmable read-only memory. A form of read-only memory that can be electrically programmed once by disruption of internal connection points.

propagation delay
In logic devices, the amount of time between when the input is stable and valid to when the output is stable and valid. The smaller the propagation delay, the faster a logic device can respond to input changes. Also known as *gate delay*. See also *latency*.

QFN
(SMD) Quad, flat, no-lead package.

QFP
(SMD) Quad flat pack, a package type.

QSOP
(SMD) Quarter small-outline package.

radio frequency (RF)
Any AC signal used for wireless communications or inadvertently emitted by a circuit (see *radio frequency interference*).

radio frequency interference (RFI)
The unintentional generation of RF signals by a circuit or device that can interfere with the operation of another another circuit or device not physically connected to the source of the interference.

RAM
See *random access memory*.

random access memory (RAM)
A form of binary data read/write memory that allows date storage and data retrieval to or from any location (address) in the memory space in any order.

RC
See *resistor-capacitor network*.

resistor-capacitor (RC) network
Refers to a network comprising of resistors and capacitors whose values have been chosen for a particular RC time constant.

reactance
The opposition to current offered by a network of capacitors and inductors in the form of a counter-electromotive force. Reactance is expressed in ohms.

read-only memory (ROM)
A form of nonvolatile memory that only allows its contents to be read, not written. ROM types include the basic design wherein the contents are part of the IC fabrication, and variants that can be erased and reprogrammed by using UV light or a specific voltage on a certain pin of the device.

rectifier
A solid-state device that allows current to flow in only one direction. Typically used to convert alternating current to pulsed direct current. See also *diode*.

relay
An electromechanical device that is basically an electrically operated switch. A relay consists of an armature that is moved by an electromagnet to operate one or more switch contacts.

request
(USB) Sent by a host to a device to request data or have the device perform an action.

resistance
A measure of the opposition to electrical flow represented by the symbol R and measured in ohms.

resistor
An electrical component used to restrict the amount of current that can flow through a circuit. Available in many formats, such as fixed, tapped, and variable.

resonance
A condition in an LC circuit that occurs when the inductive reactance (XL) equals the capacitive reactance (XC) at a particular frequency. See also *resonant circuit*.

resonant circuit
A resonant, or tuned, LC circuit that combines an inductor and capacitor to create a circuit that is responsive to a particular frequency. Depending on the configuration, the circuit can have a high or low impedance at the resonant frequency and either pass or block the input signal. If a resistive component is used, it is referred to as an *LRC circuit*.

response time
The time required for a sensor to respond to a step change in the input, or for a circuit to sense a step change and generate a response output.

RF
See *radio frequency*.

RFI
See *radio frequency interference*.

rheostat
A two-terminal resistor with an adjustable center connection and only one end connection. Similar to a potentiometer with the

wiper terminal connected to one of the end terminals.

RI
RS-232 signal: Ring Indicator.

ripple
A small amount of residual AC voltage present in the output of a DC power supply. Most power supplies are rated for some maximum amount of residual ripple.

RMS
See *root mean square*.

ROM
See *read-only memory*.

root mean square (RMS)
The value of an AC sine wave that is equivalent to some DC value in terms of power. For a sine wave, the RMS value is 70.7% of the peak AC value.

RTS
RS-232 signal: Request to Send.

RxD
RS-232 signal: Received Data.

SBGA
(SMD) Super ball-grid array.

Schottky diode
A diode made using a Schottky-barrier junction, a type of metal-semiconductor junction, rather than the P-N junction used by conventional semiconductor diodes. Schottky diodes exhibit high switching speeds and a low forward voltage drop.

SCLK
SPI Signal: Serial Clock.

SCR
See *silicon-controlled rectifier*. See also *thyristor*.

secondary cell
A device whose chemical ingredients convert chemical energy to electrical energy and in which the chemical reactions can be reversed when the device is recharged.

semiconductor
Solid crystalline substances that exhibit varying degrees of electrical conductivity depending on chemical alterations or external conditions. Examples include silicon, germanium, and gallium arsenide. These materials are the foundational material for diodes and transistors.

serial
A form of data transfer wherein bits are sent and received one at a time over a single conductor or channel. Can also refer to a sequence of action performed in a specific order with no other actions. See also *serial interface*.

serial interface
An interface in which data is sent in a single stream of bits, usually on a single wire-plus-ground, wire-pair, or single wireless channel. See also *serial*.

serial peripheral interface (SPI)
A three-wire serial interface developed by Motorola.

series circuit
A circuit with two or more loads (or sources) connected end to end, resulting in only one current path. All components in a series circuit have the same current but can have different voltages.

short circuit
An unintentional low-resistance conductive path between two points in a circuit.

siemens (S)
The unit of measurement for conductance. It is the inverse of ohms (S = 1/ohm). Named for German engineer Ernst Werner von Siemens.

signal generator
An instrument that utilizes an oscillator to generate AC signals whose frequency and amplitude can be varied.

signal-to-noise ratio (SNR)
The ratio of the amplitude of a desired signal to the amplitude of unwanted spurious noise signals in communications channel. SNR is usually expressed in dB, and the larger the number (the greater the ratio), the better.

silicon-controlled rectifier (SCR)
A three-terminal type of semiconductor switch, controlled by a gate terminal. Once the device is placed into a conducting state, it will remain in that state until the voltage drop across the anode and cathode terminals reaches zero (or close to it). An SCR will conduct current in only one direction and can be modeled as a switched rectifier. See also TRIAC and *thyristor*.

single-sided board
A printed circuit board with a copper layer on only one side and (typically) no plated-through holes.

SMD
See *surface-mount device*.

SMT
See *surface-mount technology*.

SNR
See *signal-to-noise ratio*.

snubber
A circuit or device that is used to suppress voltage transients.

SOIC
(SMD) Small-outline integrated circuit.

solar cell
A photovoltaic device that converts light energy to electrical energy. Also called a *photovoltaic cell*.

solder
A low-melting-point metal alloy, often composed of tin and lead, that is used to join metallic parts such as the leads of electronic components. New formulations are becoming common that do not contain any lead, but are made from various ratios of tin, copper, silver, bismuth, indium, zinc, antimony, and traces of other metals.

solder mask
(PCB) A coating applied using a silkscreen process that prevents solder from flowing onto any areas where it's not desired and also prevents short-circuit *bridges* from forming between closely spaced traces, pads, or vias.

solenoid
A device that uses electromagnetism to produce a linear (nonrotary) mechanical operation, typically involving a movable slug in the center of a coil of wire.

solid state
Refers to active components or circuits containing active components that use semiconductor materials in their fabrication and operation. See also *semiconductor*.

SOT
(SMD) Small-outline transistor.

SPDT
Single-pole/double-throw. A switch or relay with one conductor (the pole) and two possible positions (throw) for the conductor.

SPI
See *serial peripheral interface*.

SPST
Single-pole/single-throw. A switch or relay with one conductor (the pole) and one possible position (throw) for the conductor.

SRAM
Static RAM. A type of random-access memory that does not use a clocked signal to maintain its contents, instead relying only on voltage. See also *DRAM*.

SS
SPI Signal: Slave Select (active low).

SSOP
(SMD) Shrink small-outline package.

static electricity
A stationary electrical charge on an object created by some type of mechanical transfer (e.g., air movement, water vapor movement, suspended particles in the air, Wimshurst machine, Van de Graaff generator, removing a sweater, shuffling shoes on a carpet on a dry, cold day).

substrate
(PCB) A thin phenolic or fiberglass sheet of material on which a PCB is built.

subtractive process
A PCB fabrication process that involves removing unwanted copper laminate from a PCB substrate using either a silkscreen pattern mask or a photoresist material and acid etchant.

surface-mount device (SMD)
An electronic component that mounts on the surface of a printed circuit board without the need for leads protruding through the PCB.

surface-mount technology (SMT)
Refers to methods used to produce circuit boards that utilize surface-mounted devices. See also *surface-mount device*.

terminal
A connection point for electrical components, PCBs, connectors, or a device or instrument.

through hole
(PCB) A method for mounting components on a printed circuit board (PCB) in which the pins on the components are inserted through holes in the PCB and soldered in place on the opposite side.

thyristor
A general classification of semiconductor devices that behave as electronic switches, (e.g., DIACs, SCRs, and TRIACs).

tin whiskers
(PCB) Conductive microscopic, hair-like crystals that emanate spontaneously from pure tin surfaces. Tin whiskers are often found on electroplated tin surfaces, although a tin-lead (SnPb) alloy plating can reduce the occurrence of whiskers. Also called *metal whiskers*.

tolerance
The degree of deviation, expressed as a percentage, that defines the allowable variance in a measurement or the value of a component.

trace
(PCB) The conductive path between two points on a PCB. Also called a *track*. See also *trace width*.

trace width
(PCB) Varies according to the desired current capacity and the thickness of the copper

laminate. It is somewhat analogous to the diameter of a wire in this respect.

track
(PCB) See *trace*.

transducer
A device used to convert input energy of one form into output energy of another form.

transformer
An electromagnetic device used to transfer electric energy from one circuit to another via inductively coupled wire coils. A transformer can be used to change the voltage of alternating current, by either increasing it (step-up) or decreasing it (step-down). Some transformers are designed as an isolation component to transfer current without a change in voltage.

transistor
A basic semiconductor (solid-state) device that modulates current flow between two terminals (collector and emitter) relative to the current applied to a third terminal (the base). See also *bipolar junction transistor* and *field-effect transistor*.

TRIAC
Abbreviation for *triode for alternating current*. A type of thyristor that will conduct current in either direction. As with an SCR, once the device is placed into a conducting state, it will remain in that state until the voltage drop across the anode and cathode terminals reaches zero (or close to it).

TSOC
(SMD) Thin small-outline C-lead.

TSOP
(SMD) Thin small-outline package.

TSSOP
(SMD) Thin shrink small-outline package.

TTL
Transistor-transistor logic.

TxD
RS-232 signal: Transmitted Data.

Universal Serial Bus (USB)
A data-communication standard developed as a faster replacement for the RS-232 interface, to allow easy connection of peripheral devices (e.g., digital cameras, scanners, keyboards, and mice) to a host computer.

upstream
(USB) Looking back toward the host from the perspective of the hubs and devices in a USB network.

USB
See *Universal Serial Bus*.

VA
Volt ampere(s).

Vcc
Used to designate the V+ source for a BJT or BJT circuit. Also used with ICs. Can be used to distinguish between multiple voltage in a circuit (e.g., Vcc and Vdd are positive source voltages but may not be the same voltage).

Vdd
Used to designate the V+ source for a FET or FET circuit. Also used with ICs. Can be used to distinguish between multiple voltage in a circuit (e.g., Vcc and Vdd are positive source voltages but may not be the same voltage).

very large-scale integration (VLSI)
Refers to an IC fabrication technology used to create ICs with very complex circuits and high active component counts. Applies to devices with up to a million unique

transistors and other circuit elements. See also *large-scale integration*.

via
(PCB) A plated hole that is used to connect layers on a double-sided or multi-layer PCB.

VLSI
See *very large-scale integration*.

volt
The unit of measurement for electrical potential between two points in a circuit. Formally stated, an electrical potential of 1 volt will push 1 ampere of current through a 1-ohm resistive load. Named for Italian physicist Alessandro Volta, who invented one of the first batteries.

volt-ampere (VA)
The voltage times the current feeding into an electrical load. In a direct current system or for resistive loads, the wattage and the VA are the same. In a reactive circuit involving inductors and capacitors, the VA specification will be greater than the average wattage due to phase differences between voltage and current.

volt-ohm meter (VOM)
A (now mostly obsolete) type of passive or vacuum-tube instrument with the ability to measure both voltage and resistance. VOM devices have been largely replaced by DMM instruments.

voltage
The difference in the electrical potential between any two points in a circuit. See also *electromotive force*.

voltage divider
A series resistance circuit that presents a voltage at the junction of two resistors that is proportional to the ratio of the resistor values.

voltage drop
The difference in potential between two points in a circuit as a result of current flow.

voltage regulator
A device or circuit that accepts input current with varying voltage or ripple, and generates an output current at a constant voltage. The voltage and current capability of the output may both be variable.

voltmeter
An instrument used to measure voltage.

VOM
See *volt-ohm meter*.

Vss
Sometimes used to designate the V- (ground) return for a circuit.

watt (W)
Unit of measurement for power, where 1 watt equals the work done when 1 joule is used per second. Named for British engineer and inventor James Watt.

watt-hour (Wh)
Unit of measurement defined as the energy consumed at the rate of 1 watt (W) over a period of 1 hour. Equivalent to 3,600 joules.

Zener diode
A semiconductor device that will conduct current when the reverse voltage across the device exceeds a specific level. A Zener diode is often used as simple voltage regulator, or as a reference source for a linear regulator.

zero insertion force (ZIF)
A type of IC socket that clamps the leads of an IC between thin metal fingers after insertion when a small lever is operated on the

side of the socket. Primarily used when a part, such as an EPROM, must be removed and reinserted multiple times during development activities.

ZIF

See *zero insertion force*.

Index

Symbols
1000BASE-T Ethernet, 330
100BASE-T Ethernet, 330
10BASE-T Ethernet, 329
32-bit microcontrollers, 262
4000 series logic devices, 257
    (see also CMOS)
4N25 opto-isolator, 275
555 timer, datasheet walk-through, 206
7400 series TTL logic devices, 256
8-, 16-, or 32-bit microcontrollers, 261
8-bit microcontrollers, 262
802.11 standard, 338
    modules, 340
802.11b/g, 338
802.15.4 standard, 344
8P8C connectors, 169
~ (negation) in Boolean algebra, 250

A
A (see amperes)
AC (see alternating current)
AC (alternating current), 413
    conversion to DC by power supplies, 121
    conversion to DC via rectification, 212
    frequency, voltage, and power, 416
    phase, 417
    relationships among voltage, current, and power in AC circuits, 398
    startref =ix_ACconcepts, 418
    TRIACs for AC power control, 226
    waveforms, 414
AC ripple, 450
AC wall socket, 107
active components, 203
    datasheets, how to read, 204
    diodes and rectifiers, 212
        startref =ix_activecompdiorect, 220
    electrostatic discharge (ESD), 209
    heatsinks, 226
    integrated ciruits (ICs), 228
        startref =ix_activecompIC, 233
    packages
        using different package types, 212
    packaging
        overview, 210
        surface-mount parts, 211
        through-hole parts, 211
    passive components versus, 203
    SCR and TRIAC devices, 225
    startref =ix_activecomp, 234
    transistors, 220
        startref =ix_activecomptrans, 225
ADCs (see analog-to-digital converters)
additive process, 352
adhesives, 15, 26
    glues, epoxies, and solvents, 26
    special-purpose, 30
    utilizing bonding, 28
    working wih wood and paper, 28
    working with metal, 29
    working with plastic, 28
adjustable wrenches, 63
aliasing, 389
alkaline batteries, 109
    coin-style, 115
    discharge curves for alkaline versus NiMH batteries, 109
Allen drives, 22
Allen wrench, 19, 37
alligaator clips, 38
alphanumeric coding system for coin-style batteries, 113

alternating current (AC), 6
American wire gauge (AWG), 35
amperes (A), 7, 402
amplifiers
    transistors as, 220
amplitude, 416
amplitude modulation (AM) radio, 335
analog, 267
analog interfaces, 283, 293
    from analog to digital and back, 284
    hacking analog signals, 291
    interfacing with an analog world, 284
analog oscilloscopes, 387
analog-to-digital converters, 229, 284, 289
    for SPI and I2C interfaces, 306
    four-channel 16-bit ADC module, 289
    key points to consider, 290
    low-cost ICs with SPI or I2C interfaces, 289
    packages, 289
    resolution and quantization errors, 286
AND and OR operations in Boolean algebra, 250
AND and OR relay logic circuits, 246
AND gate, 250
ANSI
    lithium coin-style batteries, 115
ANSI (American National Standards Institute)
    UTS and bolt and screw sizes, 17
ANSI/ASME sizes
    hex wrenches, 38
antistatic foam, 210
Application Examples section (datasheets), 205
arcing, 177
    contact arcing in relays, 241
Arduino, 296
    capacitors on an Arduino Leonardo, 198
    DDS module to generate waveforms, 391
    hardware frontend for oscilloscope running on Linux, 386
    units based on AVR family of microcontrollers, 259
ARM Cortex-M3 processor design, 260
ARM microcontrollers, 259
armature relays, 235
ASCII (American Standard Code for Information Interchange), 311
assembly language, programming in, 260
asssembler, 260
asynchronous data communications, 296
asynchronous interfaces, 298
    RS-232, 311
AT&T Bell Laboratories, 220

atoms, 2
    conductive properties of, 3
    hydrogen and copper, representation of, 2
    stickiness of valence shell electrons, 7
automatic wire strippers, 35
AWG (American wire gauge), 35
AWG (American Wire Gauge) standard, 147
    common solid wire gauges, 147
    common stranded wire gauges, 148
axial packages, 178
    for diodes and rectifiers, 218

B

backshells, 171
    DB-9 connector with backshell, 162
Bakelite, 364
ball drivers, 37
ball-end wrenches, 60
ball-grid array (BGA), 210
bands, 336
bandwidth, 334
barrier terminal blocks, 157
batteries, 107
    alkaline, 109
    battery packages, 107
        common battery sizes, 107
    battery-assisted solar power systems, 128
    calculating how long a battery will last, 8, 11
    charging from a solar cell module, 127
    circuits, 119
    comparison of voltages and capacities for some battery types, 109
    lead-acid, 113
    lithium, 111
    lithium-ion (Li-ion), 112
    miniature button/coin batteries, 113
    NiCad (nickel-cadmium), 111
    NiMH (nickel-metal hydride), 112
    primary, 108
    rated output voltage, 8
    secondary, 111
    selecting criteria for, 120
    silver-oxide, 110
    soldering wires to, 118
    storage considerations, 116
    using, 117
        battery holders, 117
    zinc-air, 111
baud rate, 312
baud, misuses of, 312

Bell, Alexander Graham, 150
bipolar junction tansistors (BJTs), 220
    schematic symbols for, 221
    TTL devices based on, 253
bit rate, 312
    baud versus, 312
bit rates (I2C), 305
blades (switches), 133
blind rivet tool, 25
blind rivets, 26
Bluetooth, 340
    classes, 341
    data rates, 341
    modules, 341
Bluetooth Low Energy (BLE), 342
    IC components, 343
    modules, 343
bolts, 15
    adjustable wrench, using on a large bolt, 63
    screws versus, 16
bonding, 341
    plastic bonding, 27
    using epoxies, 27
Boolean algebra, 250
breakdown voltage, 195
bridge rectifiers, 215
Bud industries, plastic electronics enclosure, 376
buffers
    for discrete outputs, 278
burrs, 47
butanone, 28
button batteries, 108
bypass diodes, 127

C

C language, 261
C-clamps, 39
C/C++
    compiling on Linux with GCC, 262
cables
    coaxial, 152
    insulation, 149
    multi-conductor, 153
        flex cables, 155
        individual conductors nd twisted pairs, 153
        ribbon cables, 154
    shielding, 152
    twisted-pair, 151
    wires and, 146
cadmium sulphide (CdS) photoresistors, 191

cap-head screws, 20
capacitance
    determinants of, 194
    for capacitors in a parallel network, 422
    for capacitors in series network, 422
    measurement of, 195, 419
    wire-wound resistors, 183
capacitors, 194, 201, 418
    breakdown voltage, 195
    capacitative coupling, 423
    capacitative phase shift, 424
    ceramic, 196
    electrolytic, 196
    in AC circuits
        charge and discharge, 420
    plastic film, 197
    RC circuits, 424
        applications, 425
    series and parallel, 422
    startref =ix_capacitors8, 198
    surface-mount, 198
    tolerance, 176
    types of, 195
    variable, 197
    voltage rating, 177
capacity (batteries), 120
carbon film resistors, 181
    color codes, 192
carbon-composition resistors, 181
cast aluminum closures, 368
cat's whisker (diode), 213
CAT5 Ethernet cable, 152
CdS (cadmium sulphide) photoresistors, 191
center-off switch (ON-OFF-ON), 136
Centronics-style connectors, 158
ceramic capacitors, 196
chain of relays, 240
channels, 268
    ISM bands divided into, 337
chips, 228
chokes, 199
chop saws, 46
circuit breakers, 130
circuit simulation software, 413
circuits
    battery, 119
    current flow in a basic circuit, 5
    equivalent, 408
    power, 8
    resistance in, 9, 148
    single switch controlling multiple circuits, 134

slide and rotary switch, 140
clamp load, 21
clamps, 38
Classic Bluetooth, 340
CMOS (complementary metal oxide semiconductor), 253
    4000 series logic devices, 255
    4000 series numbers for ICs, 253
    closing gap with TTL logic, 253
        CMOS and TTL applications, 256
    sensitivity to static of CMOS parts, 255
    use in modern microprocessors and microcontrollers, 259
coaxial cable, 152
coaxial DC power connectors, 124
coil overheating in relays, 242
coils, 199, 200
    use by relays, 243
coin batteries, 108, 113
    alphanumeric coding system for, 113
    alternative coding system for, 114
    diameter values, 114
    numeric IDs, manufacturer-assigned, 114
    shelf life, 116
    silver-oxide cells in, 110
    size codes, 115
    type codes for, 114
combination driver sets, 32
combination wrenches, 65
combinatorial logic, 251
common-mode rejection, 150
components lists, 481, 491
conductivity
    and resistance, 403
    of atoms, 3
conjunction (AND), disjunction (OR), and negation in Boolean algebra, 250
connectors, 145, 157
    assembling, 169
        backshells, 171
        crimped terminals, 170
        Ethernet connectors, 173
        IDC connectors, 172
        soldered terminals, 169
    coaxial DC power connectors, 124
    crimped, 160
    on PCs, 157
    PCB connector placement, 355
    soldered, 159
    startref =ix_connectors, 174
    termination, 157
    terminal blocks, 157
    types of, 161
        DB connectors, 161
        Ethernet connectors, 169
        jacks and plugs, 2.5 and 3.5 mm, 166
        PCB edge connectors, 163
        pin headers and sockets, 164
        USB connectors, 167
    using crimped terminals, 36
contact arcing, relays, 241
contact bounce, 143
    in relays, 242
contact rocker, 135
contactors, 235, 236
    heavy-duty, in air conditioners, 244
contacts
    correct crimping tool for, 36
coulomb, 419
countersunk holes, 20
CPLD (complex programmable logic device), 257
CPUs (central processing units), 229
    first computer CPU built from ICs, 252
crimped contact connectors, 160, 165
    asssembling, 170
crimping tools, 36, 160
    for RJ45 connectors, 173
    using for crimped contact assembly, 171
cross-over cables, 315
cross-talk, 150
cross-threading, 59
CTS signals, 314, 315
curing, 27
current, 398, 401
    alternating current (AC), 6
    definitions in electronics, 5
    determining how much a motor draws, 272
    direct current (DC), 6
    in equivalent circuits, sources of, 408
    in parallel resistance networks, 407
    measurement, 402
    measuring with a DMM, 384
    relationship with voltage and resistance (Ohm's Law), 8
    sink buffer for discrete output, 278
    sinking and sourcing in discrete interfaces, 278
    source buffer for discrete output, 279
    voltage versus, 7
    wires, capacity for carrying, 148
current flow, 4
    in a basic circuit, 5

in diodes, 218
versus current, 5
curve tracers, 213
cut-off saws, 46
cyanoacrylates (super-glue), 26, 30

D

data books, 204
data communication interfaces, 295, 347
    basic digital communication concepts, 296
    Ethernet network communications, 328
        Ethernet ICs, modules, and USB converters, 333
    other data communication methods, 345
    parts and manufacturers, 296
    RS-232, 310
        components, 316
        DTE and DCE, 314
        handshaking, 315
    RS-232 signals, 313
    RS-232 versus RS-485, 320
    RS-485, 317
        components, 319
        line drivers and receivers, 318
        multi-drop, 318
        signals, 318
    serial and parallel, 297
    SPI and I2C, 299
    synchronous and asynchronous, 298
    USB, 320
        classes of devices, 322
        components, 327
        connections, 322
        data rates, 323
        device configuration, 325
        endpoints and pipes, 325
        hacking, 327
        hubs, 324
        terminology, 321
    wireless communications, 334, 345
        2.45 GHz short-range, 337
        802.11, 338
        Bluetooth, 340
        ISM radio bands, 336
        Zigbee, 344
data communications equipment (DCE), 314
data rates (802.11), 338
data sink, 314, 321
data source, 314, 321
data terminal equipment (DTE), 314
data transmission rate (RS-232), 312
datasheets, 178, 204
    collecting, 208
    for small-signal diodes, 214
    importance of, 234
    organization, 204
    walking through (example), 206
DB connectors, 161
    backshells, 171
    panel-mount or shell types, 162
    pins, 162
DC (see direct current)
DC (direct current)
    power ratings for passive components, 177
    power supplies, 121
        bench DC power supply, 125
        modular and internal, 125
        wall plug-in power supplies, 122
    rectifier input, 216
    relationships among voltage, current, and power in DC circuits, 398
DCD signals, 314
DCE (data communications equipment), 314
DDS (direct digital synthesis) devices, 229
debouncing, relay input, 242
deburring tools, 47
decimal notation, 283
descriptors, 321
devices, 167
devices (Ethernet), 328
devices (USB), 321
    configuration, 325
diagonal wire cutters, 34
diameter-pitch (screw and bolt sizes), 18
dies, 36, 45, 171
differential signaling, 296
digital, 267
    defined, 268
digital devices, 229
digital logic, 249, 250
    (see also logic)
digital multimeter (DMM)
    checking a power supply prior to use, 124
digital multimeters (DMMs), 381, 396
    examining the voltage on a sensor, 272
    resolution and accuracy, 383
    using, 383
        measuring current consumed by PCB circuit, 384
digital oscilloscopes, 388
digital-to-analog converters, 288, 290

for SPI and I2C interfaces, 306
hacking tips, 292
low-cost ICs with SPI or I2C interfaces, 290
output voltage range, 288
resolutions, 290
using, points to keep in mind, 291
DIO (digital or discrete input/output), 268, 268
    discrete I/O port expansion chips, SPI and I2C, 306
    pins on CY7C68013A microcontroller, 269
    using logic-level translation, 280
diodes, 204, 212, 219
    axial package types, 218
    exotic, 217
    LEDs (light-emitting diodes), 216
    packaging, 210
    part numbers beginning with 1N, 213
    ratings, 214
    small-signal, 213
        specifications for a 1N4148 diode, 214
    surface-mount packages, 219
    Zener diodes, 217
DIP (dual-inline package), 229
direct current (DC), 6
direct digital synthesis (DDS) devices, 229
discrete, 267
    defined, 268
discrete control interfaces, 267, 282
    discrete inputs, 272
        optical isolators, 274
        using a pull-up or pull-down resistor, 273
        using active input buffering, 274
        using relays with inputs, 274
    discrete interface, 268
    discrete interface applications, 269
    discrete outputs, 277
        buffering, 278
        current sinking and sourcing, 278
    hacking a discrete interface, 270
    interface-level translation components, 282
    logic-level translation, 280
        using BSS138 FET, 280
        using NTB0101 translator, 281
        using TXB0108, 281
discrete interface, 267
disjunction in Boolean algebra, 250
displays (I2C and SPI peripherals), 307
distributed systems, 295
DO prefix, part numbers for rectifiers with axial leads, 218
doping, 216

double-pole (DP) switches, 135
double-pole double-throw (DPDT) switches, 135
double-pole single-throw (DPST) switches, 135
double-throw (DT) switches, 135
downstream, 321
DPDT (double-pole double-throw) relays, 236
DPST (double-pole single-throw) relays, 236
Dream Cheeky USB missile launcher for Linux, 328
drill bits, 44
    for printed circuit boards, 44
drilling out the head of stuck fasteners, 69
drills, 43
    rotary tools versus, 42
    step drill, 48
drive types (screws and bolts), 19, 22
    combination driver set for, 32
    socket and hex drivers, 36
driver ICs, 316
driver relays, 237, 240
DSR signals, 314
DT (double-throw) switches, 135
DTE (data terminal equipment), 314
DTR signals, 314
dual-inline package (DIP), 229
dummy loads, 183
dusty or dirty environment, packaging for, 378

E

ECL (emitter-coupled logic), 253
edge connectors, 163
    with IDC termination, 164
EIA-232 (see RS-232)
electric screwdrivers
    battery-powered, 58
electrical considerations when working with logic components, 264
Electrical Specifications section (datasheets), 205
electrician's hammer, 33
electricity
    common, everyday use of the term, 2
    electric charge, 3
    electric current or current flow, 4
    formal definition, 2
electrolytic capacitors, 195, 196
    tantalum, 197
electromagnetic force (EMF), 399
electromagnetic interference (EMI)
    minimizing with shielding, 152
electromotive force (emf), 6

electronics
  basic electonics theory, topics in, 397
    AC (alternating current), 413
    current, 401
    ix_electronicsAC, 418
    power, 402
    resistance, 403
    units of measurement, 397
    voltage, current, and power, 398
  basic electronics theory, topics in
    equivalent circuits, 408
electrons, 2
  flow of
    water-flowing-in-a-pipe analogy, 1
  orbital shells, 3
electrostatic discharge (ESD), 209
  controlling when working with logic components, 264
  using prevention techniques with logic components, 264
emitter-coupled logic (ECL), 253
end cutters (nippers), 34
endpoints (USB), 321, 326
energy density
  silver-oxide batteries, 111
enumeration, 321
environment, considering for enclosures, 378
epoxide functional group, 27
epoxies, 26
  applications for, 27
  curing, 27
  defined, 27
  using with metal, 29
  using with plastics, 29
epoxy resin, 27
EPROM memory, 307
equivalent circuits, 408
  applications, 412
  lumped-parameter elements, 409
  Thévenin, 409
  voltage and current sources, 408
Ethernet, 328
  basics, 328
  example network using a hub and a switch, 332
  ICs, modules, and USB converters, 333
  network speeds, 329
Ethernet cables, 151
  assembling, 173
  connectors for wire types, 173
  crimping tools for, 36
  kits for, 174
  UTP multi-conductor cable, 154
Ethernet connectors, 169
exclusive OR (XOR), 251
exotic diodes, 217
extreme hacking, 375
extruded aluminum enclosures, 368

F
families (of logic devices), 252
farad (F), 419
farads (F), 195
fasteners, 15
  adhesives, 15, 26
    glues, epoxies, and solvents, 26
    special-purpose, 30
    working with metal, 29
    working with plastic, 28
    working with wood and paper, 28
  bonding, 27
  reference guides for, 22
  rivets, 25
  screws and bolts, 16
    drive types, 19
    head styles, 20
    selecting, 21
    self-tapping screws, 25
    sizes of, 17
    suggested inventory of screws, 17
    washers, 23
  working with, tool techniques, 56
    adjustable wrenches, 63
    dealing with stubborn fasteners, 68
    hex-head fasteners and socket wrenches, 61
    hex-socket-head fasteners and hex wrenches, 59
    rivets, 66
    screwdriver sizes and types, 56
    self-tapping screws, 59
    wrenches (spanners), 64
field-effect transistors (FETs), 221, 222
  BSS138 FET, 280
  metal-oxide-type transistors, CMOS based on, 253
filters
  passive, 436
fixed-terminal connectors, 159
fixed-value resistors, 181
flash memory, 307
flash memory sticks (thumbdrives), 323

flashlights, 4
flat washers, 23
flat-end (straight-shaft) wrenches, 60
flat-head screws, 20
flat-pack styles (ICs), 229
flex cables, 155
flexure points, 157
Fluke two-part plastic instrument enclosure, 376
flush wire cutters, 34
FM (frequency modulation) radio, 335
foiled twisted-pair (FTP) cables, 151
footprint (PCB parts), 355
forward bias, 214
forward current, 214
    LEDs (light-emitting diodes), 217
forward voltage, 214
    LEDs (light-emitting diodes), 217
FPGA (field-programmable gate array) devices, 257
frames (Ethernet), 329
Franklin, Benjamin, 6
freeze spray, 69
frequency, 6
    AC waveforms, 416
frequency modulation (FM) radio, 335
frequency ranges (bands), 336
frequency shift keying (FSK), 335
full-duplex, 300
full-wave rectifiers, 215
function (USB), 321
function generator, 391
Functional Description/Diagram section (datasheets), 205
fuses, 128
    differences between types, 129
    printed circuit board-mounted, 129
    selecting, 130
    sizes and ratings, 129

G

GAL (general array logic) devices, 257
gauge-pitch, 18
Gaussian frequency-shift keying (GFSK), 336
GCC toolchain, 262
gel cells, 113
general array logic (GAL) devices, 257
General Specifications section (datasheets), 205
Gerber files, 357
glass TTY, 314
glues, 26

cyanoacrylate, 30
    using with wood and paper, 28
grinders, 42
    right-angle, 42
grinding off the head of stuck fasteners, 69
grip depth (rivets), 66, 67

H

half-duplex, 300
ham radio
    coils, 200
hand saws, small, 45
handshake loop-back, 316
handshaking, 315
HC family of TTL-compatible devices, 254
head styles (screws and bolts), 20
    combination driver set for, 32
heatsinks, 226
    for power transistors, 226
    freestanding, caution with, 228
    heavy-duty, 228
hemostats, 38
henry (H), 199
hex drivers, 36
hex drives, 22
hex head bolts, 20
hex keys (see hex wrenches)
hex socket drives, 19
hex wrenches, 37
    working with, techniques, 59
        ball- and flat-end hex wrenches, 60
        using both ends, torque and, 59
HID class, 323
high-power packages
    resistors, 183
high-Z (or high-impedance) state, 301
hook-up wire, 147
hops, 341
host (USB), 321, 322
hosts, 167
hosts (Ethernet), 328
hubs (Ethernet), 331
hubs (USB), 321, 322, 324
humidity
    resistors sensitive to, 191
humistors, 191

I

I2C, 302

7- or 10-bit addresses, 304
7-bit addresses, reserved, 305
bidirectional signal lines, SDA and SCL, 303
Ethernet ICs and, 333
peripheral devices, 306
    ADCs and DACs, 306
    discrete I/O port expansion chips, 306
    displays, 307
    memory devices, 307
    other, 309
specification and user guide, 303
speeds (or bit rates) for I2C interfaces, 305
using with an AVR microcontroller, 304
IDCs (Insulation Displacement Connectors), 158
    assembling, 172
IEC 60086-3 alphanumeric coding standard for coin-style batteries, 113
impedance, 435
in-line connectors, 165
incandescent lamps, 292
Indium (PIN diodes), 217
inductance
    wire-wound resistors, 183
inductors, 199, 243, 428, 437
    chokes, 199
    coils, 200
    inductive kick-back, 431
    packages, 201
    series and parallel, 430
    transformers, 200
    variable, 200
infrared (IR), data communications by, 345
inner-tooth lock washers, 23
insulation
    removing for crimped contact assembly, 170
    stripping from wires or cables, 155
Insulation Displacement Connectors (IDCs), 158
integers, 283
integrated circuits (ICs), 228
    acting as current sink or source, 278
    block diagram of CY7C68013A IC, 269
    CMOS and TTL devices, 4000 and 7400 ICs, 253
    Ethernet, 333
    in through-hole packages, 211
    linear and digital types, 229
    logic ICs, origin of, 252
    package types, conventional, 229
    relay driver ICs, 243
    safe removal or installation, 264
    startref =ix_ICs, 233
    surface-mount packages, 230
    interfaces (USB), 321
    internal power supplies, 125
Internet of Things, 295
ions, 3
IP cores, 259
irrational numbers, 283
ISM bands, 336
    802.11 standards, 338
    Bluetooth, 340
    Zigbee, 344
ISO 68-1, and ISO 262 standard (screw and bolt sizes), 18
ITU-R (International Telecommunication Union-Radio), 336

J

jacks
    2.5 and 3.5 mm, 166
Japanese Industrial Standard (JIS) cross-drive screws, 57
jeweler's loupe, 51
jeweler's saw, 45
JIS (Japanese Industrial Standard) cross-drive screws, 57
JIS drive type, 22
Joint Electron Device Engineering Council (JEDEC), 208
    axial lead diode packages, 218
    diode part numbers, 213
    package types for electronic components, 211
    transistor package types, 223

K

Kirchkoff's circuit laws, 412
kits
    for enclosures, 370
    LEDs (light-emitting diodes), 233

L

lamps
    bench lamps with magnifier lenses, 51
large-scale integration (LSI), 252
laser diodes, 217
lasers, 345
    safety with, 346
latching data, 298
latching relay circuits, 245
LCD displays, 307

lead-acid batteries, 108, 113
   secondary, storage of, 116
LEDs (light-emitting diodes), 213, 216
   in an optical isolator, 276, 277
   package types, 219
   types of, 216
   using as discrete output buffer, 279
   voltage drop, 214
level shifting, 274
level shifting (logic), 253
level translators (RS-232), 316
light-dependent resistors (LDRs), 191, 292
light-emitting diodes (see LEDs)
light-sensitive resistors, 191
lightning, 7, 209
limit switches
   use of discrete interfaces, 270
line drivers and receivers (RS-485), 318
linear devices, 229
linear power supplies, 122
lineman's pliers, 33
link key, 341
lithium batteries, 111
   coin-style, size codes, 115
lithium-ion (Li-ion) batteries, 112
   storage of, 116
lithium-ion batteries, 108
load, 8
loads (machine screws), 21
lock washers, 23
logic, 249, 265
   basics of, 250
   logic building blocks, 4000 and 7400 ICs, 253
      4000 series of CMOS devices, 255
      7400 series of TTL logic devices, 256
      closing the TTL and CMOS gap, 253
      CMOS and TTL applications, 256
   logic equation as circuit diagram, 251
   logic families, 252
      CMOS and TTL, by year of introduction, 254
   microprocessors and microcontrollers, 259
      programming microcontrollers, 260
      selecting a microcontroller, 262
      types of microcontrollers, 261
   origin of logic ICs, 252
   programmable logic devices (PLDs), 257
   working with logic components, 263
      electrical considerations, 264
      electrostatic discharge control, 264
      physical handling and mounting, 264
      probing and measuring, 263
      selecting logic devices, 264
logic analyzers, 392
   CY7C68013A microcontroller, 268
logic circuits
   using relays, 245
logic-level translation, 280
   BSS138 FET, using, 280
   interface-level translation components, 282
   NTB0101 one-bit (single-channel) translator, 281
   TXB0108, using, 281
lug connectors, 160
lug crimpers, 36
lug-terminal relays, 239
lxardoscope, 386

M

mA (milliamps), 8
MAC (media access control) addresses, 329
magnetic circuit breakers, 130
magnifiers, 50
   bench lamp magnifiers, 51
   bench work holder with, 51
   pocket magnifiers, 51
mAh (milliamp-hour), 8, 109
Mass Storage class, 323
master/slave arrangement, I2C, 304
master/slave arrangement, SPI, 300
   command and response sequence for MAX7317 I/O expander IC, 302
   multiple slave devices with a single master, 302
master/slave communications (Bluetooth), 341
master/slave type bus (USB), 322
mats to prevent ESD damage, 209
maximum grip (rivets), 68
media independant interface (MII), 333
medium-scale integration (MSI), 252
memory devices
   SPI and I2C peripheral devices, 307
mesh networks, 344
message pipes, 326
metal
   using adhesives with, 29
   using C-clamps with, 39
metal case wire-wound resistor, 182
metal film resistors
   datasheet (example), 205
metal, using for packaging, 364, 378

cast aluminum enclosures, 368
extruded aluminun enclosures, 368
sheet metal enclosures, 369
steel cans, 374
metal-film resistors, 182
metal-working tools, specialty, 47
automatic punch, 47
deburring tool, 47
methyl ethyl ketone (MEK), 28, 29
metric system
hex wrenches in metric sizes, 38
screw and bolt sizes, 18
world-wide adoption of, 19
micro USB connectors, 168
microcontrollers
analog-to-digital converters, 289
ARM, sold as IP cores, 259
defined, 259
DIO terminals, 268
PCB with CY7C68013A microcontroller, 268
programming, 260
relay module for use with, 240
selecting, 262
types of, 261
UART functions, 299
USB capabilities, 320
using I2C with, 304
microcontrollers (microcontroller units, or MCUs), 230
microfarads (μF), 195
microprocesors
buying prebuilt modules, 259
defined, 259
transistor elements, 252
microprocessors
buying prebuilt modules, 259
low-cost motherboards using, 260
microscopes, 50
MIL-HDBK-60, 22
milli, 8
milliamp-hour (mAh), 8, 109
milliamps (mA), 8
millifarads (mF), 195
millihenry (mH), 199
mini USB connectors, 168
miniature screwdrivers, 32
MISO signals, 301
modems, 314, 368
modular power supplies, 125
modulation level, 335

MOSFETs (metal-oxide semiconductor field effect transistor), 207, 223
as power transistors, 222
mounting options
for switches, 142
multi-drop configuration, 319
multi-wire planar cables (see ribbon cables)

N

NAND (Not-AND) gate, 250
nanofarads (nF), 195
NASA
soldered connections specifications, 159
needle-nose pliers, 33
negation in Boolean algebra, 250
negative temperature coefficient (NTC) device, 191
negative voltage levels (RS-232 data signals), 311
neutrons, 2
NIC (network interface controller), 329
NiCad (nickel cadmium) batteries
storage of, 117
NiCad (nickel-cadmium) batteries, 111
NiMH (nickel-metal hydride) batteries, 112
discharge curve for NiMH and alkaline batteries, 109
NiMH (nickel-metal hydrite) batteries
storage of, 117
nippers, 34
NMOS (N-type metal oxide semiconductor), 253
nonshorting switches, 143
NOR (Not-OR) gate, 251
NPN- and PNP-type BJTs, 221
NTB0101 logic-level translator, 281
nucleus, 2
null modem cables, 315
null-modem adapters, 315
numbers, integer and real, 283
nuts
selecting right screw or bolt for, 23
suggested inventory of, 17
Nyquist frequency (or Nyquist limit), 286, 389

O

octal sockets, relays, 239
octets, 329
Ohm's Law, 8
Ohm's law, 404
Ohm, Georg Simon, 7

ohms, 7, 9
ON-OFF-ON (center-off) switch, 136
one-time programmable (OTP) PLDs, 257
op amps
    equivalent circuit model for, 412
open-drain lines, 303
operational amplifiers (op amps), 229, 445
    basic circuits, 446
    specialized, 447
optical isolators, 274
    4N25 device, 275
    building your own, 276
    other uses of, 277
    part numbers, 275
    types and packages, 275
    used as discrete input source, 275
    using as discrete output buffer, 279
    using with discrete outputs, 277
OR and AND relay logic circuits, 246
orbital shells, 3
oscilloscopes, 385, 396
    digital, 388
    how they work, 387
    lxardoscope running on Linux, 386
    miniature digital oscilloscope, 385
    prebuilt digital oscilloscopes, 387
    USB digital oscilloscope, 385
    using, 389
        tips and cautions, 390
    xoscope running on Linux, 385
out-of-phase signals, 418
outer-tooth lock washers, 23
oval-head screws, 20

P

P-N junctions, 233
packaging, 363, 380
    building or recycling enclosures, 371
        building plastic and wood enclosures, 371
        repurposing existing enclosures, 375
        unconventional enclosures, 373
    designing for electronics, 375
        device size and weight, 377
        environmental considerations, 378
        Fluke two-part plastic instrument enclosure, 376
        portable electronics enclosure, 376
        wearable personal pedometer, 376
    importance of, 363
    key trade-off considerations, 380
    sources for commercial enclosures, 379
    stock enclosures, 365
        cast aluminum, 368
        extruded aluminum, 368
        plastic enclosures, 366
        sheet metal, 369
    types of, 363
        metal, 364
        plastic, 364
pads, 351
    size of, 362
pairing, 341
PAL (programmable array logic) devices, 258
pan-head screws, 20
panel-mounted pushbutton switch, 139
paper
    using adhesives with, 28
parallel data, 296
    exchange of, 297
    PC parallel interfaces, 298
passive backplane, 260
passive components, 175
    capacitors, 194
        startref =ix_passivecompcap, 198
    chokes, coils, and transformers, 199
    packages, 178
    resistors, 178
        startref =ix_passivecompresist, 194
    startref =ix_passivecomp, 202
    tolerance, 176
    voltage, power, and temperature, 176
passive compponents
    active components versus, 203
passive filters, 436
PC tubing, using as an enclosure, 373
PCI (peripheral component interconnect) risers, 163
peak inverse voltage (PIV), 214
peripherals (USB), 321
phase (AC), 417
Phillips drives, 22
Phillips screw drivers, 19 ??
Phillips screwdrivers
    checking size and fit for a given screw, 57
photo-sensitive diodes, 213
photo-sensitive resistors, 178
photocells, 191
photoresistors, 191
    signal isolator using, 291
phototransistors, 275
    in an optical isolator, 276

photovoltaic power sources, 126
Physical Specifications section (datasheets), 205
pick-and-place machines, 350
picofarads (pF), 195
PIN diodes, 217
pin headers and sockets (connectors), 164
pin-grid array, 210
pins, 165, 268
    numbering for DB-9 connectors, 162
    transistor pin connections, 224
pipes (USB), 322, 326
pitch, 164, 211
pitch (threads), 18
plastic bonding, 27
plastic enclosures, 366
    building, 371
plastic film capacitors, 197
plastic, using for packaging, 364
    creating a mold, 364
    environmental considerations, 378
plastics
    identifying, 29
    using adhesives with, 28
PLDs (see programmable logic devices)
pliers, 32
    improper use with screws or bolts instead of a socket wrench, 62
    with built-in wire strippers, 34
plugs
    2.5 and 3.5 mm, 166
pocket magnifiers, 51
point to point, 330
poles, 135
    single-pole (SP) and double-pole (DP) switches, 135
    slide and rotary switches having multiple poles, schematics, 140
polyethylene
    adhesives and, 28
polyvinyl chloride (see PVC)
pop rivets, 26
portable electronics enclosure, 376, 377
ports, 268
positive temperature coefficient (PTC) device, 191
positive voltage levels (RS-232 data signals), 311
potential, 4
potential difference, 6
potential voltage, 399
potentiometers, 185, 186
    linear, or slider, styles, 187
    multi-turn, 188

single-turn, 186
surface-mount, 189
trimmer, 187
power, 8, 402
    calculating, 402, 404
        in parallel resistance network, 407
    calculating in a DC circuit, 8
    ratings for passive components, 177
    relationships among voltage, current, and power, 398
    trace widths and, on PCBs, 361
power control, use of discrete interfaces, 270
power dissipation
    calculating in series resistance network, 406
power saws, miniature, 46
power sources, 107
    batteries, 107
        alkaline batteries, 109
        battery circuits, 119
        battery packages, 107
        lead-acid batteries, 113
        lithium batteries, 111
        lithium-ion (Li-ion) batteries, 112
        miniature button/coin batteries, 113
        NiCad (nickel-cadmium) batteries, 111
        NiMH (nickel-metal hydride) batteries, 112
        primary batteries, 108
        secondary batteries, 111
        selecting batteries, 120
        silver-oxide batteries, 110
        storage considerations, 116
        using, 117
        zinc-air batteries, 111
    fuses and circuit breakers, 128
    photovoltaic, 126
    power supplies
        output specifications for PC power supply, 402
    power supply technology, 121
        bench DC power supply, 125
        modular and internal DC power supplies, 125
        wall plug-in DC power supplies, 122
    startref =ix_powersources, 132
power supplies, 12
power switching
    using relays, 244
power transistors, 222
    overheating issues, 226
precision components, 176
    metal film resistor, 182

resistors
    color codes, 192
primary batteries, 108
    shelf life, 116
primary winding, 201
printed circuit boards (PCBs), 349, 362
    basics of, 350
    connectors
        pin headers and sockets, 164
        sources of, 166
        terminal blocks, 157
    drill bits for, 44
    edge connectors, 163
    fabricating, 357
        etching process, 358
    fabrication, 352
        etching process (subtractive), 352
        silkscreen mask, 352
        solder mask, 352
    guidelines for layout, 359
        grid, 359
        grid spacing, 360
        location reference, 360
        pad size, 362
        sharp corners, 362
        silkscreen, 362
        trace separation, 361
        trace width for power, 361
        via separation, 362
        via sizes, 361
    history of, 349
    interconnect using pin and socket connectors, 165
    layout, 353
        arranging parts, 354
        creating the silkscreen, 356
        determining dimensions, 353
        generating the Gerber files, 356
        placing components, 355
        route traces on the component side, 356
        route traces on the solder side, 355
    measuring current consumed by a circuit, 384
    pads, vias, and traces, 351
    PCB with CY7C68013A microcontroller, 268
    PCB-mounted DB connectors, 162
    PCB-mounted fuses, 129
    PCB-mounted header connector, ribbon cable with, 158
    PCB-mounted slide switch, 137
    PCB-mounted USB connectors, 168
    PCB-mounted variable inductor, 200

pushbutton switch mounted on, 139
relays, 237
rotary switches mounted on, 138
soldering parts onto, 38
surface-mount components, 351
surface-mount parts, 211
surface-mount versus through-hole, 211
through-hole parts, 211
probe (digital logic), 263
programmable logic controllers (PLCs), 267
programmable logic devices (PLDs), 257
    device types, 257
programming in assembly language, 260
programming tools
    availability for microcontrollers, 262
project box, 367
protocol stack, 332
protons, 2
pull-up and pull-down resistors, 273, 274
pulse, 415
pulse generators, 391
punch, automatic, 47
punch-down blocks, 146
pushbutton switches, 139
    contact bounce, 143
PVC (polyvinyl chloride)
    insulation for wires and cables, 149
PVC tubing, using for enclosure, 378
pyRocket, 328

Q

quadratic mean, 416
quantization, 284
quantization error, 286

R

radial packages, 178
radio communications, 335
    (see also wireless communications)
radio frequency interference (RFI)
    choke, 199
radio kit
    using variable capacitor for tuning, 197
radio signals, 297
radios
    tuning coil, 200
rail-to-rail, 207
ratchet and socket set, 37
rational numbers, 283

RC (resistor-capacitor) circuits, 424
    applications, 425
RC (resistor-capacitor) filters, 125
RC (resistor-capacitor) time constant, 206
reactance, 418, 434
real numbers, 283
    challenge for electronic sensors and digital systems, 284
rechargeable batteries, 111
    use with solar cells, 127
recitifiers
    axial package types, 218
rectifiers, 212, 214
    LEDs (light-emitting diodes), 216
    packages, 215
    packaging, 210
    surface-mount packages, 219
    typical small solid-state rectifier, 215
reduced media independent interface (RMII), 333
reed relays, 236
reeds, 236
reference design, 205
regulators
    transistors as, 220
relay driver ICs, 243
relays, 235-247
    applications, 242
        controlling relays with low-voltage logic, 243
        power switching, 244
        relay logic, 245
        signal switching, 244
    armature relays, 235
    background, 235
    chain of, 240
    contactors, 236
    lug-terminal relays, 239
    packages, 237
    PCB (printed circuit board), 237
    reliability issues, 241
        coil overheating, 242
        contact arcing, 241
        contact bounce, 242
    selecting, 240
    socketed, 239
    using with discrete inputs, 274
requests (USB), 322
resistance, 7, 403
    conductivity and, 403
    in a circuit, 9
    in Ohm's law, 404
    measuring with a DMM, 383
    ohms, 9
    parallel resistance networks, 407
    relationships with voltage and current (Ohm's Law), 8
    series resistance networks, 404
    wire diameter and, 148
resistance networks, 405
resistor arrays, 179
resistor-capacitor (RC) snubbing circuit, 241
resistor-transistor logic (RTL), 252
resistors, 10, 178, 201, 403
    building a voltage divider (example), 10
    color codes and markings, 192
    fixed, 180
        carbon film resistors, 181
        carbon-composition resistors, 181
        high-power resistors, 183
        metal-film resistors, 182
        wire-wound resistors, 182
    in series, 405
    packages, 178
    physical forms, 179
        axial lead and surface-mount resistors, 179
    pull-up and pull-down, using for discrete input, 273
    special-purpose, 191
        humidity sensitive, 191
        light sensitive, 191
        temperature sensitive, 191
    startref =ix_resistors8, 194
    tolerance, 176
    variable, 185
        potentiometers, 186
        rheostats, 185
    voltage rating, 177
resources, 471, 479
RFI (radio frequency interference)
    choke, 199
rheostats, 185
RI signals, 314
ribbon cable cutters, 172
ribbon cables, 154
    IDC connectors, 172
    IDCs (Insulation Displacement Connectors), 158
right-angle grinder, 42
risers (PCI), 163
Ritchie, Dennis, 261
rivets, 25
    in electronics applications, 25

working with, techniques, 66
    installing a blind rivet, 66
RJ11 (telephone) connectors, 174
RJ45 connectors, 169
    crimp tool, 173
RJ45 Ethernet cable making kits, 36
Robosapien, 269
rocker switches, 137
    contact bounce, 143
    generic SPST miniature rocker switch, 137
root mean square (RMS), 416
rotary switches, 138
    caveats, 143
    circuits, 140
rotary tools, 41
    attachments, 41
    drills versus, 42
    grinders with, 42
round-head screws, 20
route traces
    on PCB component side, 356
    on PCB solder side, 355
RS-232, 310
    ASCII character encoding, 311
    components, 316
    data formats, 311
    DTE and DCE, 314
    handshaking, 315
    limitations of, 313
    signal voltage levels, 310
    signals, 313
    speed (or data transmission rate), 312
    versus RS-485, 320
RS-485, 317
    components, 319
    line drivers and receivers, 318
    multi-drop, 318
    RS-232 versus, 320
    signals, 318
RTL (resistor-transistor logic), 252
RTS signals, 314
RTS/CTS handshaking, 316
RxD signals, 314

S

sample rate
    digital-to-analog converter (DAC) output rate, 288
sample resolution, 286
sampling rate, 287

    effects on analog data digitization, 284
sandbox wire-wound resistor, 182
saws
    miniature power saws, 46
    small hand saws, 45
schematic symbols
    for digital logic, 250
    for NPN- and PNP-type BJTs, 221
    for SCR and TRIAC devices, 225
schematics, 455, 466
SCL signals, 303
SCLK signals, 301
screwdrivers, 31
    combination driver set, 32
    improperly using with hex-socket-head fasteners, 59
    JIS versus Phillips, 57
    miniature, 32
    Phillips, 56
    power
        avoiding damage to screws, 58
    sizes and types, 56
screws
    JIS versus Phillips heads, 57
    standard sizes (UTS/ANSI), 56
screws and bolts
    differences between, 16
    drilling or grinding off the head to remove when stuck, 69
    drive types, 19
    head styles, 20
    selecting, 21
        drive type, 22
        loads, 21
        shear strength, 22
    self-tapping screws, 25
    sizes of, 17
    suggested inventory of screws, washers, and nuts, 17
    varieties of, 16
    washers, 23
SCRs (silicon-controlled rectifiers), 225, 355
SDA signals, 303
secondary batteries, 111
    shelf life, 116
    storage of, 116
secondary winding, 201
security
    special screw and bolt drive types, 20
security systems, use of discrete interfaces, 270
self-tapping screws, 16, 22, 25

avoiding, reasons for, 25
situations for use, 21
techniques for working with, 59
semiconductor schematic symbols, 218
semiconductors, 3
  in bipolar junction transistors (BJTs), 221
sense inputs, 126
sensors
  analog, 284
  examining a sensor in a mystery device, 271
  linear potentiometers configured as position sensors, 188
  measurement resolution, 284
sequential logic, 252
serial communications protocols, 299
serial data, 296, 297
serial digital interfaces, 272
serial interface techniques, 296
serial interfaces
  asynchronous, 299
serial peripheral interface (see SPI)
series
  connection in series or parallel, 127
shear slippage, 21
shear strength
  cyanoacrylates, 30
  screws and bolts), 22
sheet metal enclosures, 369
shelf life of batteries, 116, 121
shell DB connectors, 162
shielded twisted-pair (STP) cables, 151, 153
shielding, 152
shorting switches, 143
shunt, 384
signal generators, 391
signal isolators, 291
signal switching
  using relays, 244
Signetics, 209
silicon rubber adhesive, 29
silicon-controlled rectifiers, 355 (see SCRs)
silkscreen for PCB, 356, 362
silver-oxide batteries, 110
  coin-style, 115
sine function, 415
sine waves, 414
single in-line package (SIP) resistor array, 180
single-pole (SP) switches, 135
single-pole double-throw (SPDT) switches, 135
single-pole single-throw (SPST) switches, 135
single-throw (ST) switches, 135

sink (current), 278
slaves (USB), 322
slide switches, 137
  caveats, 143
  circuits, 140
slotted screw drivers, 19
slotted screw drives, 22
slow-blow fuses, 129
small blind rivets, 25
small-outline IC (SOIC) packages, 231
small-scale integration (SSI), 252
small-signal diodes, 213
small-signal transistors, 221
SMD (see surface mount devices)
SMT (see surface mount technology)
snap-action switches, 139
  contact bounce, 143
snap-to-grid feature, PCB layout tool, 359
snubber, relay contact, 241
socket contacts, 165
socket drivers, 36
socket headers, 165
socket wrenches
  using with hex-head fasteners, 61
    direct-drive and ratchet-drive handles, 61
    flush seating, modifying socket for, 62
    too much torque, 62
socketed relays, 239
sockets
  connectors made from, 164
  using with logic components, 264
SOIC (see small-outline IC packages)
solar cells, 126
  efficiency, 127
  small, low-cost solar cell module, 127
soldered connectors, 159
soldering, 70
  connectors with solder terminals, 169
  of wires to batteries, 118
  using hemostats, 38
soldering tools, 49
  in ESD-safe designs, 210
  soldering irons, 49
  soldering stations, 49
solid state concepts, 445
solid-state components, 203
solid-state concepts, 438
solvent welding, 29
solvents
  plastics and, 29
  used in plastic bonding, 28

using to deal with stubborn fasteners, 68
source (current), 278
spade lug crimping tool, 36
spade lugs, 161
spanners, 56, 64
SPDT (single-pole double-throw) relays, 236
SPDT (single-pole double-throw) switches, 135
specs, 204
SPI (serial peripheral interface, 300
SPI (serial peripheral interface)
    clock polarity options, 301
    command and response sequence for MAX7317 I/O expander IC, 302
    data transfer timing diagram, 301
    Ethernet ICs and, 333
    master and slave devices, 301
    multiple slave devices with a single master, 302
    peripheral devices, 306
        ADCs and DACs, 306
        discrete I/O, 306
        displays, 307
        memory devices, 307
        other, 309
    signal lines, 300
    specification documents, 300
split-ring lock washers, 23
SPST (single-pole single-throw) relays, 235
SPST (single-pole single-throw) switches, 135
    generic SPST miniature rocker switch, 137
square waves, 415
SRAM (static RAM), 307
SS signals, 301, 302
ST (single-throw) switches, 135
stack (Ethernet and IP protocol), 331
star configuration (Ethernet networks), 169
steel cans, using as enclosures, 374
step drills, 48
step-down transformers, 200
step-up transformers, 200
stereo plugs, 167
STM32 ARM Cortex-M3 single-board computers, 260
stock enclosures, 365
    cast aluminum, 368
    extruded aluminum, 368
    plastic enclosures, 366
    sheet metal, 369
    sources for, 379
storage of batteries, 116
storage of components, ESD-safe, 210

straight-shaft wrenches, 60
stranded wire, 146
    AWG gauges, 148
    resistance, 148
stream pipes, 326
stud-mounted rectifiers, 215
substrate, 350
subtractive process, 352
Summary/Overview section (datasheets), 205
super-glue, 26
supercaps, 195
support chipset, 259
surface mount devices (SMD), 211
surface mount technology (SMT), 211
surface mounted components, 351
    active components, 211
    capacitors, 198
    diodes and rectifiers, 219
    fixed resistors, 184
        common packages, 184
    inductors, 201
    integrated ciruits (ICs), 230
    LEDs (light-emitting diodes), 220
    potentiometers, 189
    relays, 238
    resistors
        numerical markings, 193
    transistors, 225
surface-mount technology (SMT)
    soldering station for, 49
surface-mounted components, 31
    resistors, 179
sweep frequency, 390
switches, 133
    caveats, 143
    Ethernet, 169, 331
    one switch, multiple circuits, 134
    pushbutton, 139
    rocker, 137
    rotary, 138
    schematic symbols for common SP and DP switches, 135
    selection criteria, 141
    slide, 137
    slide and rotary switch circuits, 140
    snap-action, 139
    startref =ix_switches, 143
    toggle, 135
    transistors as, 221
    types of, 135
switching power supplies, 122

modular DC power supply, 125
switching time, 212
    diode ratings, 214
switching-mode power supply (SMPS)
    in a modern PC, 402
synchronous data communications, 296
synchronous interfaces, 298
synchronous serial interfaces, 297

T

T-handle hex wrenches, 37
tantalum eletrolytic capacitors, 197
taps, 45
telephone connectors, 174
    crimping tools for, 36
temperature
    cooling stubborn fasteners in order to loosen them, 69
    passive components in a circuit, 177
    resistors sensitive to, 191
terminal (DTE), 314
Tesla, Nikola, 176
test equipment, 381, 396
    advanced, 391
        logic analyzers, 392
        pulse and signal generators, 391
    basic, 381
        digital multimeters (DMMs), 381
        oscilloscopes, 385
    buying used and surplus equipment, 394
        caveats about used equipment, 395
thermal circuit breakers, 130
thermal considerations for packaging, 379
thermal-magnetic circuit breakers, 130
thermistors, 191
Thévenin's theorem, 409
third-hand gadgets, 38
through-hole parts, 211
    DIP packages for integrated circuits, 230
    transistors, 223
thumbdrives, 323
thyristors, 225
time-lag fuses, 129
timers
    555 timer, datasheet walk-through, 206
toggle handle (switches), 133
toggle switches, 134, 135
    center-off switch (ON-OFF-ON type), 136
    contact bounce, 143
    contact rocker or pole, 135
    SPST, SPDT, DPST, and DPDT forms, 136
    typical miniature toggle switch, 134
tolerance, 176
toolboxes, 53
tools, 31
    clamps, 38
    crimping tools, 36
    drill bits, 44
    drills, 43
    grinders, 42
    magnifiers and microscopes, 50
    miniature power saws, 46
    pliers, 32
    rotary tools, 41
    safety with, 31
    screwdrivers, 31
    small hand saws, 45
    socket and hex drivers, 36
    soldering tools, 49
    specialty metal-working tools, 47
    startref =ix_tools, 54
    storing in workspaces, 52
    taps and dies, 45
    techniques, 55
        startref =ix_tooltechnique, 106
        working with fasteners, 56
    tweezers, 49
    vises, 39
    wire cutters, 33
    wire strippers, 34
toothed lock washers, 24
torque
    using to dislodge stubborn fasteners, 68
traces, 351
transformers, 200
    packages, 201
    step-down and step-up, 200
transistor-transistor logic (see TTL)
transistors, 175, 203, 220, 233
    bipolar junction and field-effect types, 220
    field-effect, 222
    heatsinks for, 226
    in integrated circuits (ICs), 252
    one-transistor buffer for discrete outputs, 278
    one-transistor relay driver, 243
    package types, conventional, 223
    packaging, 210
    power transistors, 222
    small-signal, 221
    startref =ix_activecomptrans, 225
    surface-mount packages, 225

TTL and CMOS technology, 253
tri-wing, Y, and Torx drive types, 20
TRIACs (triodes for alternating current), 224, 225, 226
trim potentiometers, 189
trimmer potentiometers, 187
    surface-mount, 190
TTL (tansistor-transistor logic)
    5V TTL logic levels and DIO, 280
    closing gap with CMOS logic
        CMOS and TTL applications, 256
        digital TTL input corresponding to RS-485 signals, 318
    discrete outputs and, 277
TTL (transistor-transistor logic), 253
    7400 series numbers for ICs, 253
    7400 series of logic devices, 256
    closing gap with CMOS logic, 253
turnaround, 319
tweezers, 49
twist pitch, 150
twisted pairs, 149
    color code for wires, 151
    in bundles, degree of twist, 152
    in multi-conductor cables, 153
    making, 151
    wire gauge, wire type, and twist rate, 150
TXB0108 octal logic-level translator, 281
TxD signals, 314
Type A USB connectors, 168
Type B USB connectors, 168

U

UART (universal asynchronous teceiver-transmitter) devices, 299
    RS-232 components, 317
UJTs (unijunction transistors), 224
Unified Thread System (UTS), 17
unit charge, 2
units of measurement, 397
    standard value prefixes, 398
unshielded twisted-pair (UTP) cables, 151, 154
upstream, 322
USB (Universal Serial Bus), 320
    classes of devices, 322
    connections, 322
    connectors, 167
        types of, 167
    data rates, 323
    device configuration, 325

endpoints and pipes, 325
hacking, 327
hubs, 324
interface components, 327
logic analyzer, 393
oscilloscope, 385
terminology, 321
USB-to-Ethernet converters, 333, 334
USB peripherals, 321
user inputs, use of discrete interfaces, 270
utility boxes, using as enclosures, 375
UTS (Unified Thread System), 17
    screw and bolt sizes, 18

V

V (see volt)
v (disjunction) in Boolean algebra, 250
valence shell, 3
    stickiness of electrons, 7
variable capacitors, 197
variable inductors, 200
variable resistors, 185
very large-scale integration (VLSI), 252
vias, 351
    separation, 362
    sizes, 361
video connectors
    crimping tools for, 36
vises, 39
    clamp-on bench vise, 40
    large bench vise, 40
    small bench vise for electronics, 39
Vishay metal film resistor datasheet, 205
volt (V), 6
voltage, 4, 6, 398
    applied to a discrete input, 274
    building a voltage divider (example), 10
    comparison for some battery types, 109
    current versus, 7
    DC voltage over time, 400
    for batteries, 120
    forms of, 399
    in equivalent circuits, sources of, 408
    lithium battery output, 111
    measurement in a circuit, 401
    NiCad battery output, 111
    passive components, 176
    reading using a DMM, 272
    relationship with resistance and current (Ohm's Law), 8

voltage dividers
    series resistance network as multi-tap voltage divider, 405
voltage drops, 401
    for diodes, 214
    in series resistance network, 406

W

wall warts, 107
washers, 23
    suggested inventory of, 17
    varieties of, 23
water-based glues, 26
wattages
    in decimal form, for components, 179
watts, 8
WD-40
    using to loosen stubborn fasteners, 68
wearable devices, 377
WiFi, 338
WiFi modules, 339
wire cutters, 33
    in electronics applications, 34
wire gauges, 35
wire strippers, 34, 155
    combination tools with wire stripper, 35
    semiautomatic wire stripper in action, 156
wire-wound resistors, 182
wireless communications, 334, 345
    2.45 GHz short-range, 337
    802.11, 338
    Bluetooth, 340
    ISM radio bands, 336
    Zigbee, 344
wires, 145
    gauges, 147
    hook-up wire spools, 147
    insulation, 149
    shielding, 152
    stranded and solid wire, 146
    stripping insulation from, 155
    twisted pairs, 149
    wire and cable, 146
WLANs (wireless local area networks), 338
wood
    using adhesives with, 28
wood enclosures, building, 373
workbenches, 52
workspaces, 52
workstations
    based on small electronics vise, 39
WPANs (wireless personal area networks), 344
wrenches, 56
    adjustable, 63
    ball- and flat-end wrenches, 60
    spanners, 64
        combination box and open wrench, 65
        modifying to make thinner, 65
wrist straps to prevent ESD damage, 209

X

XOR (exclusive OR), 251
xoscope, 385

Z

Zener diodes, 213, 217
zero-ohm links, 193
Zigbee, 344
    data transfer rates by ISM band, 344
    modules, 345
zinc-air batteries, 111
zip-tie anchors
    using with battery holders, 119

About the Author

John M. Hughes is an embedded systems engineer with over 30 years of experience in electronics, embedded systems and software, aerospace systems, and scientific applications programming. He was responsible for the surface imaging software on the Phoenix Mars Lander and was part of the team that developed a novel synthetic heterodyne laser interferometer for calibrating the position control of the mirrors on the James Webb Space Telescope.

Over the years he has worked on digital engine-control systems for commercial and military aircraft, automated test systems, radio telescope data acquisition, 50+ gigapixel imaging systems, and realtime adaptive optics controls for astronomy. On his own time (when he has any) he likes to do cabinetry and furniture design, build microcontroller-based gadgets for use with greenhouses, bees, and backyard urban chickens, and write books.

Colophon

The cover images are public domain drawings from the U.S. patents of Nikola Tesla. The cover fonts are URW Typewriter and Guardian Sans. The text font is Scala; the heading font is Benton Sans.

Get even more for your money.

Join the O'Reilly Community, and register the O'Reilly books you own. It's free, and you'll get:

- $4.99 ebook upgrade offer
- 40% upgrade offer on O'Reilly print books
- Membership discounts on books and events
- Free lifetime updates to ebooks and videos
- Multiple ebook formats, DRM FREE
- Participation in the O'Reilly community
- Newsletters
- Account management
- 100% Satisfaction Guarantee

Signing up is easy:

1. Go to: oreilly.com/go/register
2. Create an O'Reilly login.
3. Provide your address.
4. Register your books.

Note: English-language books only

To order books online:
oreilly.com/store

For questions about products or an order:
orders@oreilly.com

To sign up to get topic-specific email announcements and/or news about upcoming books, conferences, special offers, and new technologies:
elists@oreilly.com

For technical questions about book content:
booktech@oreilly.com

To submit new book proposals to our editors:
proposals@oreilly.com

O'Reilly books are available in multiple DRM-free ebook formats. For more information:
oreilly.com/ebooks

O'REILLY®

©2014 O'Reilly Media, Inc. O'Reilly logo is a registered trademark of O'Reilly Media, Inc. 14373

Milton Keynes UK
Ingram Content Group UK Ltd.
UKHW051512220824
447250UK00007B/109